# BENZODIAZEPINE/GABA RECEPTORS AND CHLORIDE CHANNELS
## Structural and Functional Properties

# RECEPTOR BIOCHEMISTRY AND METHODOLOGY

**SERIES EDITORS**

**J. Craig Venter**
Section of Receptor Biochemistry
LNP, National Institute of Neurological and Communicative Disorders and Stroke
National Institutes of Health
Bethesda, Maryland

**Len C. Harrison**
Department of Diabetes and Endocrinology
The Royal Melbourne Hospital
Victoria, Australia

---

*Volume 1*

**Membranes, Detergents, and Receptor Solubilization**
J. Craig Venter and Len C. Harrison, *Editors*

*Volume 2*

**Receptor Purification Procedures**
J. Craig Venter and Len C. Harrison, *Editors*

*Volume 3*

**Molecular and Chemical Characterization of Membrane Receptors**
J. Craig Venter and Len C. Harrison, *Editors*

*Volume 4*

**Monoclonal and Anti-Idiotypic Antibodies as Probes for Receptor Structure and Function**
J. Craig Venter, Claire M. Fraser, and Jon Lindstrom, *Editors*

*Volume 5*

**Benzodiazepine/GABA Receptors and Chloride Channels: Structural and Functional Properties**
Richard W. Olsen and J. Craig Venter, *Editors*

# BENZODIAZEPINE/GABA RECEPTORS AND CHLORIDE CHANNELS
## Structural and Functional Properties

**Editors**

**Richard W. Olsen**
Department of Pharmacology and
Brain Research Institute
School of Medicine
University of California
Los Angeles, California

**J. Craig Venter**
Section of Receptor Biochemistry
LNP, National Institute of Neurological and
Communicative Disorders and Stroke
National Institutes of Health
Bethesda, Maryland

ALAN R. LISS, INC., NEW YORK

Address all Inquiries to the Publisher
Alan R. Liss, Inc., 41 East 11th St, New York, NY 10003



Printed in the United States of America.



**Library of Congress Cataloging in Publication Data**
Main entry under title:

Benzodiazepine/GABA receptors and chloride channels : structural and functional properties.

(Receptor biochemistry and methodology ; v. 5)
Includes index.
1. Benzodiazepines—Receptors. 2. GABA—Receptors.
3. Neural receptors. 4. Ionophores. 5. Chlorides in the body. 6. Biological transport. I. Venter, J. Craig.
II. Olsen, Richard W. III. Series.
QP801.B53B46 1986 599'.0188 85-31229
ISBN 0-8451-3704-2

# Contents

# Contributors

**Jeffery L. Barker,** Laboratory of Neurophysiology, National Institute of Neurological and Communicative Disorders and Stroke, National Institutes of Health, Bethesda, MD 20892 **[135]**

**Eric A. Barnard,** MRC Molecular Neurobiology Unit, University of Cambridge, England CB2 2QH **[261]**

**Claus Braestrup,** Novo Industri A/S, Pharmaceuticals R&D, DK-2880 Bagsvaerd, Denmark **[167]**

**Harry T. Chugani,** Division of Pediatric Neurology, School of Medicine, University of California, Los Angeles, CA 90024 **[315]**

**S.J. Enna,** Departments of Pharmacology and of Neurobiology and Anatomy, University of Texas Medical School, Houston, TX 77025 **[41]**

**E. Falch,** Department of Chemistry BC, The Royal Danish School of Pharmacy, DK-2100 Copenhagen Ø, Denmark **[73]**

**James B. Fischer,** Department of Pharmacology and Brain Research Institute, University of California, Los Angeles, CA 90024. Present address: Harvard School of Public Health, Laboratory of Toxicology, Boston, MA 02115 **[241]**

**Kelvin W. Gee,** Department of Pharmacology, Arizona Health Sciences Center, Tucson, AZ 85724 **[225]**

**Donald R. Gehlert,** Department of Pharmacology, University of Utah Medical Center, Salt Lake City, UT 84132 **[299]**

**Willy Haefely,** Pharmaceutical Research Department, F. Hoffmann-La Roche & Co., Ltd., CH-4002 Basel, Switzerland **[97]**

**P. Häring,** Pharmaceutical Research Department, F. Hoffmann-La Roche & Co., Ltd., CH-4002 Basel, Switzerland **[285]**

**Graham A.R. Johnston,** Department of Pharmacology, University of Sydney, NSW, 2006, Australia **[57]**

**E.W. Karbon,** Departments of Pharmacology and of Neurobiology and Anatomy, University of Texas Medical School, Houston, TX 77025 **[41]**

**Manfred Karobath,** Preclinical Research, Sandoz Ltd., CH-4002 Basel, Switzerland **[185]**

**P. Krogsgaard-Larsen,** Department of Chemistry BC, The Royal Danish School of Pharmacy, DK-2100 Copenhagen Ø, Denmark **[73]**

**Lowell J. Lawrence,** Department of Pharmacology, Arizona Health Sciences Center, Tucson, AZ 85724 **[225]**

**H. Möhler,** Pharmaceutical Research Department, F. Hoffmann-La Roche & Co., Ltd., CH-4002 Basel, Switzerland **[285]**

**L. Nielsen,** Department of Chemistry BC, The Royal Danish School of Pharmacy, DK-2100 Copenhagen Ø, Denmark **[73]**

**Mogens Nielsen,** Psychopharmacological Research Laboratory, Sct. Hans Mental Hospital, DK-4000 Roskilde, Denmark **[167]**

**Richard W. Olsen,** Department of Pharmacology and Brain Research Institute, School of Medicine, University of California, Los Angeles, CA 90024 **[xiii,241,299,315]**

**David G. Owen,** Laboratory of Neurophysiology, National Institute of Neurological and Communicative Disorders and Stroke, National Institutes of Health, Bethesda, MD 20892 **[135]**

**Peter Polc,** Pharmaceutical Research Department, F. Hoffmann-La Roche & Co., Ltd., CH-4002 Basel, Switzerland **[97]**

**J.G. Richards,** Pharmaceutical Research Department, F. Hoffmann-La Roche & Co., Ltd., CH-4002 Basel, Switzerland **[285]**

**Eugene Roberts,** Department of Neurobiochemistry, Beckman Research Institute of the City of Hope, Duarte, CA 91010 **[1]**

The number in brackets is the opening page number of the contributor's article.

**P. Schoch,** Pharmaceutical Research Department, F. Hoffmann-La Roche & Co., Ltd., CH-4002 Basel, Switzerland **[285]**

**Richard F. Squires,** Nathan S. Kline Institute for Psychiatric Research, Orangeburg, NY 10962 **[209]**

**C. Stähli,** Pharmaceutical Research Department, F. Hoffmann-La Roche & Co., Ltd., CH-4002 Basel, Switzerland **[285]**

**F. Anne Stephenson,** MRC Molecular Neurobiology Unit, University of Cambridge, England CB2 2QH **[261]**

**Porntip Supavilai,** Preclinical Research, Sandoz Ltd., CH-4002 Basel, Switzerland **[185]**

**B. Takacs,** Pharmaceutical Research Department, F. Hoffmann-La Roche & Co., Ltd., CH-4002 Basel, Switzerland **[285]**

**John F. Tallman,** Yale University School of Medicine, Ribicoff Research Facilities, New Haven, CT 06508 **[275]**

**Maharaj K. Ticku,** Division of Molecular Pharmacology, Department of Pharmacology, The University of Texas Health Science Center at San Antonio, San Antonio, TX 78284 **[195]**

**J. Craig Venter,** Section of Receptor Biochemistry, LNP, National Institute of Neurological and Communicative Disorders and Stroke, National Institutes of Health, Bethesda, MD 20892 **[xiii]**

**James K. Wamsley,** Departments of Psychiatry and Pharmacology, University of Utah Medical Center, Salt Lake City, UT 84132 **[299]**

**Henry I. Yamamura,** Department of Pharmacology, Arizona Health Sciences Center, Tucson, AZ 85724 **[225]**

# Contents of Volumes 1, 2, 3, and 4

## Volume 3: Molecular and Chemical Characterization of Membrane Receptors



## Volume 4: Monoclonal and Anti-Idiotypic Antibodies: Probes for Receptor Structure and Function

# Preface

In 1977, the specific binding of benzodiazepines to receptor sites in the brain was first described, followed quickly by observations that benzodiazepine receptor sites appeared to be tightly coupled to receptors for the major inhibitory neurotransmitter, $\gamma$-aminobutyric acid (GABA). The role of GABA as a neurotransmitter had gradually been established, and binding of radioactive ligands to specific GABA receptors had been established. The GABA hypothesis of benzodiazepine action had been proposed by Costa and by Haefely and their co-workers in 1974. But the discovery of "Valium receptors" in the brain stimulated such tremendous interest that this book on structural and functional properties of benzodiazepine/GABA receptors could be written less than 10 years later.

This, the fifth volume of the series on **Receptor Biochemistry and Methodology,** edited by J. Craig Venter and Len C. Harrison, is the first volume to deal exclusively with a single type of receptor. The benzodiazepine/GABA receptor-chloride channel complex is probably not the neurotransmitter receptor on which most research has been done and it is certainly not the first receptor to be characterized. However, it is an extremely interesting protein of great neurobiological importance, showing some properties in general with other membrane receptors and some properties unique to this protein. For example, the GABA receptor may be analogous to the nicotinic acetylcholine receptor in that it may contain the structure of a membrane ion channel (for anions instead of cations) and drug receptor sites which allosterically modulate the receptor ion channel coupling. Rapid progress on several levels of research employing biochemical, physiological, pharmacological, and immunological techniques has made it one of the best understood receptors, as described in this volume.

The history of GABA receptors and the "benzodiazepine connection" is described here by many of the major workers in the field. We feel that this volume meets three main objectives. First, the background material is presented in a particularly clear fashion to allow readers from outside the field to quickly grasp the main points. Second, the allosteric interactions between the various components of the benzodiazepine/GABA receptor complex are presented in detail from the points of view of several authors, so that the reader can choose his level of involvement with these details. Third, the most recent advances on the biochemical purification, characterization, and development of monoclonal antibodies for the benzodiazepine/GABA receptor protein are included. We believe that the readers will find the articles in this volume to be detailed state-of-the-subject reviews.

**Richard W. Olsen**
**J. Craig Venter**

**Benzodiazepine/GABA Receptors and Chloride Channels:
Structural and Functional Properties, pages 1–39**


# 1

# GABA: The Road to Neurotransmitter Status

**Eugene Roberts**

*Department of Neurobiochemistry, Beckman Research Institute of the City of Hope, Duarte, California 91010*

The history of research on GABA shows an intriguing pattern of the interweaving of biochemistry, physiology, pharmacology, nutrition, and clinical medicine that brought the subject of neural inhibition to front stage and center. Perhaps the subject of inhibition had languished in the wings for so many years because there was no material basis for it. Inhibitory neurons had not been identified, an inhibitory neurotransmitter had not been isolated and characterized, and postsynaptic sites for neural inhibition had not been shown [for background and history on concepts of neural inhibition, see Diamond et al., 1963; Eccles, 1969; Roberts et al., 1960; von Bekesy, 1967; von Euler et al., 1968; Roberts et al., 1976]. It is well to remember that it was not until 1952, two years after the report of the discovery of γ-aminobutyric acid (GABA) [Roberts and Frankel, 1950a], that the controversy as to whether synaptic transmission in the CNS is largely electrical or chemical was settled in favor of the latter [Eccles, 1982].

The saga of GABA began with the discovery of the unique presence of this substance in tissue of the vertebrate central nervous system (CNS). In the course of the study of free amino acids of various normal and neoplastic tissues in several species of animals by paper chromatography, relatively large amounts of an unidentified ninhydrin-reactive material were found in extracts of fresh brains of mouse, rat, rabbit, guinea-pig, human, frog, salamander, turtle, alligator, and chick. At most, only traces of this material were found in a large number of extracts of many other normal and neoplastic tissues and in urine and blood. The unknown material was isolated from suitably prepared paper chromatograms. A study of the properties of the substance in mouse brain revealed it to be GABA. The initial identification, based on the comigration of the unknown with GABA on paper chromatography in three different solvent systems, was followed by an absolute identification of the GABA in brain extracts by the isotope derivative method. Abstracts were submitted to the Federation meetings in 1950 reporting the presence of GABA in brain [Roberts and Frankel, 1950a; Udenfriend, 1950a]. An abstract reporting the presence of an "unidentified amino acid in brain only" also appeared from another laboratory in the proceedings of the same meeting [Awapara, 1950]. The first three complete papers dealing with the occurrence of GABA in brain all appeared in the same issue of the *Journal of Biological Chemistry* [Roberts and Frankel, 1950b; Udenfriend, 1950b; Awapara et al, 1950]. A more complete history of the early chemical work outlined above has been published elsewhere [Roberts, 1984a].

For several years the presence of GABA in brain remained a biochemical curiosity and a physiological enigma. My continued efforts to convince neurophysiologists to test GABA on various preparations at the end of their planned experiments met with complete failure. In the first review on the subject, in desperation, I concluded [Roberts, 1956], "Perhaps the most difficult question to answer would be whether the presence in the gray matter of the central nervous system of uniquely high concentrations of γ-aminobutyric acid and the enzyme which forms it from glutamic acid has a direct or indirect connection to conduction of the nerve impulse in this tissue." However, later in that year, the first suggestion that GABA might have an inhibitory function in the vertebrate nervous system came from studies in which it was found that topically applied solutions of GABA exerted inhibitory effects on electrical activity in the brain [Hayashi and Nagai, 1956; Hayashi and Suhara, 1956]. Then in 1957, the suggestion was made that indigenously occurring GABA might have an inhibitory function in the central nervous system from studies with convulsant hydrazides [Killam, 1957; Killam and Bain, 1957]. Also in 1957, suggestive evidence for an inhibitory function for GABA came from studies that established GABA as the major factor in brain extracts responsible for the inhibitory action of these extracts on the crayfish stretch receptor system [Bazemore et al., 1957]. Within a brief period the activity in this field increased greatly, so that the research being carried out ranged all the way from the study of the effects of GABA on ionic movements in single neurons to clinical evaluation of the role of the GABA system in epilepsy, schizophrenia, mental retardation, etc. This warranted the convocation in 1959 of the first truly interdisciplinary neuroscience conference ever held, at which were present most of the individuals who had played a role in opening up this exciting field [Roberts et al., 1960].

From the beginning, it made little difference to me, a biochemist, whether or not GABA was a neurotransmitter. My goal was the elucidation of its function in the nervous system, whatever it might be. However, the "transmitter question" agitated many physiologists. Initially, there was much skepticism with regard to the transmitter function of GABA because of the large quantities of GABA present in brain, 3–4 orders of magnitude higher than those of acetylcholine, the only proven neurotransmitter at that time. It would have been more acceptable to some if the first identified inhibitory neurotransmitter would have had a more exotic chemical structure than that of GABA. The resistance to acceptance of GABA as a naturally occurring neurotransmitter was supported by negative experiments and interpretations, many of which, in retrospect, were technically flawed. Since they are extant in the literature, largely without specific rebuttal, they will be briefly dealt with below before proceeding to discussion of some of the overwhelmingly supportive data. Some direct quotations are employed so as to transmit the flavor of the times.

At the aforementioned GABA symposium [Roberts et al., 1960], a number of statements were made that seemed to rule out a neurotransmitter role for GABA. It finally was remarked by someone that GABA entered the conference as a proud transmitter candidate and left it as a poor metabolite. For example, Florey, whose brilliant work had pioneered in the identification of a factor in brain extracts that exerted inhibitory effects on invertebrate preparations [Elliott and Florey, 1956; Florey, 1953; Florey, 1954a; Florey, 1954b; Florey, 1955; Florey, 1956a; Florey, 1956b; Florey, 1957; Florey, 1960; Florey and Biederman, 1960; Florey and Florey, 1958; Florey and McLennan, 1955a; Florey and McLennan, 1955b; Florey and McLennan, 1955c; Iwasaki and Florey, 1969; Koidl and Florey, 1975; Levin et al., 1961], stated [p. 74 of Roberts et al., 1960], "Although one likes to ascribe to neurogenic compounds an action on postsynaptic membranes, there is every possibility that they affect intraneuronal processes which in turn control the state of the membrane. At present, it is not possible to decide whether GABA normally occurs outside of nerve cells as a product of secretion (modulator substance). All available evidence speaks, however, against it playing a role as inhibitory transmitter in vertebrates." Florey made the above statement, at least in part, because at that time GABA had not yet been shown to possess most of the "classical" requirements for neurotransmitter, chief among which are proof of presence in inhibitory nerves, releasability from terminals of identified nerves, identity of postsynaptic action with that of the natural transmitter, and the presence of a rapid inactivating mechanism at synapses. The most critical among these is the first. As aptly stated by Fatt [p. 106, Roberts et al., 1960], ". . . The characterization of the post-junctional response must logically precede any attempt to identify the transmitter, since the criterion for the identification is its ability to duplicate the effect of pre-junctional nerve stimulation." Perhaps even more damaging to GABA's transmitter candidacy was Florey's failure to detect GABA in the crustacean nervous system and his report of the discovery of an incredibly more potent inhibitory substance, Factor I [p. 81 of Roberts et al., 1960]. The latter findings unquestionably resulted from a concatenation of technical errors [see Levin et al., 1961], since enormous concentrations of GABA later were found in inhibitory axons of crustacea [Kravitz et al., 1963a] and the existence of a Factor I independent of GABA has never been established [Koidl and Florey, 1975]. Yet we find the following statement by Florey [Koidl and Florey, 1975] as late as 1975: "Factor I, the inhibitory agent present in certain nerve extracts which is responsible for the inhibition of crayfish stretch receptor neurones, is up to 300 times more effective than can be accounted for by the contained $\gamma$-aminobutyric acid (GABA)." Faulty chemical work by McLennan led to similar earlier conclusions [McLennan, 1957a; McLennan, 1957b; McLennan, 1958; McLennan, 1959; McLennan, 1960]. Also in the 1959 conference, Van der Kloot, discoverer of the blocking effect of picrotoxin on peripheral inhibition [Van der Kloot et al., 1958; Robbins and Van der Kloot, 1958], noted [p. 410, Roberts et al., 1960], "Yet there is a compelling reason to deny that GABA itself is the transmitter. The inhibitory effects always persisted without diminuition after GABA was perfused into claws. If there is an enzymatic breakdown of GABA, the disappearance of the molecule is inconspicuous." The latter statement was based on the mistaken notion that the removal of all neurotransmitters would have to take place by a mechanism strictly analogous to the extremely rapid enzymatic destruction of acetylcholine by acetylcholinesterase at the vertebrate neuromuscular junction. Actually, the case of acetylcholine appears to be an exception. GABA and a variety of other neurotransmitters now are known to be removed rapidly from synaptically active sites by carrier-mediated transport into pre- and postsynaptic neuronal sites as well as into glial cells. While usually operating effectively as a GABA "vacuum cleaner" in a physiologically coordinated manner with the natural release mechanisms, the GABA transport systems can be more easily overwhelmed with an exogenous flood of GABA than can the cholinergic system with acetylcholine, because acetylcholinesterase is an enzyme with a very high turnover number. Van der Kloot's notion was derived from an incorrect assumption followed by an inappropriate experiment. Curtis and Watkins concluded that GABA was "not a specialized inhibitory substance" [p. 442, Roberts et al., 1960]. This was based on their failure to observe in neurons onto which GABA was applied iontophoretically [Curtis et al., 1959] the hyperpolarization which is known to occur during natural inhibition in cat spinal motoneurons [Brock et al., 1952; Coombs et al., 1953]. They stated, "The possible specific inhibitory character of these acids ($\beta$-alanine and GABA) is excluded by their failure to produce a hyperpolarization of motoneuronal membranes." It seems probable that the failure to observe hyperpolarization by the latter workers was attributable to the fact that, in the ingeniously devised coaxial microelectrodes used in the studies, "the electrical coupling between the barrels was uniformly so great as to preclude measuring the small potentials required to show hyperpolarization" [Werman, 1972]. Although glycine probably is the major inhibitory neurotransmitter in ventral regions of the spinal cord [Werman, 1972], numerous GABAergic synapses have been found immunocytochemically on the somata of spinal motoneurons [McLaughlin et al., 1975]. Wiersma, who had laid the groundwork for the use of crustacean sensory and neuromuscular synapses in quantitative physiological studies [Ellis et al., 1942; Wiersma and Schallek, 1948; Schallek and Wiersma, 1948; Wiersma et al., 1953; Hoyle and Wiersma, 1958a; Hoyle and Wiersma, 1958b; Hoyle and Wiersma, 1958c] also stated at the abovementioned conference [p. 70, Roberts et al., 1960], "At one time it looked as if GABA would meet the requirements of an inhibitory transmitter substance for many places, but in the last year or so it has become more doubtful that this is the case." This probably was based on the statements of Florey, who previously had been his colleague, and also on a paper published by Wiersma and Hoyle in 1958 in which is found the following statement [Hoyle and Wiersma, 1958c]: "We have applied this substance (GABA) in concentrations of $10^{-9}$–$10^{-3}$ to several preparations of crayfish and crab opener muscles. It did not produce mechanical inhibition in any of them, nor did it raise the resting potential in muscle fibers in which inhibitory action had this effect." Although never acknowledged publicly, the above result later was found to be attributable to a mislabeling of the bottle from which the substance to be tested was taken. They probably had applied $\alpha$-aminobutyric acid and not GABA! In the same symposium, on the basis of recording of cortical wave patterns, Purpura and Grundfest proposed that "the major action of GABA was the inactivation of depolarizing axondendritic synapses in the cortex" [p. 495, Roberts et al., 1960]. However, this was challenged by Jasper [p. 553, Roberts et al., 1960]: "The hypothesis of Purpura and Grundfest is a most intriguing one but, at present, there are other interpretations of the data which seem equally reasonable. It would seem to us more likely that the action of GABA on the cortical surface is not due to the selective blockade of excitatory synapses, unmasking a positive 'inhibitory' postsynaptic potential, but that it causes a selective depression of nervous structures in the surface layers of the cortex . . . ." Jasper's ideas are more congruent with today's understanding. Nonetheless, in the summary to the whole conference, while conceding the probable existence of true inhibition on ventral horn neurons in the spinal cord and on the crustacean sensory cell, George Bishop remarked [p. 589, Roberts et al., 1960], "In spite of considerable experimentation and even more extensive interpretation, the position is still tenable that little or no such inhibition, that accompanying a positively oriented impulse and a decrease of resistance of the synaptic membrane, occurs in the cortex of the mammal. Postimpulsive depression resembling refractoriness is on the contrary of very prevalent occurrence there." The curtain appeared to have been brought down on the candidacy of GABA as inhibitory neurotransmitter!

At a meeting of the Federated Societies held in the spring of 1959, I was approached by Kuffler and asked by him whether or not I thought there was still a chance that GABA might be a neurotransmitter and whether or not it would be worthwhile to hire a biochemist to work with him and other physiologists on the action of GABA in the crustacean nervous system as part of a team he was going to constitute upon his move from Johns Hopkins to Harvard. My affirmative answer encouraged him to proceed along the path he had taken as a result of the beautiful work with extracellular recordings published from his laboratory in the crayfish stretch receptor system, showing that GABA appeared to mimic the effects of natural inhibition in the several important respects that could then be measured by the techniques employed [Kuffler and Edwards, 1958; Kuffler, 1958; Dudel and

Kuffler, 1961]. The work of the latter group, together with that of the Takeuchis (see below), eventually produced the most convincing evidence for the role of GABA as an inhibitory neurotransmitter at crustacean peripheral synaptic junctions. Subsequently, much work in vertebrate nervous systems led to the conclusion that GABA probably is an inhibitory neurotransmitter in the CNS. However, even as late as 1964 the following appeared in a textbook [Davson, 1964]: "Thus GABA may be ruled out as a transmitter in those situations where its identity with the transmitter seemed best proved; it is not surprising, therefore, that in the mammalian central nervous system its depressant action is not analogous with inhibitory activity, its action being nonspecific on all responses of the motor neurone. . . ."

In the sections that follow I will try to trace a few of the paths of the GABA story, which, like the basal dendrites of cerebellar Purkinje cells, eventually led to the many branches of the GABA enterprises as they exist today. A few matters of particular personal interest also will be mentioned. The information on the GABA system, in general, now is so vast that even a series of books on the subject could not do justice to it.

## BASIC NEUROCHEMISTRY

### GABA and Its Precursor, Glutamate

An outline of the metabolism of GABA is shown in Figure 1. GABA is formed in the CNS of vertebrate organisms to a large extent, if not entirely, from L-glutamic acid. The reaction (reaction 5) is catalyzed by an L-glutamic acid decarboxylase (GAD) (L-glutamic 1-carboxyl-lyase, EC4.1.1.15), an enzyme found in mammalian organisms largely in neurons in the CNS, although there now are many reports of the occurrence of both GAD and GABA in neurons in the peripheral nervous system and also in some non-neural tissues and in body fluids. We succeeded in purifying to homogeneity neuronal GAD from mouse brain, and some of its properties were elucidated [Wu et al., 1973; Wu and Roberts, 1974; Wu, 1976; Matsuda et al., 1973a]. Brain GAD catalyzes the rapid $\alpha$-decarboxylation only of L-glutamic acid of the naturally occurring amino acids and, to a very slight extent, that of L-aspartic acid. Immunochemical studies indicated that the rat brain enzyme is most closely related to the mouse enzyme, followed in order by human, rabbit, calf, guinea pig, quail, frog, and trout [Matsuda et al., 1973a; Matsuda et al., 1973b; Saito et al., 1974]. The gene for brain GAD now has been cloned [Kaufman et al., 1985]. The reversible transamination of GABA with $\alpha$-ketoglutarate (reaction 9) is catalyzed by an aminotransferase, GABA-T (4-aminobutyrate: 2-oxoglutarate aminotransferase, EC2.6.19), which in the CNS is found chief-ly in the grey matter, but also is found in other tissues. The products of the transaminase reaction are succinic semialdehyde and glutamic acid. There is present in excess a dehydrogenase (succinic semialdehyde $NAD^+$ oxidoreductase, EC1.2.1.16) which cata-lyzes the oxidation of succinic semialdehydeto succinic acid, which in turn can be oxi-dized via the reactions of the tricarboxylic acid cycle. GABA-T was purified to homogeneity, and its properties have been studied in considerable detail [Schousboe et al., 1973; Schousboe et al., 1974]. Of the keto acids tested, only $\alpha$-ketoglutarate was an amino group acceptor. However, of a series of amino acids tested, $\beta$-alanine showed activity equal to that of GABA as substrate; $\delta$-aminovaleric and $\beta$-aminoisobutyric acids also were effective amino donors. Therefore, GABA-T may be important in several aspects of metabolism of $\omega$-amino acids and not just related to the GABA system. Antibodies to mouse brain GABA-T were found to cross-react with the enzyme from human, calf, mouse, rat, guinea pig, rabbit, and frog. Microcomplement fixation tests showed that only the enzyme from rat brain may be very similar to the mouse brain enzyme in terms of protein structure. Accordingly, this enzyme seems to be highly species-specific.

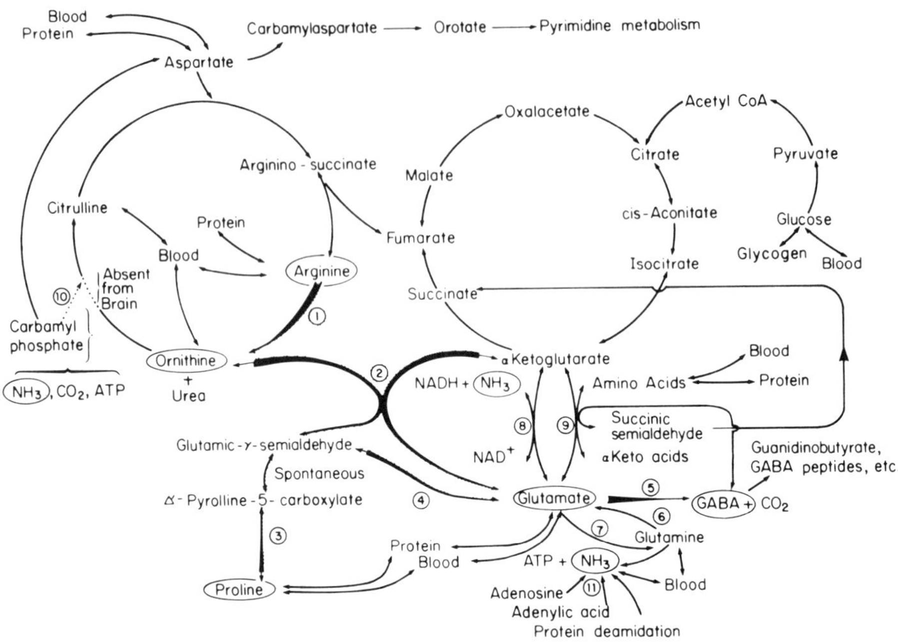


**Fig. 1.** *Metabolic relationships relevant to GABA formation and utilization.*

Steady-state concentrations of GABA in various brain areas normally are governed by the GAD activity and not by GABA-T. In many inhibitory nerves, both GAD and GABA-T are present and are found throughout the neuron, the GAD being more highly concentrated in the presynaptic terminals than elsewhere. The GABA-T is contained in mitochondria of all neuronal regions, but it seems to be richer in the mitochondria of those neuronal sites from which or onto which GABA might be liberated. Such regions are in terminals of GABAergic neurons, perikarya, and dendrites of neurons that receive GABAergic inputs and in glial and endothelial cells that are in the vicinity of GABA synapses. It became possible to visualize GAD on sections of the mammalian CNS at the light and electron microscopic levels, employing rabbit antisera to the purified enzyme and peroxidase-labeling techniques. This has led to much more definitive data than were available through cell fractionation and microdissection studies and has given a plethora of detailed information of interrelationships of various neuronal systems at the ultrastructural level (see below). Although GABA is a precursor of several substances found in nervous tissue and cerebrospinal fluid, which are listed below, no connections have yet been found between any of them and GABAergic neurotransmission: homocarnosine (GABA-histidine), homoanserine (GABA-1-methyl-histidine), γ-guanidinobutyric acid, GABA-L-cystathionine, α-(GABA)-L-lysine, GABA choline, and putreanine [(N-4-aminobutyl)-3-amino-propionic acid)].

Until recently relatively little attention was paid to the biochemical mechanisms that *normally* might be involved in controlling levels of synaptosomal GABA available for release from the terminals of GABAergic neurons. Important controls in regulation of the GABA system might be exerted at points related to the availability of glutamic acid, the substrate for GABA synthesis in nerve endings by GAD, which is the terminal enzyme in GABA formation (reaction 5). Glutamate carbon can originate from glucose through glycolysis and the Krebs cycle (upper right-hand corner of Fig. 1), from glutamine subsequent to uptake (reaction 6), and from proline (reactions 3 and 4) and ornithine (reactions 2 and 4). Ornithine (reactions 2 and 3), but not glutamate, is an effective precursor of proline in nerve terminals, a putative inhibitory neurotransmitter [Johnson and Roberts, 1984a]. Arginine can be converted in nerve terminals to ornithine (reaction 1), which in turn gives rise to glutamate (reactions 2 and 4), proline (reactions 2 and 3), and GABA (reactions 2, 4, and 5) [Johnson and Roberts, 1984a; Johnson and Roberts, 1984b; Yoneda and Roberts, 1982a; Yoneda and Roberts, 1982b; Yoneda et al., 1982; Lapinjoki et al., 1982; Shank and Campbell, 1983; Wong et al., 1982a; Wong et al., 1982b].

Quantitatively, the major overall sources of glutamate in neural tissue as a whole probably are the pathways by which ammonia is fixed via glutamate dehydrogenase (reaction 8) and by which ammonia is liberated from glutamine by glutaminase (reaction 6) [Bradford et al., 1978; Bradford and Ward, 1976; Hamberger et al., 1979; Kvamme, 1979; Morjoria and Voaden, 1979]. Although the latter two enzymes probably play major roles in the glutamate and ammonia economy of nervous tissue, from a variety of experiments it does not seem conclusive that the reactions which they catalyze furnish the glutamate that is the direct substrate for GAD (reaction 5) in presynaptic endings of GABAergic neurons. Other important sources of glutamate nitrogen may be the several amino acids that can undergo transaminations in brain, among which are arginine and ornithine. The most common nitrogen acceptor is $\alpha$-ketoglutarate (reaction 9), although transaminations also can occur with pyruvate, oxalacetate, and glyoxylate (not shown in Fig. 1).

An important link between amino acid and carbohydrate metabolism in brain occurs at the level of the ornithine-$\alpha$-ketoglutarate $\delta$-aminotransferase (reaction 2). The ornithine cycle (illustrated in the upper left-hand portion of Fig. 1), by which urea is formed in liver, is incompletely functional in brain; carbamyl transferase (reaction 10), the enzyme that catalyzes the formation of citrulline from ornithine, is absent from nervous tissue [Jones et al., 1961]. Nonetheless, arginine enters readily from the blood stream [Pardridge and Oldendorf, 1975], and urea is formed in the brain by the action of arginase (reaction 1) [Davies et al., 1961; Ratner et al., 1960]. The specific activity of arginase in brain is higher than that of the ornithine $\delta$-aminotransferase (reaction 2) [Sadasivudu and Rao, 1976], suggesting that the latter may be rate-limiting in the biosynthesis of glutamate via the linked pathway (reactions 2 and 4) or in the formation of proline from arginine (reactions 2 and 3). The rate-limiting enzyme in proline synthesis is the $\Delta^1$-pyrroline-5-carboxylate reductase, a cytosolic enzyme using NADH as cofactor, that converts $\Delta^1$-pyrroline-5-carboxylic acid to proline. Proline oxidase, a mitochondrial-bound enzyme, converts proline to $\Delta^1$-pyrroline-5-carboxylic acid (reaction 3) [Johnson and Strecker, 1962].

Glutamic-$\gamma$-semialdehyde is the common intermediate in proline and ornithine metabolism that may be transformed to glutamate by the action of a specific NAD-dependent dehydrogenase (reaction 4) [Strecker, 1960]. Although for 25 years glutamate has been considered to be a source of the carbon and nitrogen for ornithine and proline biogenesis in animal tissues, only recently the reduction of glutamate to glutamic-$\gamma$-semialdehyde has been demonstrated directly in homogenates of rat intestinal mucosa [Ross et al., 1978] and Chinese hamster ovary cells [Smith et al., 1977]. This enzymatic reaction has not yet been demonstrated in brain. It can be seen from Figure 1 that glutamic-$\gamma$-semialdehyde could give rise to ornithine and proline by reversal of the reactions by which it is formed from these amino acids. Thus, in the presence of the appropriate enzymes, glutamate, ornithine, and proline would all be potentially interconvertible. However, since the glutamate decarboxylase (reaction 5) is irreversible, it would be expected that if all of the requisite enzymes should exist in endings of GABAergic neurons, the metabolite flow would go largely in the directions arginine → ornithine → glutamate → GABA or proline → glutamate → GABA. On heuristic grounds, I originally suggested [Roberts, 1981] that the quantitatively less important metabolic pathway leading from arginine to glutamate as a source of intraterminal GABA-precursor glutamate would be more in keeping with the need for specificity of control and sequestration that generally seem to be part of neurotransmitter logistics than the more common aminotransferases (aspartate or alanine), glutaminase, or glutamic dehydrogenase. As seen from the above and considerable other data in whole brain and in nerve terminals, Arg, Orn, and Pro can give rise to Glu, which in turn may be precursor for Asp and GABA (Fig. 1). It is of particular interest that GABA exerts negative feedback on the synaptosomal biosynthetic pathway of glutamate and GABA from ornithine by acting as a competitive inhibitor of ornithine in the ornithine $\delta$-aminotransferase reaction (reaction 2) [Yoneda et al., 1982b] and that proline inhibits its own formation by inhibiting the $\Delta^1$-pyrroline-5-carboxylate reductase (reaction 3) [Yoneda and Roberts, 1982a]. In concentrations present in the CNS, GABA does not inhibit its own synthesis by GAD (reaction 5). Thus, indeed, the minor metabolic pathways, Arg → Orn → Glu and Pro → $\Delta^1$-pyrroline-5-carboxylate → Glu may be used for Glu GABA-precursor synthesis instead of, or in addition to, the more common routes employing transaminases, glutaminase, or glutamic dehydrogenase. Because of the use of mixtures of synaptosomes from many different types of neurons for the studies to date, the question still remains as to whether or not all of the necessary enzymes are present in individual terminals of GABAergic neurons. A detailed basic strategy has been outlined [Roberts, 1981] by which it may be possible to trace pertinent metabolic pathways with a combination of immunocytochemical and isotope procedures to give definitive answers to the above question. Such work is in progress in a number of laboratories, and conclusions from this work are awaited with great interest.

Meanwhile, it is of particular interest that Arg deficiency in the chick leads to motoneuron damage [Newberne et al., 1960] and in the young dog produces emesis, excess salivation, and muscle tremors [Ha et al., 1978]. Within 2 hours of a single meal of an Arg-free diet containing Glu and Pro, cats showed emesis, lethargy, vocalization, frothing at the mouth, hyperactivity, hyperesthesia, ataxia, emprosthotonos, extended limbs, and exposed claws [Morris and Rogers, 1978]. In a genetically homozygous spastic mouse, brain Arg levels were reduced by 35% by comparison with normal heterozygote controls, and urea levels were elevated to the extent of 50% as a result of a highly significant increase in brain (but not liver) arginase activity [Stewart, 1980]. In a comparison of the free amino acid content of the hippocampus obtained at postmortem from seven normal individuals and eight with Alzheimer's disease, an increase in Arg levels in the Alzheimer's brains was the only significant change found among the 24 constituents determined [Tarbit et al., 1980].

### The Vitamin $B_6$ Connection

GAD from bacteria and plants was known to require pyridoxal phosphate (PLP), a form of vitamin $B_6$, as a coenzyme [Gale, 1946]. PLP also was found to activate the brain enzyme; and semicarbazide, a carbonyl-trapping agent that reacts with the aldehyde group of PLP, inhibited the production of GABA by inhibiting GAD [Roberts and Frankel, 1951b], which subsequently was shown to be inactivated by many agents that can react with the aldehyde group of the coenzyme. PLP can be removed from the enzyme protein of GAD readily, causing loss of enzyme activity, and the lost enzymatic activity can be restored simply by the addition of the coenzyme. Pyridoxine-deficient rats showed a decrease in the degree of saturation with the coenzyme of the enzyme protein of cerebral GAD, but no decrease was found in the content of enzyme protein in the deficient animals [Roberts et al., 1951]. Brain GAD activity was restored rapidly to normal upon feeding pyridoxine to the deficient animals.

Normally, dietary forms of vitamin $B_6$ are absorbed and converted efficiently in tissues to pyridoxal phosphate or pyridoxamine phosphate, the coenzymatically active forms of the vitamin. Pyridoxal phosphate has been shown to be synthesized in brain from ATP and pyridoxal [Roberts and Frankel, 1951a]. In addition to the effects of dietary deficiency, carbonyl-trapping agents, and of pyridoxine antimetabolites, other circumstances may arise by which the tissue levels of pyridoxal phosphate may be kept too low for the optimal functioning of important rate-limiting enzymes such as GAD. The first instance of pyridoxine dependency was described in a child [Hunt et al., 1954] that showed a grossly abnormal electroencephologographic record and convulsions unless given large daily supplements of vitamin $B_6$. During the seizures in this child there was a decrease in cerebral oxygen consumption, and a return to a normal cerebral metabolic rate occurred when pyridoxine was administered [Sokoloff et al., 1959]. The defect appears to lie in an abnormally high excretion of pyridoxal and pyridoxamine in the urine, possibly because of a failure to reabsorb these substances from the blood by the kidney [Coursin, 1960]. Although not identified yet clinically, it is possible that genetic or induced defects in the handling of vitamin $B_6$ also may occur in the absorption of the dietary forms of the vitamin or in the enzymes involved in the conversions of these forms to the coenzymatically active compounds. Genetic differences in the structure of the apoenzyme of $B_6$ enzymes also may occur, so that the same enzyme in various individuals (or strains of animals) may possess differing affinities for the coenzyme and could thus achieve saturation with the coenzyme at widely different tissue levels. The finding that a seizure in an infant with a simple dietary deficiency of vitamin $B_6$ was abolished completely almost immediately after intramuscular injection of pyridoxine [Coursin, 1954] indicates that in a normal individual there is an extremely rapid rate of conversion of pyridoxine to pyridoxal phosphate and association of the coenzyme with the apoenzyme of a critical cerebral enzyme [see Tower, 1958; Tower, 1956; Roberts et al., 1964; Roberts and Eidelberg, 1960; and Roberts, 1963 for early relevant reviews].

The first indication of a physiological role for *endogenous* GABA in mammals came from studies on the mechanism of the action of convulsant hydrazides [Killam and Bain, 1957; Killam, 1957; Killam et al., 1960]. Administration of a variety of hydrazides to animals uniformly resulted in the production of repetitive seizures following a rather prolonged latent period. Seizure activity began subcortically and spread to the cortex. Since hydrazide-induced seizures can be prevented by the parenteral administration of various forms of vitamin $B_6$, it was conjectured that some enzyme system requiring PLP was being inhibited and that a decrease in activity of the enzyme was somehow related to the production of the seizures. Attention was directed to GABA and GAD, because of our previous reports of their unique occurrence in the CNS [Roberts and Frankel, 1951a; Roberts and Frankel, 1951b]; and, indeed, it was found that the seizures were accompanied by substantial decreases in levels of GABA and reductions in GAD activity of the brain areas studied. It was shown subsequently in mice that intraperitoneally administered thiosemicarbazide, one of the most potent hydrazides, significantly slowed the rate at which intracerebrally administered L-glutamic acid-U-$^{14}C$ was converted to GABA in brain [Roberts et al., 1958]. However, PLP-requiring enzymes also are involved in the metabolism of the catecholamines and the formation of serotonin. Therefore, although in some instances an excellent correlation was found to exist between the inhibition of GAD by the hydrazides and the occurrence of seizures, the results could not constitute direct proof of a causal relationship. Topically applied or intraventricularly administered GABA effectively counteracted the convulsive activities of thiosemicarbazide in cats, and electrically induced seizures in the rhinencephalon and the isocortex were shortened by application of GABA and increased by the injection of thiosemicarbazide [Killam et al., 1960]. When thiosemicarbazide, semicarbazide, or isoniazid were applied topically on segments of neuronally isolated cat cortex, in which neuronal activity did not occur spontaneously but could be elicited by electrical stimulation, there was a long latent period (10–30 min) during which time the responses to stimulation remained normal regardless of the frequency of stimulation [Grafstein, 1963]. After this period, convulsive spikes of rather large amplitude appeared. When stimulation of such preparations stopped for 10 min, the responses to stimulation returned to normal and then, after a period of stimulation, the spikes reappeared. The time required for return of the spikes was proportional to the rate of stimulation. When intervals as long as 15–20 sec were allowed between stimuli, the spikes did not reappear. The hydrazides presumably decreased the rate of formation of GABA, because at the concentrations used they are known to partially inhibit GAD. It is likely that when the rate of stimulation was slow and, therefore, rate of GABA release slow, then the generation of GABA still kept up sufficiently to prevent seizures. Increases in rates of nerve activity outstripped the availability of GABA to be released and eventually seizures occurred.

Throughout the years it was not possible to make a conclusive case for *causal* relationship between increases or decreases in seizure susceptibility in specific cerebral areas with corresponding increases or decreases in GABA levels in those areas. This was because GABA is distributed in other cellular compartments, as well as in the terminals of neurons. All of the compartments do not respond similarly to inhibition of synthesis of GABA. Loss of nerve terminal GABA may be concealed by elevated GABA levels in other cellular compartments. Thus, hundreds of papers attempting the establish such a relationship failed to do so convincingly. Recent work shows, however, that when synaptosome-enriched fractions are isolated from brain and studied under various conditions of treatment with convulsant hydrazides with and without protection by $B_6$, decreases in GABA content correlate extremely well with the onset of seizures, despite the fact that poor correlation was observed between seizure activity and whole brain tissue GABA levels [Wood, 1982; Wood et al., 1979; Geddes and Wood, 1984].

Thus, it looks like the circle has then been closed. The initial observations of the $B_6$ requirement of GAD for the formation of GABA and the decreases induced by the hydrazides and other carbonyl-trapping agents in GABA levels and GAD activities when seizures occur probably can be attributed solely to the decrease in the amount of releasable GABA in nerve terminals. Reconstitution of the ability of nerve terminals to form GABA is correlated with the elevation of seizure thresholds. The relatively firm establishment of this correlation, however, does not elucidate the neuronal mechanisms by which the seizures occur. Some aspects of the latter will be dealt with in a subsequent section.

## PHYSIOLOGY

How does the impingement of GABA on excitable membranes bring about the neural inhibitory process that counteracts the depolarizing action of excitatory processes to help maintain the polarization of a cell at an equilibrium level near that of its resting value? Elegant work with precisely defined invertebrate and vertebrate test systems employing intracellular microelectrodes has indicated that the membrane stabilization process necessary for this inhibition results from specific increases in the permeability of the postsynaptic membrane to chloride ions. Postsynaptic membranes generally are electrically unexcitable, and both excitatory and inhibitory effects usually are transmitted by chemical mediators which have effects on the specialized postsynaptic membrane.

### Invertebrate Stretch Receptors

Perhaps the earliest thoroughly studied preparations were crustacean stretch receptor organs [Alexandrowicz, 1951]. The easy accessibility of these organs for manipulation and for insertion of measuring electrodes led to a good understanding of their function. The body of the sensory neuron (SRN) lies near the receptor muscle strand, and the peripheral portions of the dendrites are imbedded in the strand. An inhibitory axon, the only source of innervation, forms synapses on the dendrites, and there also is a motor nerve supply to the muscle strand. The SRN can be activated by passive stretch or by contraction set up by the motor nerves [Wiersma et al., 1953] and can be suppressed even in the presence of stretch by stimulation of the inhibitory nerve [Kuffler and Eyzaguirre, 1955]. There are some analogies of the above system with the vertebrate muscle spindle [Kuffler, 1958]. The sequence of events leading to the firing of the SRN was summarized as follows [Kuffler, 1958]: "Stretch deformation sets up the generator potential in the peripheral portion of dendrites. By electrotonic spread other cell regions are excited and all-or-none conduction generally starts more centrally in the larger portion of dendrites, in the cell body or in the initial axon segment. The generator potential persists for the duration of the stretch, its magnitude depending on the amount of stretch deformation. This in turn controls the frequency of sensory discharge. The actual site of origin of propagated impulses has been shown to fluctuate according to intensity of depolarization. While the distal dendrite portions are clearly specialized for graded activity, the proximal parts of the dendrites, the cell body, and the axon carry fully conducted impulses." The impulses in the slowly adapting SRN of the lobster were shown to be initiated in the axon about 500 $\mu$m from the cell body [Edwards and Ottoson, 1958]. Analysis of the mechanisms of inhibition of the above system employing intracellular electrodes suggested that stimulation of the inhibitory nerve causes the liberation of a substance (or substances) that produce a specific effect on the postsynaptic membrane of the sensory cell. The inhibitory process in the SRN was believed to be attributable to increased conductance. Actually, at the equilibrium level of the inhibitory process no significant membrane potential change is set up by inhibitory impulses, although the inhibitory process is fully active [Kuffler, 1958].

The association of GABA with the crayfish stretch receptor occurred as a result of an extensive series of studies on the influence of factors present in extracts of beef brain which had inhibitory effects on the stretch receptor as well as in other physiological test systems [see Florey references, cited above]. Eventually, GABA was shown to be the major factor exerting potent inhibitory action on the stretch receptor [Bazemore et al., 1957], opening the way for many investigations of the actions of GABA in a variety of invertebrate preparations. It was found that low concentrations of GABA had effects on the crayfish stretch receptor which were remarkably similar to those found on stimulation of the inhibitory nerve [Kuffler and Edwards, 1958]. GABA concentrations 1,000 times greater than those blocking afferent discharge did not block conduction in the sensory axons. GABA action, therefore, appeared to be confined to the dendrites and cell body. The action of GABA was not localized necessarily on subsynaptic receptor sites, since it also was found to block the stretch discharge of the median receptor in the seventh thoracic segment, a structure possessing no demonstrable inhibitory synapses. GABA showed the most potent blocking action on the stretch receptor of a wide variety of related compounds tested [Edwards, 1960].

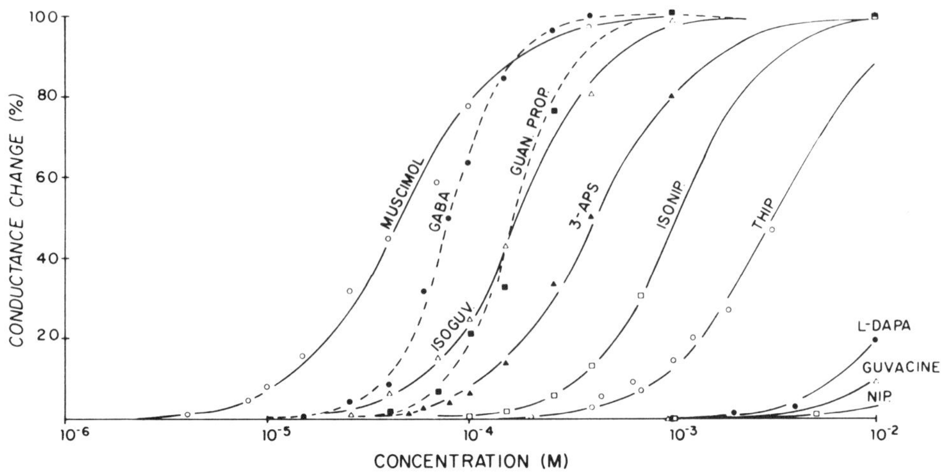


**Fig. 2.** *Concentration–conductance change relationships obtained in crayfish stretch receptor neurons for GABA and a variety of GABA analogs. Conductance change was expressed as percentage of the maximum change produced by GABA in the same neuron. The SEM for each point shown was usually less than 10% of the mean. Curves were drawn according to the equation:*

$$y = \frac{y_{max}}{1 + K_a/A^n}$$

*where y is the conductance change and $y_{max}$ the maximal conductance change produced by the compound, A is the concentration of the compound and $K_a$ is a constant, and where n = 3 for GABA and guan. prop. (β-guanidinopropionic acid); n = 2 for isoguv. (isoguvacine) and isonip. (isonipecotic acid); and n = 1.7 for muscimol, 3-APS (3-aminopropane sulfonate), and THIP (4,5,6,7-tetrahydroisoxazole-[5,4-c]pyridine-3-ol). L-DAPA-(L-α,β-diaminopropionic acid); Nip. (± nipecotic acid). (Reprinted by permission, Krause et al., 1981.)*

## Attempts to Characterize the GABA-Recognition Sites of GABA Receptors by Quantitative Correlative Studies With the Crayfish Stretch Receptors and Mouse Brain Membranes

In our laboratories were studied a variety of chemical substances, employing quantitative conductance measurements in the GABA-responding crayfish stretch receptor neuron (SRN) [Hori et al., 1978; Krause et al., 1981; Roberts et al., 1981; Swagel et al., 1973a; Swagel et al., 1973b]. The same substances also were examined using adaptations of previously published procedures with membrane preparations from mouse brain to measure GABA and/or muscimol binding to presumed GABA recognition sites of the receptor–anionophore complex [Enna and Snyder, 1975; Wang et al., 1979]. The data from widely separated animal species were compared because these were the best available test systems for comparing biological responses (crayfish) and binding (mouse) in a *quantitative* manner.

Plots of noncumulative log concentration vs. conductance change obtained in the SRN for GABA and some structural analogs of GABA are illustrated in Figure 2. The concentration–response curves were normalized by expressing the conductance change for a given concentration of agonist as a percentage of the maximal response elicited by GABA in the same neuron. *All effective agonists produced the same maximal response as did GABA.* Muscimol was the most potent compound tested. Other known receptor agonists were effective, but less potent than GABA. Nipecotic acid, guvacine and L-α,β-diaminopropionic acid(L-DAPA), all shown to be potent and selective inhibitors of GABA high affinity uptake processes in vitro [Krogsgaard-Larsen and Johnston, 1975; Roberts et al., 1978a; Lodge et al., 1977], had no effect on membrane input conductance at a concentration of $10^{-3}$ M. β-4-Chlorophenyl GABA (baclofen), a clinically relevant GABA analog with questionable GABA-like actions in vivo [Davies and Watkins, 1974; Fox et al., 1978; Waddington and Cross, 1979], but which binds to the same bicuculline-insensitive, divalent cation–dependent binding sites as GABA in rat brain [Bowery et al., 1983], had no effect on either the resting membrane conductance or on the increase in conductance produced by GABA, even at a concentration of $10^{-2}$ M.

In all cases, the slopes obtained from Hill plot analyses of the concentration–response data were greater than 1, indicating that cooperative interactions were involved in the response. However, while the value was approximately 2 for most of the agonists, the Hill slopes for GABA and for β-guanidinopropionic acid were 3.1 and 2.6, respectively, significantly greater than 2. While this finding could suggest some basic differences in mechanism of action, it was noted that of the compounds tested, these latter two are by far the most readily accumulated by cellular GABA uptake processes [Edwards and Kuffler, 1959; Iversen and Kravitz, 1968; Roberts et al., 1979a]. Muscimol, isoguvacine, 3-aminopropane sulfonate (3-APS), isonipecotic acid, and THIP (4,5,6,7-tetrahydroisoxazole-[5,4-*c*]pyridine-3-ol zwitterion) appear to be transported poorly, if at all [Krogsgaard-Larsen and Johnston, 1975; Roberts et al., 1979c; Beart and Johnston, 1973; Krogsgaard-Larsen et al., 1977; Lodge et al., 1978].

GABA uptake processes are known to exist in crustacea [Kuffler and Edwards, 1958; Iversen and Kravitz, 1968; Horwitz and Orkand, 1980; Meiners et al., 1979; Orkand and Kravitz, 1971; Sisken and Roberts, 1964] and could influence the access of susceptible GABA analogs to the postsynaptic receptors. To test whether or not this was the case, the effect of GABA on membrane input conductance was monitored in the presence and absence of nipecotic acid, a relatively potent GABA uptake blocker in crayfish [Iversen and Kravitz, 1968; Horwitz and Orkand, 1980] and mammalian nervous tissue [Krogsgaard-Larsen and Johnston, 1975]. Nipecotic acid at $10^{-3}$ M did not alter the baseline conductance of the neuron, but its presence significantly potentiated the response to a submaximal concentration of GABA. The effect of nipecotic acid on the full GABA dose–response curve is shown in Figure 3A. All data shown were obtained from the same neuron. The GABA dose–response curve was significantly shifted to the left by the presence of the uptake blocker, although the maximal response was unchanged.

The enhancing effect of nipecotic acid was greater for lower GABA concentrations, resulting in a different shape for the GABA curve, which indicated a lower apparent cooperativity for the GABA responses. The curve for muscimol, a poorly transported substance, was unaffected by the addition of nipecotic acid. Results similar to those above for nipecotic acid were obtained with another relatively potent and specific GABA uptake blocker, L-DAPA, at concentrations of the latter which, by themselves, had no direct effect on the membrane input conductance of the SRN (Fig. 3B). Guvacine ($10^{-3}$ M) also potentiated responses to GABA but was not examined in detail. The $EC_{50}$ value for GABA was not significantly altered in the presence of $10^{-3}$ M concentrations of either nipecotic acid or L-DAPA. However, the average Hill slope obtained for GABA dose-conductance data was significantly reduced from 3.20 to 2.36 by nipecotic acid and to 1.73 by $10^{-3}$ M L-DAPA. This last value was not significantly different from the average Hill slopes found for muscimol, 3-APS, and THIP, substances which are not appreciably accumulated by GABA uptake processes [Beart and Johnston, 1973; Krogsgaard-Larsen et al., 1977; Lodge et al., 1978]. When the influence of uptake processes is accounted for, it appears that all

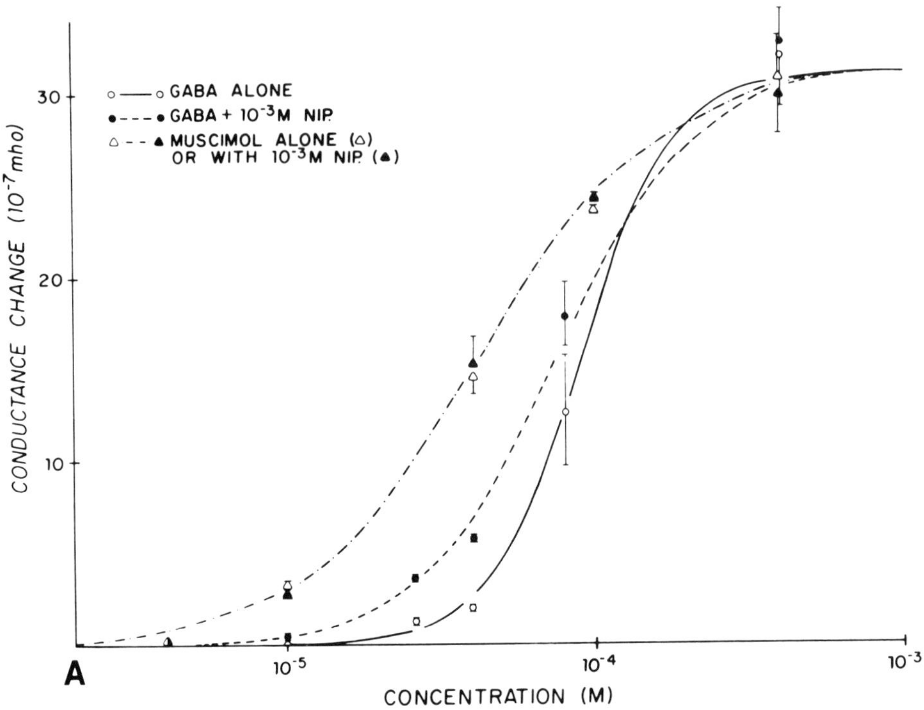


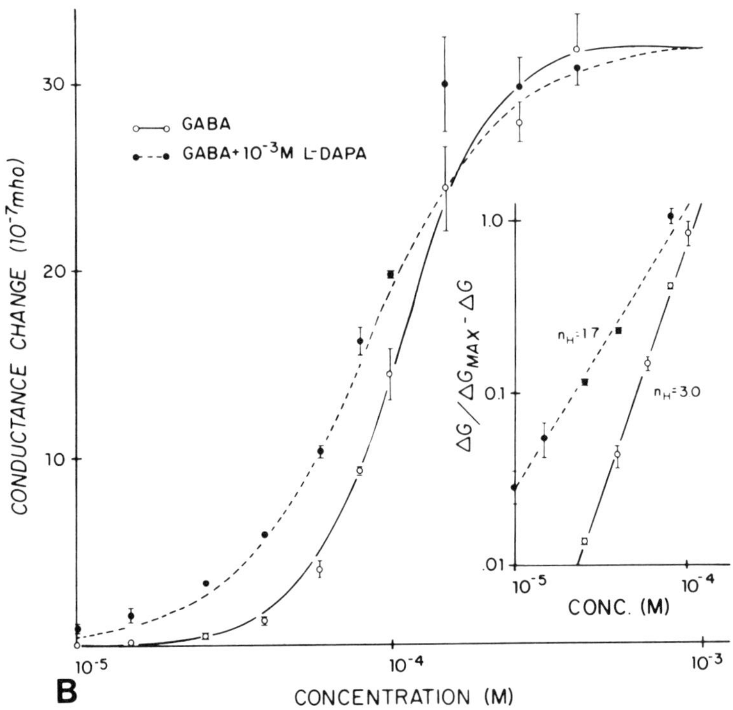


**Fig. 3.** A. *Effect of nipecotic acid on the concentration-conductance curves for GABA and muscimol. All data were obtained from the same neuron. Each point represents the mean ± S.D. for duplicate applications of test substances. Curves were drawn according to the equation given in the legend to Figure 2 with n = 3 (○), n = 2 (●) and n = 1.5 (△, ▲).* B. *Effect of L-DAPA on the concentration-conductance curve for GABA. All data were obtained from the same neuron. Each point represents the mean ± S.D. for duplicate applications of test substances. Theoretical curves were drawn using the $n_H$ values obtained from the Hill plot shown in the inset. Slopes ($n_H$) of the lines in the Hill plot were determined by linear regression analysis.*

agonists tested acted by the same cooperative mechanisms, which required at least two molecules of agonist to activate a receptor–ionophore unit.

Our quantitative conductance data for GABA agonists from the isolated SRN [Hori et al., 1978; Krause et al., 1981; Roberts et al., 1981; Swagel et al., 1973a; Swagel et al., 1973b; and see Fig. 2] were compared with quantitative receptor-related ligand binding data obtained in our laboratory with mouse brain membranes [Wang et al., 1979 and unpublished observations from this laboratory] (Fig. 4). A highly significant correlation coefficient was observed between the log of the reciprocals of the $K_I$ values obtained in the $^3$H-muscimol binding assay for various substances [Wang et al., 1979] and the log of the reciprocals of the $EC_{50}$ values determined for these compounds by conductance measurements. A similar correlation was seen with $^3$H-GABA binding data for mouse brain membranes [Roberts, 1980b, and unpublished observations], and for inhibition of $^3$H-muscimol binding by GABA analogs in crayfish muscle membranes [Meiners et al., 1979]. THIP, a rigid bicyclic analog of muscimol, showed the most obvious deviation in the correlation plot (Fig. 4) being less potent on the SRN than could be expected from the binding results. 3-APS also was somewhat less potent and ℓ-GABOB somewhat more potent on the SNR than was predicted by the binding data. *The absolute values reflecting potency in the two test systems differed, however, by several orders of magnitude between the binding and conductance assays. For example, while the $EC_{50}$ for GABA was 78 $\mu$M in the SRN conductance assay, the $K_I$ for the same substance in the membrane binding assays was 0.1 $\mu$M.*

The effects of the two stereoisomers of the GABA agonist, $\gamma$-amino-$\beta$-hydroxybutyric acid (GABOB), were studied in detail in our quantitative receptor-related assay systems [Roberts et al., 1981]. (3S)-(+)-4-amino-3-hydroxybutanoic acid (d-GABOB) was found to be about twice as potent as ℓ-GABOB in displacing $^3$H-muscimol from specific binding sites in mouse brain membrane fractions. A similar order of potency for the GABOB enantiomers was observed for the cerebrovascular GABA receptor studied in $^3$H-muscimol binding assays using bovine cerebral blood vessels [Roberts et al., 1981]. In contrast to the binding results, ℓ-GABOB was significantly more potent than d-GABOB in mimicking the postsynaptic action of GABA, which was measured as increases in membrane input conductance in the crayfish SRN assay. ℓ-GABOB also was more potent than d-GABOB when conductance measurements were made in the presence of $10^{-3}$ M nipecotic acid or DAPA, two specific GABA transport blockers, showing that differences in rates of uptake could not account for the results. The greater effectiveness of ℓ-GABOB found in the crayfish assay is congruent with work by others that showed that ℓ-GABOB, but not d-GABOB, is an effective inhibitor of induced seizure activity in cat brain and rabbit motor cortex [Aishita et al., 1978; Hayashi, 1959; Katayama, 1967; Katayama and Mori, 1977]. Thus, d-GABOB is more potent than ℓ-GABOB in membrane binding systems, while the reverse is found in physiologically responding systems. These findings suggested that the structural asymmetry of the GABA recognition sites in vitro is different from that related to activation of GABA receptors in vivo.

Another feature of the above results was that Hill plots of the effects on the SRN of d- and ℓ-GABOB gave slopes of approximately 2, consistent with the cooperativity seen for GABA responses in a variety of preparations and with a number of GABA agonists [see preceding discussion and Nistri and Constanti, 1979]. However, our binding measurements with these compounds and most previous work on $^3$H-muscimol [Wang et al., 1979] and $^3$H-GABA binding [Enna and Snyder, 1975] showed that Hill slopes of 1 were

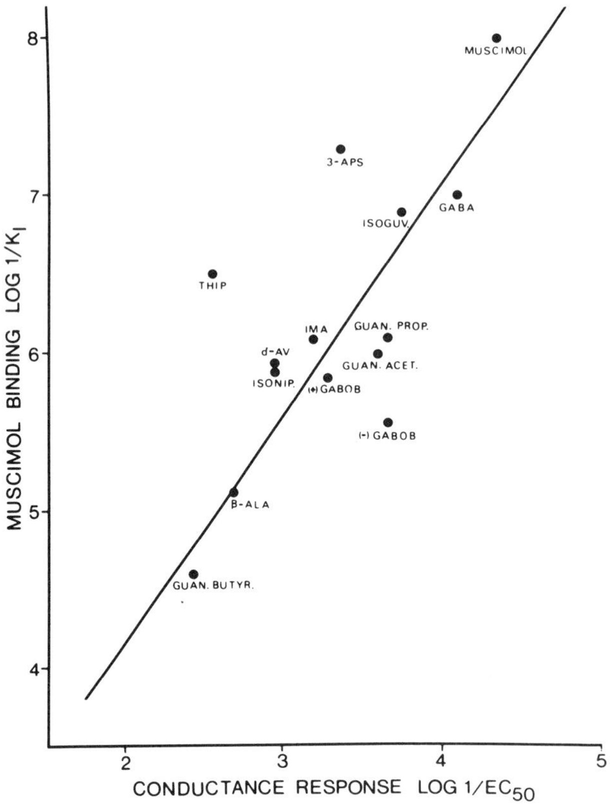


**Fig. 4.** *Correlation between conductance measurements (stretch receptor neurons) and binding measurements (mouse brain membranes), both of which are thought to reflect agonist effectiveness at GABA receptors.* $EC_{50}$ *values for agonist-induced conductance responses were taken from stretch receptor studies, and* $K_I$ *values indicating the potency of agonists in displacing [*$^3$*H]muscimol bound to GABA receptor sites in mouse brain membranes were obtained in binding assays as described in the text. The log* $1/K_I$ *values were significantly correlated with log* $1/EC_{50}$ *values with a correlation coefficient of 0.70 (*$P < 0.005$*). Omitting (—) GABOB ((3R)-(−)-4-amino-3-hydroxybutanoic acid), 3-APS, and THIP from the analysis a correlation coefficient of 0.93 (*$P < 0.001$*) was obtained, and the latter linear regression line was drawn here.*

obtained with in vitro membrane preparations. In addition, the $EC_{50}$ values found in physiological measurements of the SRN were approximately two orders of magnitude greater than those found in the in vitro binding measurements. For example, the $EC_{50}$ values for d- and ℓ-GABOB, respectively, were 550 and 230 μM in the SRN test system but were 2.8 and 6.0 μM in the receptor-related membrane binding assay employed.

Such discrepancies in affinity and degree of cooperativity between physiological and in vitro binding measurements are common for various receptor types, and many possible explanations have been discussed [Nistri and Constanti, 1979; Changeux et al., 1976; Colquhoun, 1978; Barker and Owen, this volume]. Likely possibilities are that pre- and postsynaptic and extrasynaptic membrane loci may have different binding properties and that varying proportions of these exist in different types of preparations, that the manipulations involved in making and assaying the membrane preparations alter the properties of the receptor, or that the conditions employed for the assays do not adequately reflect in vivo conditions. Cooperative interactions may be disrupted at the relevant GABA recognition sites in the membrane fragments so that individual GABA molecules are bound independently of each other. Under binding assay conditions, acetylcholine receptors have been found to shift to a desensitized state [Changeux et al., 1976].

With regard to the latter points, it is of interest to compare two in vitro studies of presumed GABA receptors [see Nistri and Constanti, 1979; De Feudis, 1978; and De Feudis, 1980 for other relevant data]. We studied the binding of $^3$H-muscimol to a membrane fraction from whole mouse brains (minus brain stems) employing $^3$H-muscimol in nanomolar concentrations [Wang et al., 1979]. Two types of sites were indicated with $K_D$ values equal to $9 \times 10^{-9}$ M and $70 \times 10^{-9}$ M, respectively, and with $B_{max}$ values of 2.1 pmoles/mg and 0.7 pmoles, respectively. The slope (n) of the Hill plot for $^3$H-muscimol binding was calculated to be 0.93, indicating no positive or negative cooperativity in the binding. The data are consistent with the presence in the membrane preparation of two different binding sites, each of which binds one molecule of muscimol (or GABA). In contrast [Giambalvo and Rosenberg, 1976], in a study of junctional complexes isolated from rat cerebellum studied with $^{14}$C-GABA in the micromolar range, a $K_D$ value of $7.4 \times 10^{-5}$ M was obtained with $B_{max}$ of 555 pmoles/mg protein. The Hill coefficient was 2.2, indicating a positive cooperativity and suggesting that two molecules of GABA may combine per site, although the pharmacological specificity expected for receptors was not demonstrated for these binding sites. Actually, in the SRN assays the $EC_{50}$ values for muscimol and GABA were $4.8 \times 10^{-5}$ M and $7.8 \times 10^{-5}$ M, respectively, and the Hill coefficients were approximately 2, close to the above values for the junctional complexes.

Based on extensive examination of CPK models of all of the agonists studied by us, in both SRN and in vitro binding systems, I now would like to present a working model consistent with current data. It has been suggested that GABA acts on its receptors in the extended form [Roberts et al., 1981; see also De Feudis, 1980]. We now propose that face 1 of the GABA molecule, as shown in Figure 5, may resemble the surface of GABA that would attach to its receptor recognition site. In keeping with considerable data on the apparent cooperativity involved with GABA responses [Krause et al., 1981; Nistri and Constanti, 1979], let us further assume that two GABA molecules fit a GABA recognition receptor cup or cavity in the extended closely packed, head-to-toe fashion, shown in Figure 6, to cooperatively activate a unit conductance change. Manipulation of CPK space–filling molecular models shows that such tightly packed dimeric arrangements are feasible. The charge neutralization that would occur as a result of the interaction of juxtaposed, oppositely charged amino and carboxyl groups would be expected to greatly decrease the degrees of hydration of the polar ends of the GABA molecules, and both coulombic and van der Waal's forces would tend to hold two molecules together in the proposed configurations. It would be expected that because of increased nonpolar properties of the dimers relative to monomers, the water solubility of such dimers would be decreased and lipid solubility increased, favoring their existence in lipidic membrane environments. Self-association of pyridine [Huyskens et al., 1977] and nicotinamide and isonicotinic hydrazide [Kopecky et al., 1978] has been demonstrated in aqueous solutions. Direct physical experiments eventually should enable us to determine whether GABA and GABA agonists can form dimers or higher associates at concentrations expected to be present at GABAergic synaptic junctions.

I now postulate that there are two equivalent, but nonidentical, membrane sites into which pairs of GABA molecules could fit (Fig. 7, A and B). These are presumed to be mirror images of each other. While these binding sites would be chirally blind insofar as GABA molecules are concerned, they would have markedly different affinities for

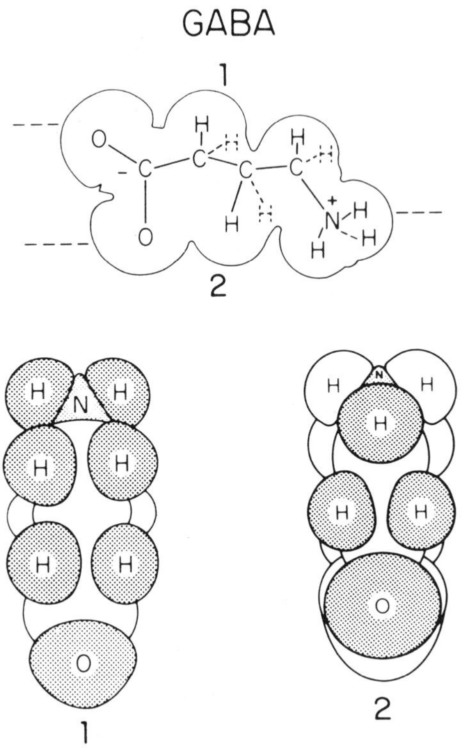


**Fig. 5.** *The faces of GABA as seen on projections of CPK space-filling models.*

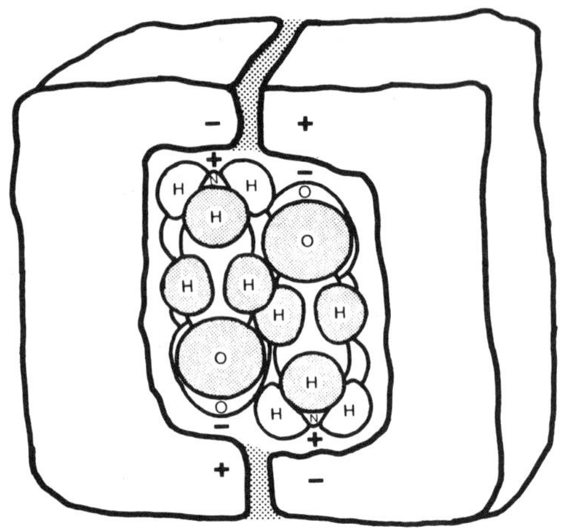

**Fig. 6.** *How two GABA molecules may fit a GABA recognition site on a GABA receptor.*

asymmetrically β-substituted derivatives of GABA, such as ℓ- or d-GABOB. As illustrated in Figure 7, the active sites, A and B, could accommodate external OH groups of the GABOB molecules, but the GABOB molecules would not pack tightly enough to fit the proposed receptor binding sites if the OH groups were in the internal positions. Thus, ℓ-GABOB would not fit the site of Figure 7A, and d-GABOB would not fit that of Figure 7B. The extra bulk and the hydrophilic character of the OH group should greatly decrease the affinity of the GABOB isomers for the proposed receptor site relative to GABA. If it is assumed that the affinities of the A site for d-GABOB and the B site for ℓ-GABOB are similar, the two-fold difference in $IC_{50}$ values found in the SRN system would suggest that there may be approximately twice as many accessible B sites as A sites. Thus, the above model schematically may represent intact receptors in responding systems and possibly those in isolated junctional complexes [Giambalvo and Rosenberg, 1976].

The destruction of cooperativity, possibly associated with desensitization of receptors, is modeled in Figure 7, C–F. Some of the manipulations required to perform in vitro binding measurements with the preparations employed to date may result in the functional (but not necessarily physical) dissociation of the subunits of the binding sites so that individual molecules of ligand can bind to them independently of each other. The probability of attachment of single zwitterionic GABA molecules in a given orientation to a doubly charged binding site would, in all likelihood, be much greater than would that of two GABA molecules to a binding site with four charges. Thus, the affinity of ligand for binding site would be greatly enhanced. In the case of the GABOB isomers, it is shown that preferential binding would take place with the OH groups facing away from the structure of the binding site and toward the aqueous environment; that is, d-GABOB would bind to subunits of the B site and ℓ-GABOB to the A sites, resulting in approximately twice as much binding being observed for the isomer that originally was

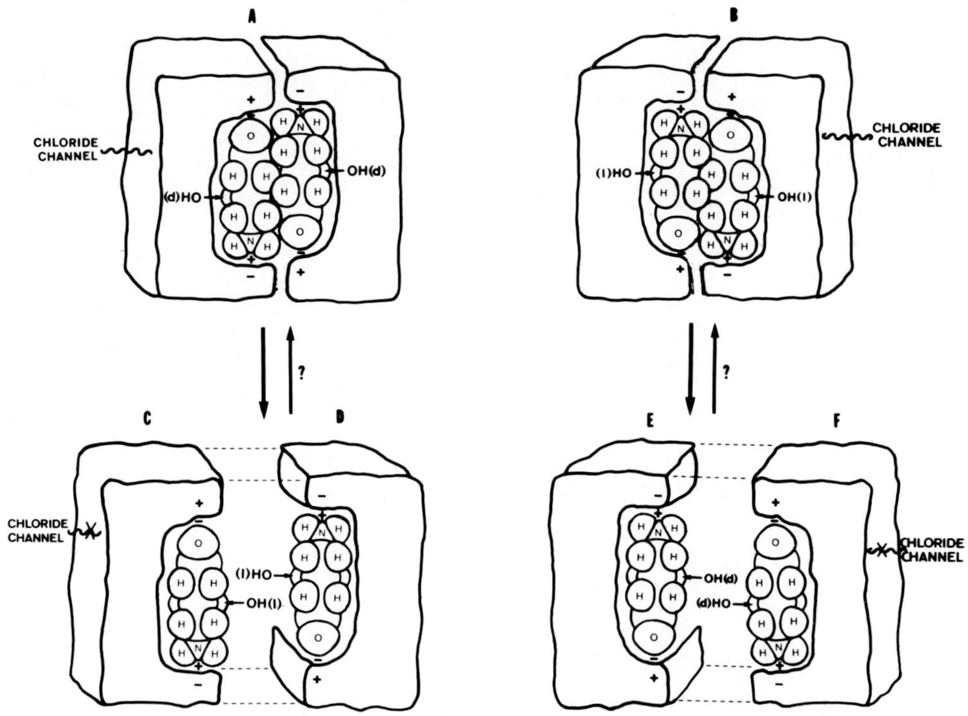


**Fig. 7.** *A model of the GABA recognition site of the GABA receptor system. It is proposed that there are two chirally different sites, A and B, that represent low-affinity states that are coupled with membrane anion gating mechanisms. Both A and B accommodate tightly packed head-to-toe dimers of GABA with approximately equal affinity, but d-GABOB attaches preferentially to A and ℓ-GABOB to B for reasons stated in the text. It is presumed that procedures for preparing brain membranes lead to a conversion to high-affinity sites that are uncoupled from anion channels (sites C–F). The possible structural reason for the increases of affinities of the latter sites for GABA and d- and ℓ-GABOB and for the reversal of affinities for the GABOB enantiomers from those found for sites A and B are discussed in the text. It is possible that the transformations suggested may take place in a reversible fashion when desensitization in responsive membranes occurs during continued action of agonist. The figure is drawn so that the reader views it from behind the presumed GABA recognition site.*

only one-half as active as the other one in producing conductance changes in physiologically responsive membranes, consistent with experimental binding data [Roberts et al., 1981].

Let us now turn to examination of the CPK models of the extended head-to-toe dimers of GABA and other pertinent GABA agonists. We searched among these compounds for the surfaces of the dimers that would resemble each other most closely. Because of the great molecular similarities between GABA and GABOB isomers, it was difficult to decide which aspects of their dimers most likely would be expected to interact with the receptor surface. Examination of the models of the total range of effective agonists studied by us in the SRN test system showed quite clearly that dimers of each of the effective substances had one structural feature in common. This is illustrated schematically for GABA in Figure 6 and in Figure 8 with CPK models for muscimol (A), GABA (B), and $\beta$-guanidinopropionate (C), three of the most potent substances on the SRN (Fig. 2). Also included in Figure 8 are models of 3-APS (D) and THIP (E), two substances with high affinity for the single binding sites of mouse brain

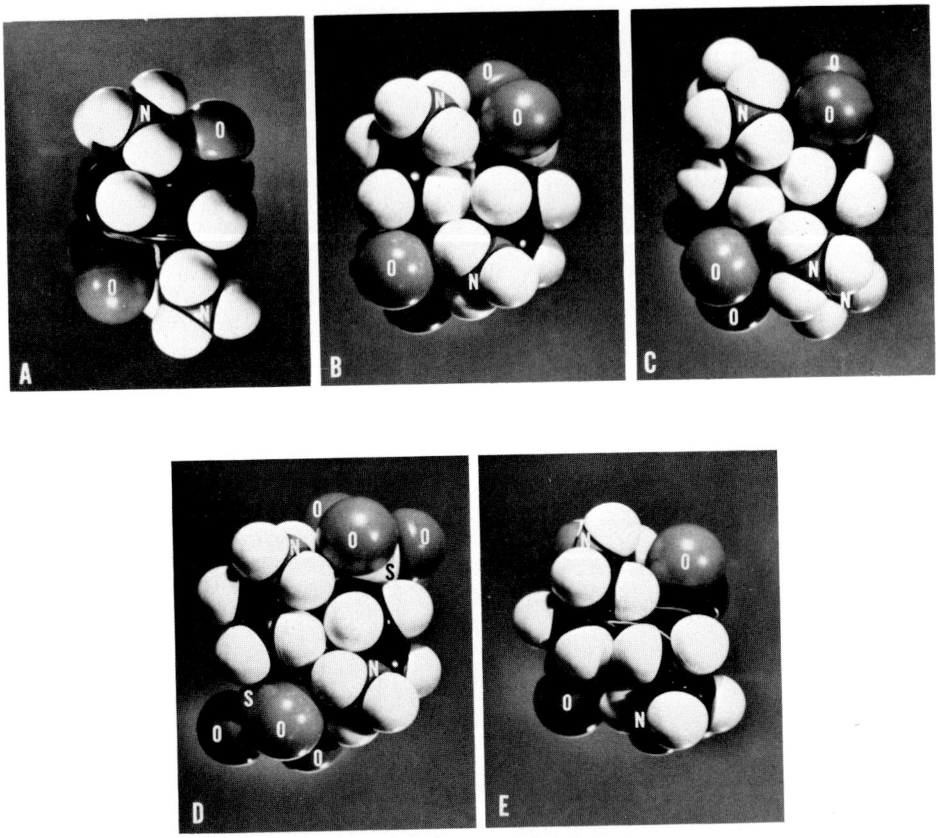


**Fig. 8.** *CPK models of head-to-toe dimers of muscimol* (A), *GABA* (B), *and β-guanidinopropionate* (C), *all potent agonists in the SRN system, and of 3-APS* (D) *and THIP* (E), *substances which are only weakly active. The marked similarities in the surfaces presented by* A, B, *and* C *and the differences between them and those of* D *and* E *should be noted.*

membranes, but with much lower efficacy on the SRN. The models of the latter two substances show that the distances between the charged groups and the flexibilities of the monomers of the latter two substances are such that it could be readily envisioned that they would bind to single sites with affinities similar to those of GABA. However, a glance at models of the head-to-toe dimers shows considerable divergence in the structure of the proposed active molecular sites from those shown by the three substances that are highly active on the SRN. Although quantitative methods of evaluation of the degree of relatedness of such surfaces to each other have not yet been established, a good correlation can be obtained between biological effectiveness and structure merely by qualitatively arranging them in order of similarity of visual "fit" to muscimol and GABA. Examination of the dimer of muscimol, the most potent agonist, showed that, in addition to the coulombic interaction of the charged groups of van der Waal's interactions, the isoxazole rings of the dimer partners would be in proper position to form a $\pi$–$\pi$ bond between one of the two double bonds of each of the rings. Thus, once formed, the dimer of muscimol would be expected to be more stable than that of GABA. It is interesting in this connection that in cultured mouse spinal neurons, muscimol opens ion channels of about the same conductance and twice the average lifetime as those activated by GABA, a finding that could explain its greater potency [Mathers and Barker, 1981].

The above is a preliminary and conjectural formulation. What is the real nature of the membrane receptor sites to which GABA molecules attach? Of course, a true understanding at the molecular level will only come when the homogeneous receptor [Sigel et al., 1983; Barnard et al., 1984] is completely characterized at a molecular level and proven to be functionally active in suitable reconstitution experiments. Such work appears to be well under way in a number of laboratories, as will be reported in detail in other papers in this volume.

The receptor structure probably consists of a protein matrix capable of interacting with other proteins, with lipids, and with a host of other cellular constituents that may modify activity of the receptor. With regard to the lipids, Watkins proposed, in general terms, that GABA could interact with phosphatidylethanolamine (PE)–protein complexes in polar discontinuities in membrane surfaces in such a way as to cause dissociation of these complexes [Watkins, 1965]. The PE–protein complexes were imagined to be formed largely by interaction of the positively and negatively charged groups of the polar head groups of PE molecules with appropriately spaced negatively charged (glutamyl or aspartyl side chains) and positively charged (basic amino acid side chains or divalent cations associated with monovalent protein anionic sites) groups, respectively, in membrane proteins. GABA could interact with the PE head groups, the protein, or with both. Support for the idea that phospholipids may interact with the GABA receptor was furnished by experiments with preparations from rat cerebellum enriched in junctional complexes [Giambalvo and Rosenberg, 1976]. It was shown that phospholipase C, which removes phospholipid head groups, increased GABA binding more than two-fold, and phospholipids were found to compete with GABA for binding to receptor protein, PE being more potent than phosphatidylcholine in this respect [Giambalvo and Rosenberg, 1976]. Homogenization of rat brain membranes (prepared with 0.5% Triton) with PE, phosphatidylcholine, or phosphatidylserine showed that PE reduced GABA binding by 50% under the conditions employed while the other phospholipids were without effect [Johnston and Kennedy, 1978]. There may be an altered relationship between phospholipids and GABA receptors in Huntington disease [Lloyd and Davidson, 1979]. However, current results cast doubt on the possibility that dissociation of PE–protein complexes [Toffano et al., 1981] might help trigger the sequence of events that would result in the increased ionic permeability that GABA produces when it acts on its receptor. Two membrane proteins, GABAmodulin [Toffano et al., 1980] and benzodiazepine receptors [Squires et al., 1980], already are known to modify the properties of GABA receptors. Certainly, others will be found. The latter subject will be discussed in a number of other contributions to this volume.

GABA neurons appear to be among the most ubiquitously distributed neurons in the vertebrate CNS. Their connectivities with each other and with neurons using other transmitters are numerous and complex. It would, indeed, be fortunate if there were only one genetically coded protein, or only a small number of such proteins, that would provide the molecular basis for responsivities of membranes to impingement of GABA. It already is certain that a variety of modifications of the GABA receptors can occur by association with other substances, and such modifications must be involved in the physiological flexibility that we already know exists in the interactions of these neurons in the CNS.

## Definitive Experiments at the Invertebrate Neuromuscular Junction

The most convincing evidence for the role of GABA as an inhibitory neurotransmitter was obtained from studies at crustacean neuromuscular junctions [see Kravitz et al., 1968; Otsuka, 1976; Potter, 1968, Takeuchi, 1976; and Atwood, 1982 for summaries]. A brilliant series of concerted chemical and physiological studies of the Harvard group started by Kuffler [Kravitz, 1962; Kravitz et al., 1962; Dudel et al., 1963; Kravitz et al., 1963a; Kravitz et al., 1963b; Kravitz et al., 1965; Otsuka et al., 1966; Otsuka et al., 1967; Hall and Kravitz, 1967a; Hall and Kravitz, 1967b; Molinoff and Kravitz, 1968; Iversen and Kravitz, 1968; Hall et al., 1970; Orkand and Kravitz, 1971] and exquisitely delicate electrophysiological studies by the Takeuchis at Juntendo University in Tokyo [Takeuchi and Takeuchi, 1965; Takeuchi and Takeuchi, 1966a; Takeuchi and Takeuchi, 1966b; Takeuchi and Takeuchi, 1967a; Takeuchi and Takeuchi, 1967b; Takeuchi and Takeuchi, 1971a; Takeuchi and Takeuchi, 1971b; Takeuchi and Takeuchi, 1975a; Takeuchi and Takeuchi, 1975b; Takeuchi, 1978; Onodera and Takeuchi, 1979] established beyond all doubt that GABA is the inhibitory neurotransmitter liberated by the inhibitory nerves. The postsynaptic action of applied GABA mimics exactly that found on stimulating inhibitory nerves. Inhibitory axons contain enormous concentrations of GABA (0.1 M), while less than 1% of these levels can be detected in excitatory nerves. GABA is released from lobster inhibitory nerves in amounts proportional to the number and frequency of stimuli applied to the nerves and is not released by stimulation of excitatory nerves. The enzyme that forms GABA from L-glutamate, GAD, is preferentially distributed in inhibitory nerves. Synaptically released GABA probably is inactivated by a $Na^+$-dependent uptake mechanism that is capable of acting against large concentration gradients, and GABA is metabolized by transamination and oxidation, as shown in Figure 1. Picrotoxin, a convulsant agent, blocks the inhibitory action of GABA and the natural inhibitory transmitter similarly. The total data adduced for GABA satisfy all of the classical requirements for the demonstration that it is a true neurotransmitter producing inhibition at the neuromuscular junction by causing selective increases in the conductance to chloride ions postsynaptically, or presynaptically at the excitatory nerve terminal. Thus, in the crustacean peripheral nervous system GABA possesses impeccable credentials as an inhibitory transmitter, perhaps more convincingly documented than in the case of any other known transmitter at any site of action.

Details of the membrane effects of GABA in crustacean systems have been reviewed [see Atwood, 1982] and experiments with voltage-clamp and patch-clamp techniques being carried out in many places will be described elsewhere in this volume.

### Vertebrate Systems

The discovery of GABA [Roberts and Frankel, 1950a] came just before there was a great increase in interest in chemical transmission [Eccles, 1982], and the devisal of the microiontophoretic techniques [Curtis and Eccles, 1958] led to the testing of the effects of substances applied directly on membranes in neurons in the CNS. With the latter technique, GABA was found to raise the membrane conductance in neurons in various regions of the mammalian CNS, e.g., cortex [Krnjevič and Schwartz, 1966; Krnjevič and Schwartz, 1967; Dreifuss et al., 1969], Deiters' nucleus [Obata et al., 1970; ten Bruggencate and Engberg, 1971], sympathetic ganglia [Adams and Brown, 1973], and in nonmammalian vertebrates as well [Martin et al., 1970; Woodward et al., 1971; Diamond et al., 1973]. In almost all instances studied, small amounts of GABA were found to have powerful, reversible inhibitory effects. A large number of studies, much more difficult to perform and necessarily less direct than can be performed in invertebrate peripheral nervous systems, permitted the conclusion that GABA typically produces an increase in membrane permeability to $Cl^-$ ions that can be measured as an increase in membrane conductance [see Krnjevič, 1976; Ito, 1976; Krnjevič, 1984 for pertinent reviews]. It is in this way that this naturally occurring transmitter can counteract the depolarizing action of excitatory processes to maintain the polarization of a cell at an equilibrium level near that of its resting value, acting essentially as a chemical voltage clamp. In most instances studied, GABA has been shown to exert a hyperpolarizing or inhibitory effect via the above mechanism. However, if high intracellular $Cl^-$ concentrations should occur, GABA can produce a decrease in membrane potential or depolarization. There are some data on release of GABA on stimulation in the CNS [Jasper et al., 1965; Jasper and Koyama, 1969; Mitchell and Srinivasan, 1969; Obata and Takeda, 1969], but these releases were observed from large aggregates of tissue and not from single inhibitory neurons. Inhibition of GABA action by the convulsants picrotoxin [Van der Kloot and Robbins, 1959] and bicuculline [Curtis et al., 1970] often has been used to indicate the presence of GABA synapses, but the pharmacology lacks the pinpoint specificity that would be required for the proof of the neurotransmittership of GABA. It is for this reason that the direct morphological demonstration of GABA neurons has been of such crucial importance in the acceptance of GABA as inhibitory neurotransmitter in the vertebrate CNS.

GABA probably is inactivated at synapses by a mechanism that involves attachment to unique membrane recognition sites, different from those for the receptor anionophores, and the subsequent removal from the synaptic junction by a $Na^+$ and $Cl^-$-dependent transport process that is similar in principle to that for many other substances [Martin, 1976; Roberts et al., 1978; Sano and Roberts, 1961; Sano and Roberts, 1963; Varon et al., 1965; Roberts et al., 1985]. Much data now question the earlier assumption that the removal of synaptically released GABA (net uptake) takes place almost entirely by reuptake into the terminals of GABAergic neurons. Evidence is accumulating that quantitatively an important part of the synaptic GABA "vacuum cleaner" also resides in $Na^+$-dependent transport sites of glial processes that invest the synapses, as well as in those of neural membranes [Campbell and Shank, 1978; Henn, 1976; Hertz et al., 1978; Schon and Kelly, 1974a; Schon and Kelly, 1975; Larsson et al., 1980]. We and others [Blaustein and King, 1976; Kuhar and Zarbin, 1978; Lasher, 1975; Kanner, 1978a; Kanner, 1978b; Kanner, 1980; Meldrum et al., 1982; Ryan and Roskoski, 1977; Roskoski, 1978; Roskoski et al., 1981] have been characterizing in detail the kinetics, ion requirements, and pharmacology of specific membrane sites related to GABA transport processes.

### Identifying GABA-Releasing Neurons in Regions of the Vertebrate CNS

The localization of GABA neurons had been inferred by correlating microchemical, electrophysiological, and iontophoretic studies with what was known of the cytoarchitecture of specific regions of brain and spinal cord. A commonly used approach was to study a particular biochemical variable in whole brain or in anatomically defined regions of brain or spinal cord. Analyses of GABA contents and GAD activities were performed in almost all identifiable brain structures and in the spinal cord. Some studies combined biochemical analyses with lesioning procedures in correlating specific neural degenerations with losses of GAD and GABA. The distributions of the components of the GABA system also were studied extensively by subcellular fractionation techniques in preparations from whole brain or selected regions. Although highly informative and necessary for range finding, interpretation of results from the above types of analyses at the cellular level suffer from the lack of definition attributable to the presence of myriad cells of different types in any dissected region; and it is difficult to make definitive conclusions about specific synaptic connectivities from them. Individual cell bodies of large neurons, such as Purkinje and Deiters' cells, and even portions thereof, have been dissected out and subjected to microanalyses. In the latter instances it was not always possible to ensure absence of contaminating presynaptic endings from other neurons or of glial cell constituents, and, therefore, their contribution to a particular measured chemical variable. The technical difficulties involved in the latter type of work usually preclude the study of large numbers of neurons in this manner.

The above approaches, while yielding fundamental and valuable data, did not make it possible to obtain a definitive understanding of how GABA neurons might participate in information processing in different parts of the vertebrate CNS. This led to the inevitable conclusion that direct visualization of components of the GABA system, particularly GABA neurons and their terminals, was necessary to obtain unequivocal proof of the existence of GABA function at specific synaptic sites in neural tissues. The most likely approaches to achieve this goal appeared to be the visualization of GAD at the light- and electron-microscopic levels. We first made many attempts to develop chemical procedures for the visualization of GAD, but all failed because of the difficulties in demonstrating histochemically the products of the enzymatic reaction, GABA and $CO_2$. The difficult alternative approach was to locate GABA neurons by immunocytochemical procedures. This required the preparation of pure GAD from brain, development of antibodies to the enzyme, and then visualization of the antibodies by a suitable labeling technique specifically at those cellular and subcellular sites where GAD, the antigen, is located. It is to this task that we dedicated a major portion of our energy for over 16 years. The basic strategy and earlier chronology of the achievement of these goals in our laboratories has been summarized [Roberts, 1975; Roberts, 1976b; Roberts, 1978a; Roberts, 1978b; Roberts, 1979; Roberts, 1980b; Ribak et al., 1981]. Much subsequent work from our own laboratories and those of others is appearing at an ever-accelerating rate, a few examples of which may be cited [Perez de la Mora et al., 1981; Houser et al., 1983; Ribak, 1981; Ribak et al., 1979; Ribak et al., 1980; Houser et al., 1980; Houser et al., 1984; Penny et al., 1983; Somogyi et al., 1983; Ohara et al., 1983; Hendrickson et al., 1983; Oertel et al., 1983; Houser et al., 1983; Barber et al., 1982]. Even GABA, itself, now can be visualized by immunocytochemical procedures [Storm-Mathisen et al., 1983].

### The Inhibited Nervous System: A New View

Although the GABA story is concerned largely with problems of neural inhibition, it is necessary to remember that neither inhibition nor the GABA system, per se, are the main issues. The question is how the GABA neurons, wherever they are demonstrated to be present, participate in information processing, together with all other neural elements, in such a way that it is possible for particular organisms to respond adaptively to their environments in a manner compatible with survival and successful reproduction. Presynaptic endings of GABA neurons on various structures in the vertebrate nervous system are illustrated by the group of light micrographs of sections of rat CNS immunocytochemically stained for GAD assembled in Figure 9. The extensive studding by GAD-positive endings on neuronal postsynaptic surfaces in the nucleus interpositus (Figs. 9a and b), globus pallidus (Fig. 9f), cortex (Fig. 9g), and hippocampus (Figs. 9h and i) is particularly noteworthy.

The ubiquity and extent of immunocytochemically visualized presynaptic endings of inhibitory GABAergic neurons on various

**Fig. 9.** *Photomicrographs of GAD-positive terminals in rat central nervous system.* a: *Deep cerebellar nucleus (nucleus interpositus). A neuron (N) is shown studded by round, GAD-positive terminals* (arrowheads) *which are probably Purkinje cell axonal terminals (magnification, ×1590).* b: *Deep cerebellar nucleus (nucleus interpositus). A grazed neuronal soma (S) showing numerous GAD-positive terminals* (arrowheads) *on the somal surface (magnification, ×1590).* c: *Cerebellar cortex. The Purkinje cell somata (P) are studded with GAD-positive terminals* (arrowheads) *which may be basket cell*

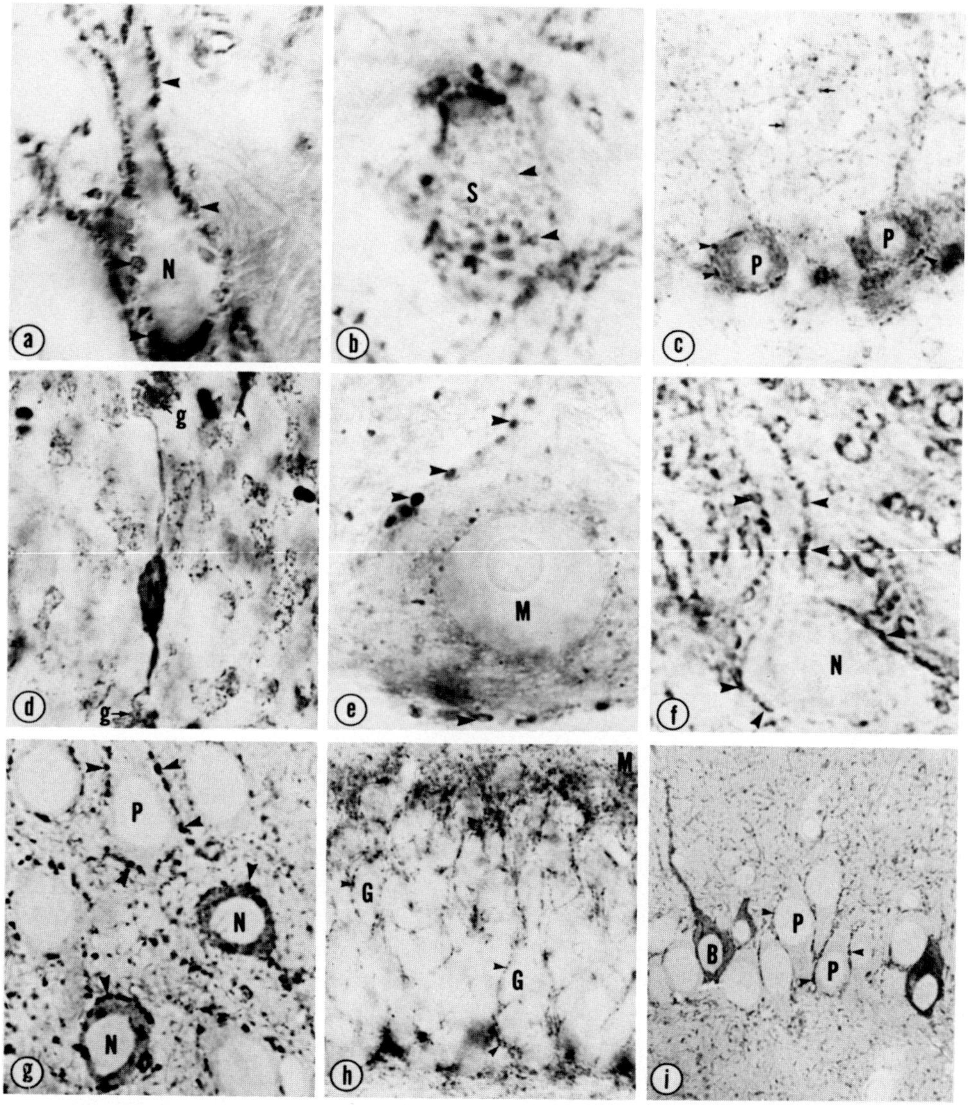


*terminals. In the molecular layer (M), there are numerous GAD-positive punctate structures* (arrows), *and in the granular layer (G), GAD-positive punctate structures (Golgi II axonal terminals) are associated with glomeruli (magnification, ×660)*. d: *Granular layer of the cerebellum from tissue treated with colchicine to block axon flow. The Golgi II neuronal soma (N) has GAD-positive cytoplasm and two of its processes are seen to terminate on glomeruli (g) (magnification, ×614)*. e: *Ventral horn of the spinal cord. A single motoneuron (M) exhibits several GAD-positive terminals* (arrowheads) *on the somal membranes (magnification, ×1571)*. f: *Globus pallidus. A neuronal profile (N) is shown which has numerous GAD-positive terminals* (arrowheads) *studding its surface. Several dendritic cross-sectional profiles (D) also are covered by punctate, GAD-positive terminals (magnification, ×1571)*. g: *Visual cortex from tissue treated with colchicine. Two neurons (N) have GAD-positive product in their cytoplasm while pyramidal cells (P) in layer V have clear cytoplasm. There are some GAD-positive terminals* (arrowheads) *on the somal surface of all of the neurons shown here (magnification, ×940)*. h: *Dentate gyrus (Ammon's horn). Numerous GAD-positive terminals* (arrowheads) *stud the profiles of granule cells (G). In the molecular layer (M), there is a heavier investment of GAD-positive terminals around apical dendrites of the granule cells (magnification, ×630)*. i: *Ammon's horn, treated with colchicine. A basket cell soma (B) exhibits cytoplasmic GAD-positive product, while pyramidal cell somata (P) do not. GAD-positive punctate structures* (arrowheads), *presumably basket cell terminals, cover the pyramidal cell somal profiles (magnification, ×560).*

structures in the vertebrate nervous system are striking. The impression is gained that one is looking at a highly restrained nervous system, the inhibitory neurons acting like reins that serve to keep the neuronal "horses" from running away [Roberts, 1976a; Roberts, 1980a; Roberts, 1984b]. In coherent behavioral sequences, innate or learned, preprogrammed circuits are *released* to function at varying rates and in various combinations. This is accomplished largely by the disinhibition of pacemaker neurons whose activities are under the dual tonic inhibitory controls of local circuit GABAergic neurons and of GABAergic projection neurons coming from neural command centers. According to this view, disinhibition is permissive, and excitatory input to pacemaker neurons would have mainly a modulatory role. Disinhibition, acting in conjunction with intrinsic pacemaker activity and often with modulatory excitatory input, is one of the major organizing principles in nervous system function. For example, cortical and hippocampal pyramidal neurons literally are studded with terminals from inhibitory GABAergic neurons. Not only are the endings of the local circuit GABAergic aspinous stellate neurons densely distributed around the somata and dendrites of the cortical pyramidal cells, but they also are located on initial axon segments, where they may act as frequency filters. In addition, GABA neurons have terminals from other GABAergic neurons impinging upon them. The pyramidal cells are tightly inhibited by local circuit inhibitory neurons which, themselves, may be inhibited by the action of other inhibitory neurons in such a way that disinhibition of the pyramidal neurons may occur. Local circuit GABAergic neurons also participate in processes that result in producing feedforward, feedback, surround, and presynaptic inhibition and presynaptic facilitation.

Both inhibition and disinhibition play key roles in information processing in all neural regions. Normally, the principal cells in particular neural sectors may be held tightly in check by constant tonic action of inhibitory neurons. Through disinhibition, neurons in one neural sector may be released to fire at different rates and sequences and, in turn, serve to release circuits at other levels of the nervous system. Communication between neural stations and substations, I believe, takes place largely by throwing of disinhibitory neural "switches." This may be the way information flows from sense organ to cerebral sensory area, through associative areas to the motor cortex, and by way of the pyramidal paths to the final motor cells of the medulla and spinal cord.

Terminals of some neurons may contain a true transmitter and also a modulator, e.g., GABA or acetylcholine and one of several neuropeptides. The frequency dependence of liberation of transmitter and modulator from inhibited nerve terminals containing both of them may not be the same. For example, a modulator that amplifies the postsynaptic efficacy of a true transmitter, or prolongs its action, may only begin to be liberated in graded amounts when the frequency of firing exceeds a certain level. Up to the latter frequency there could be a regular relationship between firing frequency, number of quanta of true transmitter released, and extent and time of postsynaptic effect. Above that frequency, the liberation of an amplifying modulator might result in a progressively increasing degree of enhancement of the postsynaptic quantal efficacy of the neurotransmitter and/or a temporal expansion of its effect. In a similar manner, it can be envisioned how a deamplifying or desensitizing modulator might operate. In the terminals of some of the primary afferents to the spinal cord (viz., pain fibers), substance P may play the role of amplifying modulator of the effects of the excitatory transmitter.

Minimally, a bifurcation of the flow of neural information from a particular sector takes place into direct through-put channels and also those leading to coordinative command centers, which at rest are believed to exert high-frequency monosynaptic tonic inhibition in various brain regions through inhibitory GABAergic projection neurons. Information arriving from several sources is integrated in the specialized neural command centers; these exert high-frequency monosynaptic tonic inhibition in various brain regions—for example, through inhibitory GABAergic projection neurons from the cerebellar cortex, the substantia nigra, the globus pallidus, and the reticular nucleus of the thalamus. Analysis of the inputs to the command centers is reflected, with variable time delays, in decreased frequencies of firing of appropriate combinations of their inhibitory output neurons, facilitating the *release* of neural activity at sites to which they project in such a manner that it becomes optimally compatible temporally and spatially with activity elsewhere in the CNS. The latter principle recently has been illustrated most convincingly in the monkey for GABAergic projection neurons from the substantia nigra pars reticulata which relay information from striatal command centers to the intermediate layers of the superior colliculus [Hikosaka and Wurtz, 1983].

Acetylcholine, the catecholamines, serotonin, neuroactive peptides, prostaglandins, and steroids in many instances may serve to optimize regional nervous system activity in relation to functional demands without themselves being involved in specific information transmittal. They exert chemical actions that influence the efficacy of the information-transmitting junctions in the mainline and command circuits. Some peptide or steroid hormones, for example, may temporarily "line-label" neural pathways, differentially facilitating behavioral options related to consummatory activities such as eating, drinking, and sex [Scheller et al., 1983]. Specific modulators may exert relatively long-lasting effects, possibly setting the gain on the efficacy of individual synapses, on specific types of synapses, or on all of the synapses in given regions or specific circuits. An inappropriate balance between availability and distribution of neural modulators with the activities in through-put neural and analyzer circuits could result in gross malfunction of the CNS. A variety of hypothalamic, pituitary, and adrenal hormones (catecholamines, peptides, and steroids) may be released into the blood stream on signals originating in the CNS, in this way altering membrane properties and metabolic states of cells in various tissues, including those of the nervous system. The cores of the neural modulator systems probably are located largely in brainstem and basal forebrain regions. Cholinergic pathways from magnocellular nuclei of the basal forebrain and noradrenergic pathways from the locus coeruleus fan out similarly to terminate in various cortical and subcortical loci, often after long traverses away from their cell bodies. Coordinated readjustments between activities of all of the above systems take place at all stages during solutions of problems facing organisms.

The points at which structural or functional lesions or both can take place that would impair the cyberneticity of such a system as that outlined above are legion. Every known transmitter candidate and every potential modulator, under one circumstance or another, has been thought to be involved in various types of neural dysfunction. Actual destruction of inhibitory GABA interneurons may be one of the major cerebral defects predisposing to seizures, at least in the case of focal epilepsy [Ribak et al., 1979; Ribak et al., 1981].

## GABA AND SEIZURES

### Consequences of Incoordination Between Inhibition and Excitation

When incoordination between the GABA system and other neurotransmitter and modulator systems persists, for whatever reason, the defect might involve a local brain region, several brain regions, or the entire CNS. A critically placed local incoordination might have drastic reverberations in the whole nervous system, as found in grand mal seizures arising from focal cortical lesions. Under relatively simple environmental conditions in individuals with such incoordinations, the nervous system could function in an apparently adequately adaptive manner, which might appear to be in the normal range. As the complexity and intensity of environmental inputs are increased, there would be a correlated increased degree of incoordination.

Let us suppose that for some reason tonically active projection and local-circuit inhibitory GABA neurons, in the entire brain or in specific regions, have considerably lower than normal effectiveness on their recipient neurons which, themselves, are normally effective; or, alternatively, that the GABA neurons are normal and the neurons they ordinarily control are less than normally responsive to GABA or are potentially hyperactive because of intrinsically arising or extrinsically produced membrane changes. With increasing excitatory input, there would be an increased tendency to release pacemaker neurons. Abnormal or inappropriate behavior might be released (e.g., unprovoked violence, mania, schizophrenia); choreic movements, seizures, or spasticity might occur; there might be hypersensitivity to visual, auditory, tactile, olfactory, gustatory, or pain stimuli. GABA neurons play important roles in control mechanisms in various hypothalamic and brainstem centers. Thus, if specific regions within these latter structures were affected in a particular situation, abnormally enhanced responses might be observed—for example, in emotional reactivity, cardiac and respiratory functions, blood pressure, food and water intake, sweating, insulin secretion, liberation of gastric acid, or motility of the colon.

The permissive element in CNS activity is postulated to be the release from local and remote tonic inhibition. In various regions of the nervous system the activity is determined largely by varying combinations of local and centrally generated disinhibitory commands. Direct excitation probably is not sufficient by itself to release neurons for activity during normal nervous system function. As the intensity of a particular input is increased from some low basal level, there would be a corresponding increase of excitatory influences, resulting in release of correspondingly greater numbers of neurons. Inputs from their excitatory axon collaterals to each other would add to the excitatory and disinhibitory influences on neurons in the same module, eventually recruiting them into synchronized discharge patterns [Johnston and Brown, 1981]. For example, there is recurrent excitation between hippocampal pyramidal cells in the CA3 region [MacVicar and Dudek, 1980]. Spikes in one pyramidal cell can evoke spikes in another pyramidal cell. This could continue until virtually all available neurons in the modules of a given neural section would be involved in responding to a pertinent input. Current data do not require the presence of abnormalities in membranes of particular neurons for this to occur. The giant EPSPs of local hippocampal seizures can result from the synchronous activity of normal neurons [Johnston and Brown, 1981; Traub and Wong, 1982].

Enhanced synchrony of neuronal firing may arise in several ways—increased rate of release of synaptic excitatory transmitters; blockade of inhibitory transmitter receptor mechanisms; desensitization of receptors to inhibitory transmitters; decreased availability of inhibitory transmitter; decreased activity of inhibitory neurons; or increased formation or activation of electrotonic junctions. Most recently the suggestion has been made that an excess of disinhibition (inhibition of GABAergic neurons) may be involved in a genetic form of epilepsy in rats [RC Roberts et al., 1984]. Decreases in rate of formation or release of inhibitory transmitter could result from decreased function or actual destruction of inhibitory neurons. There are many experimental observations relevant to one or another of the above mechanisms [Roberts, 1980]. Extensive immunocytochemical studies in many regions of the vertebrate CNS have shown that the vast majority of inhibitory neurons probably are GABAergic.

### Loss of Inhibitory, GABAergic Nerve Terminals at Sites of Focal Cortical Epilepsy

If numbers of GABA neurons were decreased, the restraints on excitatory pacemaker neurons in a given region of the CNS might be weakened and they would more easily recruit each other into the runaway activities characteristic of seizures. Identification of GABAergic neurons by immunocytochemical techniques allowed us to answer some important morphologic questions [Ribak, 1978; Roberts, 1978b; Roberts, 1980a]. The GABA-synthesizing enzyme, GAD, was observed in somata, proximal dendrites, and axon terminals of nonpyramidal neurons in the rat visual cortex in all cortical layers, including the immediately subjacent white matter [Ribak, 1978]. All GAD-positive neurons observed were nonpyramidal cells. GAD-positive terminals formed symmetric synaptic junctions most commonly with dendritic shafts and somata of pyramidal and stellate neurons and less frequently with dendritic spines and with initial axon segments of pyramidal neurons in layers III and V. The immunocytochemical localization of GAD in combination with physiologic and pharmacologic data indicates that these local circuit neurons mediate GABAergic inhibition and disinhibition in the neocortex and probably play key roles in information processing. Morphologic data in rat cortex cited above and in normal monkey sensorimotor cortex [Houser et al., 1983] are consistent with the possibility that virtually all of the types of inhibitory neurons are GABAergic. The several different morphological classes of intrinsic neurons that comprise the population of GABAergic neurons include small basket cells, spider-web cells, and chandelier cells [Houser et al., 1983].

Immunocytochemical studies of the sensorimotor cortex obtained from five epileptic monkeys treated with alumina cream correlated epileptic activity with numbers of GABAergic terminals [Ribak et al., 1979]. Highly significant reductions in numbers of GAD-positive terminals were found at the electrographically proven epileptogenic sites of alumina gel application by comparison with the contralateral nonepileptic homotopic cortex and at ipsilateral sections further away from the alumina gel. These results support the idea that a relatively selective loss of inhibitory GABAergic neurons could be responsible for susceptibility to epileptic activity observed at seizure foci. Loss of adhering GABAergic neurons by excitatory projection neurons leads to poor inhibitory control of these neurons. This allows such projection neurons to fire excessively individually and, in particular neural sectors, to engage each other in paroxysmal, seizure-like discharges.

The latter observations are particularly pertinent, since many epilepsies in human beings are caused by lesions in the brain resulting from trauma, such as head injuries, tumors, and cerebral angiomas. The above conditions may interfere with the normal blood supply to a particular region. Probably inhibitory GABAergic neurons (and glycinergic ones) have a high metabolic rate and are more sensitive to anoxia than are other types of cells [Arregui and Barer, 1980; Davidoff et al., 1967a; Davidoff et al., 1967b; Dunwiddie et al., 1980; Francis and Pulsinelli, 1982]. Their function may be disturbed differentially even by relatively slight degrees of anoxia [Dunwiddie et al., 1980]; if anoxia persists, degenerative changes and cell destruction take place rapidly. This vulnerability also may extend to anoxia resulting from excessive functional demands and the large increases in metabolic activity and changes in physical state that occur in febrile states. An extraordinarily heavy excitatory bombardment to a particular brain sector, whatever its source, might create conditions in the vicinity of the local circuit neurons, such as too great elevation of cyclic GMP level [Lolley et al., 1977] and depletion of glucose, oxygen, and ATP in the affected region, which would adversely affect the function and possibly even the existence of the GABA interneurons in the region. Pathological processes, of whatever nature, that damage the microcapillaries, alter their permeability properties, cause prolonged local vasoconstriction, or in some more direct way induce mitochondrial malfunction, can result in destruction of the inhibitory GABA neurons with the replacement of their largely axosomatic synapses on pyramidal cells by processes of reactive astrocytes. This could sensitize the affected neural sector to epileptiform discharge. Electron microscopic observations on the monkey cortical material discussed above showed there to be a marked loss of axosomatic synapses on the pyramidal cells and a replacement of synaptic appositions with astrocytic processes in the alumina cream-treated animals. However, the asymmetric, presumably excitatory, synapses on the dendrites of these pyramidal cells appeared to be largely intact [Ribak et al., 1979].

In biochemical studies of monkey cortical material complementary to the above morphological observations, a significant correlation with seizure frequency was found only with losses in GABAergic receptor–related binding and decreases in GAD activity [Bakay and Harris, 1981].

Taken together, the above observations strongly support the notion that actual destruction of inhibitory GABA interneurons is one of the major cerebral defects predisposing to seizures, at least in the case of focal epilepsy. Since the early work of Killam [Killam and Bain, 1957; Killam, 1957], it has been inferred that a decrease in GABA levels or GAD activity could give rise to seizures, and many hundreds of studies had been performed with uncertain conclusions. The immunocytochemical approach was necessary to obtain definitive results.

## CONCLUSION

The other contributions in this volume mostly will deal with "high-tech" endeavors at the cutting edge of GABA receptorology. Many of the younger scientists who will read the contributions will be in tune with them, and some even may join in the pursuit of one or another aspect of GABA receptor subunit characterization and the molecular biological manipulations that are sure to follow. I thought that it might be salutary to present some of the experiences of a "gray-beard" in the field and to acquaint the newcomers with some of the vicissitudes which I have experienced in my long association with the simple GABA molecule. It has been like peeling an onion, one problem inevitably leading to another in endless succession. I am sure that many surprises still exist. Although some of the most interesting results will come from a linear approach to a study of molecular detail, I would like to encourage those in the field to keep scanning the GABA environment for new directions and approaches. The unexpected discovery of GABA was made possible because a powerful chemical scanning technique, two-dimensional paper chromatography, was applied to brain extracts for the first time.

## ACKNOWLEDGMENTS

This work was supported in part by grants NS18858 and NS18859 from the National Institutes of Health.

The author wishes to acknowledge the invaluable help of Mrs. Betty Baker in preparing this manuscript, and of Robert Barber and Charles Ribak, who prepared Figure 9.

## REFERENCES

Adams PR, Brown DA (1973): Action of γ-aminobutyric acid (GABA) on rat sympathetic ganglion cells. Br J Pharmacol 47:639–640.

Aishita H, Akimoto A, Makita T (1978): Comparison of the inhibitory effects of GABOB optical isomers on seizure discharges elicited by electrical stimulation of the motor cortex in rabbits. IRCS Med Sci Nerv Syst Pharmacol Physiol 6:115.

Alexandrowicz JS (1951): Muscle receptor organs in the abdomen of *Homarus vulgaris* and *Palinurus vulgaris*. Quart J Microsc Sci 92:163–199.

Arregui A, Barer GR (1980): Chronic hypoxia in rats: Alterations of striato-nigral angiotensin converting enzyme, GABA, and glutamic acid decarboxylase. J Neurochem 34:740–743.

Atwood HL (1982): Synapses and neurotransmitters. In Atwood HL, Sandeman DC (eds): "The Biology of Crustacea," Vol. 3. New York: Academic Press, pp 105–150.

Awapara J (1950): Detection and identification of metabolites in tissues by means of paper chromatography. Fed Proc 9:148.

Awapara J, Landua AJ, Fuerst R, Seale B (1950): Free $\gamma$-aminobutyric acid in brain. J Biol Chem 187:35–39.

Bakay RAE, Harris AB (1981): Neurotransmitter, receptor, and biochemical changes in monkey cortical epileptic foci. Brain Res 206:387–404.

Barber RP, Vaughn JE, Roberts E (1982): The cytoarchitecture of GABAergic neurons in rat spinal cord. Brain Res 238:305–328.

Barnard EA, Stephenson FA, Sigel E, Mamalaki C, Bilbe G (1984): The purified GABA/benzodiazepine complex: Retention of multiple functions. Neuropharmacol 23:813–814.

Bazemore AW, Elliott KAC, Florey E (1957): Isolation of Factor I. J Neurochem 1:334–339.

Beart PM, Johnston GAR (1973): GABA uptake in rat brain slices: Inhibition by GABA analogues and by various drugs. J Neurochem 20:319–324.

Blaustein MP, King AC (1976): Influence of membrane potential on the sodium-dependent uptake of gamma-aminobutyric acid by presynaptic nerve terminals: Experimental observations and theoretical considerations. J Membrane Biol 30:153–173.

Bowery NG, Hill DR, Hudson AL (1983): Characteristics of $GABA_B$ receptor binding sites on rat whole brain synaptic membranes. Br J Pharmacol 78:191–206.

Bradford HF, Ward HK (1976): On glutaminase activity in mammalian synaptosomes. Brain Res 110:115–125.

Bradford HF, Ward HK, Thomas AJ (1978): Glutamine—a major substrate for nerve endings. J Neurochem 30:1453–1459.

Brock LG, Coombs JS, Eccles JC (1952): The recording of potentials from motoneurones with an intracellular electrode. J Physiol (London) 117:431–462.

Campbell GLeM, Shank RP (1978): Glutamate and GABA uptake by cerebellar granule and glial cell enriched populations. Brain Res 153:618–622.

Changeux J-P, Benedetti L, Bourgeois J-P, Brisson A, Cartaud J, Devaux P, Grünhagen H, Moreau M, Popot J-L, Sobel A, Weber M (1976): Some structural properties of the cholinergic receptor protein in its membrane environment relevant to its function as a pharmacological receptor. Cold Spring Harbor Symp Quant Biol 40:211–230.

Colquhoun D (1978): The link between drug binding and response: Theories and observations. In O'Brien RD (ed): "The Receptors: A Comprehensive Treatise." New York: Plenum Press, pp 93–142.

Coombs JS, Eccles JC, Fatt P (1953): The action of the inhibitory synaptic transmitter. Aust J Sci 16:1–5.

Coursin DB (1954): Convulsive seizures in infants with pyridoxine-deficient diet. J Am Med Assoc 154:406–408.

Coursin DB (1960): Seizures in vitamin $B_6$ deficiency. In Roberts E, Baxter C, van Harreveld A, Wiersma CAG, Adey WR, Killam KF (eds): "Inhibition in the Nervous System and $\gamma$-Aminobutyric Acid." Oxford: Pergamon Press, pp 294-301.

Curtis DR, Duggan AW, Felix D, Johnston GAR (1970): GABA, bicuculline and central inhibition. Nature (London) 226:1222–1224.

Curtis DR, Eccles RM (1958): The excitation of Renshaw cells by pharmacological agents applied electrophoretically. J Physiol (London) 141:435–445.

Curtis DR, Phillis JW, Watkins JC (1959): The depression of spinal neurones by certain acidic amino acids. J Physiol (London) 146:185–203.

Davidoff RA, Shank RP, Graham Jr LT, Aprison MH, Werman R (1967a): Association of glycine with spinal interneurons. Nature (London) 214:680–681.

Davidoff RA, Graham Jr LT, Shank RP, Werman R, Aprison MH (1967b): Changes in amino acid concentrations associated with loss of spinal interneurons. J Neurochem 14:1025–1031.

Davies J, Watkins JC (1974): The action of $\beta$-phenyl-GABA derivatives on neurones of the cat cerebral cortex. Brain Res 70:501–505.

Davies RK, Defalco AJ, Shander D, Kopelman A, Kiyasu J (1961): Urea synthesis in the living rat brain. Nature 191:288.

Davson H (1964): "A Textbook of General Physiology." Third Edition. London: Churchill, pp 806–807.

De Feudis FV (1978): Central glycine-receptors. Gen Pharmacol 9:139–144.

De Feudis FV (1980): Binding studies with muscimol; relation to synaptic $\gamma$-aminobutyrate receptors. Neuroscience 5:675–688.

Diamond S, Balvin RS, Diamond FR (1963): "Inhibition and Choice." New York: Harper and Row.

Diamond J, Roper S, Yasargil GM (1973): The membrane effects, and sensitivity to strychnine, of neural inhibition of the Mauthner cell, and its inhibition by glycine and GABA. J Physiol (London) 232:87–111.

Dreifuss JJ, Kelly JS, Krnjević K (1969): Cortical inhibition and $\gamma$-aminobutyric acid. Exp Brain Res 9:137–154.

Dudel J, Gryder R, Kaji A, Kuffler SW, Potter DD (1963): Gamma-aminobutyric acid and other blocking compounds in crustacea. I. Central nervous system. J Neurophysiol 26:721–728.

Dudel J, Kuffler SW (1961): Presynaptic inhibition at the crayfish neuromuscular junction. J Physiol (London) 155:543–562.

Dunwiddie T, Mueller A, Palmer M, Stewart J, Hoffer B (1980): Electrophysiological interactions of enkephalins with neuronal circuitry in the rat hippocampus. I. Effects on pyramidal cell activity. Brain Res 184:311–330.

Eccles JC (1969): "The Inhibitory Pathways of the Central Nervous System." Springfield, IL: Charles C. Thomas.

Eccles JC (1982): The synapse: From electrical to chemical transmission. Ann Rev Neurosci 5:325–339.

Edwards C (1960): Physiology and pharmacology of the crayfish stretch receptor. In Roberts E, Baxter CF, van Harreveld A, Wiersma CAG, Adey WR, Killam KF (eds): "Inhibition in the Nervous System and Gamma-aminobutyric Acid." Oxford: Pergamon Press, pp 386–408.

Edwards C, Kuffler SW (1959): The blocking effect of $\gamma$-aminobutyric acid (GABA) and the action of related compounds on single nerve cells. J Neurochem 4:19–30.

Edwards C, Ottoson D (1958): The site of impulse initiation in a nerve cell of a crustacean stretch receptor. J Physiol (Lond) 143:138–148.

Elliott KAC, Florey E (1956): Factor I—Inhibitory factor from brain. Assay. Conditions in brain. Simulating and antagonizing substances. J Neurochem 1:181–191.

Ellis CH, Thienes CH, Wiersma CAG (1942): The influence of certain drugs on the crustacean nerve-muscle system. Biol Bull 83:334–352.

Enna SJ, Snyder S (1975): Properties of $\gamma$-aminobutyric acid (GABA) receptor binding in rat brain synaptic membrane fractions. Brain Res 100:81–97.

Florey E (1953): Über einen nervösen Hemmungsfaktor in Gehirn und Rückenmark. Naturwiss 40:295–296.

Florey E (1954a): An inhibitory and an excitatory factor of mammalian central nervous system, and their action on a single sensory neuron. Arch Internationales de Physiologie LXII:33–53.

Florey E (1954b): Über die Wirkung von Acetylcholin, Adrenalin, Nor-adrenalin, Faktor I und anderen Substanzen auf den isolierten Enddarm des Flüsskrebses *Cambarus clarkii* girard. Zeitschriftr fur vergleichende Physiologie 36:1–8.

Florey E (1955): Untersuchungen über die Impuls-entstehung in den Streckreceptoren des Flüsskrebses. Zeitschrift fur Naturforschung Bd 10b Heft 10:591–597.

Florey E (1956a): The action of Factor I on certain invertebrate organs. Can J Biochem Physiol 34:669–681.

Florey E (1956b): Adaptationsercheinungen in den sensiblen Neuronen des Streckreceptoren des Flüsskrebses. Zeitschrift fur Naturforschung 11b(Heft 9/10):504–513.

Florey E (1957): Further evidence for the transmitter-function of Factor I. Naturwiss 44(Heft 15):424–425.

Florey E (1960): Studies on the nervous regulation of the heart beat in decapod crustacea. J Gen Physiol 46:1061–1081.

Florey E, Biederman MA (1960): Studies on the distribution of Factor I and acetylcholine in crustacean peripheral nerve. J Gen Physiol 43:509–522.

Florey E, Florey E (1958): Studies on the distribution of Factor I in mammalian brain. J Physiol 144:220–228.

Florey E, McLennan H (1955a): Die Wirkung des Hemmungsfaktors aus Gehirn und Rückenmark auf periphere und zentrale synaptische Übertragung. Naturwiss 42(Heft 2):51–52.

Florey E, McLennan H (1955b): The release of an inhibitory substance from mammalian brain, and its effect on peripheral synaptic transmission. J Physiol 129:384–392.

Florey E, McLennan H (1955c): Effects of an inhibitory factor (Factor I) from brain on central synaptic transmission. J Physiol 130:446–455.

Fox S, Krnjević K, Morris ME, Puil E, Werman R (1978): Action of baclofen on mammalian synaptic transmission. Neuroscience 3:495–515.

Francis A, Pulsinelli W (1982): The response of GABA-ergic and cholinergic neurons to transient cerebral ischemia. Brain Res 243:271–278.

Gale EF (1946): The bacterial amino acid decarboxylases. Advances in Enzymology 6:1–32.

Geddes JW, Wood JD (1984): Changes in the amino acid content of nerve endings (synaptosomes) induced by drugs that alter the metabolism of glutamate and $\gamma$-aminobutyric acid. J Neurochem 42:16–24.

Giambalvo CT, Rosenberg P (1976): The effect of phospholipases and proteases on the binding of $\gamma$-aminobutyric acid to junctional complexes of rat cerebellum. Biochim Biophys Acta 436:741–756.

Grafstein B (1963): Convulsant drug action on neuronally isolated cerebral cortex. Science 142:973–975.

Ha YH, Milner JA, Corbin JE (1978): Arginine requirements in immature dogs. J Nutrition 108:203–210.

Hall ZW, Bownds MD, Kravitz EA (1970): The metabolism of gamma aminobutyric acid in the lobster nervous system. Enzymes in single excitatory and inhibitory axons. J Cell Biol 46:290–299.

Hall ZW, Kravitz EA (1967a): The metabolism of $\gamma$-aminobutyric acid (GABA) in the lobster nervous system—I. GABA-glutamate transaminase. J Neurochem 14:45–54.

Hall ZW, Kravitz EA (1967b): The metabolism of $\gamma$-aminobutyric acid (GABA) in the lobster nervous system—II. Succinic semialdehyde dehydrogenase. J Neurochem 14:55–61.

Hamberger AC, Chiang GH, Nylen ES, Sheff SW, Cotman CW (1979): Glutamate as a CNS transmitter. I. Evaluation of glucose and glutamine as precursors for the synthesis of preferentially released glutamate. Brain Res 168:513–530.

Hayashi T (1959): The inhibitory action of $\beta$-hydroxy-$\gamma$-aminobutyric acid upon the seizure following stimulation of the motor cortex of the dog. J Physiol (Lond) 145:570–578.

Hayashi T, Nagai K (1956): Action of $\omega$-amino acids on the motor cortex of higher animals, especially $\gamma$-amino-$\beta$-oxy-butyric acid as the real inhibitory principle in brain. In "20th Int Physiol Congr Brussels, 1956. Abstracts of Reviews: Abstracts of Communications," p. 410.

Hayashi T, Suhara R (1956): Substances which produce epileptic seizures when applied on the motor cortex of dogs, and substances which inhibit the seizure directly. In "20th Int Physiol Congr Brussels, 1956. Abstracts of Reviews: Abstracts of Communications," p 410.

Hendrickson AE, Ogren MP, Vaughn JE, Barber RP, Wu J-Y (1983): Light and electron microscopic immunocytochemical localization of glutamic acid decarboxylase in monkey geniculate complex: Evidence for GABAergic neurons and synapses. J Neurosci 3:1245–1262.

Henn FA (1976): Neurotransmission and glial cells: A functional relationship? J Neurosci Res 2:271–282.

Hertz L, Wu PH, Schousboe A (1978): Evidence for net uptake of GABA into mouse astrocytes in primary cultures—its sodium dependence and potassium independence. Neurochem Res 3:313–323.

Hikosaka O, Wurtz RH (1983): Visual and oculomotor functions of monkey substantia nigra pars reticulata. IV. Relation of substantia nigra to superior colliculus. J Neurophysiol 49:1285–1301.

Hori N, Ikeda K, Roberts E (1978): Muscimol, GABA and picrotoxin: Effects on membrane conductance of crustacean neuron. Brain Res 141:364–370.

Horwitz IS, Orkand RK (1980): GABA inactivation at the crayfish neuromuscular junction. J Neurobiol 11:447–458.

Houser CR, Hendry SHC, Jones EG, Vaughn JE (1983): Morphological diversity of immunocytochemically identified GABA neurons in the monkey sensory-motor cortex. J Neurocytol 12:617–638.

Houser CR, Lee M, Vaughn JE (1983): Immunocytochemical localization of glutamic acid decarboxylase in normal and deafferented superior colliculus: Evidence for reorganization of $\gamma$-aminobutyric acid synapses. J Neurosci 3:2030–2042.

Houser CR, Vaughn JE, Barber RP, Roberts E (1980): GABA neurons are the major cell type of the nucleus reticularis thalami. Brain Res 200:341–354.

Houser CR, Vaughn JE, Hendry SHC, Jones EG, Peters A (1984): GABA neurons in the cerebral cortex. In Jones EG, Peters A (eds): "Cerebral Cortex, Vol 2, Functional Properties of Cortical Cells." New York: Plenum Press, pp 63–89.

Hoyle G, Wiersma CAG (1958a): Coupling of membrane potential to contraction in crustacean muscles. J Physiol (Lond) 143:441–453.

Hoyle G, Wiersma CAG (1958b): Excitation at neuromuscular junctions in crustacea. J Physiol (Lond) 143:403–425.

Hoyle G, Wiersma CAG (1958c): Inhibition of neuromuscular junctions in crustacea. J Physiol (Lond) 143:426–440.

Hunt AD Jr, Stokes Jr J, McCrory WW, Stroud HH (1954): Pyridoxine dependency: Report of a case of intractable convulsions in an infant controlled by pyridoxine. Pediatrics 13:140–145.

Huyskens PL, Smets RS, Mas DH (1977): Self-association of pyridine in water. Bull Soc Chim Belg 86:741–748.

Ito M (1976): Roles of GABA neurons in integrated functions of the vertebrate CNS. In Roberts E, Chase TN, Tower DB (eds): "GABA in Nervous System Function." New York: Raven Press, pp 427–448.

Iversen LL, Kravitz EA (1968): The metabolism of $\gamma$-aminobutyric acid (GABA) in the lobster nervous system—uptake of GABA in nerve-muscle preparations. J Neurochem 15:609–620.

Iwasaki S, Florey E (1969): Inhibitory miniature potentials in the stretch receptor neurons of crayfish. J Gen Physiol 53:666–682.

Jasper HH, Khan RT, Elliott KAC (1965): Amino acids released from cerebral cortex in relation to its state of activation. Science 147:1448–1449.

Jasper HH, Koyama I (1969): Rate of release of amino acids from the cerebral cortex in the cat as affected by brain stem and thalamic stimulation. Can J Physiol Pharmacol 47:889–905.

Johnson AB, Strecker HJ (1962): The interconversion of glutamic acid and proline. IV. The oxidation of proline by rat liver mitochondria. J Biol Chem 237:1876–1882.

Johnson JL, Roberts E (1984a): Arginine metabolism in mouse brain synaptosomes. J Neurochem 42:1123–1126.

Johnson JL, Roberts E (1984b): Proline, glutamate and glutamine metabolism in mouse brain synaptosomes. Brain Res 323(2):247–256.

Johnston D, Brown TH (1981): Giant synaptic potential hypothesis for epileptiform activity. Science 211:294–297.

Johnston GAR, Kennedy SME (1978): GABA receptors and phospholipids. In Fonnum F (ed): "Amino Acids as Chemical Transmitters." New York: Plenum Publishing Co., p 507–516.

Jones ME, Anderson AD, Anderson C, Hodes S (1961): Citrulline synthesis in rat tissues. Arch Biochem Biophys 95:499–507.

Kanner BI (1978a): Active transport of $\gamma$-aminobutyric acid by membrane vesicles isolated from rat brain.

Biochemistry 17:1207–1211.
Kanner BI (1978b): Solubilisation and reconstitution of the $\gamma$-aminobutyric acid transporter from rat brain. FEBS Letters 89:47–50.
Kanner BI (1980): Modulation of neurotransmitter transport by the activity of the action potential sodium ion channel in membrane vesicles from rat brain. Biochemistry 19:692–697.
Katayama Y (1976): Metabolism of 4-amino-3-hydroxybutanoic acids in the mouse organs and their inhibitory effects on penicillin-induced spike activity of the cat cerebral cortex and on electrical activity of an identified snail giant neurone. Okayama Igakkai Zasshi 88:209–221.
Katayama Y, Mori A (1977): Inhibitory action of (3R)-(-)-4-amino-3-hydroxybutanoic acid on N-amidino-benzamide induced seizure activity in cat brain. IRCS Med Sci Nerv Syst Pharmacol Physiol 5:437.
Kaufman DL, McGinnis JF, Wood TL, Tobin AJ (1985): Brain glutamate decarboxylase (GAD) CDNA cloned on the basis of its ability to produce GABA. Soc Neurosci Abstr 11:1115.
Killam KF (1957): Convulsant hydrazides 2: Comparison of electrical changes and enzyme inhibition induced by the administration of thiosemicarbazide. J Pharmacol Exp Ther 119:263–271.
Killam KF, Bain JA (1957): Convulsant hydrazides 1: In vitro and in vivo inhibition of vitamin $B_6$ enzymes by convulsant hydrazides. J Pharmacol Exp Ther 119:255–262.
Killam KF, Dasgupta SR, Killam EK (1960): Studies of the action of convulsant hydrazides as vitamin $B_6$ antagonists in the central nervous system. In Roberts E, Baxter CF, van Harreveld A, Wiersma CAG, Adey WR, Killam KF (eds): "Inhibition in the Nervous System and Gamma-aminobutyric Acid." Oxford: Pergamon Press, pp 302–316.
Koidl B, Florey E (1975): Factor I and GABA: Resolution of a long-standing problem. Comp Biochem Physiol 51C:13–23.
Kopecky F, Wojtekova M, Bednarova-Hyttnerova M (1978): Osmometric study of self-association of nicotinamide and isonicotinic acid hydrazide in aqueous solutions of 40°C and the freezing point. Coll Czech Chem Commun 43:37–46.
Krause DN, Ikeda K, Roberts E (1981): Dose-conductance relationships for GABA agonists and the effect of uptake inhibitors in crayfish stretch receptor neurons. Brain Res 225:319–332.
Kravitz EA (1962): Enzymic formation of gamma-aminobutyric acid in the peripheral and central nervous system of lobsters. J Neurochem 9:363–370.
Kravitz EA, Iversen LL, Otsuka M, Hall ZW (1968): Gamma-aminobutyric acid in the lobster nervous system; release from inhibitory nerves and uptake into nerve-muscle preparations. In von Euler C, Skoglund S, Söderberg, U (eds): "Structure and Function of Inhibitory Neuronal Mechanisms." Oxford: Pergamon Press, pp 371–376.
Kravitz EA, Kuffler SW, Potter DD (1963a): Gamma-aminobutyric acid and other blocking compounds in crustacea. III. Their relative concentrations in separated motor and inhibitory axons. J Neurophysiol 26:739–751.
Kravitz EA, Kuffler SW, Potter DD, van Gelder NM (1963b): Gamma-aminobutyric acid and other blocking compounds in crustacea. II. Peripheral nervous system. J Neurophysiol 26:729–738.
Kravitz EA, Molinoff PB, Hall ZW (1965): A comparison of the enzymes and substrates of gamma-aminobutyric acid metabolism in lobster excitatory and inhibitory axons. Proc Nat Acad Sci 54:778–782.
Kravitz EA, Potter DD, van Gelder NM (1962): Gamma-aminobutyric acid distribution in the lobster nervous system: CNS, peripheral nerves and isolated motor and inhibitory axons. Biochem Biophys Res Commun 7:231–236.
Krnjević K (1976): Inhibitory action of GABA and GABA-mimetics on vertebrate neurons. In Roberts E, Chase TN, Tower DB (eds): "GABA in Nervous System Function." New York: Raven Press, pp 269–281.
Krnjević K (1984): Neurotransmitters in cerebral cortex. In Jones EG, Peters A (eds): "Cerebral Cortex, Vol 2: Functional Properties of Cortical Cells." New York: Plenum Press, pp 39–61.
Krnjević K, Schwartz S (1966): Is $\gamma$-aminoibutyric acid an inhibitory transmitter? Nature (Lond) 211:1372–1374.
Krnjević K, Schwartz S (1967): The action of $\gamma$-aminobutyric acid on cortical neurones. Exp Brain Res 3:320–336.
Krogsgaard-Larsen P, Johnston GAR (1975): Inhibition of GABA uptake in rat brain slices by nipecotic acid, various isoxazoles and related compounds. J Neurochem 25:797–802.
Krogsgaard-Larsen P, Johnston GAR, Lodge D, Curtis DR (1977): A new class of GABA agonist. Nature (Lond) 268:53–55.
Kuffler SW (1958): Synaptic inhibitory mechanisms. Properties of dendrites and problems of excitation in isolated sensory nerve cells. Exp Cell Res Suppl 5:493–519.
Kuffler SW, Edwards C (1958): Mechanism of gamma aminobutyric acid (GABA) action and its relation to synaptic inhibition. J Neurophysiol 21:589–610.
Kuffler SW, Eyzaguirre C (1955): Synaptic inhibition in an isolated nerve cell. J Gen Physiol 39:155–184.
Kuhar MJ, Zarbin MA (1978): Synaptosomal transport: A chloride dependence for choline, GABA, glycine and several other compounds. J Neurochem 31:251–256.
Kvamme E (1979): Regulation of glutaminase and its possible implication for GABA metabolism. Adv

Exp Med Biol 123:111–138.
Lapinjoki SP, Pajunen AEI, Pulkka AE, Piha RS (1982): On the metabolism of ornithine in synaptosomal preparations. Neurochem Res 7:645–655.
Larsson OM, Hertz L, Schousboe A (1980): GABA uptake in astrocytes in primary cultures: Coupling with two sodium ions. J Neurosci Res 5:469–477.
Lasher RW (1975): Uptake of GABA by neuronal and non neuronal cells in dispersed cell cultures of postnatal rat cerebellum. J Neurobiol 6:597–608.
Levin E, Lovell RA, Elliott KAC (1961): The relation of gamma-aminobutyric acid to Factor I in brain extracts. J Neurochem 7:147–154.
Lloyd KG, Davidson L (1979): [$^3$H]GABA binding in brains from Huntington's chorea patients: Altered regulation by phospholipids? Science 205:1147–1149.
Lodge D, Curtis DR, Johnston GAR (1978): Does uptake limit the action of GABA agonists in vivo? Experiments with muscimol, isoguvacine and THIP in cat spinal cord. J Neurochem 31:1525–1528.
Lodge D, Johnston GAR, Curtis DR, Brand SJ (1977): Effects of the *Areca* nut constituents arecaidine and guvacine on the action of GABA in the cat central nervous system. Brain Res 136:513–522.
Lolley RN, Farber DB, Rayborn ME, Hollyfield JG (1977): Cyclic GMP accumulation causes degeneration of photoreceptor cells: Simulation of an inherited disease. Science 196:664–666.
MacVicar BA, Dudek FE (1980): Local synaptic circuits in rat hippocampus: Interactions between pyramidal cells. Brain Res 184:220–223.
Martin AR, Wickelgren WO, Beranck R (1970): Effects of iontophoretically applied drugs on spinal interneurones of the lamprey. J Physiol (Lond) 207:653–665.
Martin DL (1976): Carrier-mediated transport and removal of GABA from synaptic regions. In Roberts E, Chase TN, Tower DB (eds): "GABA and Nervous System Function." New York: Raven Press, pp 347–386.
Mathers DA, Barker JL (1981): GABA and muscimol open ion channels of different lifetimes on cultured mouse spinal cord cells. Brain Res 204:242–247.
Matsuda T, Wu J-Y, Roberts E (1973a): Electrophoresis of glutamic acid decarboxylase from mouse brain in sodium dodecyl sulfate polyacrylamide gels. J Neurochem 21:167–172.
Matsuda T, Wu J-Y, Roberts E (1973b): Immunocytochemical studies in glutamic decarboxylase from mouse brain. J Neurochem 21:159–166.
McLaughlin BJ, Barber R, Saito K, Roberts E, Wu J-Y (1975): Immunocytochemical localization of glutamate decarboxylase in rat spinal cord. J Comp Neurol 164:305–321.
McLennan H (1957a): A comparison of some physiological properties of Factor I and γ-aminobutyric acid. Naturwiss 44:116–117.
McLennan H (1957b): A comparison of some physiological properties of an inhibitory factor from brain (Factor I) and of γ-aminobutyric acid and related compounds. J Physiol 139:79–86.
McLennan H (1958): Absence of γ-aminobutyric acid from brain extracts containing Factor I. Nature (Lond) 181:1807.
McLennan H (1959): The identification of one active component from brain extracts containing Factor I. J Physiol 146:358–368.
McLennan H (1960): The fractionation and purification of Factor I. J Physiol 151:31–39.
Meiners BM, Kehoe P, Shaner DM, Olsen RW (1979): γ-Aminobutyric acid receptor binding and uptake in membrane fractions of crayfish muscle. J Neurochem 32:979–990.
Meldrum BS, Croucher MJ, Krogsgaard-Larsen P (1982): GABA-uptake inhibitors as anticonvulsant agents. In Okada Y, Roberts E (eds): "Problems in GABA Research." Amsterdam: Excerpta Medica, pp 182–191.
Mitchell JF, Srinivasan V (1969): Release of $^3$H-γ-aminobutyric acid from the brain during synaptic inhibition. Nature (Lond) 224:663–666.
Molinoff PB, Kravitz EA (1968): The metabolism of γ-aminobutyric acid (GABA) in the lobster nervous system—glutamic decarboxylase. J Neurochem 15:391–409.
Morjaria B, Voaden MJ (1979): The formation of glutamate, aspartate and GABA in the rat retina; glucose and glutamine as precursors. J Neurochem 33:541–551.
Morris JG, Rogers QR (1978): Ammonia intoxication in the near-adult cat as a result of a dietary deficiency of arginine. Science 199:431–432.
Newberne PM, Savage JE, O'Dell BL (1960): Pathology of arginine deficiency in the chick. J Nutrition 72:347–352.
Nistri A, Constanti A (1979): Pharmacological characterization of different types of GABA and glutamate receptors in vertebrates and invertebrates. Prog Neurobiol 13:117–235.
Obata K, Takeda K (1969): Release of γ-aminobutyric acid into the fourth ventricle induced by stimulation of the cat's cerebellum. J Neurochem 16:1043–1047.
Obata K, Takeda K, Shinozaki H (1970): Further study on pharmacological properties of the cerebellar-induced inhibition of Deiters neurones. Exp Brain Res 11:327–342.
Oertel WH, Graybiel AM, Mugnaini E, Elde RP, Schmechel DE, Kopin IJ (1983): Coexistence of glutamic acid decarboxylase- and somatostatin-like immunoreactivity in neurons of the feline nucleus reticularis thalami. J Neurosci 3:1322–1332.
Ohara PT, Lieberman AR, Hunt SP, Wu J-Y (1983): Neural elements containing glutamic acid decarbox-

ylase (GAD) in the dorsal lateral geniculate nucleus of the rat; immunohistochemical studies by light and electron microscopy. Neuroscience 8:189–211.

Onodera K, Takeuchi A (1979): An analysis of the inhibitory post-synaptic current in the voltage-clamped crayfish muscle. J Physiol (Lond) 286:265–282.

Orkand PM, Kravitz EA (1971): Localization of the sites of γ-aminobutyric acid (GABA) uptake in lobster nerve-muscle preparations. J Cell Biol 49:75–89.

Otsuka M (1976): GABA in the crustacean nervous system: A historical review. In Roberts E, Chase TN, Tower DB (eds): "GABA in Nervous System Function." New York: Raven Press, pp 245–249.

Otsuka M, Iversen LL, Hall ZW, Kravitz EA (1966): Release of gamma-aminobutyric acid from inhibitory nerves of lobster. Proc Nat Acad Sci 56:1110–1115.

Otsuka M, Kravitz EA, Potter DD (1967): Physiological and chemical architecture of a lobster ganglion with particular reference to gamma-aminobutyrate and glutamate. J Neurophysiol 30:725–752.

Pardridge WM, Oldendorf WH (1975): Kinetic analysis of blood–brain barrier transport of amino acids. Biochim Biophys Acta 401:128–136.

Penny GR, Fitzpatrick D, Schmechel DE, Diamond IT (1983): Glutamic acid decarboxylase-immunoreactive neurons and horseradish peroxidase-labeled projection neurons in the ventral posterior nucleus of the cat and *Galago senegalensis*. J Neurosci 3:1868–1887.

Perez de la Mora M, Possani LD, Tapia R, Teran L, Palacios R, Fuxe K, Hökfelt T, Ljungdahl A (1981): Demonstration of central γ-aminobutyrate-containing nerve terminals by means of antibodies against glutamate decarboxylase. Neuroscience 6:875–895.

Potter DD (1968): The chemistry of inhibition in crustaceans with special reference to gamma-aminobutyric acid. In von Euler C, Skoglund S, Söderberg U (eds): "Structure and Function of Inhibitory Neuronal Mechanisms." Oxford: Pergamon Press, pp 359–370.

Ratner S, Morell H, Carvalho E (1960): Enzymes of arginine metabolism in brain. Arch Biochem Biophys 91:280–289.

Ribak CE (1978): Aspinous and sparsely-spinous stellate neurons in the visual cortex of rats contain glutamic acid decarboxylase. J Neurocytol 7:461–478.

Ribak CE (1981): The GABAergic neurons of the extrapyramidal system as revealed by immunocytochemistry. In Di Chiara G, Gessa GL (eds): "GABA and the Basal Ganglia." New York: Raven Press, pp 23–36.

Ribak CE, Harris AB, Vaughn JE, Roberts E (1979): Inhibitory GABAergic nerve terminals decrease at sites of focal epilepsy. Science 205:211–214.

Ribak CE, Harris AB, Vaughn JE, Roberts E (1981): Immunocytochemical changes in cortical GABA neurons in a monkey model of epilepsy. In Morselli PL, Lloyd KG, Löscher W, Meldrum B, Reynolds EH (eds): "Neurotransmitters, Seizures, and Epilepsy." New York: Raven Press, pp 11–22.

Ribak CE, Vaughn JE, Barber RP (1981): Immunocytochemical localization of GABAergic neurones at the electron microscopical level. Histochemical Journal 13:555–582.

Ribak CE, Vaughn JE, Roberts E (1979): The GABA neurons and their axon terminals in rat corpus striatum as demonstrated by GAD immunocytochemistry. J Comp Neurol 187:261–284.

Ribak CE, Vaughn JE, Roberts E (1980): GABAergic nerve terminals decrease in the substantia nigra following hemitransections of the striatonigral and pallidonigral pathways. Brain Res 192:413–420.

Robbins J, Van der Kloot WG (1958): The effect of picrotoxin on peripheral inhibition in the crayfish. J Physiol (Lond) 143:541–552.

Roberts E (1956): Formation and utilization of γ-aminobutyric acid in brain. In Korey SR, Nurnberger JI (eds): "Progress in Neurobiology. I. Neurochemistry." New York: Hoeber-Harper, pp 11–25.

Roberts E (1963): γ-Aminobutyric acid (γ-ABA)-vitamin $B_6$ relationships in the brain. Am J Clin Nutr 12:291–307.

Roberts E (1975): GABA in nervous system function—An overview. In Tower DB (ed): "The Nervous System, Vol. I: The Basic Neurosciences." New York: Raven Press, pp 541–552.

Roberts E (1976a): Disinhibition as an organizing principle in the nervous system—The role of the GABA system. Application to neurologic and psychiatric disorders. In Roberts E, Chase TN, Tower DB (eds): "GABA in Nervous System Function." New York: Raven Press, pp 515–539.

Roberts E (1976b): Immunocytochemistry of the GABA system—A novel approach to an old transmitter. In Ferrendelli JA, McEwen BS, Snyder SH (eds): "Society for Neuroscience Symposia, Vol 1." Bethesda, MD: Society for Neurosciences, pp 123–138.

Roberts E (1978a): Interrelationships of GABA neurons explored by immunocytochemical techniques. In Garattini S, Pujol JF, Samanin R (eds): "Interactions Between Putative Neurotransmitters in the Brain." New York: Raven Press, pp 89–107.

Roberts E (1978b): Roles of GABA neurons in information processing in the vertebrate CNS. In Karlin A, Tennyson VM, Vogel HJ (eds): "Neuronal Information Transfer." New York: Academic Press, pp 213–239.

Roberts E (1979): New directions in GABA research. I: Immunocytochemical studies of GABA neurons. In Krogsgaard-Larsen P, Scheel-Kruger J, Kofod H (eds): "GABA-Neurotransmitters." Copenhagen: Munksgaard, pp 28–45.

Roberts E (1980a): Epilepsy and antiepileptic drugs: A speculative synthesis. In Glaser GH, Penry JK, Woodbury DM (eds): "Antiepileptic Drugs: Mechanisms of Action." New York: Raven Press, pp 667–713.

Roberts E (1980b): γ-Aminobutyric acid (GABA): A major inhibitory transmitter in the vertebrate nervous system. In Levi-Montalcini R (ed): "Nerve Cells, Transmitters and Behaviour." Rome: Pontifical Academy of Sciences, pp 163–213.

Roberts E (1981): Strategies for identifying sources and sites of formation of GABA-precursor or transmitter glutamate in brain. In DiChiara G, Gessa GL (eds): "Glutamate as a Neurotransmitter." New York: Raven Press, pp 91–102.

Roberts E (1984a): γ-Aminobutyric acid (GABA): From discovery to visualization of GABAergic neurons in the vertebrate nervous system. In Bowery NG (ed): "Actions and Interactions of GABA and Benzodiazepines." New York: Raven Press, pp 1–25.

Roberts E (1984b): GABA-related phenomena, models of nervous system function, and seizures. Ann Neurol 16(suppl):S77–S89.

Roberts E, Baxter CF, Van Harreveld A, Wiersma CAG, Adey WR, Killam KF (eds) (1960): "Inhibition in the Nervous System and Gamma-aminobutyric Acid." Oxford: Pergamon Press.

Roberts E, Chase TN, Tower DB (eds) (1976): "GABA in Nervous System Function." New York: Raven Press.

Roberts E, Eidelberg E (1960): Metabolic and neurophysiological roles of γ-aminobutyric acid. Int Rev Neurobiol 2:279–332.

Roberts E, Frankel S (1950a): γ-Aminobutyric acid in brain. Fed Proc 9:219.

Roberts E, Frankel S (1950b): γ-Aminobutyric acid in brain: Its formation from glutamic acid. J Biol Chem 187:55–63.

Roberts E, Frankel S (1951a): Glutamic acid decarboxylase in brain. J Biol Chem 188:789–795.

Roberts E, Frankel S (1951b): Further studies of glutamic acid decarboxylase in brain. J Biol Chem 190:505–512.

Roberts E, Krause DN, Wong E, Mori A (1981): Different efficacies of d- and l-γ-amino-β-hydroxybutyric acids in GABA receptor and transport test systems. J Neurosci 1:132–140.

Roberts E, Liron Z, Wong E (1985): Potentiation of $Na^+$-dependent uptake of γ-aminobutyric acid in mouse brain particles by buffer-mediated proton removal. Neurochem Res 10:1025–1046.

Roberts E, Rothstein M, Baxter CF (1958): Some metabolic studies of γ-aminobutyric acid. Proc Soc Exptl Biol Med 97:796–802.

Roberts E, Wein J, Simonsen DG (1964): γ-Aminobutyric acid (γABA), vitamin $B_6$, and neuronal function—A speculative synthesis. Vitamins and Hormones 22:503–559.

Roberts E, Wong E, Krause DN, Ikeda K, Degener P (1979): New directions in GABA research II. Some observations relevant to GABA action and inactivation. In Krogsgaard-Larsen P, Scheel-Krüger J, Kofod H (eds): "GABA Neurotransmitters." Copenhagen, Munksgaard Press, pp 46–59.

Roberts E, Wong E, Svenneby G, Degener P (1978): Sodium-dependent binding of GABA to mouse brain particles. Brain Res 152:614–619.

Roberts E, Younger F, Frankel S (1951): Influence of dietary pyridoxine on glutamic decarboxylase activity of brain. J Biol Chem 191:277–285.

Roberts RC, Ribak CE, Peterson GM, Oertel WH (1984): Increased numbers of GABAergic neurons in the inferior colliculus of the genetically epilepsy prone rat. Soc Neurosci Abstr 10:410.

Roskoski R Jr (1978): Net uptake of L-glutamate and GABA by high affinity synaptosomal transport systems. J Neurochem 31:493–498.

Roskoski R Jr, Rauch N, Roskoski LM (1981): Glutamate, aspartate, and γ-aminobutyrate transport by membrane vesicles prepared from rat brain. Arch Biochem Biophys 207:407–415.

Ross G, Dunn D, Jones ME (1978): Ornithine synthesis from glutamate in rat intestinal mucosa homogenates: Evidence for the reduction of glutamate to γ-glutamyl semialdehyde. Biochem Biophys Res Commun 85:140–147.

Ryan LD, Roskoski R Jr (1977): Net uptake of γ-aminobutyric acid by a high affinity synaptosomal transport system. J Pharmacol Exp Ther 200:285–291.

Sadasivudu B, Rao TI (1976): Studies on functional and metabolic role of urea cycle intermediates in brain. J Neurochem 27:785–794.

Saito K, Wu J-Y, Matsuda T, Roberts E (1974): Immunochemical comparisons of vertebrate glutamic acid decarboxylase. Brain Res 65:277–285.

Sano K, Roberts E (1961): Binding of γ-aminobutyric acid by brain preparations. Biochem Biophys Res Commun 4:358–361.

Sano K, Roberts E (1963): Binding of γ-aminobutyric acid by mouse brain preparations. Biochem Pharmacol 12:489–502.

Schallek W, Wiersma CAG (1948): Effects of anticholinesterases on synaptic transmission in the crayfish. Physiologia Comparata et Oecologia 1:63–67.

Scheller RH, Jackson JF, McAllister LB, Rothman BS, Mayeri E, Axel R (1983): A single gene encodes multiple neuropeptides mediating a stereotyped behavior. Cell 32:7–22.

Schon F, Kelly JS (1974a): Autoradiographic localisation of $[^3H]$GABA and $[^3H]$glutamate over satellite glial cells. Brain Res 66:275–288.

Schon F, Kelly JS (1974b): The characterisation of $[^3H]$GABA uptake into the satellite glial cells of rat sensory ganglia. Brain Res 66:289–300.

Schon F, Kelly JS (1975): Selective uptake of $[^3H]$β-alanine by glia: Association with the glial uptake system for GABA. Brain Res 86:243–257.

Schousboe A, Wu J-Y, Roberts E (1973): Purification and characterization of the 4-aminobutyrate-2-ketoglutarate transaminase from mouse brain. Biochemistry 12:2868–2873.

Schousboe A, Wu J-U, Roberts E (1974): Subunit structure and kinetic properties of 4-aminobutyrate-2-ketoglutarate transaminase purified from mouse brain. J Neurochem 23:1189–1195.
ketoglutarate transaminase purified from mouse brain. J Neurochem 23:1189–1195.

Shank RP, Campbell GLem (1983): Ornithine as a precursor of glutamate and GABA: Uptake and metabolism by neuronal and glial enriched cellular material. J Neurosci Res 9:47–57.

Sigel E, Stephenson FA, Mamalaki C, Barnard EA (1983): A γ-aminobutyric acid/benzodiazepine receptor complex of bovine cerebral cortex. J Biol Chem 258:6965–6971.

Sisken B, Roberts E (1964): Radioautographic studies of binding of γ-aminobutyric acid to the abdominal stretch receptors of the crayfish. Biochem Pharmacol 13:95–103.

Smith RJ, Downing SJ, Phang JM (1977): Enzymatic synthesis and purification of L-pyrroline-5-carboxylic acid. Anal Biochem 82:170–176.

Sokoloff L, Lassen NA, McKhann GM, Tower DB, Albers RW (1959): Effects of pyridoxine withdrawal on cerebral circulation and metabolism in a pyridoxine-dependent child. Nature 183:752–753.

Somogyi P, Smith AD, Nunzi MG, Gorio A, Takagi H, Wu JY (1983): Glutamate decarboxylase immunoreactivity in the hippocampus of the cat: Distribution of immunoreactive synaptic terminals with special reference to the axon initial segment of pyramidal neurons. J Neurosci 3:1450–1468.

Squires RF, Klepner CA, Benson DI (1980): Multiple benzodiazepine receptor complexes: Some benzodiazepine recognition sites are coupled to GABA receptors and ionophores. In Pepeu G, Kuhar MJ, Enna SJ (eds): "Receptors for Neurotransmitters and Peptide Hormones." New York: Raven Press, pp 285–293.

Stewart JA (1980): Metabolite alterations in the genetically spastic mouse. J Neurochem 36:780–783.

Storm-Mathisen J, Leknes AK, Borl AT, Vaaland JL, Edminson P, Haug F-MS (1983): First visualization of glutamate and GABA in neurones by immunocytochemistry. Nature 301:517–520.

Strecker JH (1960): The interconversion of glutamic acid and proline. III. $\Delta^1$-Pyrroline-5-carboxylic acid dehydrogenase. J Biol Chem 235:3218–3223.

Swagel MW, Ikeda K, Roberts E (1973a): Effects of GABA, imidazoleacetic acid, and related substances on conductance of crayfish abdominal stretch receptor. Nature New Biol 246:91–92.

Swagel MW, Ikeda K, Roberts E (1973b): Effects of GABA and bicuculline on conductance of crayfish abdominal stretch receptor. Nature New Biol 244:180–181.

Takeuchi A (1976): Studies of inhibitory effects of GABA in invertebrate nervous systems. In Roberts E, Chase TN, Tower DB (eds): "GABA in Nervous System Function." New York: Raven Press, pp 255–267.

Takeuchi A (1978): Kinetics of GABA action at the crayfish neuromuscular junction. In Simon P (ed): "Advances in Pharmacology and Therapeutics, Vol 2, Neurotransmitters." Oxford and New York: Pergamon Press, pp 29–32.

Takeuchi A, Takeuchi N (1965): Localized action of gamma-aminobutyric acid on the crayfish muscle. J Physiol (Lond) 177:225–238.

Takeuchi A, Takeuchi N (1966a): A study of the inhibitory action of γ-aminobutyric acid on neuromuscular transmission in the crayfish. J Physiol (Lond) 183:418–432.

Takeuchi A, Takeuchi N (1966b): On the permeability of the presynaptic terminal of the crayfish neuromuscular junction during synaptic inhibition and the action of γ-aminobutyric acid. J Physiol (Lond) 183:433–449.

Takeuchi A, Takeuchi N (1967a): Anion permeability of the inhibitory postsynaptic membrane of the crayfish neuromuscular junction. J Physiol (Lond) 191:575–590.

Takeuchi A, Takeuchi N (1967b): Electrophysiological studies of the action of GABA on the synaptic membrane. Fed Proc 26:1633–1638.

Takeuchi A, Takeuchi N (1971a): Anion interaction at the inhibitory postsynaptic membrane of the crayfish neuromuscular junction. J Physiol (Lond) 212:337–351.

Takeuchi A, Takeuchi N (1971b): Variations in the permeability properties of the inhibitory post-synaptic membrane of the crayfish neuromuscular junction when activated by different concentrations of GABA. J Physiol (Lond) 217:341–358.

Takeuchi A, Takeuchi N (1975a): Permeability changes of the crayfish muscle produced by β-guanidinopropionic acid and related substances. Neuropharmacology 14:635–641.

Takeuchi A, Takeuchi N (1975b): The structure–activity relationship for GABA and related compounds in the crayfish muscle. Neuropharmacology 14:627–634.

Tarbit I, Perry EK, Perry RH, Blessed G, Tomlinson BE (1980): Hippocampal free amino acids in Alzheimer's disease. J Neurochem 35:1246–1249.

ten Bruggencate G, Engberg I (1971): Iontophoretic studies in Deiters' nucleus of the inhibitory actions of GABA and related amino acids and the interactions of strychnine and picrotoxin. Brain Res 25:431–448.

Toffano G, Aldinio C, Balzano M, Leon A, Kirschner

G (1981): Role of membrane lipids on the regulation of GABA recognition sites: In Costa E, Di Chidra G, Gessa GL (ed): "GABA and Benzodiazepine Receptors." New York: Raven Press, pp 169–179.

Toffano G, Leon A, Massotti M, Guidotti A, Costa E (1980): In Pepeu G, Kuhar MJ, Enna SJ (eds): "Receptors for Neurotransmitters and Peptide Hormones." New York: Raven Press, pp 133–142.

Tower DB (1956): Neurochemical aspects of pyridoxine metabolism and function. Am J Clin Nutr 4:329–345.

Tower DB (1958): Pyridoxine and cerebral activity. Nutr Rev 16:161–164.

Traub RD, Wong RKS (1982): Cellular mechanism of neuronal synchronization in epilepsy. Science 216:745–747.

Udenfriend S (1950a): A micro technique for identification of organic compounds using isotopic indicators and paper chromatography. Fed Proc 9:240.

Udenfriend S (1950b): Identification of $\gamma$-aminobutyric acid in brain by the isotope derivative method. J Biol Chem 187:65–69.

Van der Kloot WG, Robbins J (1959): The effects of $\gamma$-aminobutyric acid and picrotoxin on the junctional potential and the contraction of crayfish muscle. Experientia 15:35–36.

Van der Kloot WG, Robbins J, Cooke IM (1958): Blocking by picrotoxin of peripheral inhibition in crayfish. Science 127:521–522.

Varon S, Weinstein H, Kakefuda T, Roberts E (1965): Sodium-dependent binding of $\gamma$-aminobutyric acid by morphologically characterized subcellular brain particles. Biochem Pharmacol 14:1213–1224.

von Bekesy G (1967): "Sensory Inhibition." Princeton: Princeton University Press.

von Euler C, Skoglund S, Söderberg U (1968): "Structure and Function of Inhibitory Neuronal Mechanisms." Oxford: Pergamon Press.

Waddington JL, Cross AJ (1979): Baclofen and muscimol: Behavioral and neurochemical sequelae of unilateral intranigral administration and effects on $^3$H-GABA receptor binding. Naunyn-Schmied Arch Pharmacol 306:275–280.

Wang Y-J, Salvaterra P, Roberts E (1979): Characterization of [$^3$H]muscimol binding to mouse brain membranes. Biochem Pharmacol 28:1123–1128.

Watkins JC (1965): Pharmacological receptors and general permeability phenomena of cell membranes. J Theoret Biol 9:37–50.

Werman R (1972): Amino acids as central neurotransmitters. Res Publ Association for Research in Nervous and Mental Disease (A.R.N.M.D.) 50:147–180.

Wiersma CAG, Furshpan E, Florey E (1953): Physiological and pharmacological observations on muscle receptor organs of the crayfish, *Cambarus clarkii* girard. J Exp Biol 30:136–150.

Wiersma CAG, Schallek W (1948): Influence of drugs on response of a crustacean synapse to preganglionic stimulation. J Neurophysiol 11:491–496.

Wong PT-H, McGeer EG, McGeer PL (1982a): Effects of kainic acid injection and cortical lesion on ornithine and aspartate aminotransferases in rat striatum. J Neurosci Res 8:643–650.

Wong PT-H, McGeer PL, Rossor M, McGeer EG (1982b): Ornithine aminotransferase in Huntington's disease. Brain Res 231:466–471.

Wood JD (1982): Elevation in the GABA content of nerve endings (synaptosomes) induced by anticonvulsant agents. In Okada Y, Roberts E (eds): "Problems in GABA Research." International Congress Series 565. Amsterdam: Excerpta Medica, pp 173–181.

Wood JD, Russell MP, Kurylo E, Newstead JD (1979): Stability of synaptosomal GABA levels and their use in determining the *in vivo* effects of drugs: Convulsant agents. J Neurochem 33:61–68.

Woodward DJ, Hoffer BJ, Siggins GR, Oliver AP (1971): Inhibition of Purkinje cells in the frog cerebellum. II. Evidence for GABA as the inhibitory transmitter. Brain Res 33:91–100.

Wu J-Y (1976): Purification, characterization, and kinetic studies of GAD and GABA-T from mouse brain. In Roberts E, Chase TN, Tower DB (eds): "GABA in Nervous System Function." New York: Raven Press, pp 7–55.

Wu J-Y, Matsuda T, Roberts E (1973): Purification and characterization of glutamate decarboxylase from mouse brain. J Biol Chem 248:3029–3034.

Wu J-Y, Roberts E (1974): Properties of brain L-glutamate decarboxylase: inhibition studies. J Neurochem 23:759–767.

Yoneda Y, Roberts E (1982a): A new synaptosomal biosynthetic pathway of proline from ornithine and its negative feedback inhibition by proline. Brain Res 239:479–488.

Yoneda Y, Roberts E (1982b): Synaptosomal biosynthesis of GABA from ornithine and its feedback inhibition by GABA. In Okada Y, Roberts E (eds): "Problems in GABA Research."Amsterdam: Excerpta Medica, pp 55–65.

Yoneda Y, Roberts E, Dietz GW Jr (1982): A new synaptosomal biosynthetic pathway of glutamate and GABA from ornithine and its negative feedback inhibition by GABA. J Neurochem 38:1686–1694.

Benzodiazepine/GABA Receptors and Chloride Channels:
Structural and Functional Properties, pages 41–56


# 2

# GABA Receptors: An Overview

**S.J. Enna and E.W. Karbon**

*Departments of Pharmacology and of Neurobiology and Anatomy, University of Texas Medical School, P.O. Box 20708, Houston, Texas 77025*

The development of the benzodiazepines represented a major breakthrough in psychopharmacology. Prior to their discovery, anxiety was treated with the barbiturates and meprobamate, substances which cause a significant amount of sedation in most patients. This effect led to the belief that it was necessary to sacrifice a certain amount of mental acuity to alleviate an anxious state [Enna, 1985]. The benzodiazepines radically altered this concept since it was found that, for most patients, these drugs relieved anxiety without a marked degree of sedation. Moreover, when taken orally, benzodiazepines are less complete central nervous system depressants than the barbiturates, and therefore less prone to be fatal in overdose. The terms "anxioselective" and "anxiolytic" were coined because of these findings to describe drugs which reduce anxiety without markedly depressing the central nervous system. Besides having significant clinical ramifications, this therapeutic profile raised a number of important biological questions.

Studies aimed at defining the mechanism of action of the benzodiazepines have intensified during the past decade. Two of the more important discoveries have been that these agents influence γ-aminobutyric acid (GABA) receptor function, and that the brain possesses specific benzodiazepine-binding sites [Braestrup and Nielsen, this volume; Costa and Guidotti, 1979; Tallman et al., 1978; Möhler and Okada, 1977a]. These findings suggest the existence of a specific receptor that would be responsible for regulating the level of anxiety. Therefore, to fully appreciate the proposed action of the benzodiazepines, it is necessary to have some understanding of the general properties of receptors, with particular reference to the GABAergic system.

The discovery that chemical agents influence cellular function by interacting with specific recognition sites (receptors) was one of the most important advances in the biological sciences [Enna and Strada, 1983]. Receptors may be defined as macromolecules to which a hormone, neurotransmitter, or neuromodulator attaches. Receptors are located either in association with the plasma membrane, such as those for neurotransmitters and some hormones, or with cytoplasmic or nuclear compartments, such as steroid hormone and $T_3$ receptors, respectively. Receptor activation regulates a variety of cellular activities, including the degree of ion permeability, enzyme activity, or the rate or type of transcription and translation. Thus, receptors may be viewed as the primary site for regulating cellular activity in all organs and tissues.

A number of neurotransmitter substances have been identified in the mammalian central nervous system. Among these is GABA, which appears to be, quantitatively at least, one of the most important [Iversen and Bloom, 1972]. Accordingly, agents capable of influencing GABAergic transmission may be expected to have a variety of biological effects. Indeed, given the widespread distribution of GABA and GABA receptors in the mammalian central nervous system [Enna, 1983a], it might be anticipated that their activation or inhibition would have chaotic consequences. However, GABAergic agents can be shown to have very subtle effects on brain function. This suggests that all GABA receptors may not be pharmacologically and functionally equivalent and that, as for other neurotransmitters, there may be distinct subgroups of GABA receptors which may be selectively manipulated for therapeutic gain. Therefore, the pharmacological characteristics of the benzodiazepines may be explained, in part, by their ability to influence a distinct population of GABA receptors.

The present chapter has been designed to summarize current concepts relating to GABAergic transmission, with particular emphasis on defining the different members of the GABA receptor family. The aim is to provide background information before the reader embarks on the more detailed discussions of the GABA/benzodiazepine complex. Because of the enormous literature in this area, it is impossible to reference all of the important experimental data. Readers desiring further information are urged to consult any of a number of recent monographs and reviews [Enna, 1983b; Enna and Gallagher, 1983; Olsen and Enna, 1983; Costa and Guidotti, 1979; Krogsgaard-Larsen and Christensen, 1980; Johnston, 1978].

## GENERAL CHARACTERISTICS OF THE GABAERGIC SYSTEM

Most of the GABA in the central nervous system is derived from glutamic acid through a reaction catalyzed by glutamic acid decarboxylase (GAD). Like other neurotransmitters, GABA is stored in nerve terminals and is released by a calcium-dependent process following depolarization [Ryan and Roskoski, 1975]. The liberated GABA is reaccumulated by a sodium-dependent, high-affinity uptake system located on nerve terminals and glia [Iversen and Neal, 1968]. Some GABA is metabolized to succinic semialdehyde by GABA transaminase (GABA-T). The functional significance of these biochemical events is indicated by the fact that agents which inhibit GAD significantly reduce brain GABA levels and decrease seizure threshold [Sawaya et al., 1978]. Conversely, drugs which inhibit GABA-T increase the GABA content, depress central nervous system activity, and raise the seizure threshold [Meldrum, 1978]. Likewise, drugs inhibiting GABA transport are central nervous system depressants and provide protection against chemically induced seizures [Brehm et al., 1979]. These data support the contention that GABA is an important inhibitory neurotransmitter in the brain.

Anatomical studies suggest that GABA is localized primarily in small interneurons scattered throughout the central nervous system. However, there are GABAergic projections linking the corpus striatum and nucleus accumbens to other brain regions [Scheel-Kruger, 1983]. This includes striatal GABAergic neurons that send projections to the globus pallidus, substantia nigra zona reticulata, and interpeduncular nucleus. GABAergic projections emanating from the nucleus accumbens terminate in the central portion of the globus pallidus as well as the zona compacta region of the substantia nigra and perhaps the ventral tegmental area. This distribution indicates that GABA influences a variety of central nervous system functions, including consciousness, affect, motor activity, and posture.

While the majority of GABA is confined to the central nervous system, it has been found in some peripheral tissues as well. Most evidence points to a possible role in pancreatic function and in the female reproductive system [Taniguchi et al., 1979; Erdö and Lapis, 1982; Fernandez et al., 1984; Apud et al.,

1984; Erdö, 1984; Erdö and Laszlo, 1984]. High concentrations of GABA and GABA-binding sites have been detected in the human ovary, and the GABA content varies in the rat fallopian tube during pregnancy. Moreover, GABA modulates cholinergic responses in the isolated rat oviduct, and GABA, GAD, and GABA-T have been found in the myenteric plexus [Miki et al., 1983]. GABA receptors may mediate relaxation in the rat duodenum [Maggi et al., 1984a], and influence contractions of the detrusor muscle [Maggi et al., 1984b]. There is also evidence for GABA receptors on terminals of peripheral neurons [Bowery et al., 1980], with activation of these sites inhibiting the release of neurotransmitter. However, there are as yet few data supporting a physiological role for GABA in peripheral tissue. Therefore, the significance of these findings remains to be established.

Given the extensive distribution of GABAergic neurons in brain, the selectivity of GABAergic drugs must be due to neurochemical differences rather than anatomical location. It is possible that more active GABA-ergic pathways would presumably be more sensitive to drugs inhibiting GABA synthesis, metabolism, or reuptake, with the response most likely reflecting a functional modification in these synapses. Selectivity may also occur at the level of the neurotransmitter receptor. Because GABA is a relatively flexible entity, it can assume a variety of conformations, making it conceivable that different GABA receptors recognize different molecular configurations [Krogsgaard-Larsen et al., 1983].

GABA receptors were initially defined on the basis of their sensitivity to inhibition by bicuculline and picrotoxin [Curtis et al., 1971]. More recently, evidence has accumulated that there may be a group of bicuculline-insensitive GABA receptors [Bowery et al., 1982]. These sites are defined on the basis of their sensitivity to $\beta$ -*p*-chlorophenyl GABA (baclofen), a GABA receptor agonist. Accordingly, bicuculline-sensitive, baclofen-insensitive GABA receptors are defined as $GABA_A$ sites, whereas bicuculline-insensitive, baclofen-sensitive responses are said to be mediated by $GABA_B$ receptors. $GABA_A$ receptors can be further subdivided on the basis of their sensitivity to benzodiazepines. While it is possible that all central nervous system benzodiazepine binding sites are associated with $GABA_A$ receptors, it appears that not all $GABA_A$ receptors have a benzodiazepine component [Palacios et al., 1981; Unnerstall et al., 1981]. Thus, the GABA/benzodiazepine complex represents a subgroup of the $GABA_A$ system. Together these data indicate that GABA may influence cellular activity by acting through different groups of pharmacologically and functionally distinct receptors.

## GABA RECEPTOR COMPONENTS

Ligand binding assays have provided a wealth of information about the biochemical and pharmacological characteristics of neurotransmitter receptors [Yamamura et al., 1985]. Because a number of compounds are available for labeling $GABA_A$ receptors (Table I), most work on this system has concentrated on this site. Early studies revealed that $^3$H-GABA binds with a relatively high affinity to brain membranes derived from a variety of animal species [Enna, 1983a]. These studies indicated that $^3$H-GABA binding has a distinct regional distribution in brain and that the radioligand could be displaced only by amino acids and drugs known to be electrophysiologically active at the GABA receptor, suggesting that $^3$H-GABA attaches to a component of the physiologically relevant site. Analysis of equilibrium binding data revealed both high- and low-affinity components for $^3$H-GABA binding (Table II). The two sites are most apparent in tissue that has been thoroughly rinsed or following treatment with Triton X-100 [Enna and Snyder, 1977; Olsen et al., 1981]. The higher-affinity site has an equilibrium binding constant ($K_d$) of approximately 10 nM, whereas the lower-affinity site has a binding constant in the 100

**TABLE I. Structure and Classification of Compounds Used for Labeling GABA Receptors in Brain Tissue**

| Compound | Structure | Class | Site labeled |
|---|---|---|---|
| GABA |  | $GABA_A$ & $GABA_B$ agonist | Recognition site |
| Muscimol | 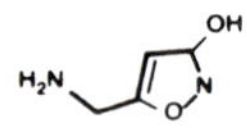 | $GABA_A$ agonist | Recognition site |
| Isoguvacine | 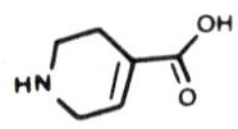 | $GABA_A$ agonist | Recognition site |
| THIP | 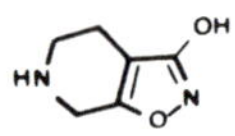 | $GABA_A$ agonist | Recognition site |
| Piperidine-4-sulphonic acid | 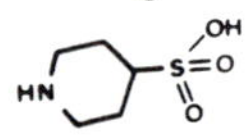 | $GABA_A$ agonist | Recognition site |
| Bicuculline | 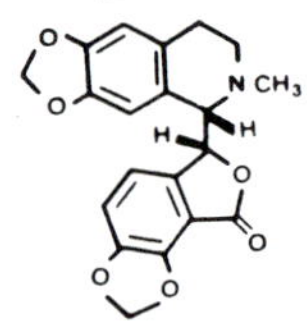 | $GABA_A$ antagonist | Recognition site |
| Dihydropicrotoxinin | 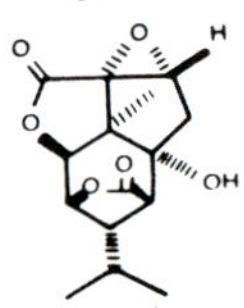 | $GABA_A$ antagonist | Chloride channel |
| t-Butylbicyclophosphorothionate |  | $GABA_A$ antagonist | Chloride channel |
| Baclofen | 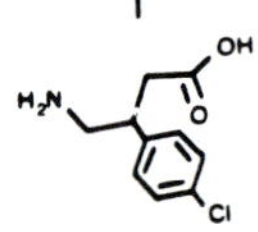 | $GABA_B$ agonist | Recognition site |

nM range. While there have been suggestions that these sites are distinct entities [Guidotti et al., 1979], it is still conceivable that they represent different states of the same receptor since there is no direct evidence indicating that they differ with regard to pharmacological selectivity or functional activity.

Several other agents have been used as ligands for labeling $GABA_A$ recognition sites (Table I). This includes muscimol, isoguvacine (1,2,3,6-tetrahydropyridine-4-carboxylic acid), THIP (4,5,6,7-tetrahydroisoxazolo [5,4-*c*]pyridin-3-ol) and piperidine-4-sulfonic acid [Beaumont et al., 1977; Falch and Krogsgaard-Larsen, 1982; Krogsgaard-Larsen et al., 1981]. The latter three compounds are synthetic analogs of GABA which have been found to possess GABA agonist properties in electrophysiological tests. The binding characteristics of these substances are very similar to those found for $^3$H-GABA itself, including multiple sites. It is interesting, however, that the number of high-affinity $^3$H-THIP sites is substantially less than that found for $^3$H-GABA. It remains to be determined whether or not this indicates that THIP is selective for a particular population of high-affinity $GABA_A$ receptors.

renergic bundle cause a selective reduction in the number of lower-affinity sites, without modifying the higher-affinity receptor [Karbon et al., 1983]. This lesion also had no effect on $GABA_A$ binding, lending further support to the notion that $GABA_A$ and $GABA_B$ binding sites represent separate receptor systems.

As with $GABA_A$ binding, $GABA_B$ sites are found throughout the central nervous system and are most enriched in the cerebellum and cerebral cortex [Wilkin et al., 1981; Karbon et al., 1983]. Unlike the $GABA_A$ system, there appears to be no relationship between $GABA_B$ receptors and the benzodiazepines. Thus, the $GABA_B$ site is differentiated from $GABA_A$ on the basis of sensitivity to baclofen, dependence on calcium ion [Majewska and Chuang, 1984], insensitivity to bicuculline, and the lack of an association with benzodiazepines. The further characterization of the $GABA_B$ binding site must await the development of selective antagonists for this receptor.

## GABA RECEPTOR FUNCTION

Mammalian GABA receptors were first identified in electrophysiological studies. Curtis and his colleagues demonstrated that GABA reduces excitatory and inhibitory postsynaptic potentials in cat spinal motoneurons [Curtis et al., 1968a & b]. Chemical excitation of Renshaw and dorsal horn cells was also inhibited by GABA, although the amino acid did not affect resting membrane potential. Krnjevic and Schwartz [1967] found that iontophoretically applied GABA causes a hyperpolarization and lowers membrane resistance in cat brain cerebral cortical neurons. This response was identical to the inhibitory postsynaptic potentials evoked by surface stimulation. GABA-mediated hyperpolarization was reversed by intracellular injections of chloride ions, suggesting that GABA is an inhibitory transmitter in the cerebral cortex and that it acts by increasing chloride flux to produce a membrane hyperpolarization. A similar hyperpolarization was observed in cat spinal motoneurons [Krnjevic, 1974]. GABA was also found to depolarize primary afferent nerve terminals in the spinal cord, resulting in a decrease in transmitter release [Schmidt, 1971]. Such presynaptic inhibition may result from an outward, rather than an inward, flow of chloride ion, although elevated levels of extracellular potassium may account for this type of depolarization as well [Nicoll and Alger, 1979]. Additional data supporting a neurotransmitter role for GABA was presented by Curtis and Johnston when they demonstrated that bicuculline selectively inhibits GABAergic responses while having little effect on the action of other inhibitory amino acids [Curtis et al., 1971]. These electrophysiological studies provided the initial evidence that GABA receptors are coupled to a chloride ion channel and that activation of the recognition site results in an alteration in membrane potential.

It has been shown that GABA is also capable of depolarizing cells which apparently lack any GABAergic innervation, such as sympathetic and dorsal root ganglia [Gallagher and Gallagher, 1983]. While the physiological significance of this response is unknown, it shares many of the features of the responses observed in brain and spinal cord neurons, making it a useful tool for studying the electrophysiological actions of GABA.

The initial data suggesting a relationship between the benzodiazepines and GABA were also derived from electrophysiological studies. Schmidt and co-workers [1967] found that diazepam potentiates presynaptic, but not postsynaptic, inhibition in the cat spinal cord. Subsequent work demonstrated that this effect is attenuated by reducing the intraneuronal concentration of GABA, suggesting that the benzodiazepines indirectly modify GABAergic function [Polc et al., 1974]. Benzodiazepines were also shown to facilitate the inhibitory feedback mechanisms in the cerebral cortex, hippocampus, and hypothalamus [Haefely and Polc, 1983], and were observed to enhance pre- and postsynaptic inhibition in the cuneate nucleus [Polc and Haefely, 1976]. Barbiturates were also found to influence inhibitory mechanisms in the spinal cord [Schmidt, 1963] and, like the benzodiazepines, enhanced GABA-mediated inhibitory transmission in a variety of brain regions [Haefely and Polc, 1983]. Unlike the barbiturates, the benzodiazepines do not normally influence neuronal activity at concentrations that have a maximal effect on GABAergic responses [Olsen, 1982]. Moreover, the benzodiazepines facilitate the frequency of chloride channel opening, while barbiturates enhance the duration of these events [Study and Barker, 1981]. Thus, while both the benzodiazepines and barbiturates facilitate GABAergic transmission, they do so by different mechanisms. These findings are in accord with the biochemical data suggesting that barbiturates directly influence the chloride channel, whereas the benzodiazepines modify recognition site sensitivity.

Studies with baclofen have revealed that it produces a hyperpolarizing response in hippocampal CA1 pyramidal cells [Newberry and Nicoll, 1984a]. This action is not sensitive to inhibition by bicuculline. The hyperpolarizing effect of baclofen appears due to an increase in potassium, rather than chloride, conductance. Stimulation of afferents to these same cells leads to a hyperpolarizing inhibitory postsynaptic potential, which is followed by a late hyperpolarizing potential [Newberry and Nicoll, 1984b]. This latter event is very similar to the response observed following baclofen administration in that it is insensitive to bicuculline and is related to an increase in potassium conductance. A similar type of late hyperpolarizing potential has been observed in granule cells of the dentate gyrus following orthodromic stimulation [Thalmann and Ayala, 1982]. Superfusion of baclofen onto neurons of the dorsolateral septum leads to a prolonged hyperpolarization, again suggesting an alteration in potassium conductance [Gallagher et al., 1984]. These studies indicate that GABA, through an action at $GABA_B$ receptors, mediates slow inhibitory postsynaptic potentials by directly stimulating an increase in potassium conductance. Clearly, the $GABA_A$ and $GABA_B$ receptors are functionally distinct entities (Table II).

In order to fully characterize GABA receptor function, it would be useful to have a biochemical measure of receptor activity. While several attempts have been made to determine whether the $GABA_A$ system influences cyclic nucleotide production, it seems that activation of this receptor has no direct effect on cyclic AMP or cyclic GMP formation. However, there have been suggestions that $GABA_A$ receptors regulate the incorporation of $^{32}Pi$ into phospholipids in rat brain, although there remains some question about the specificity of this response [Friedel et al., 1977]. Perhaps the most meaningful biochemical measure of $GABA_A$ receptor activity would be a method for studying chloride transport in vitro. Several attempts have been made to address this issue [Ticku and Olsen, 1977; Wong et al., 1984; Thampy and Barnes, 1984]. It has been reported that GABA increases chloride permeability in isolated strips of crayfish abdominal muscle through an action at bicuculline-sensitive sites, with a half-maximal response being observed at a concentration of 40 $\mu$M GABA. Wong et al. [1984] have shown that GABA increases the rate of chloride efflux from preloaded rat hippocampal slices. In this case the half-maximal response was in the high $\mu$M range. For both the abdominal muscle and hippocampal tests, the magnitude of the response and the precision of the assay were quite low, limiting the utility of these approaches. In contrast, Thampy and Barnes [1984] have shown that GABA induces a 3-fold increase in the uptake of chloride into monolayer cultures of chick brain cerebral neurons. This robust response is inhibited by both bicuculline and picrotoxin, and the $K_{0.5}$ for GABA was 8 $\mu$M, a value quite similar to the calculated $K_d$ for GABA binding to the ultralow-affinity site (Table II). Bicuculline appears to be a noncompetitive inhibitor of this response, suggesting that it does not interact at the receptor

**TABLE III. Effect of Baclofen on Norepinephrine-Stimulated Cyclic AMP Accumulation in Various Regions of the Rat Central Nervous System[a]**

| Brain region | Cyclic AMP accumulation (% conversion)[b] | | |
|---|---|---|---|
| | Basal | Norepinephrine | Norepinephrine + baclofen[c] |
| Cerebral cortex | 0.14 | 1.70 | 4.10 |
| Hippocampus | 0.16 | 2.23 | 3.81 |
| Corpus striatum | 0.09 | 2.52 | 5.18 |
| Pons-midbrain | 0.19 | 3.01 | 3.51 |
| Cerebellum | 0.25 | 1.10 | 1.00 |
| Spinal cord | 0.29 | 0.65 | 0.71 |

[a]Adapted from Karbon and Enna [1985].
[b]Cyclic AMP production was measured using a prelabeling technique and the data expressed as the percent of the total tritium pool represented by $^3$H-cyclic AMP.
[c]The influence of baclofen (100 $\mu$M) on norepinephrine (100 $\mu$M)-stimulated cyclic AMP accumulation in tissue slices obtained from various regions of the rat central nervous system.

recognition site. Picrotoxin was an uncompetitive inhibitor in this system, leading the authors to propose that this antagonist attaches only to an activated chloride channel. Continued efforts along these lines are important for perfecting a simple biochemical method for precisely defining the functional activity of GABA receptors and for examining the relationship between these receptors and the benzodiazepines.

Biochemical studies indicate that $GABA_B$ receptors influence the calcium-dependent release of other neurotransmitter agents [Bowery et al., 1980a & b, 1981]. In this scheme, $GABA_B$ receptors are located on nerve terminals such that their activation leads to an inhibition of neurotransmitter release. While the mechanism whereby $GABA_B$ receptors influence this process is unknown, the apparent relationship between $GABA_B$ binding and calcium ion suggest that these receptors may, in some way, influence the calcium component of the vesicular release process. This effect is insensitive to bicuculline and has been detected in central as well as peripheral neurons.

In contrast to $GABA_A$ receptors, $GABA_B$ receptors have been found to influence cyclic nucleotide production in the central nervous system [Karbon et al., 1984; Karbon and Enna, 1985]. These studies indicate that $GABA_B$ agonists markedly potentiate the cyclic AMP response observed during exposure of rat brain slices to hormones and neurotransmitters that directly stimulate adenylate cyclase. By themselves, the $GABA_B$ agonists have little effect on cyclic AMP production. Rather, activation of the $GABA_B$ system amplifies the second messenger response to other agents. This effect is stereoselective, resistant to inhibition by bicuculline, and is not observed with $GABA_A$ agonists. There is a distinct regional distribution for the interaction of $GABA_B$ agonists with transmitter-stimulated cyclic AMP production (Table III) [Karbon and Enna, 1985]. Baclofen enhances the second messenger response to norepinephrine in rat brain cerebral cortex, hippocampus, and corpus striatum, but has no effect in pons-midbrain, cerebellum, or spinal cord tissue. This distribution is markedly different from that found for $GABA_B$ binding [Karbon et al., 1983], indicating either that the GABA receptors involved are not identified in the $GABA_B$ binding assay, or that they represent a subclass of this receptor type. Taken together, these findings suggest that activation of $GABA_B$ receptors influences the coupling efficiency between the neurotransmitter recognition site and the other components of the second messenger system. It is unknown whether the $GABA_B$ receptors mediating this response are directly associated with the cyclic nucleotide system or whether they cause the liberation of some substance which, in turn, influences second messenger production. Because $GABA_B$ agonists alone have little effect on cyclic AMP production, it could be argued that GABA acts as a neuromodulator rather than a neurotransmitter when acting at this site.

Baclofen has also been found to have a direct inhibitory effect on adenylate cyclase in homogenates of rat brain [Wojcik and Neff, 1984]. This appears to be at variance with the data indicating that $GABA_B$ agonists potentiate second messenger production. Further studies will be required to determine which, if either, of these effects has physiological relevance.

In summary, these data indicate a clear and distinct separation between $GABA_A$ and $GABA_B$ receptors in terms of their biochemical and electrophysiological responses, supporting the concept of pharmacologically and functionally distinct receptors for this neurotransmitter. While the benzodiazepines may be viewed as enhancers of GABAergic transmission, it is evident that they are highly selective for a particular population of GABA receptors, which may account for their distinctive effects on central nervous system function.

## PURIFICATION OF GABA RECEPTORS

Receptor binding studies have suggested that $GABA_A$ receptors are part of a supramolecular complex consisting of a benzodiazepine binding site, a chloride ion channel, and the receptor recognition site. It would be useful to know whether these components are located on the same molecule or reside on different molecular species which interact under the proper conditions. The most direct way to address this issue is to purify the receptor so that the components can be characterized in isolation. Numerous investigators have undertaken such experiments [Sigel et al., 1982 & 1983; Asano et al., 1983; Asano and Ogawara, 1980; Gavish and Snyder, 1981; Chang and Barnard, 1982; Stephenson and Olsen, 1982; Kuriyama and Ito, 1983; Mernoff et al., 1983; Kuriyama et al., 1984]. The general approach has been to use detergents to extract and solubilize the receptor protein. Detergents such as deoxycholate, nonidet P-400, Triton X-100, and CHAPS have been used for this purpose. Following solubilization, the receptor is isolated using a benzodiazepine affinity column [Sigel et al., 1982 & 1983; Gavish and Snyder, 1981; Kuriyama et al., 1984] in conjunction with ion exchange chromatography [Sigel et al., 1983; Kuriyama et al., 1984]. Following elution the physicochemical properties of the protein are ascertained.

Considering the number of approaches taken to purify the GABA/benzodiazepine complex, the results are in reasonably good agreement. Evidence suggests that the $GABA_A$ and benzodiazepine binding sites reside on the same macromolecule, with a variety of receptor preparations retaining the ability to bind GABAergic drugs, benzodiazepines, and TBPS. Moreover, GABA-activated benzodiazepine binding has been demonstrated in soluble fractions [Gavish and Snyder, 1981; Mernoff et al., 1983]. As with the membrane preparation, barbiturates enhance both benzodiazepine and muscimol binding to solubilized receptors [Stephenson and Olsen, 1982; Kuriyama and Ito, 1983]. This effect is enhanced by chloride ion, suggesting that the chloride component has been isolated as well. The binding site kinetics of the solubilized receptor are very similar to those found in membrane preparations, although the type of detergent used for solubilization influences these characteristics. The ability to observe interactions between GABA, benzodiazepines, barbiturates, and chloride ion in the soluble fraction suggests that these components all reside on the same macromolecule.

The GABA/benzodiazepine receptor complex has a molecular weight in the 200,000–300,000 dalton range as determined by radiation inactivation, gel filtration, and sucrose gradient analysis [Asano et al., 1983; Chang and Barnard, 1982; Stephenson and Olsen, 1982; Kuriyama and Ito, 1983]. When subjected to electrophoresis in the presence of a denaturing agent, this complex yields one or two major stained peptide bands [Sigel et al., 1982 & 1983; Kuriyama et al., 1984]. $^3$H-Flunitrazepam labels both of these proteins, suggesting further that the two binding sites are intimately related and probably reside on a single entity [Sigel et al., 1983].

Many questions remain regarding the properties of the GABA/benzodiazepine complex. It would be useful to determine the stoichiometry of the GABA/benzodiazepine binding sites and to learn whether there are benzodiazepine sites that are not associated with $GABA_A$ receptors. Further studies may also reveal the nature of putative endogenous regulators of GABA receptors, since there is evidence that the benzodiazepines influence GABA receptor function by modifying such substances [Costa and Guidotti, 1979]. Studies are also under way to raise antibodies against the purified GABA/benzodiazepine complex [Barnard et al., 1984; Möhler and Richards, this volume]. These will be useful for localizing these sites in brain and for studying receptor regulation at the molecular level.

The GABA/benzodiazepine receptor complex has recently been synthesized in *Xenopus* oocytes following the injection of rat brain mRNA [Constanti et al., 1984]. This exciting advance should facilitate a study of the assembly of the GABA/benzodiazepine complex and may ultimately provide the tools needed for a more detailed examination of the individual components.

In contrast to $GABA_A$ receptors, few attempts have been made to isolate the $GABA_B$ site. Investigations along these lines are limited by the paucity of compounds for labeling this site. It is likely, however, that in the near future agents will be discovered which will irreversibly attach to the $GABA_B$ receptor and therefore can be used to isolate this protein.

## CLINICAL CONSIDERATIONS

The clinical impact of the benzodiazepines indicates the therapeutic potential of GABAergic drugs. While few agents specifically designed as GABA receptor agonists have been tested in humans, the available data suggest that such compounds may be anticonvulsants, analgesics, and, perhaps, antidepressants [Morselli and Lloyd, 1983; Krogsgaard-Larsen et al., 1984]. The $GABA_B$ receptor agonist baclofen has been used for some time to treat spasticity [Davidoff, 1978]. As more data are accumulated about the biochemical, pharmacological, and functional properties of the different GABA receptors, it is conceivable that a variety of structurally diverse compounds may be found efficacious in these and other conditions.

The finding that $GABA_B$ receptor activation influences transmitter-stimulated second messenger production suggests that the $GABA_B$ system may represent an important modulator of transmitter function. Manipulation of a neuromodulator should have more subtle effects on function than direct alteration of neurotransmitter activity, yielding more clinically useful agents. In the case of major psychiatric disorders such as schizophrenia, current therapy entails the use of drugs that block a number of neurotransmitter receptors, including dopamine, α-adrenergic, histamine, and cholinergic muscarinic sites [Enna and Coyle, 1983]. While dopamine receptor blockade is undoubtedly related to the therapeutic response, direct inhibition of this and other receptor systems probably accounts for many of the side-effects associated with these drugs. If it can be demonstrated that $GABA_B$ receptors regulate the activity of brain dopaminergic systems, it is theoretically possible that a $GABA_B$ receptor antagonist may be beneficial in the treatment of schizophrenia inasmuch as it should reduce dopaminergic activity. Such an agent may have fewer side-effects than contemporary drugs, since there is little evidence to support a transmitter role for GABA in the periphery. Conversely, $GABA_B$ receptor agonists might be efficacious in the treatment of depression, since this disorder is characterized by an underactive noradrenergic system [Fuller, 1981].

Given the finding that $GABA_B$ agonists amplify the second messenger response to norepinephrine, such a drug might be useful for the treatment of this disorder. Indeed, recent evidence suggests that the neurochemical response to antidepressants is facilitated when a $GABA_B$ receptor agonist is administered concomitantly [Enna et al., 1985].

Both laboratory and clinical data indicate that $GABA_A$ and $GABA_B$ agonists have analgesic potential [Kendall et al., 1982; Krogsgaard-Larsen et al., 1984; Vaught et al., 1984]. A limitation to their use, however, is the fact that the analgesic dose is similar to, or higher than, the sedative dose. Given the likelihood of pharmacologically and functionally distinct GABA receptors, it is possible that drugs may be discovered which selectively activate those receptors associated with analgesia while having little affinity for GABA receptors mediating sedation. Because GABA agonist–induced analgesia is similar to that found for the opiates, the development of a GABA analgesic could represent a significant advance in the management of pain.

Continued studies on the properties of the GABA/benzodiazepine complex may lead to the development of more selective anxiolytic and hypnotic agents. It is possible that some benzodiazepine receptors may not be associated with the GABAergic system, or that there may be differences in the benzodiazepine binding sites even among those coupled to the GABA receptor [Johnson et al., 1983 and this volume]. Exploitation of such findings could result in the discovery of anxioselective agents virtually devoid of central nervous system depressant activity. A more complete characterization of the GABA/benzodiazepine complex could therefore yield safer therapeutic agents.

## CONCLUSIONS

The discovery of the relationship between the benzodiazepines and the GABA receptor system has had a number of clinical and scientific ramifications. This finding has provided direct evidence indicating different types of GABA receptors in the mammalian central nervous system. It now appears that there are at least two types of $GABA_A$ receptors that are distinguished on the basis of their affiliation with a benzodiazepine component. While the clinical characteristics of the GABA/benzodiazepine receptor system are well established, the consequences of a selective activation of those $GABA_A$ receptors not associated with the benzodiazepines is unknown. Indeed, given the evidence suggesting a host of kinetically distinct $GABA_A$ receptors, it is conceivable that drugs may be found that will selectively manipulate these sites and thereby more precisely regulate a variety of brain functions.

The identification of $GABA_B$ binding sites supports the concept of GABA receptor diversity. As opposed to $GABA_A$ receptors, the $GABA_B$ sites are not associated with chloride ion flux, nor are they related to the benzodiazepines. Rather, among other things, these receptors appear to be linked to transmitter-stimulated second messenger production, suggesting that they may modulate responses to other transmitters in brain.

While it is certain that the benzodiazepine act through the GABAergic system, it must be borne in mind that they influence only a fraction of the total complement of GABA receptors in the central nervous system. Given the ubiquitous distribution of GABAergic neurons and GABA receptors and the diverse effects of GABA receptor activation, it seems likely that the interesting pharmacology of the benzodiazepines gives only a hint as to the clinical benefits that may be derived from manipulating this neurotransmitter system. The insights provided by the studies defining the relationship between benzodiazepines and GABA have played an important role in developing a conceptual framework for describing receptor regulation in general. For this reason, the contents of this volume will not only provide the reader with an intimate knowledge of the GABA receptor/benzodiazepine complex, but will also serve as a primer on state-of-the-art receptor research.

## ACKNOWLEDGEMENTS

Preparation of this manuscript was made possible, in part, by support from the National Science Foundation, National Institute of Mental Health, and Bristol-Myers Corporation. We thank Mrs. Doris Thornton for her excellent secretarial assistance.

## REFERENCES

Apud JA, Tappaz ML, Celotti F, Negri-Cesi P, Masotto C, Racagni G (1984): Biochemical and immunohistochemical studies on the GABAergic system in the rat fallopian tube and ovary. J Neurochem 43:120–125.

Asano T, Ogasawara N (1980): Solubilization of γ-aminobutyric acid receptor from rat brain. Life Sci 26:1131–1137.

Asano T, Yamada Y, Ogasawara N (1983): Characterization of the solubilized GABA and benzodiazepine receptors from various regions of bovine brain. J Neurochem 40:209–214.

Barnard EA, Stephenson FA, Sigel E, Mamalaki C, Bilbe G, Constanti A, Smart TG, Brown DA (1984): Structures and properties of the brain GABA/benzodiazepine receptor complex. In Kito S, Segawa T, Kuriyama K, Yamamura HI, Olsen RW (eds): "Neurotransmitter Receptors: Mechanism of Action and Regulation." New York: Plenum Press, pp 235–254.

Beaumont K, Chilton WS, Yamamura HI, Enna SJ (1977): Muscimol binding in rat brain: Association with synaptic GABA receptors. Brain Res 148:153–162.

Bowery NG, Doble A, Hill DR, Hudson AL, Shaw J, Turnbull MJ (1980a): β -Chlorophenyl GABA (Baclofen) is a selective ligand for a novel GABA receptor on nerve terminals. Brain Res Bull 5:497–502.

Bowery NG, Doble A, Hill DR, Hudson AL, Shaw JS, Turnbull MJ, Warrington R (1981): Bicuculline-insensitive GABA receptors on peripheral autonomic nerve terminals. Eur J Pharmacol 73:53–70.

Bowery NG, Hill DR, Hudson AL (1982): Bicuculline-insensitive $GABA_B$ receptors in mammalian brain: Specific binding of $^3$H-GABA and $^3$H-baclofen. In Okada Y, Roberts E (eds): "Problems in GABA Research." Amsterdam: Excerpta Medica, pp 302–310.

Bowery NG, Hill DR, Hudson AL, Doble A, Middlemiss DN, Shaw J, Turnbull M (1980): (-)Baclofen decreases neurotransmitter release in the mammalian CNS by an action at a novel GABA receptor. Nature 283:92–94.

Brehm L, Krogsgaard-Larsen P, Jacobsen P (1979): GABA upake inhibitors and structurally related "pro-drugs." In Krogsgaard-Larsen P, Scheel-Kruger J, Kofod H (eds): "GABA-Neurotransmitters". Copenhagen: Munksgaard, pp 247–262.

Browner M, Ferkany JW, Enna SJ (1981): Biochemical identification of pharmacologically and functionally distinct GABA receptors in rat brain. J Neurosci 1:514–518.

Chang L-R, Barnard EA (1982): The benzodiazepine/GABA receptor complex: Molecular size in brain synaptic membranes in solution. J Neurochem 39:1507–1518.

Constanti A, Houamed KM, Smart TG, Bilbe G, Brown DA, Barnard EA (1984): Functional brain GABA and glycine receptors in mRNA-injected *Xenopus* oocytes. Neuropharmacology 23:817–818.

Costa E, Guidotti A (1979): Molecular mechanisms in the receptor action of benzodiazepines. Ann Rev Pharmacol Toxicol 19:531–545.

Curtis DR, Duggan AW, Felix D, Johnston GAR (1971): Bicuculline, an antagonist of GABA and synaptic inhibition in the spinal cord of the cat. Brain Res 32:69–96.

Curtis DR, Hösli L, Johnston GAR (1968a): A pharmacological study of the depression of spinal neurones by glycine and related amino acids. Exp Brain Res 6:1–8.

Curtis DR, Hösli L, Johnston GAR, Johnston IH (1968b): The hyperpolarization of spinal motoneurones by glycine and related amino acids. Exp Brain Res 5:235–258.

Davidoff RA (1978): Pharmacology of spasticity. Neurology 28:46–51.

Enna SJ (1983a): GABA receptors. In Enna SJ (ed): "The GABA Receptors." Clifton NJ: Humana Press, pp 1–23.

Enna SJ (ed) (1983b): "The GABA Receptors." Clifton NJ: Humana Press.

Enna SJ (1985): γ-Aminobutyric acid (GABA) pharmacology and neuropsychiatric illness. In Coyle JT (ed): "Psychiatry Update." Vol. IV. New York: American Psychiatric Association Press, pp 67–82.

Enna SJ, Andree T (1983): GABA receptor heterogeneity: Relationship to benzodiazepines. In Usdin E, Skolnick, P, Tallman, J, Greenblatt, D, and Paul, SM (eds): "Pharmacology of Benzodiazepines." New York: Macmillan Press, pp 121–132.

Enna SJ, Coyle JT (1983): Neuroleptics. In Coyle JT, Enna SJ (eds): "Neuroleptics: Neurochemical, Behavioral and Clinical Perspectives." New York: Raven Press, pp 1–14.

Enna SJ, Gallagher JP (1983): Biochemical and electrophysiological characteristics of mammalian GABA receptors. Int Rev Neurobiol 24:181–212.

Enna SJ, Karbon EW, Duman RS (1985): $GABA_B$ agonists and imipramine-induced modifications in rat brain β-adrenergic receptor binding and function. In Morselli PL, Lloyd KG (eds): "GABA and Mood Disorders." New York: Raven Press (in press).

Enna SJ (1985): γ-Aminobutyric acid (GABA) pharmacology and neuropsychiatric illness. In Coyle JT (ed): "Psychiatry Update." Vol. IV. New York: American Psychiatric Association Press, pp 67–82.

Enna SJ, Strada SJ (1983): Postsynaptic receptors: Recognition sites, ion channels and second messengers. In Rosenberg R, Grossman R, Schochet S, Heinz ER, Willis W (eds): "Clinical Neurosciences," Vol. V. New York: Churchill Livingstone, pp 145–170.

Erdö SL (1984): Alteration of GABA levels in ovary and fallopian tube of the pregnant rat. Life Sci 34:1879–1884.

Erdö SL, Lapis E (1982): Bicuculline-sensitive GABA receptors in rat ovary. Eur J Pharmacol 85:243–246.

Erdö SL, Laszlo A (1984): High specific γ -aminobutyric acid binding to membranes of the human ovary. J Neurochem 42:1464–1467.

Falch E, Krogsgaard-Larsen P (1982): The binding of the GABA agonist $^{3}$H-THIP to rat brain synaptic membranes. J Neurochem 38:1123–1129.

Fernandez I, Orensanz LM, de Ceballos ML (1984): GABA modulation of cholinergic transmission in rat oviduct. Life Sci 35:357-364.

Friedel RO, Slotnick RN, Bombardt PA (1977): Effects of γ-aminobutyric acid on $^{32}P_i$ incorporation into rat brain phospholipids in vivo. Life Sci 20:235–242.

Fuller RW (1981): Enhancement of monoaminergic neurotransmission by antidepressant drugs. In Enna SJ, Malick JB, Richelson E (eds): "Antidepressants: Neurochemical, Behavioral and Clinical Perspectives." New York: Raven Press, pp 1–12.

Gallagher JP, Gallagher P (1983): Electrophysiological characteristics of GABA-receptor complexes. In Enna SJ (ed): "The GABA Receptors." Clifton, New Jersey: Humana Press, pp 25–61.

Gallagher JP, Stevens DR, Gallagher P (1984): Actions of GABA and baclofen on neurons of the dorsolateral septal nucleus (DLSN) *in vitro*. Neuropharmacology 23:825–826.

Gavish M, Snyder SH (1981): γ-Aminobutyric acid and benzodiazepine receptors: Co-purification and characterization. Proc Natl Acad Sci USA 78:1939–1947.

Guidotti A, Gale K, Suria A, Toffano G (1979): Biochemical evidence for two classes of GABA receptors in rat brain. Brain Res 172:566–571.

Haefely W, Kulcsar A, Möhler H, Pieri L, Polc P, Schaffner R (1975): Possible involvement of GABA in the central actions of benzodiazepines. Adv Biochem Psychopharmacol 14:131–151.

Haefely W, Polc P (1983): Electrophysiological studies on the interaction of anxiolytic drugs with GABAergic mechanisms. In Malick JB, Enna SJ, Yamamura HI (eds): "Anxiolytics: Neurochemical, Behavioral and Clinical perspectives." New York: Raven Press, pp 113–146.

Iversen LL, Bloom FE (1972): Studies on the uptake of $^{3}$H-GABA and $^{3}$H-glycine in slices and homogenates of rat brain and spinal cord by electron microscopic autoradiography. Brain Res 41:131–143.

Iversen LL, Neal MJ (1968): The uptake of $^{3}$H-GABA by slices of rat cerebral cortex. J Neurochem 15:114–1149.

Johnson RW, Tallman JF, Squires R, Yamamura HI (1983): Benzodiazepine receptor heterogeneity and regulation. In Malick JB, Enna SJ, Yamamura HI (eds): "Anxiolytics: Neurochemical, Behavioral and Clinical Perspectives." New York: Raven Press, pp 93–112.

Johnston GAR (1978): Neuropharmacology of amino acid inhibitory transmitters. Annu Rev Pharmacol Toxicol 18:269–289.

Johnston GAR (1983): Regulation of GABA receptors by barbiturates and by related sedative-hypnotic and anticonvulsant drugs. In Enna SJ (ed): "The GABA Receptors." Clifton, New Jersey: Humana Press, pp 107–128.

Karbon EW, Duman RS, Enna SJ (1983): Biochemical identification of multiple $GABA_B$ binding sites: Association with noradrenergic terminals in rat forebrain. Brain Res 274:393–396.

Karbon EW, Duman RS, Enna SJ (1984): $GABA_B$ receptors and norepinephrine-stimulated cAMP production in rat brain cortex. Brain Res 306:327–332.

Karbon EW, Enna SJ (1985): Characterization of the relationship between γ -aminobutyric acid B ($GABA_B$) agonists and transmitter-coupled cyclic nucleotide generating systems in rat brain. Mol Pharmacol 27:53–59.

Kendall DA, Browner M, Enna SJ (1982: The antinociceptive effect of GABA agonists: evidence for a cholinergic involvement. J Pharmacol Exp Ther 220:482–487.

Krnjević K (1974): Chemical nature of synaptic transmission in vertebrates. Physiol Rev 54:418–540.

Krnjević K, Schwartz S (1967): The action of γ-aminobutyric acid on cortical neurons. Exp Brain Res 3:320–336.

Krogsgaard-Larsen P, Christensen AV (1980): GABA agonists and antagonists. Annu Rep Med Chem 15:41–50.

Krogsgaard-Larsen P, Falch E, Christensen AV (1984): Chemistry and pharmacology of the GABA agonists THIP (Gaboxadol) and isoguvacine. Drugs of the Future 9:597–618.

Krogsgaard-Larsen P, Jacobsen P, Falch E (1983): Structure–activity requirements of the GABA receptor. In Enna SJ (ed): "The GABA Receptors." Clifton, New Jersey: Humana Press, pp 149–176.

Krogsgaard-Larsen P, Snowman A, Lummis SC, Olsen

RW (1981): Characterization of the binding of the GABA agonist $^3$H-piperidine-4-sulphonic acid to bovine brain synaptic membranes. J Neurochem 37:401–409.

Kuriyama K, Ito Y (1983): Some characteristics of solubilized and partially purified cerebral GABA and benzodiazepine receptors. In Mandel P, DeFeudis FV (eds): "CNS Receptors–From Molecular Pharmacology to Behavior." New York: Raven Press, pp 59–70.

Kuriyama K, Yonda Y, Taguchi J, Takahoshi M, Ohkuma S (1984): Properties of a purified $\gamma$-aminobutyric acid (GABA) receptor and modulation of GABA receptor binding by membrane phospholipids. Neuropharmacology 23:839–840.

Maggi CA, Manzini S, Meli A (1984a): Evidence that GABA receptors mediate relaxation of rat duodenum by activating intramural nonadrenergic-noncholinergic neurons. J Auton Pharmacol 4:77–85.

Maggi CA, Santicioli P, Meli A (1984b): $GABA_B$ receptors mediate inhibition of field stimulation induced contractions of detrusor muscle strips from newborn rats. J Auton Pharmacol 4:45–51.

Majewska MD, Chuang D-M (1984): Modulation by calcium of $\gamma$-aminobutyric acid (GABA) binding to $GABA_A$ and $GABA_B$ recognition sites in rat brain. Mol Pharmacol 25:352–359.

Meldrum BS (1978): $\gamma$-Aminobutyric acid and the search for new anticonvulsant drugs. Lancet 2:304–306.

Mernoff S, Cherwinski HM, Becker JW, de Blas AL (1983): Solubilization of brain benzodiazepine receptors with a zwitterionic detergent: Optimal preservation of their functional interaction with the GABA receptors. J Neurochem 41:752–758.

Miki Y, Taniyama K, Tanaka C, Tobe T (1983): GABA, glutamic acid decarboxylase, and GABA transaminase levels in the myenteric plexus in the intestine of humans and other mammals. J. Neurochem 40:861–865.

Möhler H, Okada T (1977a): Benzodiazepine receptor: Demonstration in the central nervous system. Science 198:849–851.

Möhler H, Okada T (1977b): GABA receptor binding with $^3$H-(+)-bicuculline methiodide in rat CNS. Nature 267:65–67.

Möhler H, Okada T (1978): Properties of $\gamma$-aminobutyric acid receptor binding with (+)- $^3$H-bicuculline methiodide in rat cerebellum. Mol Pharmacol 14:256–265.

Morselli PL, Lloyd KG (1983): Clinical Pharmacology of GABA agonists. In Enna SJ (ed): "The GABA Receptors." Clifton, New Jersey: Humana Press, pp 305–336.

Newberry, NR, Nicoll RA (1984a): Direct hyperpolarizing action of baclofen on hippocampal pyramidal cells. Nature 308:450-452.

Newberry NR, Nicoll RA (1984b): A bicuculline-resistant inhibitory post-synaptic potential on rat hippocampal pyramidal cells *in vitro*. J Physiol 348:239–254.

Nicoll RA, Alger BE (1979): Presynaptic inhibition: Transmitter and ionic mechanisms. Int Rev Neurobiol 21:219–258.

Olsen RW (1982): Drug interactions at the GABA receptor-ionophore complex. Annu Rev Pharmacol Toxicol 22:245–277.

Olsen RW, Bergman MO, Van Ness PC, Lummis SC, Watkins AE, Napias C, Greenlee DV (1981): Gamma-aminobutyric acid receptor binding in mammalian brain. Heterogeneity of binding sites. Mol Pharmacol 19:217–227.

Olsen RW, Enna SJ (1983): GABA and anxiolytics. In Malick JB, Enna SJ, Yamamura HI (eds): "Anxiolytics: Neurochemical, Behavioral and Clinical Perspectives." New York: Raven Press, pp 55–76.

Olsen RW, Snowman AM (1982): Chloride-dependent enhancement by barbiturates of $\gamma$-aminobutyric acid receptor binding. J Neurosci 2:1812–1823.

Olsen RW, Snowman AM (1983): $^3$H-Bicuculline methochloride binding to low-affinity $\gamma$-aminobutyric acid receptor sites. J Neurochem 41:1653–1663.

Palacios JM, Unnerstall JR, Young WS, Kuhar MJ (1981): Radiohistochemical studies of benzodiazepine and GABA receptors and their interactions. In Costa E, DiChiara G, Gessa GL (eds): "GABA and Benzodiazepine Receptors." New York: Raven Press, pp 53–76.

Polc P, Haefely W (1976): Effect of benzodiazepines, phenobarbitone and baclofen on synaptic transmission in the cat cuneate nucleus. Arch Pharmacol 294:121–131.

Polc P, Möhler H, Haefely W (1974): The effect of diazepam on spinal cord activities: Possible sites and mechanisms of action. Naunyn Schmiedeberg's Arch Pharmacol 284:319–337.

Ramanjaneyulu R, Ticku MK (1984): Binding characteristics and interactions of depressant drugs with $^{35}$S-t-butylbicyclophosphorothionate, a ligand that binds to the picrotoxinin site. J Neurochem 42:221–229.

Ryan LD, Roskoski R (1975): Selective release of newly synthesized and newly captured GABA from synaptosomes by potassium depolarization. Nature 248:254–256.

Sawaya C, Horton R, Meldrum B (1978): Transmitter synthesis and convulsant drugs: Effects of pyridoxal phosphate antagonists and allylglycine. Biochem Pharmacol 27:475–481.

Scheel-Kruger J (1983): The GABA receptor and animal behavior. In Enna SJ (ed): "The GABA Receptors." Clifton, New Jersey: Humana Press, pp 215–256.

Schmidt RF (1963): Pharmacological studies on the

primary afferent depolarization of the toad spinal cord. Pflugers Arch 277:325–346.
Schmidt RF (1971): Presynaptic inhibition in the vertebrate central nervous system. Ergebn Physiol. 63:20–101.
Schmidt RF, Vogel E, Zimmerman M (1967): Die Wirkung von Diazepam auf die prasynaptische Hemmung und Andere Ruckenmarksreflexe. Naunyn Schiedeberg's Arch Pharmacol 258:69–82.
Sigel E, Mamalaki C, Barnard EA (1982): Isolation of a GABA receptor from bovine brain using a benzodiazepine affinity colum. FEBS Lett 147:45–48.
Sigel E, Stephenson FA, Mamalaki C, Barnard EA (1983): A γ-aminobutyric acid/benzodiazepine receptor complex of bovine cerebral cortex. J Biol Chem 258:6965-6971.
Squires RF, Casida JE, Richardson M, Saederup E (1983): $^{35}$S-t-Butylbicyclophosphorothionate binds with high affinity to brain-specific sites coupled to γ-aminobutyric acid-A and ion recognition sites. Mol Pharmacol 23:326–336.
Stephenson FA, Olsen RW (1982): Solubilization by CHAPS detergent of barbiturate-enhanced benzodiazepine-GABA receptor complex. J Neurochem 39:1579–1586.
Study RE, Barker J (1981): Diazepam and (−) pentobarbital: fluctuation analysis reveals different mechanisms for potentiation of GABA responses in cultured central neurons. Proc Natl Acad Sci USA 78:7180–7184.
Supavilai P, Karobath M (1984): $^{35}$S-t-Butylbicyclophosphorothionate binding sites are constituents of the γ-aminobutyric acid benzodiazepine receptor complex. J Neurosci 4:1193–1200.
Tallman JF, Thomas JW, Gallagher DW (1978): GABAergic modulation of benzodiazepine binding site sensitivity. Nature 24:383-385.
Taniguchi H, Okada Y, Seguchi H, Shimada C, Seki M, Tsutou A, Baba S (1979): High concentration of gamma-aminobutyric acid in pancreatic beta cells. Diabetes 28:629–633.
Thalmann RH, Ayala GF (1982): A late increase in potassium conductance follows synaptic stimulation of granule neurons of the dentate gyrus. Neurosci Lett 29:243–248.
Thampy KG, Barnes EM (1984): γ-Aminobutyric acid-gated chloride channels in cultured cerebral neurons. J Biol Chem 259:1753–1757.
Ticku MK (1983): Benzodiazepine-GABA receptor-ionophore complex. Neuropharmacology 22:1459–1470.
Ticku MK, Olsen RW (1977): γ-Aminobutyric acid-stimulated chloride permeability in crayfish muscle. Biochem Biophys Acta 464:519–529.
Ticku MK, Van Ness PC, Haycock JW, Levy WB, Olsen RW (1978): Dihydropicrotoxin binding sites in rat brain: Comparison to GABA receptors. Brain Res 150:642–647.
Unnerstall Jr, Kuhar MJ, Niehoff DL, Palacios JM (1981): Benzodiazepine receptors are coupled to a subpopulation of γ-aminobutyric acid (GABA) receptors: Evidence from a quantitative autoradiographic study. J Pharmacol Exp Ther 218:797–804.
Vaught J, Pelley K, Costa LG, Settler P, Enna SJ (1985): A comparison of the antinociceptive responses to the GABA receptor agonists THIP and baclofen. Neuropharmacology 24:211–216.
Wilkin GP, Hudson AL, Hill DR, Bowery NG (1981): Autoradiographic localization of GABA receptors in rat cerebellum. Nature 274:584–587.
Wojcik WJ, Neff NH (1984): γ-Aminobutyric acid B receptors are negatively coupled to adenylate cyclase in brain, and in the cerebellum these receptors may be associated with granule cells. Mol Pharmacol 25:24–28.
Wong EHF, Leeb-Lundberg LMF, Teichberg VI, Olsen RW (1984): γ-Aminobutyric acid activation of Cl- flux in rat hippoccampal slices and its potentiation by barbiturates. Brain Res 303:267–275.
Yamamura HI, Enna SJ, Kuhar MJ (eds) (1985): "Neurotransmitter Receptor Binding," 2nd edition. New York: Raven Press.

Benzodiazepine/GABA Receptors and Chloride Channels:
Structural and Functional Properties, pages 57–71


# 3

# Multiplicity of GABA Receptors

Graham A.R. Johnston

*Department of Pharmacology, University of Sydney, NSW, 2006, Australia*

GABA, in common with most of the established neurotransmitters with the possible exception of glycine, is capable of interacting with more than a single class of receptor sites on neuronal membranes. There is much evidence for two broad classes of GABA receptors designated $GABA_A$ and $GABA_B$, while recently a third class ($GABA_C$) has been proposed.

What can be the function of apparently different receptors for a single agonist? Firstly, different receptors for a single agonist might be necessary for different responses to receptor activation. For example, a different molecular mechanism might be involved for an activated receptor complex to initiate a change in the ionic permeability of a neuronal membrane as is envisaged in the activation of certain $GABA_A$ receptors, compared to the mechanism needed to initiate an increased rate of production of an intracellular messenger such as cyclic GMP which may be the result of activation of certain $GABA_B$ receptors. Thus the linking of neurotransmitter receptors to different classes of molecular consequences may involve different types of receptors.

Secondly, receptors of different affinity for the same agonist may cooperate in the regulation of responses. For example, low-affinity presynaptic receptors ("autoreceptors") may mediate feedback inhibition of transmitter release when the concentration of transmitter in the synaptic environment exceeds that required to saturate high-affinity postsynaptic receptors. Receptors of different affinity on postsynaptic membranes may interact to regulate the response characteristics to changing agonist concentrations. Such interaction could be in the form of positive cooperativity resulting in a more rapid change in response to a small change in agonist concentration over a key concentration range (a "trigger response"), or negative cooperativity and a slow change in response to a large change in agonist concentration. Such allosteric interactions between different receptors may be similar to those well documented between enzyme subunits. High- and low-affinity binding sites for GABA have been described, and certain GABA responses appear to reflect positive cooperativity.

Thirdly, receptor properties may be changed by modulators ("allosteric effectors") or localization. The changes may be in affinity and/or availability and may mean that some of the receptor populations classified as "different" really represent the same gene products in different environments. A variety of endogenous substances termed GABARINS (GABA-*Re*ceptor-*IN*hibitor*S*), which may include GABA itself, peptides, phospholipids, purines, and chloride ions, appear to be able to alter the apparent properties of certain GABA receptors, as do some classes of centrally active drugs including the benzodiazepines and barbiturates.

How should we investigate receptors for neurotransmitters such as GABA? Clearly a multidisciplinary approach is necessary. The consequences of receptor activation may be measured using electrophysiological techniques, chemical measurements, and behavioral observations. Structure–activity studies are probably essential to the characterization of receptors, and in general the wider the range of agonists and antagonists available, the better is the characterization (and usually the larger the number of proposed different receptors!). Ligand binding studies are becoming increasingly important in investigations of GABA receptor affinity and density, and their modulation by GABARINS and environmental factors, and in aiding the purification of binding proteins to homogeneity. In recent years, two techniques have produced important new findings regarding GABA receptors. Noise analysis and patch-clamp studies are providing information on the dynamics of GABA receptor activation. These two techniques have afforded, for example, some further insight into the action of barbiturates on certain $GABA_A$ receptors: pentobarbital increases affinity of GABA binding to rat brain membranes by decreasing the rate of dissociation of GABA, consistent with the observed increased lifetime of GABA-activated chloride channels in spinal cultured neurons in the presence of pentobarbital. A further technique shows great promise for the study of GABA receptors particularly concerning the variety and nature of the gene products that appear to constitute certain GABA–receptor–ionophore supramolecular complexes. This technique involves the incorporation of messenger RNAs, coding for specific brain proteins, into amphibian oocytes which subsequently develop GABA receptors that can be investigated by electrophysiological methods. Coupled with quantitative structure–activity studies of GABA agonists and antagonists, and using highly purified messenger RNAs, these incorporation studies should provide a truly molecular understanding of the different types of GABA receptors, their activation, and their modulation.

## GABA ANTAGONISTS

### Bicuculline

The most widely used GABA antagonist is the phthalide isoquinoline alkaloid bicuculline (Fig. 1). Its GABA antagonist properties were discovered in 1970 [Curtis et al., 1970] as a result of a systematic study of convulsant alkaloids. Bicuculline is usually studied as the hydrochloride, or as the more stable and more water-soluble methochloride [Johnston et al., 1972]. Consistent with the widespread function of GABA as an inhibitory neurotransmitter, bicuculline hydrochloride on systemic administration antagonizes synaptic inhibitions in most, if not all, areas of the central nervous system (CNS) [Johnston, 1978a]. Bicuculline methochloride is excluded by the blood–brain barriers, thus enabling studies of peripheral bicuculline-sensitive GABA receptors without complications from central actions.

On local administration by iontophoresis, bicuculline hydrochloride and methochloride antagonize the neuronal action of GABA on most CNS neurons and some neurons in the periphery. While a variety of neurochemical data does indicate that GABA is involved in most bicuculline-sensitive inhibitions [Johnston, 1978a], it should be kept in mind that other bicuculline-sensitive neuronal inhibitors are found in the CNS; these inhibitors include β-alanine, δ-aminolevulinic acid, ethylenediamine, imidazole-4-acetic acid, and taurine. There are regional and species differences in the sensitivity of agonists to antagonism by bicuculline. For example, bicuculline antagonizes the inhibitory action of taurine in cat cerebral and cerebellar cortices but not in cat spinal cord or medulla [Johnston, 1978a].

Bicuculline (0.1–1 $\mu$M) is a competitive antagonist ($pA_2 = 5.35$) at $GABA_A$ receptors in the rat cuneate nucleus [Simmonds, 1978],

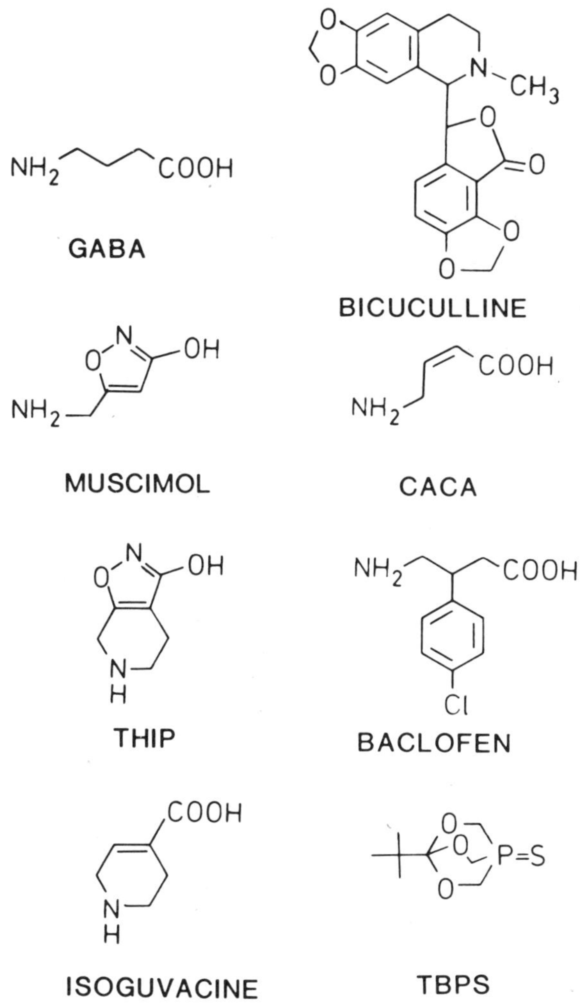


**Fig. 1.** *The structures of some of the compounds used in studies of multiple GABA receptors.*

and concentrations as low as 10 nM have been found to antagonize GABA-induced inhibitions in explants of rat cerebellum [Gahwiler, 1975]. Concentrations of bicuculline in excess of 1 $\mu$M are needed to displace radioactive GABA bound to rat brain membranes, but the potency is enhanced to submicromolar levels in the presence of thiocyanate and certain other anions [Enna and Snyder, 1977]. This effect of thiocyanate has been interpreted on the basis of exposure of a hydrophobic accessory binding site which enhances the interaction of bicuculline with low-affinity GABA binding sites [Maksay and Ticku, 1984]. Radioactive bicuculline methochloride binds to high-affinity sites ($K_D$ 30 nM) on rat membranes in the presence of thiocyanate, and these sites may coincide with the low-affinity sites ($K_D$ 1$\mu$M) for GABA that are implicated in many of the physiological actions of GABA [Olsen and Snowman, 1983] and benzodiazepines [Skerritt et al., 1982]. These observations are consistent with the proposal, based on structural similarities between bicuculline and certain conformationally restricted GABA agonists, that GABA and bicuculline share some common receptors [Andrews and Johnston, 1979]. These appear to be low-affinity $GABA_A$ binding sites. High concentrations of thiocyanate, perchlorate, or chloride ions may increase the affinity of bicuculline by favoring an antagonist-preferring conformational state [Möhler and Okada, 1978; Olsen and Snowman, 1983], with an exposed hydrophobic accessory binding site [Maksay and Ticku, 1984]. Autoradiographic studies on rat cerebellum indicate that bicuculline methochloride binding sites mimic the distribution of benzodiazepine binding sites, but not that of the GABA agonist muscimol binding to high-affinity GABA binding sites [Olsen et al., 1984]. This indicates that bicuculline and muscimol preferentially label different subtypes of $GABA_A$ sites, with the bicuculline-preferring subtypes being associated with benzodiazepine binding sites [Olsen et al., 1984].

#### Picrotoxinin and TBPS

The convulsant picrotoxinin acts as a noncompetitive antagonist of GABA acting at different sites to bicuculline [Simmonds, 1980]. Binding studies using radioactive dihydropicrotoxinin indicate that picrotoxinin acts on GABA-activated chloride channels rather than on GABA recognition sites [Olsen, 1981]. The displacement of GABA agonist binding by picrotoxinin and related convulsants is dependent on the presence of chloride or related anions [Supavilai et al., 1982]. Recently a new ligand, [$^{35}$S]-*t*-butyl bicyclophosphorothionate (TBPS; Fig. 1), has been introduced to study binding sites associated with GABA-activated chloride channels [Squires et al., 1983]. TBPS is one of a series of convulsants that have a cagelike structure and appear to act as GABA antagonists as the result of interaction with chloride channels [Bowery et al., 1976; Olsen, 1981].

TBPS in the presence of halide ions binds with high affinity to brain-specific binding sites with properties similar, but not identical to dihydropicrotoxinin binding sites [Squires et al., 1983].

TBPS binding is inhibited by $GABA_A$ agonists, but not by selective $GABA_B$ or $GABA_C$ agonists [Squires et al., 1983; Johnston et al., 1984]. Benzodiazepine agonists enhance while benzodiazepine inverse agonists inhibit TBPS binding [Supavilai and Karobath, 1984]. TBPS binding sites can be inactivated separately from the binding sites for $GABA_A$ ligands or those for benzodiazepines [Supavilai and Karobath, 1984]. A wide variety of centrally active drugs influence TBPS binding [Squires et al., 1983]. Barbiturates have biphasic effects on TBPS binding, enhancing at low and inhibiting at high concentrations [Supavilai and Karobath, 1984]. Convulsant $\gamma$-butyrolactones and caprolactams inhibit TBPS binding [Weissman et al., 1984; Skerritt et al., 1984]. There is a direct correlation between inhibition of TBPS by a series of 15 caprolactam derivatives and the modulation of muscimol binding to rat brain membranes in the presence of chloride ions [Skerritt et al., 1984].

A variety of insecticides influence TBPS binding. Studies with 37 pyrethroids showed absolute correlation between inhibition of TBPS binding and neurotoxicity on intracerebral injection into mice [Lawrence and Casida, 1983]. These studies indicate that the mechanism of toxicity of such insecticides may involve an interaction with GABA-activated chloride channels by binding to sites closely associated with, but not identical to TBPS sites. Some DDT-pyrethroid hybrid insecticides are potent enhancers of TBPS binding to rat brain membranes interacting at sites distinct from sites that bind GABA or benzodiazepines [Johnston et al., 1984].

The biphasic actions of etazolate and pentobarbital on TBPS binding with Hill coefficients greater than 1 for their inhibitory effects would be consistent with a multiplicity of

TBPS binding sites capable of interacting with positive cooperativity [Supavilai and Karobath, 1984]. It seems clear that TBPS binding sites on rat brain membranes can be influenced in a highly structurally specific manner suggestive of a supramolecular complex of $GABA_A$ binding sites coupled to chloride channels with specific sites for several classes of drugs. These classes include: (1) benzodiazepines, (2) barbiturates, (3) picrotoxinin and related convulsants, (4) TBPS and related cage convulsants, (5) γ-butyrolactones, (6) caprolactams, and (7) insecticides related to pyrethroids. The sites for some of these drug classes may overlap. Noncompetitive GABA antagonism may result from the activation of certain of these sites on the supramolecular complex. The presence or absence of these sites which are distinct from the GABA recognition sites may lead to the observed multiplicity of GABA-receptor-ionophore complexes.

## GABA AGONISTS

A large number of analogues of GABA are available for the study of the structural requirements for the interaction of GABA with a variety of neuronal receptors, transport carriers, and enzymes [Allan and Johnston, 1983]. Of particular interest are the analogues of restricted conformation, the study of which leads to knowledge of the particular active conformations of the flexible molecule of GABA interacting with particular recognition sites of receptors, transport carriers, and active sites of enzymes [Johnston et al., 1979].

On the basis of sensitivity to bicuculline as a selective antagonist, GABA agonists may be classified into two groups: bicuculline-sensitive and bicuculline-insensitive (Fig. 2). Bicuculline-sensitive agonists have been termed $GABA_A$ agonists to distinguish them from the bicuculline-insensitive agonist baclofen, which is termed a $GABA_B$ agonist [Hill and Bowery, 1981; Bowery et al., 1983]. Not all bicuculline-insensitive GABA agonists act on receptors sensitive to baclofen, and a third class of agonists, termed $GABA_C$ agonists, may interact with bicuculline-insensitive sites on neuronal membranes.

### $GABA_A$ Agonists and Bicuculline-sensitive Actions

Extensive structure–activity studies on GABA analogues that activate bicuculline-sensitive receptors in vivo indicate that GABA

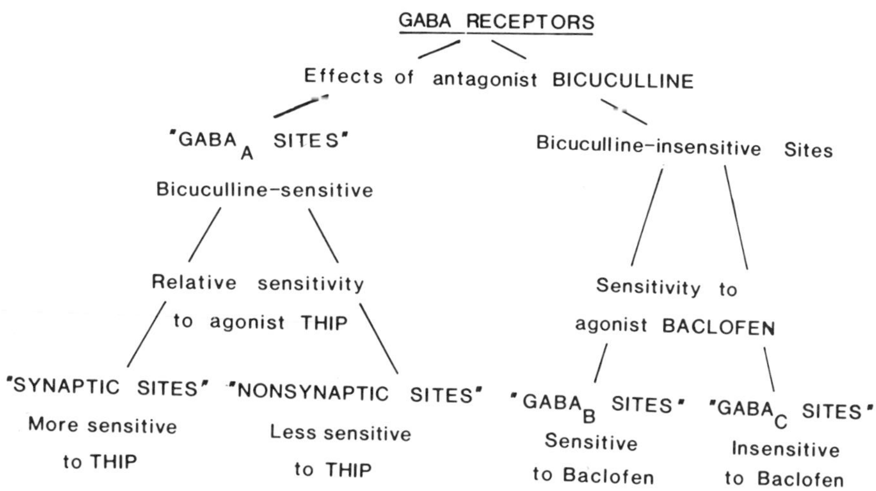


**Fig. 2.** *Outline of a possible classification of GABA receptors/binding sites.*

interacts with such receptors in partially extended and almost planar conformations [Krogsgaard-Larsen et al., 1977]. Analogues which act as $GABA_A$ agonists include muscimol, THIP, and isoguvacine (Fig. 1). The bicyclic GABA analogue THIP is a key compound, since its semirigid structure appears to endow particular specificity on its receptor interactions [Krogsgaard-Larsen et al., 1977; Krogsgaard-Larsen and Falch, 1981].

Bicuculline-sensitive actions of GABA are observed widely in the mammalian CNS and less frequently in the periphery. Bicuculline-sensitive receptors may occur as widely on nerve terminals ("presynaptic") exhibiting similar substrate specificity in each location as judged by studies in the cat spinal cord [Curtis et al., 1980]. There is evidence for some regional and species differences in bicuculline-sensitive receptors. Thus the substrate specificity for bicuculline-sensitive responses in the cat neurohypophysis differs from that in the cat spinal cord [Mathison and Dreifuss, 1980], and the relative potency of GABA and muscimol differs in the cat cerebral cortex from that in the rat spinal cord [Biscoe et al., 1972]. A number of bicuculline-sensitive receptors may exist for inhibitory amino acids with regional and species differences, as indicated by the following findings about the actions of taurine: (1) a bicuculline-sensitive, strychnine-insensitive action in the rat cerebellum; (2) a bicuculline-sensitive, strychnine-sensitive action in the cat cerebral cortex; and (3) a bicuculline-insensitive, strychnine-sensitive action in the cat spinal cord [Curtis et al., 1971; Frederickson et al., 1978]. In each of these regions GABA has a bicuculline-sensitive, strychnine-insensitive action. Thus it appears as though not all bicuculline-sensitive neuronal receptors are linked to GABA receptors.

GABA has "nonsynaptic" or "extrasynaptic" actions (for example, on nerve fibers which are devoid of synapses) that can be antagonized by bicuculline [Brown, 1979]. These nonsynaptic actions have different structure–activity properties from synaptic actions as judged by studies on a range of conformationally restricted GABA analogues as depressants of neuronal activity in the hemisected spinal cord from immature rats (a "synaptic" action) and as depolarizers of dorsal root fibers from the same animals (a "nonsynaptic" action) [Allan et al., 1980c]. These two classes of bicuculline-sensitive receptors may be the basis of the biphasic action of GABA on CA1 pyramidal cells in the rat hippocampus producing depolarization and hyperpolarization at different sites with different agonist and antagonist profiles [Alger and Nicoll, 1982]. The hyperpolarizing responses may arise from "synaptic" GABA receptors on the cell soma, while the depolarizing responses may arise from "nonsynaptic" responses on the dendrites. THIP appears to be relatively much more potent as a GABA agonist on "synaptic" than on "nonsynaptic" receptors, both in the spinal cord/dorsal root studies [Allan et al., 1980a] and in the hippocampal studies [Alger and Nicoll, 1982]; thus THIP may provide a basis for dividing $GABA_A$ receptors into subclasses of "synaptic" and "nonsynaptic" receptors (Fig. 2).

Fluctuation analyses of current responses in neurons cultured from mouse spinal cord and dorsal root ganglia indicate that GABA and $GABA_A$ agonists activate chloride ion channels of very similar conductance but significantly different duration in the lifetime of the activated channels [Barker and Mathers, 1981]. THIP has a relatively short duration of action (13 ms) in such studies compared to that of GABA (30 ms) and muscimol (76 ms); this indicates that predominantly "nonsynaptic" receptors are involved, since THIP and GABA exhibit similar potency on "synaptic" $GABA_A$ receptors in the cat spinal cord [Krogsgaard-Larsen et al., 1977]. There is some evidence from electrophysiological studies for positive cooperativity in the activation of $GABA_A$ receptors [Johnston, 1978b; Krause et al., 1981].

Functional $GABA_A$ receptors have been expressed in Xenopus oocytes injected with rat brain mRNA [Houamed et al., 1984]. Responses to GABA can be antagonized by bicuculline and picrotoxin and can be enhanced by benzodiazepines and barbiturates. Too few GABA analogues have been tested thus far to judge as to whether the expressed receptors show agonist profiles similar to "synaptic" or "nonsynaptic" $GABA_A$ receptors.

### High- and Low-Affinity $GABA_A$ Binding Sites

Binding studies have been carried out with a variety of $GABA_A$ agonists as ligands, including muscimol [DeFeudis, 1980], isoguvacine [Morin and Wasterlain, 1980], piperidine-4-sulfonic acid [Krogsgaard-Larsen et al., 1981] and THIP [Falch and Krogsgaard-Larsen, 1982]. Like GABA, these ligands exhibit at least biphasic binding kinetics being analyzed in terms of high- and low-affinity binding sites [Olsen, 1981]. There is no evidence of cooperactivity between high- and low-affinity GABA binding sites [Horng and Wong, 1979]. Indeed, these kinetically different sites appear to be functionally distinct differing in rates of heat inactivation and sensitivity to changes in pH, temperature, and ionic environment [Olsen et al., 1981].

The agonist specificity of high-affinity GABA binding to brain membranes resembles that observed in electrophysiological studies for "nonsynaptic" rather than "synaptic" GABA receptors [Allan et al., 1980c]. It may be that the methods used in preparing brain membranes for binding studies alter the organization of GABA binding sites such that a majority of the isolated sites behave as if they were "nonsynaptic." Different membrane preparations are known to favor somewhat different agonist specificities [Krogsgaard-Larsen et al., 1979].

Low-affinity GABA binding sites are known to interact with benzodiazepine sites [Tallman et al., 1978; Karobath and Sperk, 1979; Skerritt et al., 1982; Burch et al., 1983]. GABA is known to enhance benzodiazepine agonist binding in concentrations appropriate to low-affinity GABA binding sites. Benzodiazepine agonists increase the affinity of GABA binding to low-affinity sites in appropriately prepared membrane preparations. The nexus between benzodiazepine agonist sites and low-affinity $GABA_A$ sites is not simple: enhancement of benzodiazepine agonist binding by GABA is a relatively robust phenomenon, whereas enhancement of low-affinity $GABA_A$ binding by benzodiazepines involves endogenous modulators [Johnston and Skerritt, 1984].

Electrophysiological studies show that benzodiazepine agonists can enhance the neuronal effects of GABA and of $GABA_A$ agonists such as muscimol and THIP. Binding studies show that GABA, muscimol, and THIP can enhance benzodiazepine agonist binding under appropriate conditions [Supavilai and Karobath, 1980] and that the binding of GABA and muscimol, but not THIP, can be enhanced by benzodiazepine agonists [Skerritt and Johnston, 1983; Skerritt and MacDonald, 1984]. It has been suggested that a low-affinity binding site, $K_D$ of 20–50 $\mu$M, for THIP, which is difficult to detect in THIP binding studies, may be involved in the enhancement of THIP action by benzodiazepines [Skerritt and MacDonald, 1984].

Treatment of brain membranes with silver nitrate alters the binding of GABA agonists such as muscimol from biphasic kinetics to a single high-affinity system [Supavilai et al., 1982]. In such membranes, muscimol is able to enhance benzodiazepine agonist binding by a high-affinity interaction, while TBPS binding sites are undetectable [Supavilai and Karobath, 1984]. These observations show that high-affinity states of GABA agonist binding sites can be functionally linked to benzodiazepine agonist binding sites under conditions where TBPS binding sites are perturbed. This is consistent with purification studies on GABA and benzodiazepine binding sites, which show that GABA binding sites of high-affinity copurify with benzodiazepine sites as part of the same macromolecular complex [Gavish and Snyder, 1981; Stephenson and Olsen, 1982; Lo and Snyder, 1983]. These studies indicate that the high-affinity GABA binding sites functionally linked to benzodiazepine sites in purified solubilized membranes, or in membranes treated with silver nitrate, are different from the high-affinity GABA binding sites observed with relatively intact membrane preparations. Diazotization of brain membranes with *p*-diazobenzenesulfonic acid selectively eliminates low-affinity GABA binding and the ability of GABA to enhance benzodiazepine binding and decreases potency of antagonists to displace agonist binding [Burch et al., 1983; Maksay and Ticku, 1984]. On the other hand, thiocyanate reversibly reduces the availability of high-affinity GABA binding sites, increasing the potency of antagonists and decreasing the potency of agonists to displace bound GABA. These results have been interpreted on the basis of the existence of distinct high- and low-affinity GABA binding sites which can exist in interconvertible agonist and antagonist conformations [Maksay and Ticku, 1984].

The agonist specificity of low-affinity $GABA_A$ binding sites may be assessed by studying the ability of substances to enhance benzodiazepine agonist binding [Tallman et al., 1978] or displace the binding of bicuculline methochloride [Olsen and Snowman, 1983]. Low-affinity $GABA_A$ sites assessed on the basis of enhancement of benzodiazepine binding appear to have a quite different agonist specificity from high-affinity $GABA_A$ sites [Johnston et al., 1982]. Recent studies have indicated that *cis*-3-[(aminoiminomethyl)thio]propenoic acid may be a selective ligand for low-affinity $GABA_A$ sites being more potent than GABA or muscimol in enhancing benzodiazepine binding [Allan et al., 1984; Johnston et al., 1984].

### Structure–Activity Studies on $GABA_A$ Agonists

Computer modeling indicates that $GABA_A$ agonists show definite structural similarities with bicuculline, consistent with the observations that these substances show some common binding sites [Andrews and Johnston, 1979]. Such modeling has suggested that GABA may interact with $GABA_A$ receptors in a rather narrow range of "active conformations" from the "bicuculline conformation" (defined on the basis of the structure of bicuculline) to the "muscimol" conformation (defined on the basis of the potent $GABA_A$ agonist, muscimol). Within this range, the agonists muscimol, THIP, *trans*-4-aminocrotonic acid, (+)-*cis*-3R-aminocyclopentane-1S-carboxylic acid, and *cis*-3-[(aminoiminomethyl)thio]pentenoic acid may adopt conformations close to the "muscimol conformation," whereas 4,5-dihydromuscimol and (+)-*trans*-3S-aminocyclopentane-1S-carboxylic acid may adopt the "bicuculline conformation." These $GABA_A$ agonists may form the basis for distinguishing between different classes of bicuculline-sensitive receptors. There is already evidence that THIP preferentially activates synaptic rather than nonsynaptic $GABA_A$ receptors and that *cis*-3-[(aminoiminomethyl)thio]pentenoic acid preferentially activates low- rather than high-affinity $GABA_A$ binding sites as noted above. Computer-assisted design may lead to the development of more selective agonists and antagonists with which to distinguish different bicuculline-sensitive GABA receptors.

### Baclofen as a $GABA_B$ Agonist

Baclofen (Lioresal, β-(p-chlorophenyl)-GABA) (Fig. 1) was originally designed as a lipophilic derivative of GABA able to cross the blood–brain barriers [Bowery, 1983a]. It has proved to be a relatively selective GABA agonist for a class of receptors ($GABA_B$) that is not antagonized by the "classic" $GABA_A$ antagonist bicuculline [Bowery et al., 1980; Hill and Bowery, 1981; Bowery et al., 1983]. The bicuculline-insensitive depressant action of baclofen on the firing of CNS neurons in vivo in the cat spinal cord and cerebral cortex was first described in 1974 [Curtis et al., 1974; Davies and Watkins, 1974] and subsequently a number of conformationally restricted analogues of GABA were found to have similar actions, e.g., *cis*-4-aminocrotonic acid (CACA; Fig. 1). [Johnston et al., 1975]. The structural similarities between certain of these analogues led to the proposal that they constitute another class of GABA agonist acting on bicuculline-insensitive receptors [Johnston, 1976, 1978b]. The possibility of bicuculline-insensitive receptors for GABA had also been proposed on the basis of other considerations by Ryall [1975]. GABA agonists were considered to act on bicuculline-sensitive GABA receptors in relatively extended conformations and on bicuculline-insensitive GABA receptors in relatively folded conformations. Recent binding studies using radioactive (−)-baclofen have shown that these conformationally restricted analogues that have bicuculline-insensitive depressant actions on neurons do not interact with binding sites on brain membranes for (−)-baclofen [Johnston et al., 1984]. Furthermore, there is electrophysiological evidence for two pharmacologically distinct bicuculline-insensitive actions of GABA in the hippocampus, only one of which is shared by baclofen [Ault and Nadler, 1983]. Thus it appears that GABA may be able to activate more than one class of bicuculline-insensitive receptors while baclofen has a selective action activating only what have been designated $GABA_B$ receptors by Hill and Bowery [1981].

The concept of baclofen as a selective $GABA_B$ agonist came from studies on the peripheral nervous system, where GABA was shown to have a bicuculline-insensitive inhibitory action on the release of noradrenaline from atria and vas deferens [Bowery and Hudson, 1979]. Baclofen had a similar effect to GABA, but the classic GABA agonists, such as isoguvacine and muscimol, were inactive [Bowery et al., 1980]. This action of baclofen was stereospecific, with (−)-baclofen being at least 100 times more potent than (+)-baclofen consistent with other pharmacological actions of baclofen [Olpe et al., 1978].

Baclofen-sensitive, bicuculline-insensitive receptors have been found extensively in the peripheral nervous system and appear to be widespread in the central nervous system [Bowery, 1983a]. These $GABA_B$ receptors do not appear to be linked to chloride channels, and there is no evidence for their modulation by benzodiazepines or barbiturates. They may be coupled to calcium ion fluxes, and the binding of baclofen to neuronal membranes is a calcium-dependent process. GTP reduces the affinity of GABA for $GABA_B$ binding sites without influencing $GABA_A$ sites.

Binding studies on $GABA_B$ sites may be carried out using radioactive GABA or baclofen, preferably the recently introduced [$^3$H]-(−)-baclofen [Weatherby et al., 1984], in the presence of calcium ions to promote $GABA_B$ site binding and of isoguvacine to antagonize $GABA_A$ site binding [Bowery et al., 1983]. In some instances the same results are obtained with either ligand, but in other cases (−)-baclofen does not displace all of the specifically bound GABA under these conditions [Drew et al., 1984] consistent with GABA binding to sites other than those labeled by baclofen.

Autoradiographic studies indicate that $GABA_B$ sites are distinct from $GABA_A$ sites in the cerebellum and spinal cord. In the cerebellum, $GABA_B$ sites are localized in the molecular cell layer while $GABA_A$ sites are mainly in the granule cell layer [Wilkin et al., 1981]. $GABA_B$ sites in the spinal cord are localized to Rexed laminae II and III of the dorsal horn, while $GABA_A$ sites are relatively uniformly distributed throughout the gray matter [Price et al., 1984]. The $GABA_B$ sites localized by autoradiography of bound [$^3$H]GABA may also reflect some $GABA_C$ sites [Johnston et al., 1984] and thus [$^3$H]-(−)-baclofen should be the ligand of choice at present for further localization studies. It is not the ideal ligand, however, since not all of the pharmacological actions of (−)-baclofen are necessarily related to activation of $GABA_B$ binding sites [Johnston et al., 1981; Lanthorn et al., 1981; Yanagisawa et al., 1981].

Structure–activity studies on $GABA_B$ agonists, as assessed by ability to displace bound radioactive baclofen, indicate that a *substituted* phenyl substituent in the $\beta$-position of the GABA molecule is necessary for a selective interaction with $GABA_B$ sites [Bowery et al., 1983]. The order of potency for the substituted β-phenyl-GABA derivatives is *p*-chloro > *o*-chloro > *p*-fluoro > *m*-chloro. It is of particular interest that β-phenyl-GABA itself is inactive as a displacer of bound radioactive baclofen, though this derivative is able to displace radioactive GABA in the presence of calcium ions and isoguvacine [Bowery et al., 1983]. The active stereoisomer of baclofen has the R(−)-configuration [Chang et al., 1982] and steric hindrance resulting from such a substitution at the β-position of the GABA carbon chain may prevent interaction with $GABA_A$ and $GABA_C$ receptor sites.

δ-Aminovaleric acid has been reported to be an antagonist at some $GABA_B$ sites, whereas it acts as an agonist at $GABA_A$ sites [Muhyaddin et al., 1982a,b]. δ-Aminovaleric acid is a relatively weak displacer of bound radioactive baclofen [Bowery et al., 1983].

### $GABA_C$ Agonists?

A number of conformationally restricted analogues of GABA have been found to have bicuculline-insensitive depressant actions on the firing on CNS neurons: *cis*-4-aminocrotonic acid (CACA; Fig. 1) [Johnston et al., 1975], 5-hydroxy-3-(1-aminopropyl)isoxazole [Krogsgaard-Larsen et al., 1975], *trans*-2- and *trans*-3-aminocyclohexane carboxylic acid [Johnston, 1976], *cis*-2-(aminomethyl)cyclopropane carboxylic acid [Allan et al., 1980a], and *trans*-3-aminocyclobutane carbolic acid [Allan et al., 1980b]. Certain of these analogues show structural similarities which led to the proposal that they may constitute a class of GABA agonist acting on bicuculline-insensitive receptors [Johnston, 1976, 1978b]. The effects of a variety of these conformationally restricted GABA analogues have been examined on the binding of $^3$H-(−)-baclofen and of $^3$H-GABA to synaptic membranes from rat cerebellum under conditions reported to favor binding to bicuculline-insensitive sites, i.e., in the presence of calcium ions with bicuculline-sensitive binding being inhibited by the selective agonist isoguvacine [Hill and Bowery, 1981; Bowery et al., 1983]. While most of these analogues inhibited GABA binding, they were ineffective as inhibitors of $^3$H-(−)-baclofen binding [Drew et al., 1984; Johnston et al., 1984]. This has led to the hypothesis of a third major class of GABA receptors, designated $GABA_C$ sites, which are insensitive to either the $GABA_A$ antagonist bicuculline or the $GABA_B$ agonist baclofen (Fig. 2).

Analogues of GABA in folded conformations are able to displace radioactive GABA bound to rat cerebellar membranes in the presence of calcium ions and isoguvacine; in the case of *cis*-4-aminocrotonic acid and *cis*-2-(aminomethyl)cyclopropane carboxylic acid, they are more potent inhibitors of GABA bound under these conditions than in the absence of calcium ions and isoguvacine (i.e., conditions favoring binding to $GABA_A$ sites) [Drew et al., 1984; Johnston et al., 1984]. The results obtained with *cis*-4-aminocrotonic acid (CACA) indicate that this GABA analogue may be a selective ligand for a hitherto undescribed class of GABA binding sites. CACA is inactive at 100 μM as a displacer of $^3$H-(−)-baclofen bound to $GABA_B$ sites, and relatively weak ($IC_{50}$ 25 μM, which could result from a less than 1% contamination with the *trans*-isomer) as a displacer of $^3$H-GABA bound to $GABA_A$ sites, but relatively potent ($IC_{50}$ 1 μM, 10-fold more potent than the *trans*-isomer) as a displacer of $^3$H-GABA bound in the presence of the selective $GABA_A$ agonist isoguvacine [Drew et al., 1984; Johnston et al., 1984]. β-Phenyl-GABA may also act as a selective $GABA_C$ agonist, since it inhibits ($IC_{50}$ 10 μM) the binding of GABA in the presence of isoguvacine but is inactive at 100 μM against $GABA_A$ or baclofen binding to rat brain membranes [Bowery et al., 1980].

Ault and Nadler [1983] have provided electrophysiological evidence using rat hippocampal slices for a bicuculline-resistant action of bath-applied GABA on synaptic responses in the fascia dentata, which is not shared by baclofen. Their results indicate two pharmacologically distinct, bicuculline-insensitive actions of GABA in the hippocampus, which might be explained on the basis of the activation of both $GABA_B$ and $GABA_C$ binding sites by GABA but of only $GABA_B$ sites by baclofen.

These observations of course do not provide evidence for any physiological significance of a class of GABA sites that are distinct from $GABA_A$ and $GABA_B$ sites, but they may provide a stimulus for further investigation and development of more selective ligands for the various possible classes of GABA receptors. *cis*-4-Aminocrotonic acid may be an important lead compound for the investigation of this proposed new class of GABA binding sites. Interactive computer graphics indicate that the molecular structures of *cis*-4-aminocrotonic acid and *cis*-2-(aminomethyl) cyclopropane carboxylic acid can be closely superimposed with respect to their "GABA backbones," their amino and carboxyl functions, whereas *trans*-3-aminocyclobutane carboxylic acid (which is some 60 times less active than the two former compounds in displacing GABA) can be superimposed much less closely, particularly with respect to its carboxyl group. Isoguvacine, the selective $GABA_A$ ligand, matches poorly to *cis*-4-aminocrotonic acid with the possibility for steric hindrance from one of the piperidine methylene groups. This may explain why isoguvacine and related ligands such as THIP and piperidine-4-sulfonic acid are selective for $GABA_A$ sites.

5-Hydroxy-3-(1-aminopropyl)isoxazole has been described as a moderately potent, bicuculline-insensitive neuronal depressant [Krogsgaard-Larsen et al., 1975] which does not inhibit the binding of $^3$H-($\pm$)-baclofen [Krogsgaard-Larsen et al., 1984] or the bicuculline-sensitive binding of $^3$H-GABA [Krogsgaard-Larsen and Johnston, 1978]. This isoxazole is also inactive at 100 $\mu$M as an inhibitor of the binding of $^3$H-GABA to the proposed $GABA_C$ sites in the present studies, suggesting that it may interact with yet another class of sites in its depressant action on neuronal firing and supporting the concept of a multiplicity of bicuculline-insensitive binding sites for GABA.

## CLASSIFICATION OF GABA RECEPTORS

How should receptors for neurotransmitters be classified? A review entitled, "Is it possible to integrate dopamine receptor terminology?" [Offermeier and van Rooyan, 1982] highlights the difficulties of "the prodigious classification of dopamine receptors" that has resulted from the many different approaches to this problem which extends to many neurotransmitters. The classical approach of "shorthand" codes which started with muscarinic and nicotinic cholinoceptors has a limited usefulness, but to whom but an expert in the field do the codes D1, D2, etc., communicate anything of significance? I have attempted (unsuccessfully!) to use an even more complicated code to classify GABA receptors: e.g., AH-145 was to describe postsynaptic hyperpolarizing receptors sensitive to bicuculline and barbiturates, but not to benzodiazepines [Johnston, 1981]. This classification was intended to reflect the anatomical localization of the receptors, the effect of their activation on neuronal properties, and their pharmacological profile. The total number of theoretically possible combinations was 48; in "clarifying and consolidating these numerous possibilities," Gallagher and Shinnick-Gallagher (1983) concluded, "Thus, the classification of GABA receptor complexes may be even more diverse than that proposed by Johnston [1981]"! Also commenting on this classification, Bowery (1983b) concluded, "As Johnston [1981] points out, there may be many subtypes of bicuculline-sensitive GABA sites pre- or postsynaptic, linked or not linked, to benzodiazepine and/or barbiturate sites. All we need are the definitive baskets in which to put our eggs." To do this we need more selective agents with which to investigate the possible different types of GABA receptors, and these will take many years to develop and critically evaluate. Until a more definitive classification can be developed, I would suggest that investigators should describe at least the positive information about the receptors under study. Thus studies of GABA receptor activation should describe the anatomical localization and the physiological and pharmacological properties, while studies of GABA binding should decribe the source of the binding membranes and their major biochemical properties. It should be borne in mind that certain GABA receptors are parts of complex domains and that an apparent multiplicity of receptors may well reflect the same gene product for the GABA recognition site in different environments.

The present alphabetical classification of $GABA_A$ sites as bicuculline-sensitive and $GABA_B$ sites as baclofen-sensitive and bicuculline-insensitive is a useful shorthand description; the possible emergence of $GABA_C$ sites that are bicuculline-insensitive and baclofen-insensitive indicates that it can be expanded to encompass new findings. Classifying and understanding the interrelationships of different GABA receptors is a very complex problem, as indicated by the likelihood of $GABA_A$, $GABA_B$, and $GABA_C$ sites being found on the same terminals of primary afferent fibers in the spinal cord!

## ACKNOWLEDGMENTS

The author is grateful to his colleagues Dr. R.D. Allan, Ms. A.D. Benton, Ms. S. Chen Chow, Mrs. C.A. Drew, Dr. R. Kazlauskas, Ms. H. Rzezniczak, and Dr. R.P. Weatherby for helpful discussions. The financial support of the Australian National Health and Medical Research Council is gratefully acknowledged.

## REFERENCES

Alger BE, Nicoll RA (1982): Pharmacological evidence for two kinds of GABA receptor on rat hippocampal pyramidal cells studies in vitro. J Physiol 328:125–141.

Allan RD, Curtis DR, Headley PM, Johnston GAR, Lodge D, Twitchin B (1980a): The synthesis and activity of *cis*- and *trans*-2-(aminomethyl)cyclopropane carboxylic acid as conformationally restricted analogues of GABA. J Neurochem 34:651–655.

Allan RD, Curtis DR, Headley PM, Johnston GAR, Kennedy SME, Lodge D, Twitchin B (1980b): Cyclobutane analogs of GABA. Neurochem Res 5:43–50.

Allan RD, Evans RH, Johnston GAR (1980c): γ-Aminobutyric acid agonists: an in vitro comparison between depression of spinal synaptic activity and depolarization of spinal root fibres in the rat. Br J Pharmacol 70:609–615.

Allan RD, Johnston GAR (1983): Synthetic analogs for the study of GABA as a neurotransmitter. Med Res Rev 3:91–118.

Allan RD, Hiern BP, Johnston GAR, Kazlauskas R, Rzezniczak H (1984): Isothiouronium salts as GABA agonists. Clin Exp Pharmacol Physiol (suppl.) 8:113.

Andrews PR, Johnston GAR (1979): GABA agonists and antagonists. Biochem Pharmacol 28:2697–2702.

Ault B, Nadler JV (1983): Physiological evidence for two pharmacologically distinct, bicuculline-insensitive actions of GABA in the rat hippocampal slice. Soc Neurosci Abs 9:411.

Barker JL, Mathers DA (1981): GABA analogues activate channels of different duration on cultured mouse spinal neurons. Science 212:358–361.

Biscoe TJ, Duggan AW, Lodge D (1972): Antagonism between bicuculline, strychnine, and picrotoxin and depressant amino acids in the rat nervous system. Comp Gen Pharmacol 3:423–433.

Bowery NG (1983a): Baclofen: 10 years on. Trends Pharm Sci 3:400–403.

Bowery NG (1983b): Classification of GABA receptors. In Enna SJ (ed.) "The GABA Receptors." Clifton, New Jersey: Humana Press, pp 177–213.

Bowery NG, Collins JF, Hill RG, Pearson S (1976): GABA antagonism as a possible basis for the convulsant action of a series of bicyclo phosphorus esters. Br J Pharmacol 57:435P–436P.

Bowery NG, Hudson AL (1979): γ-Aminobutyric acid reduces evoked release of $^3$H-noradrenaline from sympathetic nerve terminals. Br J Pharmacol 66:108P.

Bowery NG, Hill DR, Hudson AL, Doble A, Middlemiss DN, Shaw J, Turnbull MJ (1980): (−)-Baclofen decreases neurotransmitter release in the mammalian CNS by an action at a novel GABA receptor. Nature 283:92–94.

Bowery NG, Hill DR, Hudson AL (1983): Characteristics of $GABA_B$ receptor binding sites on rat whole brain synaptic membranes. Br J Pharmacol 78:191–206.

Brown DA (1979): Extrasynaptic GABA systems. Trends Neurosci 2:271–273.

Burch TP, Thyagarajan R, Ticku MK (1983): Group selective reagent modification of the benzodiazepine-γ-aminobutyric acid receptor-ionophore complex reveals that low-affinity γ-aminobutyric acid receptors stimulate benzodiazepine binding. Mol Pharmacol 23:52–59.

Chang CH, Yang DSC, Yoo CS, Wang BC, Pletcher J, Sax M, Terrence CF (1982): Structure and absolute configuration of (R)-baclofen monohydrochloride. Acta Cryst B38:2065–2067.

Curtis DR, Duggan AW, Felix D, Johnston GAR (1970): GABA, bicuculline and central inhibition. Nature 226:1222–1224.

Curtis DR, Duggan AW, Felix D, Johnston GAR, McLennan H (1971): Antagonism between bicuculline and GABA in the cat brain. Brain Res 33:57–73.

Curtis DR, Game CJA, Johnston GAR, McCulloch RM (1974): Central effects of β-(p-chlorophenyl)-γ-aminobutyric acid. Brain Res 70:493–499.

Curtis DR, Bornstein JC, Lodge D (1980): In vivo analysis of GABA receptors on primary afferent terminations in the cat. Brain Res 194:255–258.

Davies J, Watkins JC (1974): The action of β-phenyl-GABA derivatives on neurones of the cat cerebral cortex. Brain Res 70:501–505.

DeFeudis FV (1980): Binding studies with muscimol; relation to synaptic γ-aminobutyrate receptors. Neuroscience 5:675–688.

Drew CA, Johnston GAR, Weatherby RP (1984) Bicuculline-insensitive GABA receptors: studies on the binding of (−)-baclofen to rat cerebellar membranes. Neurosci Lett 52:317–321.

Enna SJ, Snyder SH (1977): Influences of ions, enzymes, and detergents on γ-aminobutyric acid receptor binding in synaptic membranes. Mol Pharmacol 13:442–453.

Falch E, Krogsgaard-Larsen P (1982): The binding of the specific GABA agonist [$^3$H]THIP to rat brain synaptic membranes. J Neurochem 38:1123–1129.

Frederickson RCA, Neuss M, Morzorati SL, McBride WJ (1978): A comparison of the inhibitory effects of taurine and GABA on identified Purkinje cells and other neurons in the cerebellar cortex of the rat. Brain Res 145:117–126.

Gahwiler BH (1975): The effects of GABA, picrotoxin and bicuculline on the spontaneous bioelectric activity of cultured cerebellar Purkinje cells. Brain Res 99:85–95.

Gallagher JP, Shinnick-Gallagher P (1983): Electrophysiological characteristics of GABA-receptor complexes. In Enna SJ (ed): "The GABA Receptors." New Jersey: Humana Press, pp 25–61.

Gavish M, Snyder SH (1981): γ-Aminobutyric acid and benzodiazepine receptors: copurification and characterization. Proc Natl Acad Sci USA 78:1939–1942.

Hill DR, Bowery NG (1981): $^3$H-baclofen and $^3$H-GABA bind to bicuculline-insensitive $GABA_B$ sites in rat brain. Nature 290:149–152.

Horng JS, Wong DT (1979): γ-Aminobutyric acid receptors in cerebellar membranes of rat brain after a treatment with Triton X-100. J Neurochem 32:1379–1386.

Houamed KM, Bilbe G, Smart TG, Constanti A, Brown DA, Barnard EA, Richards BM (1984): Expression of functional GABA, glycine and glutamate receptors in Xenopus oocytes injected with rat brain mRNA. Nature 310:318–321.

Johnston GAR (1976): Physiologic pharmacology of GABA and its antagonists in the vertebrate nervous system. In Roberts E, Chase TN, Tower DB (eds): "GABA in Nervous System Function." New York: Raven Press, pp 394–411.

Johnston GAR (1978a): Neuropharmacology of amino acid transmitters. Annu Rev Pharmacol Toxicol 18:269–289.

Johnston GAR (1978b): Amino acid receptors. In Smythies JR, Bradley RJ (eds): "Receptors in Pharmacology." New York: Marcel Dekker, pp 295–333.

Johnston GAR (1981): GABA receptors. In Lombardini JB, Kenny AD (eds): "The Role of Peptides and Amino Acids as Neurotransmitters." New York: Alan R. Liss, pp 1–17.

Johnston GAR, Beart PM, Curtis DR, Game CJA, McCulloch RM (1972): Bicuculline methochloride as a GABA antagonist. Nature New Biol 240:219–220.

Johnston GAR, Curtis DR, Beart PM, Game CJA, McCulloch RM, Twitchin B (1975): *Cis* and *trans*-4-aminocrotonic acid as GABA agonists of restricted conformation. J Neurochem 24:157–160.

Johnston GAR, Allan RD, Kennedy SME, Twitchin B (1979): Systematic study of GABA analogues of restricted conformation. In Krogsgaard-Larsen P, Scheel-Kruger J, Kofod H (eds): "GABA-Neurotransmitters." Copenhagen: Munksgaard, pp 149–164.

Johnston GAR, Hailstone MH, Freeman CG (1981): Baclofen: stereoselective inhibition of excitant amino acid release. J Pharm Pharmacol 32:230–231.

Johnston GAR, Willow M, Davies LP, Allan RD (1982): Enhancement of diazepam binding by GABA analogues. Neurosci Lett (Suppl.) 8:S60.

Johnston GAR, Skerritt JH (1984): GABARINS and the nexus between GABA and benzodiazepine receptors. In Bowery NG (ed): "Actions and Interactions of GABA and Benzodiazepines." New York: Raven Press, pp 179–189.

Johnston GAR, Allan RD, Benton AD, Chen Chow S, Drew CA, Hiern BP, Holan G, Kazlauskas R, Rzezniczak H, Weatherby RP (1984): GABA receptors and related binding sites. Proc. 9th IUPHAR Congress, Vol. 3, pp 179–183.

Karobath M, Sperk G (1979): Stimulation of benzodiazepine receptor binding by γ-aminobutyric acid. Proc Natl Acad Sci USA 76:1004–1006.

Krause DN, Ikeda K, Roberts E (1981): Dose-conductance relationships for GABA agonists and the effect of uptake inhibitors in crayfish stretch receptor neurons. Brain Res 225:319–332.

Krogsgaard-Larsen P, Johnston GAR, Curtis DR, Game CJA, McCulloch RM (1975): Structure and biological activity of a series of conformationally restricted analogues of GABA. J Neurochem 25:803–809.

Krogsgaard-Larsen P, Johnston GAR, Lodge D, Curtis DR (1977): A new class of GABA agonist. Nature 268:53–55.

Krogsgaard-Larsen P, Johnston GAR (1978): Structure–activity studies on the inhibition of GABA binding to rat brain membranes by muscimol and related compounds. J Neurochem 30:1377–1382.

Krogsgaard-Larsen P, Hjeds H, Curtis DR, Lodge D, Johnston GAR (1979): Dihydromuscimol, thiomuscimol and related heterocyclic compounds as GABA analogues. J Neurochem 32:1717–1724.

Krogsgaard-Larsen P, Falch E (1981): GABA agonists—development and interactions with the GABA-receptor complex. Mol Cell Biochem 38:129–146.

Krogsgaard-Larsen P, Snowman A, Lummis SC, Olsen RW (1981): Characterization of the binding of the GABA agonist [$^3$H]piperidine-4-sulphonic acid to bovine brain synaptic membranes. J Neurochem 37:401–409.

Krogsgaard-Larsen P, Jacobsen P, Falch E (1984): Structure-activity requirements of the GABA receptor. In Enna SJ (ed): "The GABA Receptors." Clifton, New Jersey, Humana Press, pp 149–176.

Lanthorn TH, Cotman CW (1981): Baclofen selectively inhibits excitatory synaptic transmission in the hippocampus. Brain Res 25:171–178.

Lawrence LJ, Casida JE (1983): Stereospecific action of pyrethroid insecticides on the $\gamma$-aminobutyric acid receptor-ionophore complex. Science 221:1399–1401.

Lo MMS, Snyder SH (1983): Two distinct solubilized benzodiazepine receptors: differential modulation by ions. J Neurosci 3:2270–2279.

Maksay G, Ticku MK (1984): Diazotization and thiocyanate differentiate agonists from antagonists for the high- and low-affinity receptors of $\gamma$-aminobutyric acid. J Neurochem 43:261–268.

Mathison RD, Dreifuss JJ (1980): Structure–activity relationships of a neurohypophysial GABA receptor. Brain Res 187:476–480.

Möhler H, Okada T (1978): Properties of $\gamma$-aminobutyric acid receptor binding with (+)-[$^3$H]bicuculline methiodide in rat cerebellum. Mol Pharmacol 14:256–265.

Morin AM, Wasterlain C (1980): The binding of $^3$H-isoguvacine to mouse brain synaptic membranes. Life Sci 26:1239–1245.

Muhyaddin M, Roberts PJ, Woodruff GN (1982a): Presynaptic $\gamma$-aminobutyric acid receptors in the rat anococcygeus muscle and their antagonism by 5-aminovaleric acid. Br J Pharmacol 77:163–168.

Muhyaddin M, Roberts PJ, Woodruff GN (1982b): GABA reduces the evoked neuronal release of $^3$H-noradrenaline from rat anococcygeus muscle: Antagonism by 5-aminovalerate. Br J Pharmacol 77:suppl. 409P.

Offermeier J, van Rooyan JM (1982): Is it possible to integrate dopamine receptor terminology? Trends Pharmacol Sci 3: 326–330.

Olpe HR, Demieville H, Baltzer V, Bencze WL, Koella WP, Wolf P, Haas HL (1978): The biological activity of d- and l-baclofen (lioresal). Eur J Pharmacol 52:133–136.

Olsen RW (1981): GABA-benzodiazepine-barbiturate receptor interactions. J Neurochem 37:1–13.

Olsen RW, Bergman MO, Van Ness PC, Lummis SC, Watkins AE, Napias C, Greenlee DV (1981): $\gamma$-Aminobutyric acid receptor binding in mammalian brain: heterogeneity of binding sites. Mol Pharmacol 19:217–227.

Olsen RW, Snowman AM (1983): [$^3$H]bicuculline methochloride binding to low affinity GABA receptor sites. J Neurochem 41:1653–1663.

Olsen RW, Snowhill EW, Wamsley JK (1984): Autoradiographic localization of low affinity GABA receptors with [$^3$H]bicuculline methochloride. Eur J Pharmacol 99:247–248.

Price GW, Wilkin GP, Turnbull MJ, Bowery NG (1984): Are baclofen-sensitive $GABA_B$ receptors present on primary afferent terminals of the spinal cord? Nature 307:71–74.

Ryall RW (1975): Amino acid receptors in CNS: I. GABA and glycine in spinal cord. Handbk Psychopharmacol 4:83–128.

Simmonds MA (1978): Presynaptic actions of $\gamma$-aminobutyric acid and some antagonists in a slice preparation of cuneate nucleus. Br J Pharmacol 63:495–502.

Simmonds MA (1980): Evidence that bicuculline and picrotoxin act at separate sites to antagonise $\gamma$-aminobutyric acid in rat cuneate nucleus. Neuropharmacology 19:39–45.

Skerritt JH, Willow M, Johnston GAR (1982): Diazepam enhancement of low affinity GABA binding to rat brain membranes. Neurosci Lett 29:63–66.

Skerritt JH, Johnston GAR (1983): Diazepam stimulates the binding of GABA and muscimol but not THIP to rat brain membranes. Neurosci Lett 38:315–320.

Skerritt JH, MacDonald RL (1984): Diazepam enhances the action but not the binding of the GABA analog, THIP. Brain Res 297:181–186.

Skerritt JH, Johnston GAR, Chen Chow S, Macdonald RL, Prager RH, Ward AD (1984): Differential modulation of GABA receptors by caprolactam derivatives with CNS depressant or convulsant activity. Brain Res 331:225–233.

Squires RF, Casida JE, Richardson M, Saederup E (1983): [$^{35}$S]t-Butylbicyclophosphorothionate binds with high affinity to brain specific sites coupled to γ-aminobutyric acid-A and ion recognition sites. Mol Pharmacol 13:326–336.

Stephenson FA, Olsen RW (1982): Solubilization by CHAPS detergent of barbiturate-enhanced benzodiazepine-GABA receptor complex. J Neurochem 39:1579–1586.

Supavilai P, Karobath M (1980): The effect of temperature and chloride ions on the stimulation of [$^3$H]flunitrazepam binding by the muscimol analogues THIP and piperidine-4-sulfonic acid. Neurosci Lett 19:337–341.

Supavilai P, Karobath M (1984): [$^{35}$S]-t-Butylbicyclophosphorothionate binding sites are constituents of the γ-aminobutyric acid benzodiazepine receptor complex. J Neurosci 4:1193–1200.

Supavilai P, Mannonen A, Karobath M (1982): Anion-dependent modulation of $^3$H-muscimol binding and of GABA stimulation of $^3$H-flunitrazepam binding by picrotoxin and related CNS convulsants. Eur J Pharmacol 81:687–691.

Tallman JF, Thomas JW, Gallager DW (1978): GABAergic modulation of benzodiazepine binding site sensitivity. Nature 274:383–385.

Weatherby RP, Allan RD, Johnston GAR (1984): Resolution of the stereoisomers of baclofen by high performance liquid chromatography. J Neurosci Methods 10:23–28.

Weissman BA, Burke TR, Rice KC, Skolnick P (1984): Alkyl-substituted γ-butyrolactones inhibit [$^{35}$S]TBPS binding to a GABA linked chloride ionophore. Eur J Pharmacol 105:195–196.

Wilkin GP, Hudson AL, Hill DR, Bowery NG (1981): Autoradiographic localisation of $GABA_B$ receptors in rat cerebellum. Nature 274:584–587.

Yanagisawa M, Akagi H, Otsuka M (1981): A further study on the mechanism of action of baclofen in the spinal cord. Proc 8th Int Congr Pharmacol, Tokyo:604.

Benzodiazepine/GABA Receptors and Chloride Channels:
Structural and Functional Properties, pages 73–95


# 4

# The Active Site of the GABA Receptor

**P. Krogsgaard-Larsen, L. Nielsen, and E. Falch**

*Department of Chemistry BC, The Royal Danish School of Pharmacy, Universitetsparken 2, DK-2100 Copenhagen Ø, Denmark*

The amino acid γ-aminobutyric acid (GABA) is an inhibitory neurotransmitter concerned with the control of neuronal activity in the mammalian central nervous system (CNS) [Curtis and Johnston, 1974; Krnjević, 1974; Roberts et al., 1976]. Impaired transmission at such GABA-operated synapses appears to be implicated in certain neurologic and psychiatric disorders [Johnston, 1978; Enna, 1981; Krogsgaard-Larsen, 1981; Meldrum, 1982], and GABA is known to be involved in the central regulation of a variety of physiological functions, including: cardiovascular mechanisms [DeFeudis, 1981; Williford et al., 1980]; the regulation of endocrine functions [Grandison and Guidotti, 1979; Enna, 1981]; food intake [Blavet et al., 1982]; and the sensation of pain [Kendall et al., 1982; DeFeudis, 1982]; and anxiety [Cananzi et al., 1980; Olsen and Enna, 1982; Hoehn-Saric, 1983]. These findings have brought agents with direct or indirect GABA-stimulating effects into focus as potential therapeutic agents.

## SUBSTRATE AND AGONIST SPECIFICITY OF THE GABA SYNAPTIC MECHANISMS

The GABA neurotransmission process can be stimulated directly by receptor agonists and indirectly using inhibitors of the GABA uptake mechanisms or the GABA-metabolizing enzyme GABA:2-oxoglutarate aminotransferase (GABA-transaminase). Based on studies of the structure–activity relations (SAR) for conformationally restrained GABA analogues, it has been established that the GABA receptors, the GABA transport systems, and GABA-transaminase have similar but distinctly different structural specificities (Krogsgaard-Larsen, 1978; Johnston et al., 1979; Krogsgaard-Larsen, 1981; Palfreyman et al., 1981], indicating that the pharmacology of these synaptic mechanisms can be studied selectively. These findings probably reflect that GABA adopts different conformations during the interaction with the biomolecules concerned.

The structures of a number of cyclic amino acid analogues of GABA with specific or highly selective effects on different GABA synaptic mechanisms are shown in Figure 1. While gabaculine is a very potent $k_{cat}$ inhibitor of GABA-transaminase [Palfreyman et al., 1981], its saturated analogue (1*SR*,3*RS*)-3-aminocyclohexanecarboxylic acid (ACHC) is a selective inhibitor of neuronal GABA uptake [Neal and Bowery, 1977]. Isonipecotic acid is a potent GABA agonist with no affinity for the GABA uptake systems [Krogsgaard-Larsen and Falch, 1981]. As emphasized in Figure 1, all of these compounds contain a GABA structure element. On the other hand, nipecotic acid and (3*RS*,4*SR*)-4-hydroxypiperidine-3-carboxylic acid (*cis*-4-OH-nipecotic acid) have no detectable affinity for the GABA receptors, but these compounds are

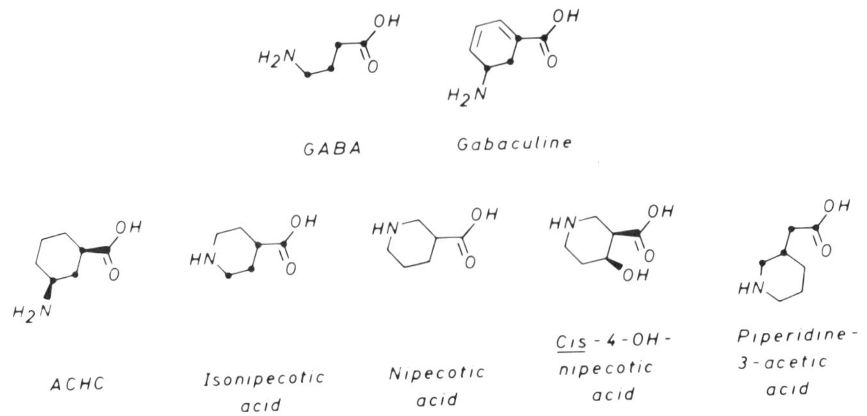


**Fig. 1** The structures of GABA, the GABA-transaminase inhibitor gabaculine, the GABA agonist isonipecotic acid, the GABA uptake inhibitors aminocyclohexane-carboxylic acid, nipecotic acid and *cis*-4-OH-nipecotic acid, and the inactive GABA analogue piperidine-3-acetic acid.

potent GABA uptake inhibitors, *cis*-4-OH-nipecotic acid being a selective inhibitor of and substrate for the glial uptake system [Schousboe et al., 1979; Krogsgaard-Larsen, 1980; Schousboe et al., 1983], but none of these amino acids are GABA analogues in the proper sense of the word. It is interesting to note that piperidine-3-acetic acid, which is a structural analogue of GABA (Fig. 1), does not interact significantly with any of the GABA synaptic recognition sites [Johnston et al., 1979].

## CLASSIFICATION OF THE GABA RECEPTORS BASED ON STRUCTURE–ACTIVITY STUDIES

GABA neurons are ubiquitous in the mammalian CNS, and virtually all neurons are sensitive to GABA, rendering treatment of diseases with regioselective degeneration of GABA neurons by GABA agonists difficult. However, accumulating evidence derived from studies on GABA analogues suggests the existence of different types of GABA receptors, but in the absence of specific agonists and antagonists for most of these proposed subpopulations of receptors the exact number of physiologically relevant GABA receptors is unknown.

Based on electrophysiological and receptor-binding studies, the GABA receptors can be divided into two apparently heterogeneous groups:

1) GABA receptors sensitive to the competitive antagonists (+)-bicuculline and bicuculline methochloride (BMC); and
2) bicuculline- and bicuculline methochloride-insensitive GABA receptors.

### Bicuculline-Sensitive GABA Receptors

This class of GABA receptors ($GABA_A$ receptors) includes *presynaptic* (axo-axonic) and *postsynaptic* (axo-somatic and axo-dendritic) receptors, which have very similar pharmacological characteristics [Curtis, 1978; Curtis et al., 1982]. *Extrasynaptic* GABA receptors, which are also sensitive to bicuculline, apparently have agonist specificities different from those of synaptic GABA receptors [Curtis, 1978; Brown, 1979; Alger and Nicoll, 1982].

There is some evidence supporting the view that the $GABA_A$ receptors can be further subdivided. The agonist selectivity of GABA receptors in the cat cerebral cortex and spinal cord are dissimilar (Curtis et al., 1971a, 1971b), and while 5,6,7,8-tetrahydro-4*H*-isoxazolo[3,4-*d*]azepin-3-ol (iso-THAZ) (Fig.

2) has a GABA antagonist profile after injection into the rat substantia nigra [Arnt and Krogsgaard-Larsen, 1979], this compound is an antagonist at glycine receptors in the feline spinal cord [Krogsgaard-Larsen et al., 1982b]. These differences between spinal and supraspinal GABA receptors have recently been further elucidated. While muscimol, isoguvacine, and GABA (Fig. 2) are agonists with a decreasing order of potency at $GABA_A$ receptors in the cat spinal cord [Krogsgaard-Larsen et al., 1977; 1979b], these compounds seem to activate bicuculline-sensitive GABA receptors which regulate the release of glutamic acid in rat brains, with a different order of potency [Mitchell, 1982]. At these receptors muscimol is only a moderately potent agonist, weaker than GABA, whereas isoguvacine and 3-aminopropanesulfonic acid (3APS) (Fig. 2) are the most potent agonists so far tested.

### Bicuculline-Insensitive GABA Receptors

The GABA analogue baclofen (Fig. 3) is a bicuculline-insensitive depressant of neuronal firing [Curtis et al., 1974; Davies and Watkins, 1974], and baclofen interacts with a population of binding sites, which bind GABA but not bicuculline or various bicuculline-sensitive GABA agonists including isoguvacine [Hill and Bowery, 1981]. While these receptors ($GABA_B$ receptors) seem to modulate the release of monoamines in the CNS [Bowery et al., 1980], their relevance to the therapeutic effect of baclofen is unclear. Baclofen potently reduces monosynaptic excitation in the cat spinal cord, an effect which probably is not mediated by GABA receptors [Curtis et al., 1981].

A number of GABA analogues, such as kojic amine (Fig. 3) and 3-hydroxy-4-aminobutyric acid (3-OH-GABA) interact with the $GABA_B$ receptor sites, but these compounds are far from being specific $GABA_B$ agonists. Kojic amine also is a moderately potent $GABA_A$ agonist [Yarbrough et al., 1979; Atkinson et al., 1979], and whereas the (*S*)-(+)-isomer of 3-OH-GABA interacts slightly more effectively with the $GABA_A$ receptors than the (*R*)-(−)-isomer (Fig. 3) [Krogsgaard-Larsen et al., 1983], these optical isomers are equally effective at the $GABA_B$ receptor sites [N.G. Bowery, unpublished]. The heterocyclic GABA analogue 5-(3-aminopropyl) tetrazole (Fig. 3) has recently been shown to interact weakly with $GABA_A$ as well as $GABA_B$ receptor sites in vitro [Schlewer et al., 1984]. 5-Aminovaleric acid (DAVA) has a very weak antagonist profile at peripheral $GABA_B$ receptors [Muhyaddin et al., 1982], but 5-aminovaleric acid does not affect significantly central $GABA_B$ receptors, and it interacts more or less effectively with all recognition sites at synapses containing $GABA_A$ receptors [Krogsgaard-Larsen, 1980]. A detailed analysis of the physiological and pharmacological importance of the $GABA_B$ receptors must await the development of specific $GABA_B$ agonists and, in particular, potent and specific antagonists at these receptors.

Other groups of bicuculline methochloride–insensitive GABA receptors may exist in the mammalian CNS, although the evidence is still indirect and very incomplete [Johnston and Allan, 1984]. *cis*-4-Aminocrotonic acid, *cis*-2-aminomethylcyclopropanecarboxylic acid (CAMP), 5′-ethylmuscimol, and 5′-pro-

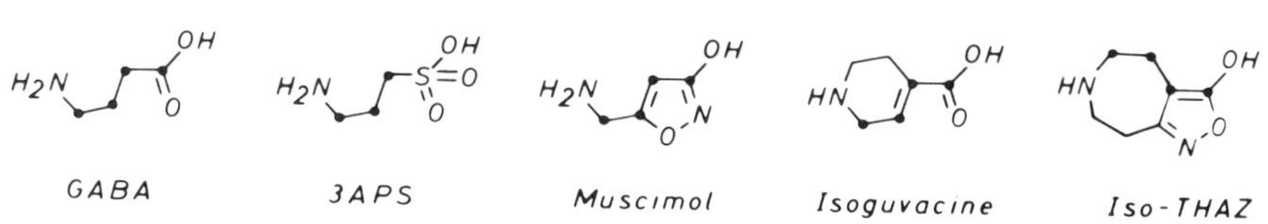


**Fig. 2** The structures of GABA, the GABA agonists 3-aminopropanesulfonic acid (3APS), muscimol, and isoguvacine, and the GABA/glycine antagonist iso-THAZ.

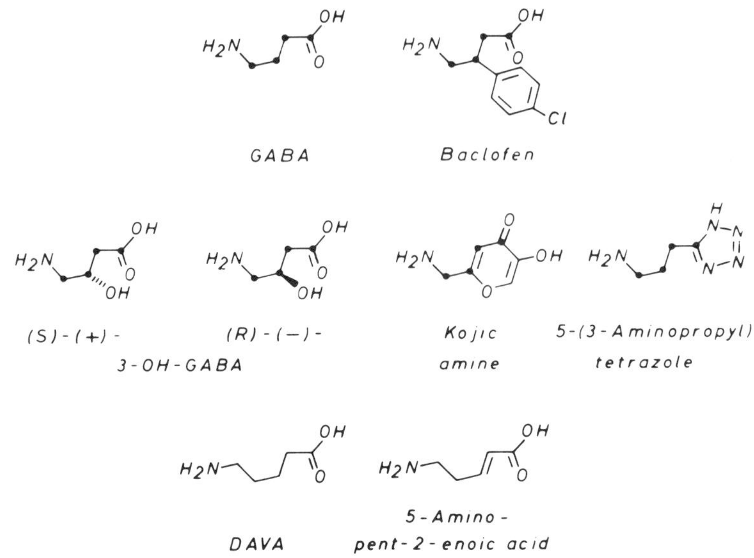


**Fig. 3** The structures of GABA and baclofen and a number of GABA analogues with effects on the $GABA_A$ and $GABA_B$ receptors, and the structures of the $GABA_B$ antagonist 5-aminovaleric acid (DAVA) and the inactive analogue 5-aminopent-2-enoic acid.

pylmuscimol (Fig. 4) are moderately potent neuronal depressants insensitive to bicuculline methochloride as well as to strychnine [Johnston et al., 1975; Allan et al., 1980; Krogsgaard-Larsen et al., 1975]. The detection of GABA binding sites, different from $GABA_A$ and $GABA_B$ receptor sites, to which *cis*-4-aminocrotonic acid appears to bind with some selectivity, has recently been reported [Johnston and Allan, 1984]. Such binding sites, which have tentatively been named $GABA_C$ sites [G.A.R. Johnston, this volume], may be the sites of action of some of the bicuculline methochloride–insensitive GABA-like neuronal depressants shown in Figure 4.

## BICUCULLINE-SENSITIVE GABA ($GABA_A$) AGONISTS

Electrophysiological methods, which offer the most direct approach to studies of receptor mechanisms, as well as receptor-binding techniques, have inherent limitations. Electrophysiological methods allow studies of alterations of neuronal membranes as a result of activation or blockade of the GABA receptors, but such experiments normally are not sufficiently precise to obtain quantitative data. The receptor sites studied in binding experiments [Enna and Snyder, 1975], on the other hand, probably represent more or less disin-

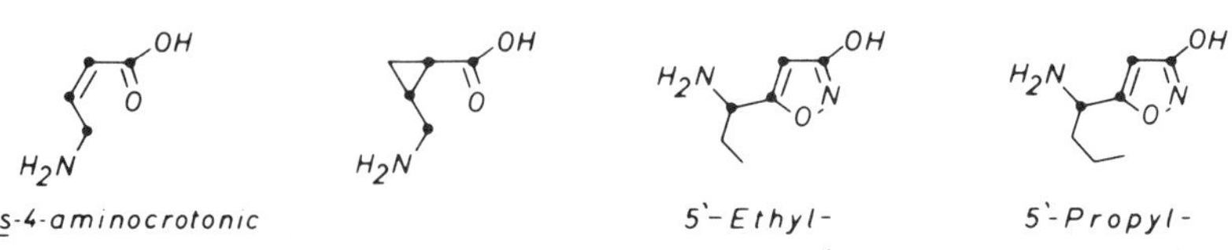


**Fig. 4** The structures of some GABA analogues with bicuculline-insensitive neuronal depressant activities.

tegrated forms of the physiological receptor(s), and interpretation of binding data in terms of receptor mechanisms must be performed very guardedly. However, using these techniques in combination, the interaction of a number of GABA analogues with the GABA receptors has been studied, and based on such studies the pharmacological characteristics of the $GABA_A$ receptors in particular have been quite extensively elucidated [Krogsgaard-Larsen and Falch, 1981; Curtis et al., 1982].

## $GABA_A$ Agonist–Receptor Interactions

**Structural and conformational factors.** A wide variety of GABA analogues has been developed by systematic alteration of the structural parameters characterizing the molecule of GABA. A necessary, but not sufficient, condition for GABA agonist activity is the presence of a negatively and a positively charged group. Figure 5 illustrates a number of zwitterionic GABA analogues in the unionized form.

Some analogues, in which the carboxyl group of GABA has been replaced by conformationally frozen and flattened functionalities, like muscimol, dihydromuscimol, and thiomuscimol, possess higher apparent affinity for and intrinsic activity at the GABA receptors than GABA itself [Krogsgaard-Larsen et al., 1979b]. Quite surprisingly, these effects are shared by 3-aminopropanesulfonic acid [Curtis and Watkins, 1965; Enna and Snyder, 1975; Krogsgaard-Larsen et al., 1981a], which contains the bulky tetrahedrally arranged sulfonate group. Replacement of the carboxylate group of GABA by other acid groups does, however, result in reduced GABA agonist activity. The 3-hydroxy-4*H*-pyran-4-one nucleus of kojic amine is to some extent accepted by the GABA receptors [Yarbrough et al., 1979], but the 5-isoxazolol analogue of muscimol, isomuscimol, is much weaker than muscimol as a GABA agonist [Krogsgaard-Larsen et al., 1979b] (Fig. 5). The boronic and phosphonic acid analogues of GABA are virtually devoid of GABA agonist activity [Curtis and Watkins, 1965], possibly because of the presence of two acid groups in the acid moieties of these compounds. Similarly, the heterocyclic GABA analogues 5-aminomethyl-1,2,4-triazol-3-ol (B) [Armstrong et al., 1982], 5-(2-aminoethyl)hydantoin (C) [Johnston et al., 1979], and 5-aminomethyl-3-hydroxyfuran-2(5*H*)-one (D) [Allan et al., 1983] have no detectable affinity for GABA receptor sites.

Within certain limits the primary amino group of GABA can be replaced by other basic functionalities without substantial loss of GABA agonist activity, imidazole-4-acetic acid and *cis*-3-[(aminoiminomethyl)thio]propenoic acid (A) (Fig. 5) being quite potent GABA agonists [Nistri and Constanti, 1979; Johnston et al., 1979; Johnston and Allan, 1984]. While the *N*-methyl derivatives of muscimol and thiomuscimol are GABA agonists, although much weaker than the parent compounds [Krogsgaard-Larsen et al., 1982b; D.R. Curtis and P. Krogsgaard-Larsen, unpublished], *N*-methyl-GABA is virtually inactive, and *N,N*-dimethyl-GABA is totally inactive [Curtis and Watkins, 1960; Krogsgaard-Larsen and Johnston, 1978; Krogsgaard-Larsen et al., 1982b] (Fig. 5).

As illustrated in Fig. 6, the cyclic GABA analogues isonipecotic acid and isoguvacine interact with the $GABA_A$ receptors without affecting GABA uptake (Krogsgaard-Larsen et al., 1977; 1979b), whereas the GABA analogue homo-β-proline interacts quite effectively with both of these GABA synaptic mechanisms [Larsson et al., 1981; Falch et al., 1985]. The structure-activity studies presented in Figure 6 clearly illustrate that the GABA uptake mechanisms do not accept aminosulfonic acid analogues of GABA. Thus, not only the sulfonic acid analogues of isonipecotic acid and isoguvacine, piperidine-4-sulfonic acid (P4S) and 1,2,3,6-tetrahydropyridine-4-sulfonic acid (DH-P4S), respectively, do not affect GABA uptake in vitro, but also the aminosulfonic acid analogues of homo-β-proline and nipecotic acid have little or no effect on GABA uptake [Krogsgaard-Larsen et al., 1980a; Falch et al., 1985].

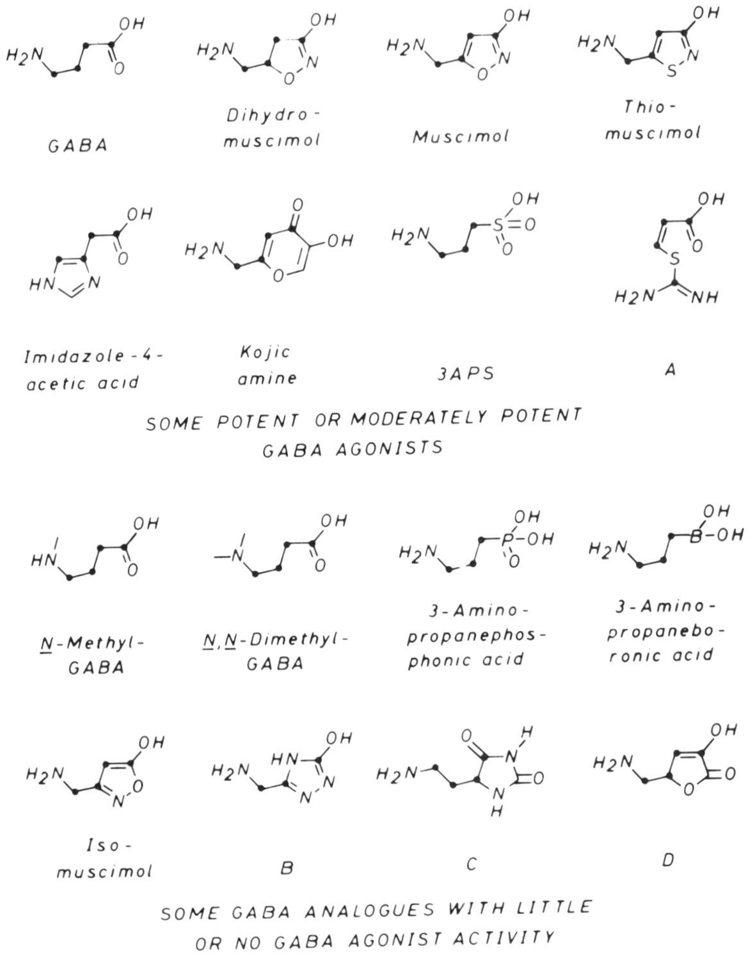


**Fig. 5** The structures of a number of GABA agonists and some GABA analogues with little or no affinity for the GABA receptors.

**Electronic factors.** All of the GABA analogues discussed in this chapter have zwitterionic structures as exemplified in Figure 7. It is normally difficult to map out precisely the electronic structures of these zwitterions, including the degree of delocalization of their charges in vitro, and it is impossible to determine the exact electronic structures of GABA agonists during their interaction with the GABA receptors. Nevertheless, it is necessary to consider the approximate charged structures of GABA analogues in SAR studies on GABA agonists [Krogsgaard-Larsen et al., 1979a; Armstrong et al., 1982].

The pronounced potency of dihydromuscimol as a $GABA_A$ agonist is consistent with its structural similarity with GABA, and the much lower activity of isomuscimol is readily understandable on the basis of its very different charged structure [Krogsgaard-Larsen et al., 1979b] (Fig. 7). The electronic structure of the sulfonate group of 3-aminopropanesulfonic acid is different from those of the negatively charged moieties of GABA and dihydromuscimol, which may explain why the mechanism of action of 3-aminopropanesulfonic acid and other aminosulfonic acid GABA agonists apparently is somewhat different from

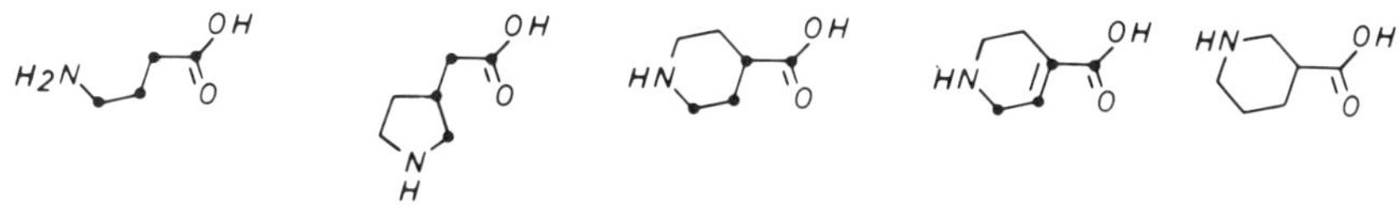


| | GABA | Homo-β-proline | Isonipecotic acid | Isoguvacine | Nipecotic acid |
|---|---|---|---|---|---|
| A | --- | --- | --- | ---- | 0 |
| B | 0.03 | 0.3 | 0.3 | 0.04 | >100 |
| U | 2.0 | 3.0 | >300 | >300 | 6.0 |

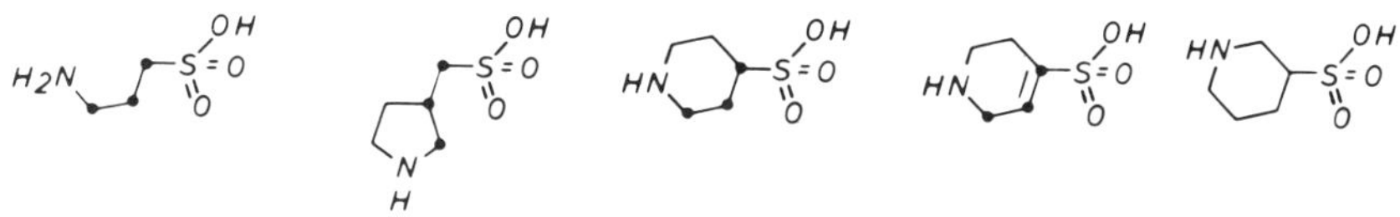


| | 3APS | PMSA | P4S | DH-P4S | P3S |
|---|---|---|---|---|---|
| A | ---- | --- | ---- | --- | 0 |
| B | 0.08 | 0.3 | 0.03 | 0.3 | >100 |
| U | 300 | 90 | >300 | >300 | >300 |

**Fig. 6** Structures and effects on the $GABA_A$ receptors and GABA uptake of some amino acids and aminosulfonic acids structurally related to GABA. *A*: bicuculline methochloride-sensitive GABA agonist activity relative to that of GABA (---) as determined microelectrophoretically [Curtis et al., 1971a]; *B*: Inhibition of $GABA_A$ receptor binding in vitro ($IC_{50}$, $\mu$M) using rat brain membranes [Falch and Krogsgaard-Larsen, 1982]; *U*: Inhibition of neuronal GABA uptake in vitro ($IC_{50}$, $\mu$M) using a crude synaptosomal fraction isolated from rat brain [Falch et al., 1985].

that of GABA and muscimol analogues [Krogsgaard-Larsen et al., 1983].

As illustrated in Figure 7, the degree of charge delocalization of the imidazole and guanidino groups of imidazole-4-acetic acid, β-guanidinopropionic acid, and A is different from that of the amino group of GABA. An increasing degree of delocalization of the positive charges seems, to some extent, to be tantamount to reduced GABA agonist activity. The protolytic properties of the basic groups concerned have not been included into these SAR considerations, but the GABA receptors are known to accept GABA agonists possessing amino groups with quite different base strengths.

**Stereochemical factors.** Stereo-SAR studies have disclosed different effects of the optical isomers of chiral GABA analogues on the binding of GABA and various radioactive GABA agonists, although the degree of stereoselectivity depends on the structure of the asymmetric GABA analogues [Krogsgaard-Larsen, 1981; Falch and Krogsgaard-Larsen, 1982]. While these and other studies [Johnston et al., 1978] indicate stereoselectivity of the GABA receptor recognition sites, stereoselectivity of the GABA receptors with respect to GABA agonists is problematic [Krogsgaard-Larsen et al., 1983] (Fig. 8). (*S*)−(+)-3-OH-GABA is more potent than the *R*-form with respect to inhibition of GABA binding, but these optical antipodes are equipotent as depressants of neuronal firing. This apparent equipotency of (*S*)−(+)- and (*R*)-(−)-3-OH-GABA may, however, be explained by the findings that the former antipode is the more effective inhibitor of, and thus probably the better substrate for, the GABA uptake mechanisms (Fig. 8). Consequently, the *S*-isomer

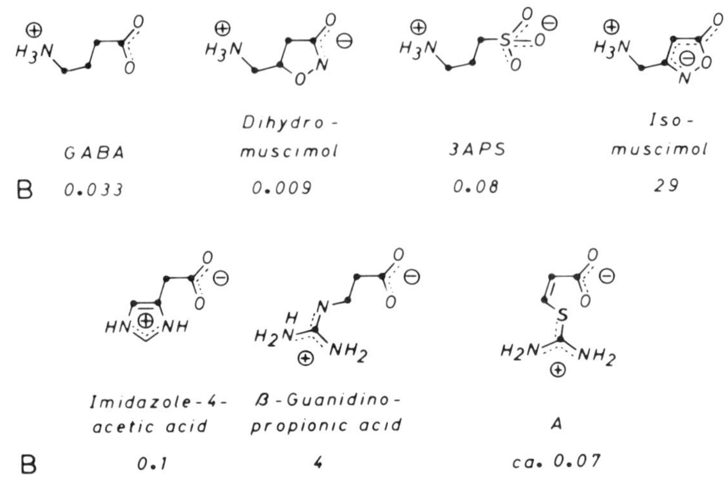


**Fig. 7** The approximate charged structures of some GABA agonists and their affinities for the $GABA_A$ receptor sites in vitro. *B*: See legend for Figure 6.

may be more effectively removed from the receptor environments than the *R*-form after microelectrophoretic application near neurons.

Such factors can hardly explain the apparent equipotency of (*S*)-(−) and(*R*)-(+)- 5′-methylmuscimol as $GABA_A$ agonists [Krogsgaard-Larsen et al., 1983]. The possibility that the *S*-isomer, which is the more potent inhibitor of GABA binding, is taken up more effectively than the *R*-isomer in vivo is not very likely, since neither isomer affects significantly GABA uptake in vitro (Fig. 8). Furthermore, (*S*)-(−)-*trans*-4-methyl-4-aminocrotonic acid (4-Me-*trans*-ACA) is a specific $GABA_A$ agonist, whereas the (*R*)-(+)-isomer of this conformationally restrained GABA analogue is a GABA uptake inhibitor without significant affinity for the GABA receptor sites [Schousboe et al., 1979]. Thus, based on arguments from analogy it is unlikely that (*S*)-(−)-5′-methylmuscimol is taken up more effectively than the (*R*)-isomer in vivo. The relationship between structure and activity of the four optical isomers of 3-aminocyclopentanecarboxylic acid emphasizes the different stereochemical requirements of the GABA receptors and uptake mechanisms (Fig. 8) [Johnston et al., 1978].

It may be concluded from the stereo-SAR studies on the optical isomers of 4-methyl-*trans*-aminocrotonic acid and 5′-methylmuscimol that the GABA receptor recognition sites, as measured in binding experiments, and the ion channel-coupled GABA receptor, exhibit different degrees of stereoselectivity. These findings emphasize the limitations of binding studies in investigations of GABA receptor mechanisms.

## GABA RECEPTOR ANTAGONISTS: STRUCTURE–ACTIVITY STUDIES

As illustrated in Figure 10, the classical antagonists (1*S*,9*R*)-(+)-bicuculline and picrotoxinin [Curtis and Johnston, 1974] have different sites of action in the GABA receptor complex. Both bicuculline and bicuculline methochloride appear to interact directly with the GABA receptor [Frere et al., 1982; Nowak et al., 1982], possibly with an "antagonist conformational state" of the receptor [Möhler and Okada, 1977]. It has recently been demonstrated that bicuculline methochloride selectively binds to low-affinity sites of the GABA receptors [Olsen and Snowman,

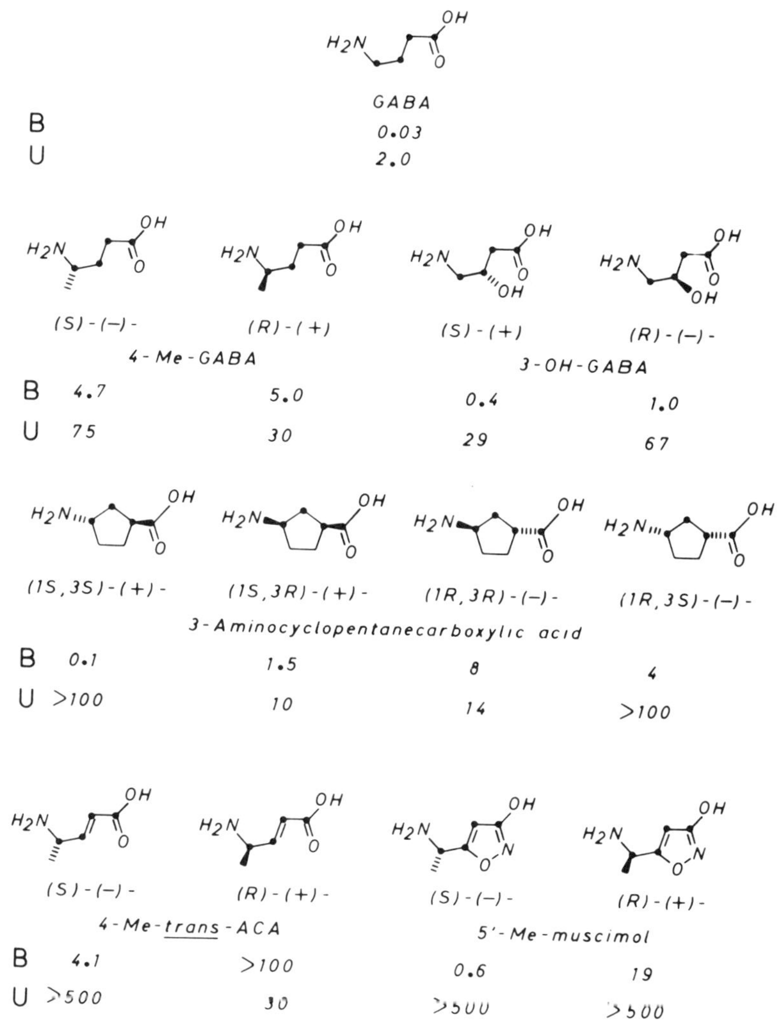


**Fig. 8** The relationship between structures and effects in vitro on the $GABA_A$ receptor sites and the neuronal GABA uptake system of a number of GABA analogues with known absolute configuration. *B* and *U*: See legend for Fig. 6.

1983; Olsen et al., 1984]. In agreement with the apparent stereoselectivity of the activation of bicuculline methochloride-sensitive GABA receptors in the mammalian CNS (Fig. 8), these receptors are blocked in a stereoselective manner [Enna et al., 1977; Möhler and Okada, 1977]. Thus (1*R*,9*S*)-(-)-bicuculline has very weak GABA antagonist properties, and it does not affect significantly the binding of radioactive bicuculline methochloride. Although narcotine has the same absolute stereochemistry as bicuculline (Fig. 9), it is ten times weaker than bicuculline as a GABA antagonist [Enna et al., 1977]. The methoxy group on the tetrahydroisoquinoline ring of narcotine may force the molecule to adopt a conformation different from that, in which bicuculline binds to the GABA receptor. The (1*R*,9*S*)- and (1*S*,9*S*)-isomers of narcotine have negligible GABA antagonist properties, and neither (1*R*,9*R*)-, (1*R*,9*S*)-, nor (1*S*,9*S*)-hydrastine, none of which has the same ab-

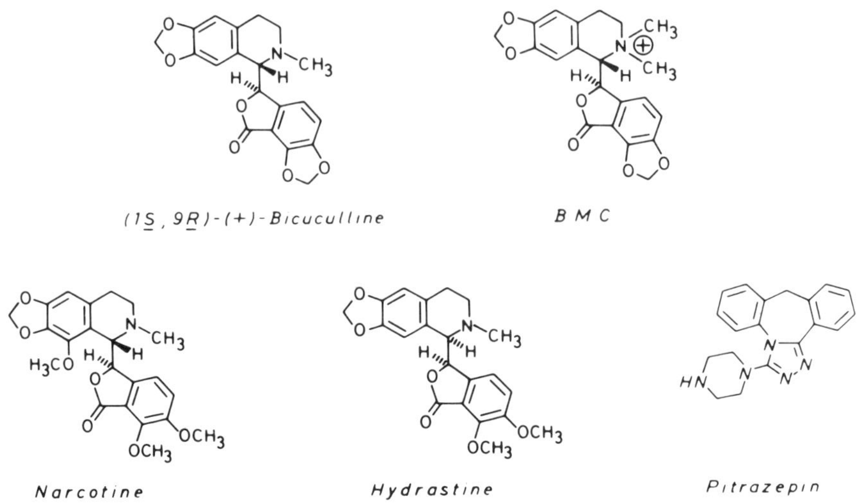


**Fig. 9** The structures of the GABA receptor antagonist bicuculline and some related phthaleideisoquinoline alkaloids and the novel GABA receptor antagonist pitrazepin [Gähwiler et al., 1984].

solute configuration as bicuculline (Fig. 9), have significant GABA antagonist properties [Enna et al., 1977; Bowery et al., 1979].

## THE POSTSYNAPTIC GABA$_A$ RECEPTOR COMPLEX

As a result of electrophysiological and receptor binding studies along different lines, a picture of the postsynaptic GABA$_A$ receptor complex begins to emerge [Olsen, 1981; Skolnick and Paul, 1981] (Fig. 10). The chloride channel, which is the central unit of this complex, is controlled by the GABA$_A$ receptor consisting of two [Enna and Snyder, 1977; Olsen et al., 1981] or possibly three GABA binding sites [Falch and Krogsgaard-Larsen, 1982; Olsen and Snowman, 1982], although the physiological relevance of these multiple binding sites, which can be detected in vitro under different experimental conditions, is unclear. The receptor appears to be regulated by various additional units, which can be detected in vitro as distinct binding sites for the benzodiazepines [Squires and Braestrup, 1977], picrotoxinin [Ticku et al., 1978], and the avermectins, including avermectin B$_{1a}$ [Supavilai and Karobath, 1981]. Recent studies have disclosed intimate contacts between these additional sites of the GABA receptor complex [Pong and Wang, 1982].

The physiological role of these binding sites, which have ligand specificities different from that of the GABA receptor sites, is unknown, but the allosteric interactions between these sites in vitro may reflect certain aspects of the physiological GABA receptor mechanism [Olsen, 1981]. At the present time these sites are most appropriately described as the *pharmacological receptors* for a variety of therapeutic agents and pharmacological tools as exemplified in Fig. 10 [Nielsen et al., 1981; Squires et al., 1983]. The situation is further complicated by the discovery of different subtypes of benzodiazepine binding sites and by the findings that not all GABA receptors are coupled to benzodiazepine sites and vice versa [Squires et al., 1980]. Future studies undoubtedly will reveal an even higher degree of complexity of this receptor complex.

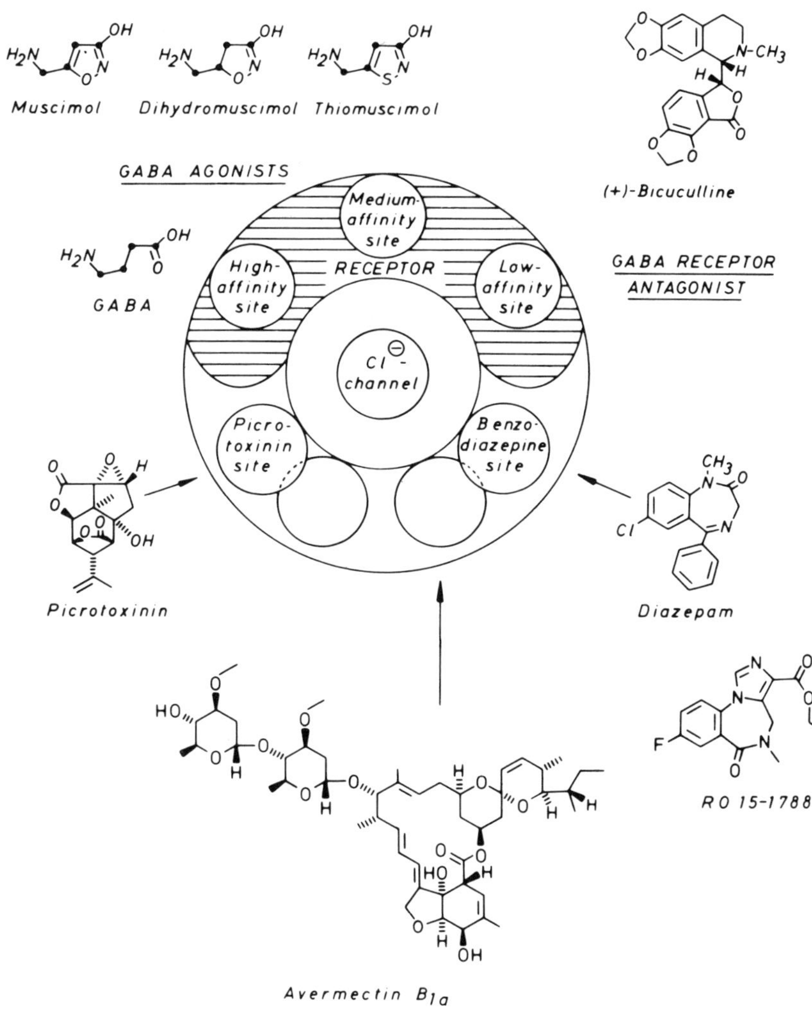


**Fig. 10** The structure of the $GABA_A$ receptor complex and the sites of action of some GABA agonists, GABA antagonists, and various drugs.

## MOLECULAR PHARMACOLOGY OF $GABA_A$ AGONISTS

There generally is a positive correlation between the potency of bicuculline-sensitive GABA agonists, having little or no affinity for the GABA uptake systems, and their ability to displace radioactive GABA from GABA receptor sites in vitro [Krogsgaard-Larsen and Falch, 1981; Krogsgaard-Larsen et al., 1981a; Falch and Krogsgaard-Larsen, 1982]. However, these molecular pharmacological studies have not shed much light on the mechanisms underlying the sequence of events between the binding of the GABA agonist to the receptor recognition site and the resulting alterations of the postsynaptic membrane. Important steps in this process probably are conformational changes of the receptor macromolecule(s), activation of a coupling mechanism between the receptor and ion channel, and transformation of this channel from an inactive into an active stage with subsequent influx of chloride ions [Krogsgaard-Larsen and Arnt, 1980; Skolnick and Paul, 1981].

### The Multiplicity of $GABA_A$ Binding Sites

As part of our interest in the mechanism of action of GABA agonists, we have labeled a number of specific GABA agonists with different structural characteristics including isoguvacine, piperidine-4-sulfonic acid and 4,5,6,7-tetrahydroisoxazolo[5,4-*c*]pyridin-3-ol (THIP), and compared the binding of these agonists and GABA [Falch and Krogsgaard-Larsen, 1982]. Analyses of the binding data for all three ligands revealed that a receptor model with three independent binding sites was statistically preferable to receptor models containing one or two sites (Fig. 10). The apparent equilibrium dissociation constants ($K_D$) and the maximum density of the binding sites ($B_M$) for GABA and THIP are listed in Figure 11. As indicated these data could not be determined precisely for the low-affinity site. The most conspicuous difference between these agonist ligands is the low density of the high-affinity binding sites for $[^3H]$THIP. Although these observations have been made under nonphysiological conditions, it is interesting to compare the binding data for $[^3H]$GABA and $[^3H]$THIP with the results of some recent electrophysiological studies of Alger and Nicoll [1982]. While GABA was shown to activate synaptic as well as extrasynaptic hippocampal GABA receptors, of which the latter class of receptors was highly sensitive to bicuculline methochloride, THIP proved to activate selectively synaptic GABA receptors which were less sensitive to bicuculline methochloride. Since the membranes used in binding studies probably contain both synaptic and extrasynaptic GABA receptor sites, it is therefore tempting to assume that the medium-affinity binding sites (Fig. 10) represent synaptic GABA receptors, and the high-affinity sites extrasynaptic receptors. On the other hand, the results obtained from electrophysiological and binding studies on cultured cortical and spinal neurons led to the proposal that high-affinity GABA binding sites represent postsynaptic GABA receptors, whereas low-affinity sites reflect GABA receptors of presynaptic nature [Frere et al., 1982; Nowak et al., 1982]. In light of the above-mentioned in vitro and in vivo effects of THIP, it is interesting to note that this specific GABA agonist is a highly active pharmacological agent in animals and humans [Krogsgaard-Larsen, 1981; Christensen et al., 1982; Krogsgaard-Larsen, 1984; Krogsgaard-Larsen et al., 1984].

### Effects of $GABA_A$ Agonists on Benzodiazepine Binding

The molecular mechanisms underlying the interaction between the GABA receptor, the chloride channel, and the benzodiazepine site are unknown [Olsen, 1981]. However, the availability of specific $GABA_A$ agonists with different structural and physicochemical properties have stimulated studies of these coupling mechanisms, which manifest themselves as allosteric interactions in vitro [Tallman et al., 1978; Karobath et al., 1979; Braestrup et al., 1979; Supavilai and Karobath, 1980]. Like GABA, GABA agonists having aminomethyl groups with a high degree of rotational freedom, such as dihydromuscimol and muscimol (Fig. 10), stimulate the binding of benzodiazepines under certain in vitro conditions. The ability of GABA agonists with more rigid structures such as homo-$\beta$-proline, isonipecotic acid, and isoguvacine (Fig. 6) to increase benzodiazepine binding is slightly reduced, an effect which appears also to be a function of the presence of secondary amino groups in these GABA agonists [Krogsgaard-Larsen and Arnt, 1980; Falch et al., 1985]. The almost rigid GABA agonist THIP has virtually no effect on benzodiazepine binding at low temperatures and in the absence of chloride [Maurer, 1979; Karobath et al., 1979]. Like piperidine-4-sulfonic acid [Braestrup et al., 1979], 1,2,3,6-tetrahydropyridine-4-sulfonic acid, and (RS)-

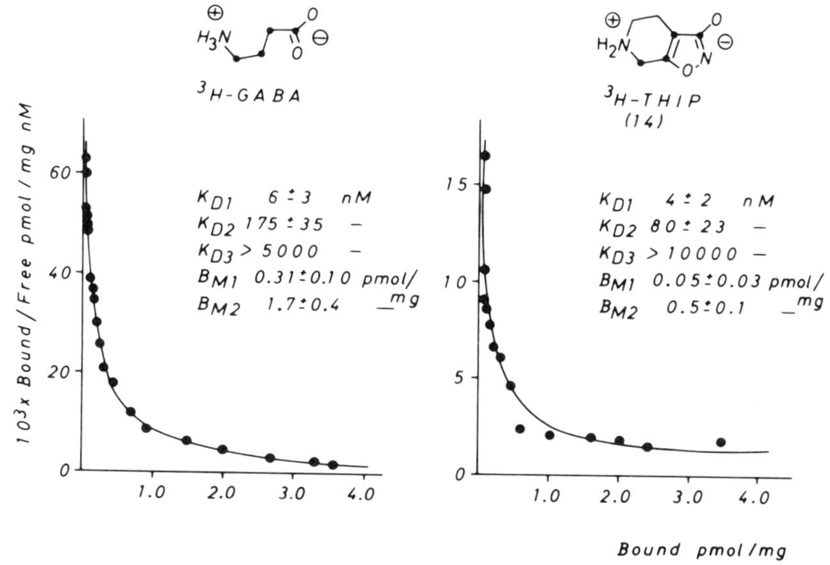


**Fig. 11** Scatchard plots of the specific binding of radioactive GABA and THIP to rat brain synaptic membranes. The binding parameters are derived from computer-fitted nonlinear regression analysis of the data [Falch and Krogsgaard-Larsen, 1982].

3-pyrrolidinemethanesulfonic acid (PMSA) all of which contain sulfonate groups (Fig. 6), are deactivators of benzodiazepine binding when tested at 0°C and in the absence of chloride ions [Falch et al., 1985]. At higher temperatures and in the presence of chloride these aminosulfonic acids as well as THIP actually do stimulate benzodiazepine binding, although the maximal levels of stimulation by these GABA agonists are different and lower than those observed for GABA and muscimol [Supavilai and Karobath, 1980; Falch et al., 1985].

The concentrations of GABA agonists required for 50% stimulation of benzodiazepine binding are typically about two orders of magnitude higher than the $IC_{50}$ values for the compounds as inhibitors of the receptor binding of GABA and various GABA agonist ligands [Karobath et al., 1979; Braestrup et al., 1979; Krogsgaard-Larsen and Falch, 1981; Falch et al., 1985]. These observations, supported by studies using membranes in which the high-affinity GABA binding sites are selectively destroyed without affecting the ability of the GABA receptor sites to stimulate benzodiazepine binding [Browner et al., 1981], strongly suggest that the GABA receptor-benzodiazepine coupling is mediated by low-affinity GABA receptor sites.

As mentioned earlier the aminosulfonic acids piperidine-4-sulfonic acid, 1,2,3,6-tetrahydropyridine-4-sulfonic acid and pyrrolidinemethanesulfonic acid are potent agonists at $GABA_A$ receptors in the cat spinal cord [Krogsgaard-Larsen et al., 1980a; Falch et al., 1985], and the binding characteristics of piperidine-4-sulfonic acid and GABA are similar [Krogsgaard-Larsen et al., 1981a; Falch and Krogsgaard-Larsen, 1982]. Nevertheless, there is evidence to suggest that the mechanism of interaction of piperidine-4-sulfonic acid with the GABA receptors is somehow different from those of GABA, muscimol, isoguvacine, and THIP [Krogsgaard-Larsen et al., 1983]. While a number of optically active GABA analogues have similar *relative* potencies as inhibitors of the binding of radioactive GABA, muscimol, isoguvacine, THIP, and piperidine-4-sulfonic acid, their *absolute* potencies as inhibitors of piperidine-4-sulfonic acid binding are typically an order of magnitude higher than those determined using the other GABA agonist ligands [Krogsgaard-Larsen, 1981; Falch and Krogsgaard-Larsen, 1982]. Comparative studies on isonipecotic acid and piperidine-4-sulfonic acid and their respective unsaturated analogues isoguvacine and 1,2,3,6-tetrahydropyridine-4-sulfonic acid support the proposed different mechanism of action of amino acid and aminosulfonic acid GABA agonists. While isoguvacine is substantially more effective than isonipecotic acid as a GABA agonist, 1,2,3,6-tetrahydropyridine-4-sulfonic acid is proportionally weaker than piperidine-4-sulfonic acid [Falch et al., 1984] (Fig. 6). Under certain experimental conditions the binding of piperidine-4-sulfonic acid has proved to be more sensitive than the binding of GABA to stimulation by barbiturates [Olsen and Snowman, 1982]. Finally, it is interesting to note that piperidine-4-sulfonic acid is virtually inactive at GABA receptors in central neurons of the invertebrates *Periplaneta* and *Limulus*, which otherwise have agonist specificities similar to those of $GABA_A$ receptors in the mammalian CNS [Roberts et al., 1981].

## PHARMACOKINETIC ASPECTS OF $GABA_A$ AGONISTS

All compounds so far known with specific actions on GABA receptors have zwitterionic structures, and the early stage of the pharmacology of $GABA_A$ agonists reflects the difficulties in developing such compounds with satisfactory pharmacokinetic properties. Small, and frequently negligible, fractions of amino acids exist as unionized molecules in solution, the ratio between the concentrations of ionized and unionized molecules (I/U ratio) being a function of the difference between the $pK_a$I- and II- values [Edsall and Wyman, 1958; Krogsgaard-Larsen, 1981; Krogsgaard-Larsen and Falch, 1981]. A great difference between the $pK_a$ values of neutral amino acids is tantamount to high I/U ratios for the compounds.

Since amino acids are likely to penetrate the blood–brain barrier in the unionized form, it is of pharmacological interest to develop analogues of GABA with small differences in the $pK_a$ values, and thus lower I/U ratios, compared to GABA [Krogsgaard-Larsen et al., 1982a]. Like GABA, isoguvacine and in particular piperidine-4-sulfonic acid have high I/U ratios (Fig. 12), and none of these compounds penetrate the blood–brain barrier [Krogsgaard-Larsen, 1981; Krogsgaard-Larsen et al., 1981b]. THIP and muscimol, on the other hand, have much lower I/U ratios [Krogsgaard-Larsen et al., 1981b]. Approximately 0.1% and 0.2%, respectively, of doses of THIP or muscimol exist as unionized molecules in aqueous solution, and these values can explain why THIP and muscimol enter the brain very easily after peripheral administration in mice and rats [Krogsgaard-Larsen et al., 1981b; Moroni et al., 1982] and in humans [Krogsgaard-Larsen et al., 1984].

The acid strength of the 3-isothiazolol nucleus is much less pronounced than that of the 3-isoxazolol unit, and as a result of the small difference between the $pK_a$ values of the 3-isothiazolol zwitterion thiomuscimol, this compound has a very low I/U ratio (Fig. 12). Thiomuscimol is approximately equipotent with muscimol as a GABA agonist (Krogsgaard-Larsen et al., 1979b), and in light of its low I/U ratio, behavioral pharmacological studies of thiomuscimol obviously are of great interest. Such studies are in progress.

Muscimol not only lacks specificity as a GABA agonist [Krogsgaard-Larsen et al., 1979b] (Fig. 13), it is relatively toxic [Christensen et al., 1982], and it is rapidly metabolized after peripheral administration [Maggi and Enna, 1979; Baraldi et al., 1979; Moroni et al., 1982]. This instability in vivo has recently been shown to be a consequence of muscimol being a substrate for GABA transaminase [Fowler et al., 1983]. THIP, on the other hand, is well tolerated by various animal species [Krogsgaard-Larsen and Christensen, 1980; Christensen et al., 1982]. THIP is active after oral administration, and it is excreted unchanged and to some extent in a

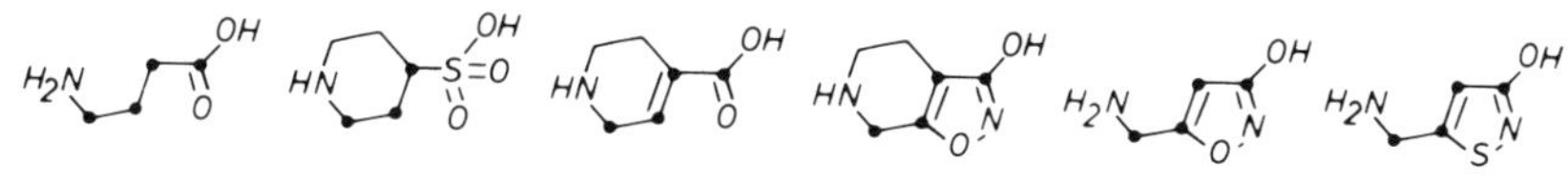


| | GABA | P4S | Isoguvacine | THIP | Muscimol | Thiomuscimol |
|---|---|---|---|---|---|---|
| $pK_A$ | 4.0;10.7 | <1;10.3 | 3.6;9.8 | 4.4;8.5 | 4.8;8.4 | 6.1;8.9 |
| I/U | 800,000 | >1,000,000 | 200,000 | 1,500 | 900 | 13 |
| Penetration of the BBB | No | No | No | Yes | Yes | Not tested |

**Fig. 12** The structures, $pK_A$ values, and I/U ratios of some $GABA_A$ agonists and their ability to penetrate the blood–brain barrier.

conjugated form in the urine from animals and humans [Schultz et al., 1981]. THIP is at present the subject of investigations in the human clinic.

In light of the potency and specificity of THIP as a $GABA_A$ agonist, this compound has been subjected to quite extensive structural modifications with the aim of developing compounds with similar pharmacological profiles. However, as exemplified in Figure 13, even minor structural changes of the molecule of THIP result in complete loss of affinity for the $GABA_A$ receptors [Krogsgaard-Larsen and Johnston, 1978; Krogsgaard-Larsen and Falch, 1981; Haefliger et al., 1984; P. Krogsgaard-Larsen, unpublished], emphasizing the remarkable agonist specificity of the $GABA_A$ receptors.

## PHARMACOLOGICAL AND CLINICAL STUDIES ON THIP

### Effects on Seizures and Ethanol Withdrawal Symptoms

There is an overwhelming amount of indirect evidence, derived from experimental models of epilepsy, supporting the view that pharmacological stimulation of GABA neurotransmission may have therapeutic interest in epilepsy [Meldrum, 1982]. The anticonvulsant effects of THIP and muscimol have been compared in a variety of animal models. THIP typically is 2–5 times weaker than muscimol in suppressing seizure activities. In mice [Löscher, 1982] and in gerbils with genetically determined epilepsy [Löscher et al., 1983], systemically administered THIP has proved very effective in suppressing seizure activity, and THIP is capable of reducing audiogenic seizures in DBA/2 mice [Meldrum and Horton, 1980]. However, THIP failed to protect baboons with photosensitive epilepsy against photically induced myoclonic responses [Meldrum and Horton, 1980]. In acute as well as chronic studies, THIP is capable of reducing seizures in mice induced by various agents [Minchin and Nutt, 1983]. An analysis of these effects [Minchin and Nutt, 1983] and the effects of THIP on ponto-geniculo-occipital (PGO) activity in cats [Neal and Bond, 1983], seems to indicate an as-yet-unclarified involvement of the serotonin system. THIP has been subjected to a single-blind controlled trial in patients with epilepsy, in which THIP was added to the concomitant antiepileptic treatment [Petersen et al., 1983]. Under these conditions no significant effects of THIP were detected, although a trend was observed for lower seizure frequency during a period of submaximal doses of THIP.

The effects of THIP on ethanol withdrawal symptoms in rats have recently been studied [Frye et al., 1983]. While intracisternally administered THIP proved effective in reducing audiogenic clonic-tonic seizures, no effects on forelimb tremor were observed. These selective effects may have clinical interest.

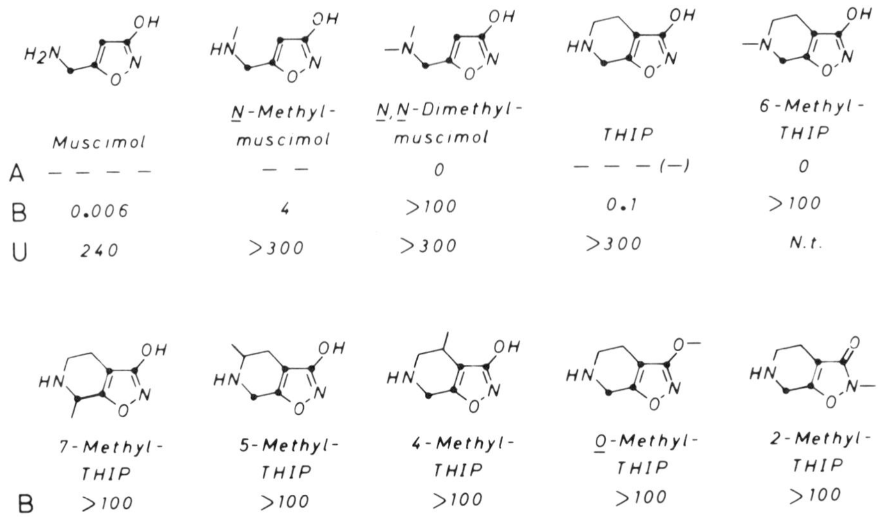


**Fig. 13** The relationship between structures and effects in vivo and in vitro on the $GABA_A$ receptors and the neuronal GABA uptake system in vitro of some analogues of muscimol and THIP. *A*,*B*, and *U*: See legend for Figure 6. N.t. = not tested.

## Analgesic and Anxiolytic Effects

In recent years the involvement of the central GABA system in pain mechanisms has been the subject of intensive studies [DeFeudis, 1982; Andree et al., 1983]. Studies of the clinical prospects of GABA-mediated analgesia have been made possible by the demonstration of very potent analgesic effects of THIP in different animal models [Christensen and Larsen, 1982; Hill et al., 1981]. THIP-induced analgesia is insensitive to naloxone [Kendall et al., 1982; Hill et al., 1981; Grognet et al., 1983], indicating that the effect is not mediated by the opiate receptors. Quite surprisingly, THIP analgesia cannot be reversed by bicuculline [DeFeudis, 1982; Kendall et al., 1982; Hill et al., 1981], which may indicate the involvement of a distinct class of GABA receptor. On the other hand, THIP-induced analgesia can be reduced by atropine [Kendall et al., 1982; Grognet et al., 1983] and potentiated by cholinergics such as physostigmine, reflecting as-yet-unclarified functional interactions between GABA and acetylcholine neurons and the central opiate systems rather than a direct action of THIP on muscarinic receptors.

THIP and morphine are approximately equipotent as analgesics, although their relative potencies are dependent on the animal species and experimental models used [Christensen and Larsen, 1982; Grognet et al., 1983]. Acute injection of THIP potentiates morphine-induced analgesia, and chronic administration of THIP produces a certain degree of functional tolerance to its analgesic effects [Andree et al., 1983]. In contrast to earlier findings [Christensen and Larsen, 1982] the results of recent studies have been interpreted in terms of some cross-tolerance between THIP and morphine [Andree et al., 1983]. In contrast to morphine, THIP does not cause respiratory depression [Lindeburg et al., 1983]. Clinical studies on postoperation patients and patients with chronic pain of malignant origin [Kjaer and Nielsen, 1983] have disclosed potent analgesic effects of THIP in the latter group of patients at doses of 5–30 mg (im) of THIP. In these cancer patients and also in patients with chronic anxiety [Hoehn-Saric, 1983], the desired effects of THIP were accompanied by side-effects, notably sedation, nausea, and in a few cases euphoria. The side-effects of THIP have been described as mild and similar in quality to those of other GABA-mimetics [Hoehn-Saric, 1983]. These undesirable effects of THIP may to a certain extent be ascribed to the nonoptimal pharmacokinetics of THIP [Hoehn-Saric, 1983], emphasizing the need for a sustained-release preparation of THIP.

It is assumed that the postsynaptic GABA receptor complex is mediating the anxiolytic effects of the benzodiazepines [Paul et al., 1981], and, consequently, it is of interest to see whether GABA agonists have anxiolytic effects. Muscimol has proved effective in conflict tests, though with a pharmacological profile different from that of diazepam [Cuomo et al., 1981], and in humans muscimol in low doses was found to sedate and calm some schizophrenic patients [Tamminga et al., 1979]. In a recent study on a number of patients with chronic anxiety, the effects of THIP were assessed on several measures of anxiety [Hoehn-Saric, 1983]. Although these effects were accompanied by side effects [Hoehn-Saric, 1983], the combination of analgesic and anxiolytic effects of THIP would seem to have therapeutic prospects.

## Cardiovascular Effects

There is very strong evidence that GABA is involved in the regulation of cardiovascular mechanisms [DeFeudis, 1981; Antonaccio and Taylor, 1977]. While intracerebroventricularly administered THIP reduced blood pressure as well as heart rate, systemically administered THIP did not affect these functions significantly [Persson, 1983; Mesdjian et al., 1983]. On the other hand, systemically administered GABA and isoguvacine had significant cardiovascular effects, which could be blocked by bicuculline [Mesdjian et al., 1983]. Since THIP, but not GABA and isoguvacine, is capable of penetrating the blood–brain barrier [Schultz et al., 1981; Krogsgaard-Larsen et al., 1982a], these results seem to indicate that peripheral GABA receptors are involved in the regulation of cardiovascular functions [Mesdjian et al., 1983]. It is interesting to note that GABA and a number of GABA agonists produce a dose-dependent dilation of isolated cat and dog artery segments, apparently by activation of GABA receptors with pharmacological characteristics similar to those of central postsynaptic GABA receptors [Edvinsson and Krause, 1979].

## Effects on Spasticity

The results of pharmacological studies on the spastic mouse are consistent with a role of GABA in spasticity [Biscoe and Fry, 1982]. Systemic administration of the GABA agonists muscimol, THIP, and isoguvacine affected spinal cord activities in the cat [Polc, 1979]. Since isoguvacine does not readily penetrate the blood–brain barrier [Krogsgaard-Larsen et al., 1981b; 1982a], its pharmacological effects in this animal model [Polc, 1979] may suggest that some parts of the spinal cord are not effectively protected by a blood–brain barrier. THIP has been studied recently in spastic patients [Mondrup and Pedersen, 1983]. At oral doses of 15–25 mg, THIP clearly reduced the monosynaptic T-reflexes without affecting the flexor threshold significantly [Mondrup and Pedersen, 1983].

## Interactions of THIP With Dopaminergic and Cholinergic Systems in the CNS

Studies in recent years have disclosed very complex interactions between different neurotransmitter systems in the basal ganglia, notably the dopamine, GABA, and cholinergic systems [Chase and Walters, 1976; Christensen et al., 1982; Di Chiara and Gessa, 1981]. These interactions have been studied extensively with the intention of getting better insight into the mechanisms underlying schizophrenia, Parkinson disease, and different dyskinetic syndromes and, furthermore, of developing new strategies for the treatment of these severe diseases. Much interest has been focused on the nigrostriatal dopamine neurons, which form part of the nigrostriatal "feedback" pathway, of which the striatonigral GABA neurons terminate within the substantia nigra pars reticulata, possibly on cholinergic neurons. The dopamine neurons of this system originating in substantia nigra pars compacta as well as the mesolimbic dopamine neurons involved in an analogous "feedback" loop are assumed to be under inhibitory GABA control [Chase and Walters, 1976; Christensen et al., 1982; Di Chiara and Gessa, 1981]. These aspects have opened up the prospects of using GABA-stimulating therapies in the treatment of schizophrenic patients, and, consequently, THIP has been studied quite extensively in different animal models.

While activation of GABA receptors in substantia nigra pars reticulata of rats have dramatic behavioral consequences [Arnt et al., 1979], the dopamine neurons in substantia nigra pars compacta and the mesolimbic dopamine neurons are much less sensitive. Direct application of THIP in the respective brain areas actually has weak inhibitory effects on both types of dopamine neurons, whereas systemically administered THIP weakly stimulates these dopamine neurons [Waszczak and Walters, 1980; Waszczak et al., 1980]. No simple explanation of these apparently self-contradictory observations has, so far, been forwarded. The behavioral effects of acute and chronic administration of THIP have been studied [Christensen et al., 1982]. Dopamine agonist–induced locomotor activity and stereotypies are altered by simultaneous treatment with GABA agonists, the former activity being depressed and the latter intensified [Christensen et al., 1979]. From a clinical point of view, the interactions between THIP and neuroleptics may be particularly interesting. Most neuroleptic drugs inhibit dopamine-induced stereotypy and induce catalepsy in animals, the former effect being related to clinical antipsychotic effects and the latter to extrapyramidal side-effects of neuroleptics [Arnt and Christensen, 1981]. Since THIP, and also scopolamine, antagonize the antistereotypic effects of some neuroleptics, it has been tentatively concluded that GABA agonists such as THIP would probably not potentiate the antipsychotic effect of neuroleptics but rather antagonize it [Arnt and Christensen, 1981].

The interactions between THIP and the central dopamine systems have also been studied in monkeys [Bjoerndal et al., 1983]. Analyses of the complex pharmacological profile in this animal of THIP, which to some extent was similar to that of diazepam, led to the conclusion that THIP would probably have limited therapeutic effect in different kinds of dyskinesia, and THIP actually has proved ineffective in reducing the symptoms of dyskinetic patients [Korsgaard et al., 1982].

## SUMMARY

In agreement with the antinociceptive actions of THIP in a variety of animal models, clinical studies in different types of patients have disclosed very potent analgesic actions of THIP. The analgesic effects of THIP clearly are of nonopioid nature. While some of the side-effects of morphine, such as nausea, are also seen after administration of THIP to patients, THIP does not, in contrast to morphine, affect respiration or blood pressure. THIP is a relatively weak anxiolytic in humans, and it has proved quite effective in suppressing some of the symptoms in spastic patients. These effects together are interesting in regard to different clinical situations. The clinical studies on THIP have disclosed side-effects of THIP, notably sedation, ataxia, and euphoria. In some groups of patients these unwanted effects have been described as mild, but in some clinical situations they have been found unacceptable. In light of the potent anticonvulsant effects of THIP in a variety of animal models, its very weak effects in epileptic patients are somewhat surprising. It must, however, be emphasized that in this clinical situation as well as in those involving patients with neurological disorders, THIP was administered concomitantly with other traditionally used drugs to patients who had been under various kinds of drug treatment for long periods of time.

## ACKNOWLEDGMENTS

This work was supported by grants from the Danish Medical Research Council. The secretarial and technical assistance of Mrs. B. Hare and Mr. S. Stilling, respectively, is gratefully acknowledged.

## REFERENCES

Alger BE, Nicoll RA (1982): Pharmacological evidence for two kinds of GABA receptors on rat hippocampal pyramidal cells studied in vitro. J Physiol (London) 328:125–141.

Allan RD, Curtis DR, Headley PM, Johnston GAR, Lodge D, Twitchin B (1980): The synthesis and activity of cis- and trans-2-(aminomethyl)cyclopropanecarboxylic acids as conformationally restricted analogues of GABA. J Neurochem 34:652–654.

Allan RD, Johnston GAR, Kazlauskas R, Tran H (1983): Synthesis of analogues of GABA. IX. 5-(Aminomethyl)-3-hydroxyfuran-2(5*H*)-one. Aust J Chem 36:977–981.

Andree T, Kendall DA, Enna SJ (1983): THIP analgesia: cross tolerance with morphine. Life Sci 32:2265–2272.

Antonaccio MJ, Taylor DG (1977): Involvement of central GABA receptors in the regulation of blood pressure and heart rate of anesthetized cats. Eur J Pharmacol 46:283–287.

Armstrong DR, Breckenridge RJ, Suckling CJ (1982): Muscimol and some aza analogues: molecular orbital calculations relating to their potency as GABA mimetics. J Theor Biol 97:267–276.

Arnt J, Christensen AV (1981): Differential reversal by scopolamine and THIP on the antistereotypic and cataleptic effects of neuroleptics. Eur J Pharmacol 69:107–111.

Arnt J, Krogsgaard-Larsen P (1979): GABA agonists and potential antagonists related to muscimol. Brain Res 177:395–400.

Arnt J, Scheel-Krüger J, Magelund G, Krogsgaard-Larsen P (1979): Muscimol and related GABA receptor agonists: the potency of GABAergic drugs *in vivo* determined after intranigral injection. J Pharm Pharmacol 31:306–313.

Atkinson JG, Girard Y, Rokach J, Rooney CS, McFarlane CS, Rackham A, Share NN (1979): Kojic amine — a novel $\gamma$-aminobutyric acid analogue. J Med Chem 22:99–106.

Baraldi M, Grandison L, Guidotti A (1979): Distribution and metabolism of muscimol in the brain and other tissues of the rat. Neuropharmacology 18:57–62.

Biscoe TJ, Fry JP (1982): Some pharmacological studies on the spastic mouse. Br J Pharmacol 75:23–35.

Bjoerndal N, Gerlach J, Casey DE, Christensson E (1983): Effect of apomorphine and GABAergic drugs in monkeys pretreated with haloperidol. Psychopharmacology 79:220–225.

Blavet N, DeFeudis FV, Clostre F (1982): THIP inhibits feeding behaviour in fasted rats. Psychopharmacology 76:75–78.

Bowery NG, Collins JF, Cryer G, Inch TD, McLaughlin NJ (1979): The GABA receptor: stereospecificity and structure–activity studies. In Mandel P, DeFeudis FV (eds): "GABA-Biochemistry and CNS Functions." New York: Plenum Press, pp 339–353.

Bowery NG, Hill DR, Hudson AL, Doble A, Middlemiss DN, Shaw J, Turnbull M (1980): (−)-Baclofen decreases neurotransmitter release in the mammalian CNS by an action at a novel GABA receptor. Nature 283:92–94.

Braestrup C, Nielsen M, Krogsgaard-Larsen P, Falch E (1979): Partial agonists for brain GABA-benzodiazepine receptor complex. Nature 280:331–333.

Brown DA (1979): Extrasynaptic GABA systems. Trends Neurosci 2:271–273.

Browner M, Ferkany JW, Enna SJ (1981): Biochemical identification of pharmacologically and functionally distinct GABA receptors in rat brain. J Neurosci 1:514–518.

Cananzi AR, Costa E, Guidotti A (1980): Potentiation by intraventricular muscimol of the anticonflict effect of benzodiazepines. Brain Res 196:447–453.

Chase TN, Walters JR (1976): Pharmacological approaches to the manipulation of GABA mediated synaptic function in man. In Roberts E, Chase TN, Tower DB (eds): "GABA in Nervous System Function." New York: Raven Press, pp 497–513.

Christensen AV, Arnt J, Scheel-Krüger J (1979): Decreased antistereotypic effect of neuroleptics after additional treatment with a benzodiazepine, a GABA agonist or an anticholinergic compound. Life Sci 24:1395–1402.

Christensen AV, Larsen JJ (1982): Antinociceptive and anticonvulsive effects of THIP, a pure GABA agonist. Pol J Pharmacol Pharm 34:127–134.

Christensen AV, Svendsen O, Krogsgaard-Larsen P (1982): Pharmacodynamic effects and possible therapeutic uses of THIP, a specific GABA-agonist. Pharm Week Sci Ed 4:145–153.

Cuomo V, Cortese I, Siro-Brigiani G (1981): Effects of muscimol alone and in combination with diazepam on shuttle-box performance in rats. Arzneimittel-

forsch 31:1724–1726.
Curtis DR (1978): Pre- and non-synaptic activities of GABA and related amino acids in the mammalian nervous system. In Fonnum F (ed): "Amino Acids as Chemical Transmitters." New York: Plenum Press, pp 55–86.
Curtis DR, Johnston GAR (1974): Amino acid transmitters in the mammalian central nervous system. Ergebn Physiol 69:97–188.
Curtis DR, Watkins JC (1960): The excitation and depression of spinal neurones by structurally related amino acids. J Neurochem 6:117–141.
Curtis DR, Watkins JC (1965): The pharmacology of amino acids related to gamma-aminobutyric acid. Pharmacol Rev 17:347–391.
Curtis DR, Duggan AW, Felix D, Johnston GAR (1971a): Bicuculline, an antagonist of GABA and synaptic inhibition in the spinal cord of the cat. Brain Res 32:69–96.
Curtis DR, Game CJA, Johnston GAR, McCulloch RM (1974): Central effects of β-(p-chlorophenyl)-γ-aminobutyric acid. Brain Res 70:493–499.
Curtis DR, Lodge D, Bornstein JC, Peet MJ (1981): Selective effects of (−)-baclofen on spinal synaptic transmission in the cat. Exp Brain Res 42:158–170.
Curtis DR, Duggan AW, Felix D, Johnston GAR, McLennan H (1971b): Antagonism between bicuculline and GABA in the cat brain. Brain Res 33:57–73.
Curtis DR, Lodge D, Bornstein JC, Peet MJ, Leah JD (1982): The dual effects of GABA and related amino acids on the electrical threshold of ventral horn group Ia afferent terminations in the cat. Exp Brain Res 48:387–400.
Davies J, Watkins JC (1974): The action of β-phenyl-GABA derivatives on neurones of the cat cerebral cortex. Brain Res 70:501–505.
DeFeudis FV (1981): GABA and "neuro-cardiovascular" mechanisms. Neurochem Int 3:113–122.
DeFeudis FV (1982) γ-Aminobutyric acid and analgesia. Trends Pharmacol Sci 3:444–446.
Di Chiara G, Gessa GL (1981):"GABA and the Basal Ganglia." New York: Raven Press.
Edsall JT, Wyman J (1958): "Biophysical Chemistry," Vol. I. New York: Academic Press, pp 485–486.
Edvinsson L, Krause DN (1979): Pharmacological characterization of GABA receptors mediating vasodilation of cerebral arteries in vitro. Brain Res 173:89–97.
Enna SJ (1981): GABA receptor pharmacology. Functional considerations. Biochem Pharmacol 30:907–913.
Enna SJ, Snyder SH (1975): Properties of γ-aminobutyric acid (GABA) receptor binding in rat brain synaptic membrane fractions. Brain Res 100:81–97.
Enna SJ, Snyder SH (1977): Influences of ions, enzymes, and detergents on γ-aminobutyric acid-receptor binding in synaptic membranes of rat brain. Mol Pharmacol 13:442–453.
Enna SJ, Collins JF, Snyder SH (1977): Stereospecificity and structure–activity requirements of GABA receptor binding in rat brain. Brain Res 124:185–190.
Falch E, Jacobsen P, Krogsgaard-Larsen P, Curtis DR (1985): GABA-mimetic activity and effects of diazepam binding of aminosulphonic acids structurally related to piperidine-4-sulphonic acid. J Neurochem 44:68–75.
Falch E, Krogsgaard-Larsen P (1982): The binding of the GABA agonist [$^3$H]THIP to rat brain synaptic membranes. J Neurochem 38:1123–1129.
Fowler LJ, Lovell DH, John RA (1983): Reaction of muscimol with 4-aminobutyrate aminotransferase. J Neurochem 41:1751–1754.
Frere RC, Macdonald RL,Young AB (1982): GABA binding and bicuculline in spinal cord and cortical membranes from adult rat and from mouse neurons in cell culture. Brain Res 244:145–153.
Frye GD, McCown TJ, Breese GR (1983): Differential sensitivity of ethanol withdrawal signs in the rat to γ-aminobutyric acid (GABA)mimetics: blockade of audiogenic seizures but not forelimb tremors. J Pharmacol Exp Ther 226:720–725.
Gähwiler BH, Maurer R, Wüthrich HJ (1984): Pitrazepin, a novel $GABA_A$ antagonist. Neurosci Lett 45:311–316.
Gillis RA, Williford DJ, Souza JD, Quest JA (1982): Central cardiovascular effects produced by the GABA receptor agonist drug, THIP. Neuropharmacology 21:545–547.
Grandison L, Guidotti A (1979): γ-Aminobutyric acid receptor function in rat anterior pituitary: evidence for control of prolactin release. Endocrinology 105:754–759.
Grognet A, Hertz F, DeFeudis FV (1983): Comparison of the analgesic actions of THIP and morphine. Gen Pharmacol 14:585–589.
Haefliger W, Révész L, Maurer R, Römer D, Büscher HH (1984): Analgesic GABA agonists. Synthesis and structure–activity studies on analogues and derivatives of muscimol and THIP. Eur J Med Chem 19:149–156.
Hill DR, Bowery NG (1981): $^3$H-Baclofen and $^3$H-GABA bind to bicuculline-insensitive $GABA_B$ sites in rat brain. Nature 290:149–152.
Hill RC, Maurer R, Buescher HH, Roemer D (1981): Analgesic properties of the GABA-mimetic THIP. Eur J Pharmacol 69:221–224.
Hoehn-Saric R (1983): Effects of THIP on chronic anxiety. Psychopharmacology 80:338–341.
Johnston GAR (1978): Neuropharmacology of amino acid inhibitory transmitters. Annu Rev Pharmacol Toxicol 18:269–289.
Johnston GAR, Allan RD (1984): GABA agonists. Neu-

ropharmacology 23:831–832.

Johnston GAR, Allan RD, Kennedy SME, Twitchin B (1979): Systematic study of GABA analogues of restricted conformation. In Krogsgaard-Larsen P, Scheel-Krüger J, Kofod H (eds): "GABA-Neurotransmitters. Pharmacochemical, Biochemical and Pharmacological Aspects." Copenhagen: Munksgaard, pp 149–164.

Johnston GAR, Allan RD, Andrews PR, Kennedy SME, Twitchin B (1978): Stereospecific actions of GABA analogues. In Simon P (ed): "Advances in Pharmacology and Therapeutics," Vol. 2. Oxford: Pergamon Press, pp 11–18.

Johnston GAR, Curtis DR, Beart PM, Game CJA, McCulloch RM, Twitchin B (1975): *Cis*- and *trans*-4-aminocrotonic acid as GABA analogues of restricted conformation. J Neurochem 24:157–160.

Karobath M, Placheta P, Lippitsch M, Krogsgaard-Larsen P (1979): Is stimulation of benzodiazepine receptor binding mediated by a novel GABA receptor? Nature 278:748–749.

Kendall DA, Browner M, Enna SJ (1982): Comparison of the antinociceptive effect of GABA agonists: evidence for a cholinergic involvement. J Pharmacol Exp Ther 220:482–487.

Kjaer M, Nielsen H (1983): The analgesic effect of the GABA-agonist THIP in patients with chronic pain of malignant origin. A phase-1-2 study. Br J Clin Pharmacol 16:477–485.

Korsgaard S, Casey DE, Gerlach J, Hetmar O, Kaldan B, Mikkelsen LB (1982): The effect of 4,5,6,7-tetrahydroisoxazolo[5,4-*c*]pyridin-3-ol (THIP) in tardive dyskinesia: A new gamma-aminobutyric acid agonist. Arch Gen Psychiatry 39:1017–1021.

Krnjević K (1974): Chemical nature of synaptic transmission in vertebrates. Physiol Rev 54:418–540.

Krogsgaard-Larsen P (1978): GABA agonists and uptake inhibitors of restricted conformations: structure–activity relations. In Fonnum F (ed): "Amino Acids as Chemical Transmitters." New York: Plenum Press, pp 305–321.

Krogsgaard-Larsen P (1980): Inhibitors of the GABA uptake systems. Mol Cell Biochem 31:105–121.

Krogsgaard-Larsen P (1981): $\gamma$-Aminobutyric acid agonists, antagonists, and uptake inhibitors. Design and therapeutic aspects. J Med Chem 24:1377–1383.

Krogsgaard-Larsen P (1984): THIP, a specific and clinically active GABA agonist. Neuropharmacology 23:837–838.

Krogsgaard-Larsen P, Arnt J (1980): Pharmacological studies of interactions between benzodiazepines and GABA receptors. Brain Res Bull (Suppl 2) 5:867–872.

Krogsgaard-Larsen P, Christensen AV (1980): GABA agonists and antagonists. Annu Rep Med Chem 15:41–50.

Krogsgaard-Larsen P, Falch E (1981): GABA agonists. Development and interactions with the GABA receptor complex. Mol Cell Biochem 38:129–146.

Krogsgaard-Larsen P, Johnston GAR (1978): Structure–activity studies on the inhibition of GABA binding to rat brain membranes by muscimol and related compounds. J Neurochem 30:1377–1382.

Krogsgaard-Larsen P, Falch E, Christensen AV (1984): Chemistry and pharmacology of the GABA agonists THIP (gaboxadol) and isoguvacine. Drugs of the Future 9:597–618.

Krogsgaard-Larsen P, Honorè T, Thyssen K (1979a): GABA receptor agonists: design and structure–activity studies. In Krogsgaard-Larsen P, Scheel-Krüger J, Kofod H (eds): "GABA-Neurotransmitters. Pharmacochemical, Biochemical and Pharmacological Aspects." Copenhagen: Munksgaard, pp 201–216.

Krogsgaard-Larsen P, Falch E, Mikkelsen H, Jacobsen P (1982a):Development of structural analogues and pro-drugs of GABA agonists with desirable pharmacokinetic properties. In Bundgaard H, Hansen AB, Kofod H (eds):"Optimization of Drug Delivery." Copenhagen, Munksgaard, pp 225–234.

Krogsgaard-Larsen P, Johnston GAR, Lodge D, Curtis DR (1977): A new class of GABA agonist. Nature 268:53–55.

Krogsgaard-Larsen P, Snowman A, Lummis SC, Olsen RW (1981a): Characterization of the binding of the GABA agonist [$^3$H]piperidine-4-sulfonic acid to bovine brain synaptic membranes. J Neurochem 37:401–409.

Krogsgaard-Larsen P, Falch E, Peet MJ, Leah JD, Curtis DR (1983): Molecular pharmacology of the GABA receptors and GABA agonists. In Mandel P, DeFeudis FV (eds): "CNS Receptors — From Molecular Pharmacology to Behaviour." New York: Raven Press, pp 1–13.

Krogsgaard-Larsen P, Falch E, Schousboe A, Curtis DR, Lodge D (1980a): Piperidine-4-sulphonic acid, a new specific GABA agonist. J Neurochem 34:756–759.

Krogsgaard-Larsen P, Hjeds H, Curtis DR, Lodge D, Johnston GAR (1979b): Dihydromuscimol, thiomuscimol, and related heterocyclic compounds as GABA analogues. J Neurochem 32:1717–1724.

Krogsgaard-Larsen P, Hjeds H, Curtis DR, Leah JD, Peet MJ (1982b): Glycine antagonists structurally related to muscimol, THIP, or isoguvacine. J Neurochem 39:1319–1324.

Krogsgaard-Larsen P, Jacobsen P, Brehm L, Larsen JJ, Schaumburg K (1980b): GABA agonists and uptake inhibitors designed as agents with irreversible actions. Eur J Med Chem 15:529–535.

Krogsgaard-Larsen P, Johnston GAR, Curtis DR, Game CJA, McCulloch RM (1975): Structure and biological activity of a series of conformationally restricted analogues of GABA. J Neurochem 25:803–809.

Krogsgaard-Larsen P, Schultz B, Mikkelsen H, Aaes-Jørgensen T, Bøgesø KP (1981b): THIP, isoguvacine, isoguvacine oxide, and related GABA agonists. In DeFeudis FV, Mandel P (eds): "Amino Acid Neurotransmitters." New York, Raven Press, pp 69–76.

Larsson OM, Thorbek P, Krogsgaard-Larsen P, Schousboe A (1981): Effect of homo-β-proline and other heterocyclic GABA analogues on GABA uptake in neurons and astroglial cells and on GABA receptor binding. J Neurochem 37:1509–1516.

Lindeburg T, Foelsgaard S, Sillesen H, Jacobsen E, Kehlet H (1983): Analgesic, respiratory and endocrine responses in normal man to THIP, a GABA agonist. Acta Anaesthesiol Scand 27:10–12.

Löscher W (1982): Comparative assay of anticonvulsant and toxic potencies of sixteen GABAmimetic drugs. Neuropharmacology 21:803–810.

Löscher W, Frey HH, Reiche R, Schultz D (1983): High anticonvulsant potency of γ-aminobutyric acid (GABA)mimetic drugs in gerbils with genetically determined epilepsy. J Pharmacol Exp Ther 226:839–844.

Maggi A, Enna SJ (1979): Characteristics of muscimol accumulation in mouse brain after systemic administration. Neuropharmacology 18:361–366.

Maurer R (1979): The GABA agonist THIP, a muscimol analogue, does not interfere with the benzodiazepine binding site on rat cortical membranes. Neurosci Lett 12:65–68.

Meldrum B (1982): Pharmacology of GABA. Clin Neuropharmacol 5:293–316.

Meldrum B, Horton R (1980): Effects of the bicyclic GABA agonist, THIP, on myoclonic and seizure responses in mice and baboons with reflex epilepsy. Eur J Pharmacol 61:231–237.

Mesdjian E, DeFeudis FV, Jadot G, Valli M, Brugerolle B, Bouyard P (1983): Comparison of the cardiovascular effects of intravenously administered GABA, THIP, and isoguvacine in the anaesthetized rat. Drug Dev Res 3:311–318.

Minchin MCW, Nutt DJ (1983): Acute and chronic effects of THIP on seizure threshold. Proc Br Pharmacol Soc, Cambridge, 6–8 April, p 11.

Mitchell R (1982): Interactions of agonists and antagonists with a novel type of GABA receptor. Biochem Pharmacol 31:2684–2686.

Mondrup K, Pedersen E (1983): The acute effect of the GABA-agonist, THIP, on proprioceptive and flexor reflexes in spastic patients. Acta Neurol Scand 67:48–54.

Moroni F, Forchetti MC, Krogsgaard-Larsen P, Guidotti A (1982): Relative disposition of the GABA agonists THIP and muscimol in the brain of the rat. J Pharm Pharmacol 34:676–678.

Muhyaddin M, Roberts PJ, Woodruff GN (1982): Presynaptic γ-aminobutyric acid receptors in the rat anococcygeus muscle and their antagonism by 5-aminovaleric acid. Br J Pharmacol 77:163–168.

Möhler H, Okada T (1977): GABA receptor binding with $^3$H-(+)-bicuculline-methiodide in rat CNS. Nature 267:65–67.

Neal H, Bond A (1983): The influence of GABAergic drugs on PGO activity in the cat. Neuropharmacology 22:881–886.

Neal MJ, Bowery NG (1977): Cis-3-aminocyclohexanecarboxylic acid: a substrate for the neuronal GABA transport system. Brain Res 138:169–174.

Nielsen M, Schou H, Braestrup C (1981): [$^3$H]Propyl β-carboline-3-carboxylate binds specifically to brain benzodiazepine receptors. J Neurochem 36:276–285.

Nistri A, Constanti A (1979): Pharmacological characterization of different types of GABA and glutamate receptors in vertebrates and invertebrates. Prog Neurobiol (Oxford) 13:117–235.

Nowak LM, Young AB, Macdonald RL (1982): GABA and bicuculline actions on mouse spinal cord and cortical neurons in cell culture. Brain Res 244:155–164.

Olsen RW (1981): GABA-benzodiazepine-barbiturate receptor interactions. J Neurochem 37:1–13.

Olsen RW, Enna SJ (1983): GABA and anxiolytics. In Malick JB, Enna SJ, Yamamura HI (eds): "Anxiolytics: Neurochemical, Behavioral and Clinical Perspectives. New York: Raven Press, pp 55–76.

Olsen RW, Snowman AM (1982): Chloride-dependent enhancement by barbiturates of GABA receptor binding. J Neurosci 12:1812–1823.

Olsen RW, Snowman AM (1983):[$^3$H]Bicuculline methochloride binding to low-affinity γ-aminobutyric acid receptor sites. J Neurochem 41:1653–1663.

Olsen RW, Ticku MK, Greenlee D, Van Ness P (1979): GABA receptor and ionophore binding sites: interaction with various drugs. In Krogsgaard-Larsen P, Scheel-Krüger J, Kofod H (eds): "GABA-Neurotransmitters. Pharmacochemical, Biochemical and Pharmacological Aspects." Copenhagen: Munksgaard, pp 165–178.

Olsen RW, Bergman MO, Van Ness PC, Lummis SC, Watkins AE, Napias C, Greenlee DV (1981): Gamma-aminobutyric acid receptor binding in mammalian brain: Heterogeneity of binding sites. Mol Pharmacol 19:217–227.

Olsen RW, Snowhill EW, Wamsley JK (1984): Autoradiographic localization of low affinity GABA receptors with [$^3$H]bicuculline methochloride. Eur J Pharmacol 99:247–248.

Palfreyman MG, Schechter PJ, Buckett WR, Tell GP, Koch-Weser J (1981): The pharmacology of GABA-transaminase inhibitors. Biochem Pharmacol 30:817–824.

Paul SM, Marangos PJ, Skolnick P (1981): The benzodiazepine-GABA-chloride ionophore receptor complex: Common site of minor tranquilizer action. Biol

Psychiatry 16:213–229.

Persson B (1983): Cardiovascular effects of THIP in the rat. J Pharm Pharmacol 35:759–761.

Petersen HR, Jensen I, Dam M (1983): THIP: a single-blind controlled trial in patients with epilepsy. Acta Neurol Scand 67:114–117.

Polc P (1979): Effects of GABA-mimetic agents on the cat spinal cord. Prog Neuro-Phychopharmacol 3:345–352.

Pong SS, Wang CC (1982): Avermectin $B_{1a}$ modulation of γ-aminobutyric acid receptors in rat brain membranes. J Neurochem 38:375–379.

Roberts CJ, Krogsgaard-Larsen P, Walker RJ (1981): Studies on the actions of GABA, muscimol and related compounds on *Periplaneta* and *Limulus* central neurones. Comp Biochem Physiol 69C:7–11.

Roberts E, Chase TN, Tower DB (eds) (1976): "GABA in Nervous System Function." New York, Raven Press.

Schlewer G, Wermuth CG, Chambon JP (1984): Analogues tétrazoliques d'agents GABA-mimétiques. Eur J Med Chem 19:181–186.

Schousboe A, Larsson OM, Wood JD, Krogsgaard-Larsen P (1983): Transport and metabolism of γ-aminobutyric acid in neurons and glia: Implications for epilepsy. Epilepsia 24:531–538.

Schousboe A, Thorbek P, Hertz L, Krogsgaard-Larsen P (1979): Effects of GABA analogues of restricted conformation on GABA transport in astrocytes and brain cortex slices and on GABA receptor binding. J Neurochem 33:181–189.

Schultz B, Aaes-Jørgensen T, Bøgesø KP, Jørgensen A (1981): Preliminary studies on the absorption, distribution, metabolism, and excretion of THIP in animal and man using $^{14}C$-labelled compound. Acta Pharmacol Toxicol 49:116–124.

Skolnick P, Paul SM (1981): Benzodiazepine receptors Annu Rep Med Chem 16:21–29.

Squires RF, Braestrup C (1977): Benzodiazepine receptors in rat brain. Nature 266:732–734.

Squires RF, Klepner CA, Benson DI (1980): Multiple benzodiazepine receptor complexes: some benzodiazepine recognition sites are coupled to GABA receptors and ionophores. In Pepeu P, Kuhar MJ, Enna SJ (eds): "Receptors for Neurotransmitters and Peptide Hormones." New York: Raven Press, pp 285–293.

Squires RF, Casida JE, Richardson M, Saederup E (1983): $^{35}S$-*t*-Butyl-bicyclophosphothionate binds with high affinity to brain specific sites coupled to GABA-A and ion recognition sites. Mol Pharmacol 23:326–336.

Supavilai P, Karobath M (1980): The effect of temperature and chloride ions on the stimulation of [$^{3}H$] flunitrazepam binding by the muscimol analogues THIP and piperidine-4-sulphonic acid. Neurosci Lett 19:337–341.

Supavilai P, Karobath M (1981): *In vitro* modulation by avermectin $B_1a$ of the GABA/benzodiazepine receptor complex of rat cerebellum. J Neurochem 36:798–803.

Tallman JF, Thomas JW, Gallager DW (1978): GABAergic modulation of benzodiazepine binding site sensitivity. Nature 274:383–385.

Tamminga CA, Crayton JW, Chase TN (1979): Improvement of tardive dyskinesia after muscimol therapy. Arch Gen Psychiatry 36:595–598.

Ticku MK, Van Ness PC, Haycock JW, Levy WB, Olsen RW (1978): Dihydropicrotoxinin binding sites in rat brain: Comparison to GABA receptors. Brain Res 150:642–647.

Waszczak BL, Hruska RE, Walters JR (1980): GABAergic actions of THIP *in vivo* and *in vitro*: a comparison with muscimol and GABA. Eur J Pharmacol 65:21–29.

Waszczak BL, Walters JR (1980): Effects of GABAergic drugs on single unit activity of A9 and A10 dopamine neurons. Brain Res Bull(Suppl 2) 5:465–470.

Williford DJ, Hamilton BL, Souza JD, Williams TP, DiMicco JA, Gillis RA (1980): Central nervous system mechanisms involving GABA influence arterial pressure and heart rate in the cat. Circ Res 47:80–88.

Yarbrough GG, Williams M, Haubrich DR (1979): The neuropharmacology of a novel γ-aminobutyric acid analog, kojic amine. Arch Int Pharmacodyn 241:266–279.

Benzodiazepine/GABA Receptors and Chloride Channels:
Structural and Functional Properties, pages 97–133


# 5

# Physiology of GABA Enhancement by Benzodiazepines and Barbiturates

**Willy Haefely and Peter Polc**

*Pharmaceutical Research Department, F. Hoffmann-La Roche & Co., Ltd., CH-4002 Basel, Switzerland*

A view frequently encountered is that our knowledge about the involvement of GABA in the pharmacological actions of benzodiazepines was initiated by the identification of specific high-affinity binding sites for these drugs in the CNS and by the finding that GABA modulates their binding to receptors. The fact is that electrophysiological studies had clearly identified the GABAergic synapse as the site of action of benzodiazepines [Haefely et al., 1975; Polc and Haefely, 1976] well before the description of specific high-affinity benzodiazepine sites [Möhler and Okada, 1977; Squires and Braestrup, 1977] and prior to the findings of an increased binding of benzodiazepines in the presence of GABA [e.g., Tallman et al., 1980] and of an increased binding of GABA in the presence of benzodiazepines [Costa et al., 1978], both pointing to the physical coupling of the two receptors. Similarly, barbiturates were shown to enhance GABAergic synaptic transmission [Nicoll et al., 1975] and the effect of exogenous GABA [Ransom and Barker, 1975] well before the allosteric interaction between GABA and barbiturate binding sites on the GABA receptor–ionophore complex was found [see Olsen, 1982].

Electrophysiological techniques are the only ones currently at hand for studying *the single GABA receptor–ionophore complex in action.* As will be shown, experiments using these techniques have led to the conclusion that benzodiazepines act mainly by increasing the probability of single channel openings in response to GABA [Study and Barker, 1981; Bormann and Sakmann, 1984], while barbiturates have a much more complex action, acting in the lower active dose-range predominantly by prolonging the mean open time of the GABA-activated chloride channel [Study and Barker, 1981] and, in higher concentrations, opening directly the GABA-gated chloride channels [Mathers and Barker, 1980].

In this chapter we review the physiological consequences of the modulatory action of benzodiazepine receptor ligands and barbiturates on GABAergic neurotransmission. We first discuss the basic changes which these drugs induce in the GABA receptor–ionophore function. Aspects of the interaction of benzodiazepine receptor ligands with the receptor–ionophore complex which are frequently neglected, such as intrinsic efficacy and receptor reserve, are emphasized, because they seem to determine to a large extent which of the countless GABAergic synapses in the CNS are affected to which degree by given doses of these agents. We then review the effects of benzodiazepine receptor ligands and barbiturates on GABA-mediated synaptic actions in the various regions of the CNS and discuss the possible involvement of these distinct GABAergic synapses in the different drug effects in the intact CNS.

## EARLY INVESTIGATIONS ON THE SYNAPTIC PHARMACOLOGY OF BARBITURATES AND BENZODIAZEPINES

The phenomenon of so-called presynaptic inhibition (Fig. 5) of spinal motoneurons (see below) was intensively studied by neurophysiologists in the early sixties. The effects of a variety of drugs on presynaptic inhibition were investigated in the hope of finding evidence for or against the involvement of the same neurotransmitter in pre- and postsynaptic inhibition. Pentobarbital, already found by Eccles and Malcolm [1946] to prolong the duration of dorsal root potentials, and other barbiturates were shown to increase the intensity and time course of presynaptic inhibition [Schmidt, 1963]. Diazepam was reported by Schmidt et al. [1967] to increase the amplitude of dorsal root potentials and to intensify presynaptic inhibition of the monosynaptic excitation of motoneurons by Ia (muscle spindle) afferents. The effect of barbiturates and diazepam on presynaptic inhibition in the spinal cord remained an unexplained curiosity, proposed to account, in part at least, for the central muscle relaxant activity of these drugs [Schmidt et al., 1967], until in the early seventies the role of GABA as neurotransmitter in the axo-axonal synapses on primary afferent endings became increasingly substantiated [Curtis et al., 1971; Miyata and Otsuka, 1972; Bell and Anderson, 1972; Barker and Nicoll, 1972; Davidoff, 1972]. This suggested the possibility that these drugs interact rather specifically and directly with the GABAergic transmission between GABAergic interneurons and primary afferent axon endings. Pharmacological interaction studies performed in the cat spinal cord provided results consistent with, but of course not proving, a specific action of diazepam at the GABAergic synapse [Polc et al., 1974]. It was the extension of similar studies to other identified GABAergic synapses in various parts of the CNS that led to the proposal of a GABA-related mode of action of benzodiazepines [Haefely et al., 1975; Polc and Haefely, 1976]. While

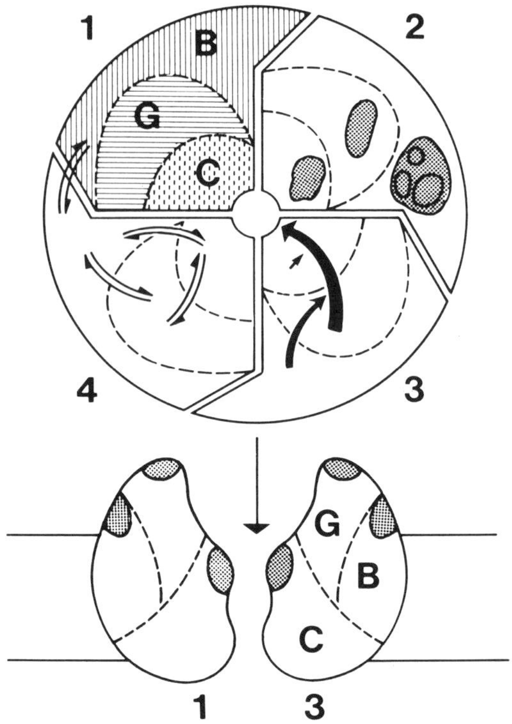


**Fig. 1.** *Hypothetical model of the benzodiazepine–GABA receptor–ionophore complex. The globular oligomeric protein is seen from the extracellular space* (top) *and in a cross-section through the membrane* (bottom). *For explanation see text.*

these early studies did not allow for a distinction between a presynaptic (e.g., facilitatory effect on GABA release) and a postsynaptic site of action (changes in GABA receptors or postreceptor events) of benzodiazepines at the GABAergic synapse, subsequent studies in vivo and in vitro concentrated on the interaction of benzodiazepines and barbiturates with the membrane effects of exogenous GABA. The main results are discussed in the next paragraph.

## MODEL OF THE BENZODIAZEPINE–GABA RECEPTOR–IONOPHORE COMPLEX; TERMINOLOGY OF ITS LIGANDS

The current view on the composition and function of the benzodiazepine receptor–GABA receptor ionophore complex [see

Haefely et al., 1985] is condensed in the schematic diagram of Figure 1 (in this chapter only the GABA receptor and the neuronal or central type of benzodiazepine receptor are considered). This complex is an oligomeric glycoprotein, probably a tetramer. One monomer, of a molecular mass of slightly above 50 kD, contains both the GABA receptor and the benzodiazepine receptor. The precise assembly of the subunits in the complex is not yet known; in particular, it is still uncertain whether the anion channel is a subunit of the complex distinct from the monomers acting as both GABA receptor and benzodiazepine receptor. In the diagram it is assumed for the sake of simplicity that four identical or very similar monomeric subunits are assembled in such a way that they can also function as the anion channel. Accordingly, each subunit is assumed to have three structurally and functionally different domains (indicated on subunit 1): the channel domain (C), the GABA receptor (G), and the benzodiazepine receptor (B). Each domain has at least one binding site for ligands (indicated on subunit 2): a binding site for "channel agents" (barbiturates, many convulsants, and a variety of other agents); a binding site for GABA mimetics and GABA antagonists; and a binding area for benzodiazepine receptor agonists, antagonists, and inverse agonists. The basic function of the complex [Simmonds, 1984] is the gating of the anion channel (indicated on subunit 3). Activation of the GABA receptor, which is transmitted intramolecularly to the channel domain, results in its assuming the channel-open conformation *(largest arrow)*. Agonistic and inverse agonistic ligands of the benzodiazepine receptor allosterically modulate the gain of this function of the GABA receptor–ionophore *(second largest arrow)*, agonists by facilitating this function (either via an increase in the affinity of the GABA receptor for GABA, a facilitated isomerization of the ligand-associated GABA receptor, an improved coupling between GABA receptor and channel domain, or via any combination of these mechanisms), inverse agonists by dampening this ionophore-gating function. Channel agents acting on the channel domain also modulate the gating function in the positive or negative direction *(small arrow)* or directly induce the open state of the channel. A ligand-induced conformational change in any one of the three domains is capable of altering the functions of the other two *(indicated by bidirectional arrows on subunit 4)*.

The terminology used here for the ligands of the three domains of the benzodiazepine receptor–GABA receptor–ionophore complex is presented in Table 1. For the ligands of the GABA receptor, GABA mimetics, and GABA receptor blockers, no problems arise. As for the ligands of the benzodiazepine receptor, the terminology proposed by Polc et al. [1982] is used. *Benzodiazepine receptor agonists* are those acting like the classic benzodiazepine tranquilizers to increase the gain of the GABA receptor–ionophore complex (positive intrinsic efficacy, positive allosteric modulation); *inverse benzodiazepine receptor agonists* are those reducing this gain (negative intrinsic efficacy, negative allosteric modulation). Other names for inverse agonists are: convulsant antagonists, anxiogenic antagonists, antagonists with negative efficacy (as opposed to neutral antagonists), active antagonists, contragonists. *Antagonists at the benzodiazepine receptor* ideally have no intrinsic efficacy but block the effects of both agonists and inverse agonists. *Partial agonists* and *partial inverse agonists* have submaximal positive or negative intrinsic efficacies, respectively. The situation for ligands of the chloride ionophore may turn out to be similar to that just proposed for ligands of the benzodiazepine receptor, with anticonvulsant and anesthetic barbiturates having a positive intrinsic efficacy, and convulsants of the picrotoxinin type displaying negative intrinsic efficacy. Antagonists with no intrinsic efficacy at the ionophore binding sites have not yet been found; they would, however, be a prerequisite for accepting an analogous basic function of the channel agent receptor and the benzodiazepine receptor.

**TABLE 1. Overview on the Various Classes of Ligands of the Benzodiazepine–GABA Receptor–Ionophore Complex**

| Ligands of the benzodiazepine–GABA receptor–ionophore complex |
|---|
| $GABA_A$ receptor |
| Agonists or GABA mimetics (GABA, muscimol) |
| Partial agonists (THIP?) |
| Antagonists (bicuculline) |
| Benzodiazepine receptor |
| Benzodiazepine receptor agonists (benzodiazepine derivatives, cyclopyrrolones, β-carbolines, pyrazoloquinolines, imidazopyridines) |
| Benzodiazepine receptor partial agonists (benzodiazepine derivatives, triazolopyridazines, β-carbolines, pyrazoloquinolines) |
| Benzodiazepine receptor blockers (antagonists) (benzodiazepine derivatives, β-carbolines, pyrazoloquinolines) |
| Partial inverse benzodiazepine receptor agonists (benzodiazepine derivatives, β-carbolines, pyrazoloquinolines) |
| Inverse benzodiazepine receptor agonists (β-carbolines) |
| Ligands of the chloride channel or ionophore ("channel agents") |
| Anticonvulsant and convulsant barbiturates |
| Picrotoxinin and "cage convulsants" (e.g., TBPS) |
| Tetrazoles (e.g., pentylenetetrazole) |
| Convulsant benzodiazepines (e.g., Ro 5-3663, Ro 5-4864) |
| Pyrazolopyridines |
| Clomethiazole |
| Avermectin $B_{1a}$ |
| Etomidate |

## THE BASIC CHANGES INDUCED BY BENZODIAZEPINES AND BARBITURATES IN THE GABA RECEPTOR–IONOPHORE FUNCTION

The basic pharmacological aspects under which GABA potentiation by benzodiazepines and barbiturates has to be considered are: (1) selectivity for GABA (as compared with the interaction with other neurotransmitters); (2) type of potentiation (changes in apparent affinity and/or efficacy of GABA); (3) dose-dependency of the potentiating action (mono- or biphasic, relation with receptor binding affinities); (4) relative activities (potency, efficacy) of potentiating agents and the relation of these activities with all or some distinct in vivo pharmacological effects; and (5) molecular mechanism of potentiation (improved electrosecretory coupling in GABA release, inhibition of GABA inactivation, increase of GABA receptor affinity, changes in receptor–effector coupling, increase in GABA receptor availability, increase in the driving force for chloride ion flux, changes in single-channel properties).

Some of these basic pharmacological questions can be answered by studying multicellular or single-cell activities in vivo or in vitro. Most of these problems, however, require in vitro experiments for the control of drug and transmitter concentrations, as well as intracellular recording for the control of membrane conductance and ion fluxes; a few require the study of single-membrane channels.

### Selectivity

The potentiation of GABA effects by *benzodiazepine receptor agonists* is highly selective. None of the monoamines, nor the other established or putative inhibitory amino acid neurotransmitters (glycine, β-alanine, taurine), was affected by benzodiazepines in concentrations that increase the effect of GABA [Nistri et al., 1980; Choi et al., 1981a; White et al., 1981; Skerritt et al., 1984]. This is particularly important, as some of these other amino acids, e.g. glycine, activate anion channels with similar properties as the channels gated by GABA [Hamill et al., 1983]. No effect of benzodiazepine receptor agonists on excitatory neurotransmitters has been found in relevant concentrations [White et al., 1981; Skerritt et al., 1984]. The selectivity for GABA of *inverse agonists at benzodiazepine receptors* has not yet been investigated, but the specific antagonism of their effects by Ro 15-1788 suggests that at least $\beta$-carboline inverse agonists selectively decrease the effect of GABA.

The situation is quite different with *barbiturates*. While the selectivity was found to be rather good as far as inhibitory amino acids are concerned, the effects of excitatory transmitters (glutamate, aspartate, acetylcholine) start to be depressed by concentrations of barbiturates similar to or only slightly higher than those producing potentiation of GABA effects [Evans, 1979; Ransom and Barker, 1975, 1976; Barker and Ransom, 1978; Macdonald and Barker, 1978, 1979a,b; Nicoll and Wojtowicz, 1980; Lodge and Curtis, 1978; Connors, 1981: Headley et al., 1981]. While the chloride conductance increase and the GABA potentiation induced by barbiturates can be blocked by picrotoxin, the depression of glutamate effects was found to be insensitive to picrotoxin [Barker and Ransom, 1978].

Selectivity should also be considered with respect to drug-induced changes in transmitter effects versus drug-induced changes in nonsynaptic membrane properties of neurons. The available information on this point is not as comprehensive as for transmitter selectivity. Whereas benzodiazepines have mostly been found not to alter neuronal membrane properties unrelated to GABA-induced conductance changes in concentrations producing GABA potentiation [Choi et al., 1981a,b], some authors have described effects on spontaneous and evoked membrane events unrelated to GABA receptor–mediated effects [MacDonald and Barker, 1979b, 1982; Carlen et al., 1983]. Most of the latter findings were obtained with flurazepam; multiple evidence suggests that this molecule, which is essentially a prodrug, has properties not shared by other benzodiazepines. There is a vast literature on the effects of barbiturates on nonsynaptic neuronal (and nonneuronal) membrane phenomena [Nicoll, 1980].

## Type of GABA Potentiation

The dose–response curve for the conductance increasing effect of GABA in cultured spinal cord cells was found to be shifted to the left by the *benzodiazepines* chlordiazepoxide and diazepam without a change in its maximum [Choi et al., 1981a,b; Farb et al., 1984; Bormann and Sakmann, 1984; Skerritt and Macdonald, 1984a]. Similar findings were reported with flurazepam, des-chloro-diazepam [Nistri and Constanti, 1978] and midazolam [Nistri and Berti, 1984a,b] for the potentiation of the depolarizing effect of GABA on primary afferent endings in the frog hemisected spinal cord preparation, and with chlordiazepoxide and flurazepam in the rat isolated superior cervical ganglion [Little, 1984]. The potentiating action of midazolam in the frog spinal cord was reversed by the benzodiazepine receptor antagonist Ro 14-7437 [Nistri and Berti, 1983]. Benzodiazepine receptor agonists can, therefore, be said to increase the apparent potency of GABA, but not to increase its apparent efficacy. Although this statement is very simple, it is illustrated schematically in Fig. 2, because inspection of this benzodiazepine receptor agonist–induced shift of the GABA dose–response curve tells us a great deal about the quantitative aspects of GABA potentiation by these drugs at the innumerable GABAergic synapses in the intact CNS. If the potentiating action of benzodiazepine receptor agonists is tested using a single dose of GABA (as is often done), the increase of the GABA effect will depend critically on the selected dose of GABA. Per-

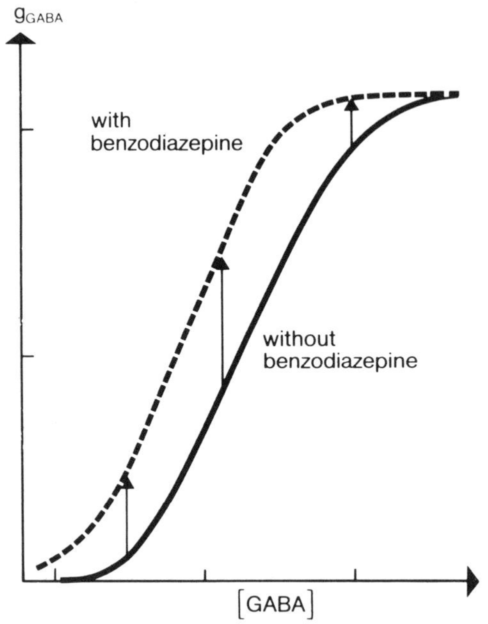


**Fig. 2.** *Schematic diagram of the effect of BZR agonists on the dose-response curve of GABA. The magnitude of the (chloride) conductance induced by GABA* (ordinate) *is plotted against concentrations of GABA* (abscissa). *In the presence of a BZR agonist the GABA dose-response curve is shifted in parallel to the left; the maximum conductance increase of GABA is not altered.*

centagewise, this increase will be several hundred-fold, if the selected dose produces less than about 20% of the GABA maximum. The higher the doses of GABA, the smaller the increase of its effect by benzodiazepine receptor agonists will be in percent of the preagonist effect of GABA, and the augmentation will virtually disappear at nearly maximal doses of GABA. There will be no increase of the effect of GABA at a dose that saturates the available benzodiazepine receptors. The impact of this simple pharmacological principle on the determination of the intrinsic efficacy of benzodiazepine receptor ligands, as well as the functional consequences of these drugs on the efficiency of GABAergic synaptic transmission, will become clear in later discussions.

The effect of *barbiturates* is more complex. In isolated cat spinal ganglion cells, phenobarbital at $10^{-4}$ M was found to produce a parallel 3-fold shift to the left of the dose-depolarizaton curve for GABA and, in contrast to benzodiazepines, to increase the GABA maximum [Gallagher et al., 1981]. A similar increase in the GABA maximum was found with pentobarbital [Higashi and Nishi, 1982]. In rat cuneate nucleus slices, the curve for the depolarizing action of muscimol on dorsal column fiber endings was shifted to the left by a factor of 6.8 by $10^{-4}$ M pentobarbital, which seemed to be about 13 times more potent than phenobarbital [Simmonds, 1981]. Higher concentrations of barbiturates produce very complex changes in the GABA dose–response curve because of their bicuculline- and picrotoxin-sensitive direct activation of GABA receptors or chloride channels at concentrations ≧ 1 mM [Nicoll, 1975b; Barker and Ransom, 1978; Macdonald and Barker, 1978; Gallagher et al., 1981; Simmonds, 1981; Higashi and Nishi, 1982; Nicoll and Madison, 1982]. Hence, barbiturates differ from benzodiazepine receptor agonists quite markedly in being able to increase the effect of a receptor-saturating dose of GABA. In addition, barbiturates may increase $K^+$ conductance [Nicoll and Madison, 1982]. The various anesthetic and anticonvulsant barbiturates seem to differ markedly both qualitatively and quantitatively in their effects on GABA; however, no precise structure–activity relation has been elaborated. The available data are not sufficient for a correlative study of binding to "chloride channel binding sites" and modification of GABA effects.

## Dose-Dependency of the GABA-Potentiating Activity

Dose-response curves for the GABA-potentiating action of benzodiazepine receptor agonists have been determined on spinal cord neurons in primary cell culture. Choi et al. [1981a,b] found chlordiazepoxide to produce a concentration-dependent shift to the left of the GABA dose–response curves in concentrations between $10^{-6}$ M and $10^{-4}$ M, reflecting the weak affinity of this benzodiazepine for its receptors. Midazolam was extremely potent in potentiating the depolarizing effect of GABA on primary afferents within the frog spinal cord, a half-maximum shift being obtained at $10^{-9}$ M and the maximum shift being seen at 2–2.5 × $10^{-8}$ M; flurazepam and des-chloro-diazepam were much less potent than midazolam [Nistri and Constanti, 1978; Nistri and Berti, 1984a,b]. It should be noted that the experiments with midazolam were performed at 4°C, which may account for its very high potency.

The relationship between the potency for enhancing the effect of GABA and the binding affinity for specific benzodiazepine receptors was studied by Farb et al. [1984] on spinal neurons from chicken embryos in monolayer cell culture. Flunitrazepam and clonazepam were markedly more potent than diazepam and chlordiazepoxide. The presently available data seem to support the view that, overall, the binding affinity and the GABA-potentiating potency in vitro are well correlated. However, the agreement was not as good in absolute terms: the $ED_{50}$ values (for GABA potentiation) were nearly 40 times higher than the $K_i$ values (for inhibition of $^3$H-flunitrazepam binding). A major part of this discrepancy may be due to the fact that the binding experiments were carried out on isolated membranes at 4°C, whereas the physiological studies were performed at 37°C on intact neurons. $K_D$ values for benzodiazepine receptor ligands are known to be 5–10 times higher at 37°C than in the cold [Speth et al., 1979]. The remaining discrepancy might become smaller if binding were determined on intact cells instead of membrane preparations, suggesting a fairly good absolute correlation between receptor affinity and physiological potency. The extremely high potency of midalozam as a GABA potentiator in the frog spinal cord raises another question, namely, that of the temperature dependency of the benzodiazepine receptor–GABA receptor function. Using rat brain membranes, thermodynamic considerations suggested that the initial binding of benzodiazepine receptor agonists and antagonists was similar at low temperature, but that the receptor isomerization underlying receptor activation by agonists may occur only at higher temperatures [Möhler and Richards, 1981]. It will be of interest to compare benzodiazepine receptor–GABA receptor function of homeothermic and poikilothermic animals at low temperature.

The GABA potentiating effect of some benzodiazepine receptor agonists was found to decrease and to be even reversed at high concentrations. With midazolam in the frog spinal cord, this was observed at concentrations above $10^{-7}$ M [Nistri and Berti, 1984a,b], while flurazepam depressed the effect of GABA at $10^{-4}$–$10^{-3}$ M [Choi et al., 1981a]. This depression of GABA effects at higher concentrations of benzodiazepine receptor agonists is not apparent from the results of Farb et al. [1984].

Comparative dose–response studies of the GABA potentiating action of *barbiturates* are not available.

### Efficacy of Benzodiazepine Receptor Ligands

The intrinsic efficacy [Furchgott, 1966; Ruffolo, 1982; Kenakin, 1984] of benzodiazepine receptor ligands as modulators of the GABA receptor function can be assessed in a precise way only by determining the changes that occur in the GABA dose–response curves in the presence of varying concentrations of ligands.

**Agents with positive intrinsic efficacy (agonists and partial agonists).** The first experiments of this kind were performed by Choi et al. [1981a,b] on spinal cord cells of chicken embryos in primary cell culture. Chlordiazepoxide was shown to produce a concentration-dependent left-hand shift of the GABA dose–response curve. The maximum shift, obtained at $10^{-4}$ M, was about 2-fold. A similar value was obtained in the frog spinal cord for the much more potent midazolam [Nistri and Berti, 1984a,b]. The very laborious procedure for obtaining such data seems to be the main reason for the scarcity of similar studies. Farb et al. [1984] and Chan and Farb [1985] have used a simpler way to estimate the efficacy of various benzodiazepine receptor ligands on spinal neurons in culture. They applied a concentration of GABA producing approximately 10% of the maximum membrane conductance increase that can be obtained with GABA. Increasing concentrations of ligands were then added, and the effectiveness of the test dose of GABA was measured. When the percent increase of the GABA effect (GABA-induced conductance in the presence of a drug versus GABA-induced conductance in its absence) was plotted against the concentration of the benzodiazepine receptor ligands, flunitrazepam and diazepam were found to have approximately the same efficacy, while clonazepam, flurazepam, and chlordiazepoxide had only one-half or one-third the efficacy of flunitrazepam. The benzodiazepine receptor blocker Ro 15-1788, as well as the triazolopyridazine CL 218 872, had only about 15% of the efficacy of flunitrazepam. The low efficacy found with clonazepam is rather surprising but has also been reported by Skerritt and Macdonald [1984a] on mouse spinal cord neurons. Whether the experiments of Farb et al. [1984] and Chan and Farb [1985] reflect peculiarities of the benzodiazepine receptors in chicken neurons or whether the method used is responsible for these unexpected efficacy values (only one compound was tested on a given neuron), is a very important issue for our future understanding of pharmacological profiles of benzodiazepine receptor agonists and partial agonists on the basis of intrinsic efficacies. The results of these authors would indicate that flunitrazepam and diazepam are full agonists, and that clonazepam, flurazepam, and chlordiazepoxide are partial agonists with medium efficacy, whereas CL 218, 872 and Ro 15-1788 would be partial agonists with very low intrinsic efficacy.

When flunitrazepam was covalently bound to benzodiazepine receptors by exposure to ultraviolet light, it was found that the resting chloride conductance was unaltered, but the GABA-potentiating effect of chlordiazepoxide was reduced [Farb et al., 1984]. This would suggest that photoaffinity labeling of the benzodiazepine receptor with flunitrazepam does not fix the receptor in its activated form, but blocks the positive modulatory function of agonists, leaving the negative modulatory function of inverse agonists unaffected.

**Agents with negative intrinsic efficacy.** Farb et al. [1984] and Chan and Farb [1985] used the same technique to study the effects of the two $\beta$-carboline-3-carboxylate derivatives, $\beta$-CCM and DMCM, on the GABA-induced increase of membrane conductance. Both compounds markedly reduced, but did not completely abolish, the effect of GABA, thus showing negative efficacy [Braestrup et al., 1983; Jensen and Lambert, 1983] and confirming earlier suggestions that they are inverse agonists (DMCM) and partial inverse agonists ($\beta$-CCM). Skerritt and Macdonald [1984b] found a GABA depressant effect of DMCM, but not of $\beta$-CCE and $\beta$-CCM; all $\beta$-carbolines blocked the GABA-facilitating effect of diazepam.

A comparison of the efficacy of barbiturates as GABA potentiators has not yet been published.

## Molecular Mechanisms of the GABA-Potentiating Action of Benzodiazepines and Barbiturates

The effect of *benzodiazepines* on the GABA dose–response curve is, in general pharmacological terminology, an increase of the apparent potency of GABA. The lack of an increase of the maximum effect of the dose–response curve strongly suggests that benzodiazepines do not increase the number of functional GABA receptors. An increase of the driving force for the GABA-induced chloride flux can be excluded because benzodiazepines have been found not to alter the resting membrane potential or the reversal potential for the GABA-induced potential change [Choi et al., 1981a; Segal and Barker, 1984]. The remaining theoretical alternative mechanisms are an increase of the conductivity and a change in the ion selectivity of single channels, an increase of the affinity of the GABA receptor(s) for GABA, a facilitation of the isomerization of the ligand-associated GABA receptor into the activated receptor conformation, or a facilitated conformational transition of the closed chloride channel into the open form in response to GABA receptor activation (facilitated coupling of the GABA receptor with the chloride channel, facilitated gating mechanism). These possible mechanisms were investigated by spectral analysis (fluctuation or noise analysis) of the GABA-induced membrane current noise (the random miniature jumps in the current flow representing single channel-gating events) in voltage-clamped single cells of mouse spinal cord or rat hippocampus in dissociated cell culture [Study and Barker, 1981; Segal and Barker, 1984; Barker and Owen, this volume]. These studies showed that diazepam increases the probability of single-channel opening events with no change in the single-channel conductivity and little, if any, increase in the mean single-channel open time. In contrast to benzodiazepine receptor agonists, but in a similar manner as bicuculline and picrotoxin, the two $\beta$-carboline inverse agonists, $\beta$-CCM and DMCM, reduced the frequency of single channel opening events without affecting the average channel lifetime [Barker et al., 1984]. $\beta$-CCM and DMCM were observed to reduce ($EC_{50}$ 18 $\mu$M and 10 $\mu$M, respectively) the conductance increase of GABA without direct effects on the neuronal membrane; maximal inhibition of the effect of the GABA test dose was 53% and 70%, respectively. The GABA depression was not surmountable by increasing the concentration of GABA; however, clonazepam reversed the depressant effect of DMCM [Farb et al., 1984]. These authors also found that DMCM inhibited to the same extent the chloride conductance induced by a low dose of GABA in the absence of picrotoxin and that induced by a high dose of GABA in the presence of picrotoxin, indicating that DMCM and picrotoxin act on different sites.

The most direct approach to the study of the molecular mechanism of action of benzodiazepines has been the recording of single GABA-gated chloride channels in cultured spinal cord neurons using the patch-clamp technique. A preliminary report by Bormann and Sakmann [1984] indicates that the only effect of diazepam is to increase the frequency of channel openings in response to GABA: the single channel response to GABA in the presence of benzodiazepines was found to be identical with a response to a higher concentration of GABA in the absence of the drug. These findings are compatible with either an increase in the GABA receptor affinity, a facilitated GABA receptor activation, or a facilitated gating function of the receptor–ionophore complex, or any combination of these three single individual mechanisms. Whereas the two latter mechanisms cannot yet be studied, radioligand binding studies suggest that benzodiazepines are able to increase the affinity of the GABA receptor [Costa, 1983; Johnston and Skerritt, 1983].

Put in the simplest term, benzodiazepines seem to increase the gain of the GABA receptor–ionophore function. In the presence of a benzodiazepine, less GABA is required to induce a given chloride conductance than in its absence. This effect observed on single channels as well as on a whole single cell is fundamental to an understanding of the functional changes that a benzodiazepine receptor agonist will produce in the various GABAergic synapses in the intact CNS.

Fluctuation analysis has shown that *barbiturates* have a clearly different molecular mechanism of action than benzodiazepine receptor agonists [Study and Barker, 1981]. Indeed, (−)−pentobarbital was found to prolong the mean open time of GABA receptor–gated chloride channels in cultured mouse spinal cord neurons and to reduce the frequency of single channel openings. No change in the single channel conductance was observed. The GABA-independent increase of chloride conductance by higher concentrations of (−)−pentobarbital was found to be due to the activation of channels that had a similar conductance as those activated by

GABA, but whose average lifetime was five times longer than that of channels activated by GABA [Mathers and Barker, 1980; Barker and Mathers, 1981].

## FROM "SINGLE-CHANNEL BEHAVIOR" TO BEHAVIOR

### Benzodiazepine Receptor Agonists

The basic effect of benzodiazepine receptor agonists, as studied on single GABA-gated anion channels, is to increase the probability of channel-opening events on exposure to GABA. On the level of a single channel, the response to a given concentration of GABA in the presence of a benzodiazepine receptor agonist is identical with the response to a higher GABA concentration in the absence of the drug. This nicely explains the parallel left-hand shift by benzodiazepine receptor agonists of the GABA dose–response curve obtained on single cells or on cell populations. What does this basically simple mechanism teach us about the complex effects of benzodiazepine receptor ligands in the intact CNS?

Assuming that this basic mechanism is operating on all neurons throughout the CNS at all subsynaptic membranes facing a GABAergic nerve terminal, a first important prediction can be made: *The magnitude of the facilitatory effect of a benzodiazepine receptor agonist at any single GABAergic synapse will depend critically on the actual concentration of GABA in the synaptic cleft.* Due to the parallel left-hand shift of the GABA dose–response curve induced by benzodiazepine receptor agonists, these agents will produce a dramatic (percent) increase in the effect of a small concentration of GABA; this percent increase will rapidly become smaller as the GABA concentration increases towards the maximally effective concentration, when the drugs will be apparently ineffective. In functional terms, i.e., with respect to inhibition of excitatory membrane events, the potentiation of the GABA effect by benzodiazepine receptor agonists would be most significant at GABA concentrations close to its $IC_{50}$. Indeed, a several-hundred-fold percent increase by benzodiazepine receptor agonists of the chloride conductance induced by a threshold concentration of GABA would result in only a very small increase in synaptic inhibition, and near-maximal inhibition produced by high GABA concentrations will be further increased only immaterially by the drugs. Unfortunately, no information exists for any GABAergic synapse to indicate which degree of saturation of subsynaptic GABA receptors is achieved by the intrasynaptic concentration of GABA that occurs after the single impulse in the GABAergic input cell.

Hence, because benzodiazepine receptor agonists increase the apparent potency of GABA, *the functional effect of these drugs will not be quantitatively the same in all synapses of the CNS* (even if we assume a uniform drug concentration throughout), but will depend on the concentration of GABA in the individual synapses. This synaptic concentration fluctuates as a function of the activity of GABAergic input neurons. As indicated in Fig. 3, the GABA concentration is likely to be similar within all synapses formed by the input neuron $GABA_1$ on target neurons A and B, and, therefore, the functional consequence of a benzodiazepine receptor agonist is presumably similar in these synapses. At a given moment the GABA concentration in synapses formed by input neuron $GABA_2$ may be quite different from that in synapses formed by neuron $GABA_1$, and the functional consequence of a benzodiazepine receptor agonist will, accordingly, also be different. We should not expect, therefore, that a given dose of a benzodiazepine receptor agonist, which is, in principle, capable of producing a similar shift of the GABA dose–response curve at all GABAergic synapses, actually potentiates ongoing GABAergic transmission to the same extent at all these synapses in the intact CNS.

An evenly distributed potentiation by benzodiazepine receptor agonists of ongoing GABAergic transmission in all synapses throughout the CNS and, hence the induction of an inhibitory state of similar intensity in all neu-

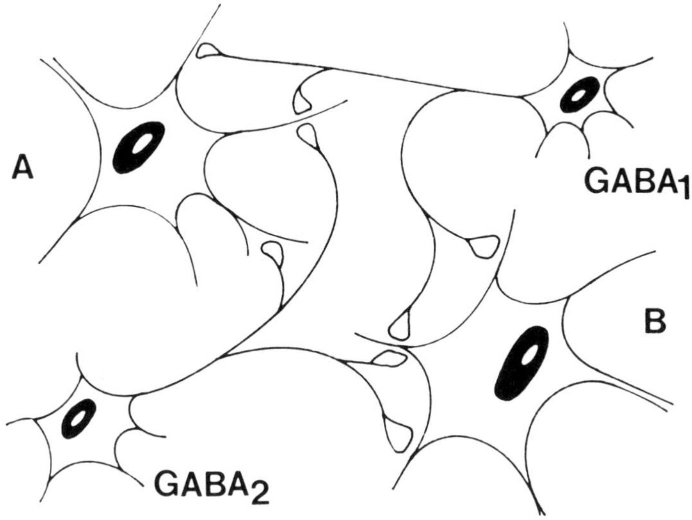


**Fig. 3.** *Schematic diagram to illustrate the role of the GABAergic input activity to a neuron in the greatly varying potentiating effect of benzodiazepines at different GABAergic synapses. The GABAergically innervated neurons A and B receive their (convergent) GABAergic input from neurons $GABA_1$ and $GABA_2$. The concentration of GABA in the synapses formed by the neuron $GABA_1$ are probably similar whether these synapses are on target neuron A or neuron B. The potentiating effect of a BZ agonist will, therefore, also be similar at these synapses. The concentration of GABA is likely to be similar in all synapses formed by $GABA_2$, but may be quite different from that in the synapses formed by $GABA_1$. Accordingly, the potentiating effect of a benzodiazepine may be qute different in synapses formed by neurons $GABA_1$ and $GABA_2$.*

rons of the CNS, is unlikely for an additional reason. Although all central neurons examined so far were found to contain GABA receptors, the absolute and relative (to other transmitter synapses) numbers (density) of GABAergic synapses on the various neurons probably differ markedly. Accordingly, *the inhibitory effectiveness of the total GABAergic input onto various neurons will vary from very slight to extremely marked.*

The possible existence of a functional *receptor reserve for the GABA–benzodiazepine receptor* is a further complicating factor when extrapolating single-receptor-channel data to the intact CNS. Some neurons may contain such a high density of GABA receptors and such a high number of GABAergic inputs that occupation of only a fraction of all receptors ("effective fraction") will be required for a maximum increase of the membrane permeability to chloride. That fraction of GABA receptors that does not have to be occupied for this maximum effect is called "spare receptor" fraction; even when these spare receptors are activated by GABA, the channels they open will not further increase chloride conductance over that produced by the "effective fraction." The receptor reserve of a cell may be variable rather than being a constant property: indeed, considering as an example the hyperpolarizing action of GABA, less chloride channels would be required to be activated in order to shift the membrane potential towards the chloride equilibrium potential in a resting or only slightly excited neuron, but a higher chloride conductance increase (a greater number of activated channels) would be necessary to obtain the same hyperpolarizing endpoint in a neuron which is under a strong excitatory bombardment.

## Partial Agonists at Benzodiazepine Receptors

The phenomenon of receptor reserve is likely to have a great impact on the activity profile of partial agonists at benzodiazepine receptor, in a similar way as it is known to affect the organ specificity and, hence, activity profile of other drugs (e.g., agents with activity on adrenoreceptors) [Ruffolo, 1982]. Full agonists are ligands whose binding to the receptor results always or in a great proportion of interactions in the activation of the receptor. Full agonists have a high ratio of "activated receptor–ligand complex" number to "inactive receptor-ligand complex" number. Partial agonists form ligand–receptor complexes that often fail to activate the receptor; they need a higher "fractional receptor occupancy" for a given intensity of effect than full agonists. Antagonists virtually always fail to activate receptors and only block them; the higher their "fractional receptor occupancy," the more marked is their antagonistic effect. Partial agonists have been described recently for benzodiazepines and nonbenzodiazepines [Braestrup et al., 1984; Haefely, 1984]. They behave as "mixed agonists–antagonists," mimicking agonists for some effects (e.g., anxiolytic activity) and antagonizing other effects of agonists (e.g., sedation and muscle relaxation). Such profiles have been tentatively explained by postulating multiple subtypes of benzodiazepine receptors. Partial agonism (submaximal intrinsic efficacy) and differences in the receptor reserves among various central neurons would seem to provide a straightforward alternative explanation for the mixed agonist–antagonist property without the need for postulating heterogeneity of benzodiazepine receptor species. As shown schematically in Fig. 4, three neurons with differing GABA–benzodiazepine receptor reserves may be assumed to exist. Drugs A, B, C, and D represent four ligands of benzodiazepine receptor with full and partial agonistic properties. The full agonist A needs to occupy only a small fraction (about 10%) of all receptors present on a neuron with 90% receptor reserve in order to produce the maximum shift of the GABA dose–response curve which can be achieved through benzodiazepine receptor activation. The same agonist still produces a full response in a neuron with no receptor reserve, although in this case the full effect requires 100% fractional receptor occupancy. Agonist A will produce the maximum shift of the GABA dose–response curve in all three types of neurons, albeit at a different fractional receptor occupancy and, hence, at different doses. Drugs B and C are partial agonists (have submaximal intrinsic efficacy); they produce a full potentiation of GABA effects in neurons with 90% receptor reserve. However, in neurons with less or no receptor reserve they produce only a small shift of the GABA dose–response curve. Ligand D, which has such a low intrinsic efficacy as to produce only a half-maximum potentiation of GABA in neurons with high receptor reserve, behaves as an almost pure antagonist at neurons with low or no receptor reserve. If we assume, in gross oversimplification, that those neurons, which are involved in creating the emotion of anxiety, its associated inhibition of behavior, and its increase of vigilance, have a high benzodiazepine–GABA receptor reserve, while neurons mediating sedation and muscle relaxation have only a low or no receptor reserve at all, the profile of partial benzodiazepine receptor agonists (full anxiolytic effectiveness, reduced sedative and muscle-relaxant/ataxic effectiveness) can be understood easily.

## Inverse Agonists at Benzodiazepine Receptors

With inverse agonists, which depress the effect of GABA by reducing the probability of channel openings in response to GABA [Barker et al., 1984], the most likely situation is opposite to that induced by benzodiazepine receptor agonists: With increasing doses their depression of GABA receptor function can be expected to occur first at neurons with no or low receptor reserve, and only at higher doses will they depress GABAergic transmission in neurons with high receptor reserve. There is evidence that the vigilance state is increased by doses of inverse agonists which are lower than those producing anxiety and epileptiform

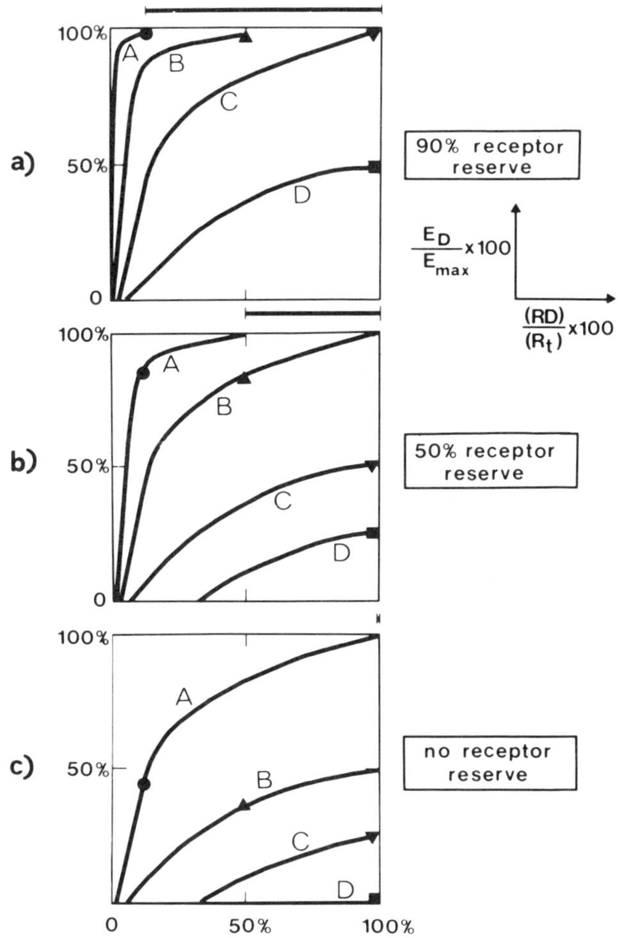


**Fig. 4** *Schematic diagram to explain the "mixed agonistic–antagonistic" property of partial BZR agonists based on the existence of GABA sensitive target cells with differing GABA-benzodiazepine receptor reserves. The three cells a,b, and c have a receptor reserve of 90%, 50%, or 0%, respectively. The ligands A, B, C, and D, differ from each other by their intrinsic efficacy. Fractional receptor occupancy on the abscissa is plotted against intensity of effect on the ordinate (in percent of the maximum that can be generated by the receptor). The horizontal bars on top of the three graphs indicate the fraction of spare receptors. $E_D$ = maximal effect produced by a drug ligand; $E_{max}$ = maximal effect that can be generated by the receptor; (RD) = number of receptor–ligand complexes; ($R_t$) = total number of receptors.*

activity [Ongini, 1983]. It appears that of the presently available inverse agonists, DMCM has the highest negative intrinsic efficacy; however, a complete suppression of GABA receptor function does not seem to be achieved even with this agent.

## Anticonvulsant and Anesthetic Barbiturates

In contrast to benzodiazepine receptor agonists, anticonvulsant and anesthetic barbiturates produce a shift to the left of the GABA dose–response curve as well as an increase of its maximum [Gallagher, 1981; Higashi and Nishi, 1982]. On the molecular level, these barbiturates have been shown to increase the mean open time of the GABA-activated chloride channels [Study and Barker, 1981]. In addition, barbiturates at higher concentrations are capable of opening chloride channels in the absence of GABA, most probably by a

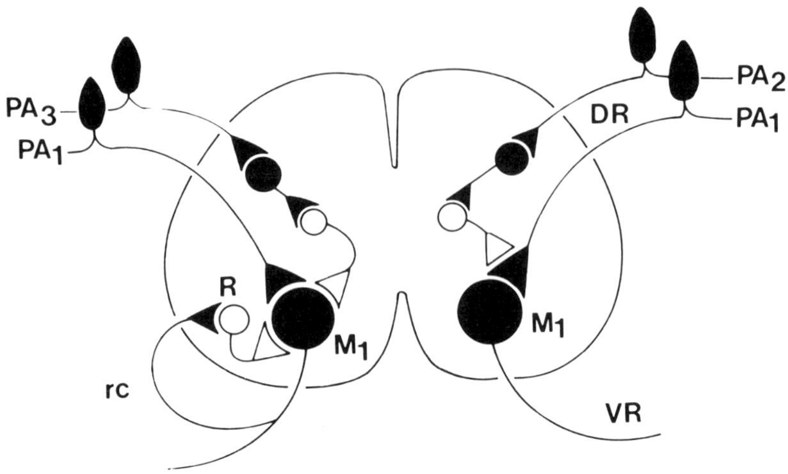


**Fig. 5.** *Schematic diagram of three inhibitory pathways depressing the monosynaptic excitation of motoneurons in the spinal cord. Excitatory neurons are depicted in black, inhibitory neurons in white. M1 are two motoneurons innervating a muscle whose muscle spindles activate the primary afferent fibers PA1 (M1 and PA1 are the efferent and afferent neurons forming the reflex arch for the stretch reflex). Primary afferent fiber PA3 (from a muscle group antagonistic at the joint to the muscle innervated by M1 motoneurons), in addition to exciting monosynaptically a M3 motoneuron (not shown), innervates disynaptically an inhibitory interneuron which impinges on M1 (feedforward, antagonistic, postsynaptic inhibition). R is an inhibitory interneuron (Renshaw cell) activated by a recurrent collateral (rc) of the motoneuron axon (feedback inhibition, Renshaw inhibition). PA2 is a primary afferent from a heteronymous muscle (acting on another joint than M1) or a cutaneous nerve; it activates disynaptically a GABAergic interneuron which forms an axo-axonal synapse with PA1 endings (feedforward presynaptic inhibition). DR = dorsal root, VR = ventral root.*

direct effect on the ionophore, since they can be blocked by picrotoxin [Nicoll, 1975b; Barker and Ransom, 1978; Macdonald and Barker, 1979b; Higashi and Nishi, 1982; Nicoll and Madison, 1982]. Therefore, while benzodiazepine receptor agonists cannot induce a more intense degree of the inhibitory state in a given neuron than can be reached by maximal activation of the total GABAergic input in the absence of a drug, barbiturates in higher concentrations will induce a much higher inhibitory state than can be achieved without drug. It is easily understandable that many neurons will become completely unresponsive to excitatory synaptic inputs and hence completely silent in the presence of barbiturate concentrations that maximally enhance the effect of GABA and, in addition, directly induce chloride channel opening. This more severe state of neuronal depression that can be achieved by barbiturates may explain the complete surgical anesthesia that is induced by barbiturates, but not by benzodiazepine receptor agonists, and the relatively marked effect of barbiturates on electroshock-induced seizures in the absence of such an effect with several, otherwise very potent, benzodiazepine receptor agonists.

## MODIFICATION BY BENZODIAZEPINE RECEPTOR LIGANDS AND BARBITURATES OF ENDOGENOUS AND EXOGENOUS GABA EFFECTS IN VARIOUS REGIONS OF THE CNS

GABAergic synapses are present in all major areas of the central gray matter, albeit at a greatly varying regional density. The distribution of specific, high-affinity, central-type binding sites for benzodiazepines roughly corresponds to that of GABA receptors

[Schoch et al., 1985]. It is of interest to briefly review those neurons and neuronal pathways in the various CNS regions, where benzodiazepine receptor ligands and barbiturates have been studied in vivo or in vitro for their effect on endogenous and/or exogenous GABA. The basic aspects of the interaction of these agents with the function of the GABA receptor–ionophore complex have already been elaborated, and it is against this background that the findings in the various CNS regions will be discussed. The general physiology of GABA-mediated synaptic inhibition has been treated in previous chapters and is summarized in the following three paragraphs.

## Spinal Cord

**GABAergic pre- and postsynaptic inhibition.** Because of the relative accessibility of neurons and the advanced knowledge of many neuronal pathways within the spinal cord, this region has been used most extensively in electrophysiological studies of synaptic excitation and inhibition [Eccles, 1964]. The *concept of two types of synaptic inhibition, postsynaptic and presynaptic inhibition,* has been elaborated on the basis of findings obtained on spinal motoneurons (Fig. 5). Motoneurons (M) can be excited monosynaptically by primary afferent neurons ($PA_1$) which have their cell bodies in spinal ganglia, whose peripheral axon branches carry impulses produced in stretch receptors in the muscle spindles of the same muscle group, and whose central axon branches enter the spinal cord through dorsal roots and form synapses on the perikaryon and dendrites of motoneurons (so-called stretch reflex). Monosynaptic excitation of motoneurons by these muscle afferents can be depressed (inhibited) by the activation of various pathways, three of which are depicted in Fig. 5. First, stimulation of primary afferent fibers from muscle groups ($PA_3$) antagonistic to the group innervated by $M_1$motoneurons (e.g., extensor versus flexor muscles) produces hyperpolarization of $M_1$ motoneurons. Second, each action potential fired by a motoneuron is followed by hyperpolarization, produced by an inhibitory interneuron called Renshaw cell (R), which is excited by recurrent collaterals (rc) from the motoneuron axon. These two types of synaptic inhibition are called postsynaptic inhibition because excitatory transmission through the primary afferent–motoneuron synapse is depressed by an action of the inhibitory transmitter on the membrane of the motoneuron and, hence, on the postsynaptic side of the excitatory synapse. Antagonistic inhibition is an example of *feedforward inhibition*; recurrent Renshaw inhibition is an example of *feedback inhibition*. Thirdly, monosynaptic excitation of motoneurons is also inhibited by stimulation of primary afferents ($PA_2$) from heteronymous muscles (acting on another joint) or by activation of cutaneous afferents. With this type of inhibition, no change in the membrane properties of the motoneuron can be detected. The reason for this is that the GABAergic inhibitory interneuron involved forms axo-axonal synapses with axon terminals of the $PA_1$ afferents and depresses monosynaptic excitation by depressing the release of the excitatory transmitter from the $PA_1$ ending. The action of the inhibitory transmitter is on the presynaptic side of the excitatory synapse and, therefore, this type is called presynaptic inhibition. Of course, GABA released from the inhibitory interneuron acts on the membrane of the primary afferent ending, which is postsynaptic with respect to this GABAergic axo-axonal synapse. Presynaptic inhibition does not reduce the response of the motoneurons to excitatory input neurons other than those which are inhibited by the GABAergic interneuron. Presynaptic inhibition allows for the selective depression of distinct excitatory inputs to the motoneurons, whereas postsynaptic inhibition reduces its response to all types of excitatory inputs.

The GABA receptors and associated anion channels, through which GABA mediates pre- and postsynaptic inhibition, appear to be very similar, if not identical. GABA receptor activation results in a conductance increase of the membrane, which is mainly due to an increased permeability for chloride anions. While this conductance increase occurs in all GABA-sensitive neurons, the resulting change in membrane potential is a hyperpolarization in most cells but a depolarization in particular types of neurons, e.g., the primary sensory neurons and sympathetic ganglion cells. The reason for these differences is primarily the existence of different electrochemical chloride gradients; while the intracellular chloride concentration in most neurons is considerably lower than the extracellular one, an inwardly directed chloride pump mechanism results in a smaller concentration gradient in some types of neurons. Neurons, which under physiological conditions respond with a membrane hyperpolarization to GABA, can change their response into depolarization when the intracellular chloride concentration is experimentally increased by leakage out of chloride from an intracellular pipette containing this anion.

Both the increased membrane conductance (decreased membrane resistance) and the membrane potential change (hyperpolarization, depolarization) probably contribute to the inhibitory effect of GABA [Nicoll and Alger, 1979; Levy, 1980]. The increased chloride conductance shunts the ion fluxes (primarily sodium) that underly excitatory postsynaptic potentials and the upstroke of action potentials; excitatory postsynaptic potentials may therefore not reach threshold depolarization for the generation of an action potential, and the electrotonic propagation of local depolarization is impeded. In primary afferent endings GABA produces a depolarization (primary afferent depolarization); this and the underlying chloride conductance increase reduce the amplitude of action potentials and may inhibit their active propagation into the fine terminal arborization. Reduction of action potential amplitude, partial failure of action potential invasion of terminal arborization ("branch point failure") and, perhaps, GABA effects unrelated to chloride conductance, may contribute jointly in depressing transmitter release from primary afferent endings.

**Modification by benzodiazepines and barbiturates of presynaptic inhibition and of GABA effects on primary sensory neurons.** Presynaptic inhibition can be resolved in three processes accessible to electrophysiological investigation.

1. Inhibition of the monosynaptic motoneuron reflex may be assessed by recording the motoneuron monosynaptic reflex response (recording of single-cell or of mass action potential amplitude in ventral rootlets) in the absence of and during activation of the presynaptic inhibitory pathway; typical "inhibition curves" are thereby obtained by plotting the size of the monosynaptic response in percent of the noninhibited response against time after activation of the presynaptic inhibitory pathway; recovery of transmission takes from 200–500 ms.

2. Primary afferent depolarization is usually recorded as a negative deflection in a dorsal rootlet, called dorsal root potential; this dorsal root potential is the mass depolarization generated in primary afferent terminals, which propagates electrotonically (passively and decrementally) over the primary afferent axons from the intraspinal site of generation antidromically towards the cell body in spinal ganglia. The time course of the dorsal root potential roughly reflects that of the presynaptic inhibition of the monosynaptic excitation of motoneurons.

3. Depolarization of intraspinal primary afferent axons should decrease their electric stimulation threshold, i.e., increase their electrical excitability. To test this, a fine electrode is positioned in the spinal cord at sites where primary afferent endings are concentrated and constant electric shocks are applied. They induce antidromic action potentials in a number of primary afferent axons terminating in the vicinity of the electrode tip. During primary afferent depolarization the amplitude of the antidromic mass action potential recorded in a dorsal rootlet increases because more primary afferents are now being excited by the constant test stimulus. Since the very endings of

primary afferents are not easily accessible to direct recording, in particular intracellular recording, spinal ganglion cells have been used, assuming that the membrane of endings and of the cell body do not differ essentially in their responsiveness to GABA.

The early findings of Schmidt et al. [1967] with diazepam have been repeated, confirmed, and extended by many investigators with an increasing number of *benzodiazepine receptor agonists* both of the benzodiazepine family [see Haefely et al., 1981; Leah et al., 1983; Polc, unpublished results] and of other chemical classes, e.g., cyclopyrrolones [Boireau et al., 1984] and triazolopyridazines [Polc, 1981]. They dose-dependently increase the amplitude and/or the duration of dorsal root potentials and accentuate and prolong the presynaptic inhibition of monosynaptic reflex responses. The *benzodiazepine receptor blockers* Ro 15-1788 and Ro 15-3505 reverse the effect of benzodiazepine receptor agonists on dorsal root potentials, but not the effect of barbiturates [Polc et al., 1981; Polc, unpublished]. The benzodiazepine receptor agonists have only a moderate effect on the time constant of decay at dorsal root potential except at high doses [Schaffner, unpublished], when they may slow down transmission in the polysynaptic pathway mediating presynaptic inhibition. Although these methods allow for a rather accurate determination of a threshold dose, they detect only large differences in the intrinsic efficacy of various agents, as is the case with certain partial agonists [Polc, unpublished]. Benzodiazepine receptor agonists also enhance presynaptic inhibition of motoneurons induced by pathways descending from the pontine and medullary reticular formation [Polc et al., 1974; Polzin and Barnes, 1976]. These drugs differ from GABA mimetics, such as GABA and muscimol, in not producing primary afferent depolarization directly in the absence of activated presynaptic inhibitory pathways [Polc, 1978]. In the cat spinal cord, potentiation of presynaptic inhibition was virtually completely abolished by thiosemicarbazide; diazepam did not overcome this effect of the glutamic acid decarboxylase inhibitor [Polc et al., 1974]. Leah et al. [1983] could demonstrate an increase of primary afferent depolarization of single dorsal root fibers by systematically administered midazolam. Microionophoretically applied midazolam in the areas of presumed primary afferent endings enhanced synaptically induced primary afferent depolarization as well as depolarization induced by local application of GABA and piperidine-4-sulfonate. The effect of midazolam was readily reversed by Ro 15-1788, which by itself did not affect primary afferent depolarization. When dissociated chick spinal cord neurons and dorsal root ganglion cells were cocultured, spontaneous depolarizing potentials could sometimes be recorded in dorsal root ganglion cells, which seemed to be generated in axo-axonal synapses on their terminals [Choi et al., 1981a,b]; these presumed spontaneous primary afferent depolarizations were resistant to strychnine, abolished by bicuculine, but enhanced by chlordiazepoxide.

The two β-carboline derivatives with *inverse agonistic* activity at benzodiazepine receptors, β-CCM, and β-CCE, depressed dorsal root potentials in the cat spinal cord; their effect was reversed by Ro 15-1788 [Polc et al., 1982]. Similar findings were obtained with DMCM [Polc, unpublished].

The depolarizing effect of GABA on primary afferents in the frog hemisected spinal cord was enhanced by flurazepam, midazolam, and chlordiazepoxide [Nistri and Constanti, 1978; Nistri and Berti, 1983, 1984a,b]. The dose–response curve for the depolarizing effect of GABA was shifted to the left in a dose-dependent manner; the maximum shift was about 2 fold. No GABA-potentiating effect of diazepam was found in rat spinal ganglion cells, which are the cell bodies of primary afferents [Désarménien et al., 1980; Lamour, 1979].

*Barbiturates* were found in various studies [see Haefely and Polc, 1983; Lodge and Anis, 1984] to enhance presynaptic inhibition.The effect is almost exclusively an increase of the duration of dorsal root potentials or of monosynaptic reflex depression. Whereas this prolongation of presynaptic inhibition increases dose-dependently, the amplitude of dorsal root potentials may be increased at low doses but drastically decreased at higher doses. The real effect of barbiturates on presynaptic inhibition may be difficult to assess accurately because additional effects, in particular depression of excitatory transmission to interneurons, interfere in a complex manner.

The depolarizing effect of GABA, but not of glycine, on primary afferents in the rat hemisected spinal cord was augmented by pentobarbital [Evans, 1979]. Higher concentrations than used in this study produced a marked intrinsic depolarization of primary afferents in the frog hemisected spinal cord, which was antagonized by picrotoxin [Nicoll, 1975a]. In rat isolated lumbar dorsal root ganglion cells, pentobarbital in concentrations from $10^{-5}$ to $2 \times 10^{-4}$ M enhanced the GABA-induced depolarization and conductance increase without intrinsic effect on resting membrane potential or conductance; at $10^{-3}$ M the barbiturate depolarized about half of the cells and blunted the GABA-induced conductance increase [Connors, 1981; Headley et al., 1981; Désarménien et al., 1980]. Both effects of pentobarbital were abolished by picrotoxin. On isolated dorsal root ganglion of the cat, phenobarbital, pentobarbital, and thiopental were also found to shift the dose–response curve of GABA for depolarization to the left with an increase of the maximum; at this concentration ($10^{-4}$ M) phenobarbital had no effect on membrane conductance and on the reversal potential of the GABA effect [Gallagher et al., 1981; Higashi and Nishi, 1982]. A direct depolarizing effect with complex changes in the effect of GABA was observed with millimolar concentrations of pentobarbital and thiopental [Higashi and Nishi, 1982].

**Benzodiazepine and barbiturate modification of postsynaptic (antagonistic and recurrent) inhibition.** Until recently postsynaptic inhibition in the spinal cord (Fig. 5) was believed to be mediated by a neurotransmitter different from GABA, most probably glycine, because this type of inhibition is sensitive to blockade by the glycine antagonist strychnine [Curtis and Johnston, 1974].

The inability of diazepam to affect antagonistic inhibition [Schmidt et al., 1967] was a further indication that GABA was not the mediator at inhibitory synapses on motoneurons. Recurrent Renshaw inhibition was also generally believed to be mediated by glycine. Inconsistent with the irrelevance of GABA in postsynaptic inhibition of motoneurons, however, were the immunohistochemical demonstration of glutamic acid–decarboxylase (GAD)–positive nerve terminals forming synapses with spinal motoneurons [McLaughlin et al., 1975; Roberts, 1979] and the occurrence of spontaneous inhibitory postsynaptic potentials in cultured spinal cord cells that were resistant to strychnine, blocked by bicuculline, and enhanced by chlordiazepoxide [Choi et al., 1981b]. Cullheim and Kellerth [1981] recently suggested that recurrent Renshaw inhibition may be mediated by both GABAergic and glycinergic neurons, because the late phase of this synaptic depression of motoneuron excitability was blocked by bicuculline and picrotoxin, but resistant to strychnine, while the opposite drug sensitivity was found with the early phase of Renshaw inhibition. This pharmacological heterogeneity of Renshaw inhibiton was confirmed by Polc and Haefely [1982], who in addition showed that benzodiazepines selectively enhance its late, presumably GABAergic phase. Assuming that most neurons in dissociated cell culture studied [Choi et al., 1981a,b; Study and Barker, 1981; Skerritt and Macdonald, 1984a,b; Skerritt et al., 1984] were motoneurons, their response to GABA, which was enhanced by agonistic benzodiazepines, also indicates the presence in the spinal cord of GABAergic synapses other than those on primary afferent endings. Nicoll and Wojtowicz [1980] had observed a potentiation by chlordiazepoxide of GABA-induced hyperpolarization of frog spinal motoneurons and Nistri et al. [1980] obtained similar results with flurazepam in cat spinal motoneurons in situ.

The effect of *barbiturates* on postsynaptic inhibition of spinal motoneurons is controversial: absence [Weakly et al., 1969] as well as presence of potentiation [Larson and Major, 1970] have been reported. According to the latter authors, hexobarbital markedly increases the duration of inhibitory postsynaptic potentials evoked in cat spinal motoneurons by the recurrent Renshaw pathway. The potentiation by various barbiturates of GABA effects on motoneurons in vitro [Nicoll, 1975b; Nicoll and Wojtowicz, 1980] and on presumable motoneurons in dissociated cell culture [Macdonald and Barker, 1978, 1979a,b; Ransom and Barker, 1975, 1976; Schulz and Macdonald, 1981; Barker and Mathers, 1981] are strong supportive evidence of a barbiturate–GABA interaction on the membrane of motoneurons. Interesting differences in the GABA-potentiating activity were found between enantiomers of optically active barbiturates [Huang and Barker, 1980]. In concentrations of $10^{-4}$ M and higher, some, but not all barbiturates induced a hyperpolarization of motoneurons in the absence of GABA, which could be blocked by picrotoxin, penicillin, and bicuculline but not by strychnine [Nicoll and Wojtowicz, 1980; Ransom and Barker, 1975; Macdonald and Barker, 1978]. Barbiturates that easily produce complete surgical anesthesia were much more potent in producing direct hyperpolarization than those particularly useful in the treatment of epilepsy and less effective and ineffective as general anesthetics [Macdonald and Barker, 1979a,b, Schulz and Macdonald, 1981].

*Spinal cord interneurons,* both unidentified and presumable Renshaw cells, were found to be inhibited by microionophoretic GABA. This effect, but not that of glycine, was enhanced by pentobarbital [Curtis and Lodge, 1977; Lodge and Curtis, 1978]. The benzodiazepine midazolam enhanced GABA-mediated inhibition of dorsal horn interneuron activity in a Ro 15-1788–sensitive manner [Leah et al., 1983].

Benzodiazepine receptor agonists and barbiturates depress mono- and polysynaptic excitation of spinal motoneurons. It is possible that benzodiazepine receptor agonists produce this depression of reflex activity, as well as the reduction of spontaneous and evoked activity of $\gamma$-motoneurons and Renshaw cells via enhancement of pre- and postsynaptic GABAergic inhibition. Indeed, their effect is opposite to that of bicuculline and picrotoxin as well as of inverse benzodiazepine receptor agonists, and is blocked by Ro 15-1788 [Polc et al., 1981, 1982; unpublished results]. Barbiturates have a similar depressant effect on motoneuron reflexes; their multiple actions, potentiation of GABAergic inhibition, direct activation of chloride conductance, and depression of excitatory synaptic transmission all likely contribute to reflex depression.

It should be realized that we have acquired a considerable knowledge of the effects of benzodiazepine receptor ligands and of barbiturates on synaptic transmission in sensory–motor circuits within the spinal cord. In contrast, information is virtually lacking on the effects of these drugs on transmission in sensory pathways projecting towards the brain. Very strikingly, neither benzodiazepines nor barbiturates have antinociceptive activity. Disturbances of sensory perception seen with overdoses of these drugs are usually ascribed to changes in the level of vigilance and not to modification of transmission in sensory pathways on the level of the spinal cord. Could this mean that primary afferent depolarization of C fibers [Calvillo et al., 1982], to which nociceptive primary afferents belong, is mediated by another transmitter than GABA?

## Dorsal Column Nuclei

In contrast to the spinal cord, where GABA is probably the transmitter mediating primary afferent depolarization, and, thereby, presynaptic inhibition, and where glycine seems to be the transmitter of most inhibitory synapses on motoneurons [Curtis and Johnston, 1974], GABA is believed to mediate both pre- and postsynaptic inhibition of the relay cells in the dorsal column nuclei induced by primary afferents ascending in the dorsal columns [Curtis and Johnston, 1974].

It was an important finding that *benzodiazepines* enhance pre- as well as postsynaptic inhibition in the cuneate nucleus of decerebrate cats [Polc and Haefely, 1976], because this was strong evidence that these drugs selectively enhance those synaptic inhibitions that are mediated by GABA. Pharmacological manipulation of the GABA system clearly demonstrated that the effects of benzodiazepines depended on ongoing activity of GABAergic interneurons, since in cats pretreated with the glutamic acid–decarboxylase inhibitor thiosemicarbazide diazepam failed to restore the reduced pre- and postsynaptic inhibition.

The interaction of benzodiazepines with the effect of GABA on primary afferent fibers of the rat cuneate nucleus slices was investigated by Simmonds [1981]. Lorazepam and flurazepam ($10^{-6}$ M) were found to slightly enhance the depolarizing responses to muscimol and to reduce the potency of picrotoxin as a GABA antagonist. The noncompetitive nature of the picrotoxin–benzodiazepine interaction is in line with the view that the sites of action of these drugs on the GABA receptor complex are not identical [Olsen, 1982]. In the same slice preparation the diazepam analogue Ro 5-4864, which has a high affinity for nonneuronal benzodiazepine binding sites [Richards et al., 1982] and which has convulsant activity, antagonized the potentiating effect of flurazepam on muscimol responses in a low concentration ($10^{-7}$ M); in higher concentrations ($3 \times 10^{-6}$ M), Ro 5-4864 enhanced the effect of picrotoxin and antagonized the pentobarbital-induced enhancement of muscimol responses. This may be taken as the physiological confirmation of the view derived from binding studies that Ro 5-4864 acts on a site of the GABA receptor complex that is close to or identical with the picrotoxinin binding site [Pellow et al., 1984]. Strikingly similar results were obtained with two phenylquinolines reported to bind to benzodiazepine receptors in vitro [Le Fur et al., 1981], PK 8165 and PK 9084 ($10^{-4}$ M). Either of these enhanced the potency of picrotoxin as an antagonist of muscimol. PK 8165 reduced the flurazepam-potentiation of muscimol. PK 9084 slightly antagonized muscimol responses, an effect which was reversed by flurazepam [File and Simmonds, 1984].

*Barbiturates* also enhance pre- and postsynaptic inhibition in the cuneate nucleus. Pentobarbital administered intravenously was found to increase primary afferent depolarization in the cat cuneate nucleus and to antagonize the depressant effect of picrotoxin on primary afferent depolarization [Banna and Jabbur, 1969]. When applied ionophoretically, pentobarbital was reported to depress relay cells and interneurons by a postsynaptic mechanism [Galindo, 1969]. Phenobarbital was found by Polc and Haefely [1976] to slightly enhance and markedly prolong both pre- and postsynaptic inhibition at 10 mg/kg i.v. At 30 mg/kg the potentiation of pre- and postsynaptic inhibition became less marked than at the lower dose, most probably due to the concomitant depression of excitatory transmission from primary afferents to GABAergic interneurons. Since phenobarbital did not affect the resting excitability of cuneate relay cells and primary afferents, a direct effect on GABA receptors or associated chloride channels at these doses appeared to be absent [Polc and Haefely, 1976].

An interesting difference in the mechanism of interaction of pentobarbital and phenobarbital with the GABA system was revealed by Simmonds [1981] in slices of the rat cuneate nucleus. Pentobarbital ($10^{-4}$ M), as a prototype of anesthetic barbiturates, potentiated muscimol effects on primary afferents, and reduced the potency of picrotoxin and bicuculline as GABA antagonists. In contrast, phenobarbital ($10^{-4}$ M) reduced the potency of picrotoxin, but not that of bicuculline, to inhibit muscimol responses and only slightly facilitated muscimol effects. It was suggested that anticonvulsant barbiturates interact with the picrotoxinin binding site on the GABA receptor complex, whereas the anesthetic barbiturates, in addition, directly evoke GABA-like responses.

### Unidentified Reticular Neurons in the Medulla Oblongata

The reticular formation of the brain stem (medulla, pons, mesencephalon) is a diffuse system of cell aggregates and of a vast number of nerve fibers running in all directions. Reticular neurons receive inputs from the periphery as well as from higher centers. Of the many functions ascribed to the reticular formation their roles in determining the level of consciousness and the tone of the skeletal muscle and of vegetative activities are of primary interest in the context of drugs that affect alertness, muscle tone, and vegetative functions. Although in the brain stem the density of benzodiazepine receptors is relatively low and the location of GABAergic neurons in the neuronal network is largely unknown, benzodiazepines and barbiturates have long been suggested to produce some of their effects through an action within the reticular formation.

Iontophoretically applied chlordiazepoxide and flurazepam depressed the spontaneous neuronal activity but did not affect GABA-produced inhibitions of firing in urethane-anesthetized rats. However, since bicuculline antagonized the flurazepam-induced depression, it was suggested that benzodiazepines interact with GABA-mediated processes [Dray and Straughan, 1976]. In a subsequent study, a partial antagonism of the bicuculline-evoked abolition of the GABA depression was observed with iontophoretically applied chlordiazepoxide and intravenously administered clonazepam [Bowery and Dray, 1978].

Similar effects to those observed with benzodiazepine were found after iontophoretic or intravenous barbiturates [Bowery and Dray, 1978]. The only difference was the much clearer reversal by barbiturates than by benzodiazepines of the bicuculline-induced reversal of GABA effects.

There is good evidence to believe that part of the depressant effect of benzodiazepines and barbiturates on spontaneous and evoked activity of reticular neurons [see Haefely et al., 1981] may, at least in part, be due to an enhancement of GABAergic transmission. This view is supported by the finding that pentylenetetrazole, which produces effects that are in all respects opposite to tranquilizers, increases the responsiveness of reticular neurons to sensory stimuli of different modality [Faingold, 1978, 1980]. The GABA receptor blocker bicuculline also reduces response thresholds in reticular neurons [Faingold et al., 1983].

### Lateral Vestibular Nucleus

In the lateral vestibular nucleus of Deiters, primary afferents from the vestibular apparatus activate second-order neurons that project mainly to the cerebellum (impinging as so-called mossy fibers on granule cells) and to various motor nuclei. These secondary vestibular neurons receive postsynaptic inhibition from the GABAergic Purkinje cells in the cerebellar cortex. The control loop formed by the circuitry: secondary vestibular neuron–granule cell—Purkinje cell—secondary vestibular neuron modulates the various vestibular reflexes involved, e.g., in equilibrium and eye movements.

Small doses of diazepam (0.2 mg/kg i.v.) were shown to depress in unanesthetized decerebrate cats the spontaneously active second-order vestibular neurons that respond to linear motions, and, in addition, to antagonize the excitation of these neurons elicited by pentylenetetrazole [Kirsten and Schoener, 1972]. Even smaller doses of diazepam (0.02–0.1 mg/kg i.v.) reduced the sensitivity of vestibular neurons to angular accelerations about the vertical and longitudinal axes in unanesthetized rabbits [Barmack and Pettorossi, 1980]. Since diazepam had no consistent effect on primary vestibular afferents, it was proposed that diazepam depresses this vestibular pathway by enhancing a tonic inhibition of the secondary vestibular neurons. A facilitation by benzodiazepines of the GABAergic inhibition from Purkinje neurons has also been proposed by Haefely et al. [1978] to account for the observation that diazepam antagonized the depressant effect of bicuculline on the component of a complex evoked potential elicited in Deiters' nucleus by stimulation of Purkinje cell axons of cats.

A depression of the excitability of secondary vestibular neurons by enhancement of the GABAergic inhibitory input from the cerebellar cortex may explain a beneficial effect of benzodiazepines on some forms of vertigo; it may also contribute to the ataxia produced by higher doses of these drugs.

### Dorsal Raphe Nucleus

The raphe nuclei are aggregates of cell bodies of serotonergic neurons. These neurons receive a GABAergic input of unknown origin [Belin et al., 1979; Nanopoulos et al., 1982; Nishikawa and Scatton, 1983]. The investigation of the effects of benzodiazepine receptor ligands on these 5-HT neurons is of particular interest because a depression of their activity was at some time suggested to underlie the anxiolytic effect of benzodiazepine receptor agonists [Thiébot et al., 1982].

A selective potentiation of the inhibitory effect of iontophoretically applied GABA was observed in chloral hydrate–anesthetised rats by iontophoretically administered chlordiazepoxide and flurazepam [Gallager, 1978]. Since the spontaneous firing rate of single raphe neurons was reduced by benzodiazepines only after pretreatment with the GABA-transaminase inhibitor aminooxyacetic acid (AOAA), it was suggested that a sufficient GABAergic input must be present on the raphe cells in order to observe the depressant effect of benzodiazepines. However, in recent studies using unanesthetized encéphale isolé rats, intravenous injections of benzodiazepines were shown to depress the spontaneous activity of dorsal raphe neurons in a dose-dependent manner [Laurent et al., 1983]. It is very likely that the use of an anesthetic precluded the observation of a depressant effect of benzodiazepines by Gallager [1978], because chloral hydrate by itself was found to reduce multiunit activity in unanesthetized rats and because this depression was not further increased by diazepam [Laurent et al., 1983]. A similar conclusion was drawn by Trulson and Trulson [1983]. The depressed effect of various benzodiazepines was antagonized by the benzodiazepine antagonist Ro 15-1788 [Laurent et al., 1983]. In unrestrained, freely moving cats, benzodiazepines dose-dependently reduced the spontaneous activity as well as the behavior-related increase of activity of single raphe neurons [Trulson et al., 1982].

An interesting observation, suggesting a role of the GABA-benzodiazepine receptor complex in the development of tolerance to benzodiazepine receptor agonists, is the recent finding by Gallager et al. [1984] that a three-week pretreatment with diazepam markedly decreased the potency of GABA to inhibit single raphe neuron activity. Very surprisingly, an acute dose of Ro 15-1788 reversed this subsensitivity of raphe neurons to GABA: The authors do not indicate whether the antagonist altered the spontaneous firing rate of the neurons. Additional binding experiments revealed an impairment by chronic diazepam treatment of the ability of GABA to enhance $^{3}$H-diazepam binding [Gallager et al., 1984].

The depressant effect of benzodiazepine receptor agonists on raphe neurons is consistent with their decrease of 5-HT turnover. It is, however, very questionable whether this depression of serotonergic neuron activity is directly involved in their anxiolytic action [Thiébot et al., 1984].

## Locus Coeruleus

The locus coeruleus is the main site of origin of central noradrenergic neurons and is known to contain GABAergic terminals of unknown origin. Benzodiazepine receptor agonists are known to reduce the turnover of norepinephrine, in particular under conditions of stress-induced activation of locus coeruleus neurons [see Haefely et al., 1981]; this depressant effect of benzodiazepines on noradrenergic activity has been postulated to contribute to their sedative and/or anxiolytic effects [Gray, 1982]. Recent lesion experiments do not support an essential link between noradrenergic locus coeruleus neurons and anxiolytic effect of benzodiazepines [Koob et al., 1984].

Various benzodiazepine receptor agonists were found to depress multiunit activity in the locus coeruleus of encéphale isolé rats, and Ro 15-1788 reversed this effect [Laurent et al., 1983]. Diazepam (i.v.) and flurazepam (microionophoretically) were reported to decrease the spontaneous firing rate of locus coeruleus neurons in rats anesthetized with chloral hydrate; picrotoxin reversed the effect of diazepam [Sanghera and German, 1983].

## Substantia Nigra

The substantia nigra of the midbrain has a high density of GABA receptors and benzodiazepine receptors [Richards et al., 1985]. The pars compacta contains the cell bodies of the dopaminergic output neurons which project mainly to the neostriatum; a considerable part of the dendritic tree of these neurons extends into the pars reticulata. The output neurons of the pars reticulata are probably exclusively GABAergic and project to the thalamus and the midbrain colliculi. Among the afferents to the substantia nigra are GABAergic and substance P–containing bundles from the neostriatum. GABAergic synapses can, therefore, be found on both dopaminergic and GABAergic output cells of the substantia nigra (Fig. 6).

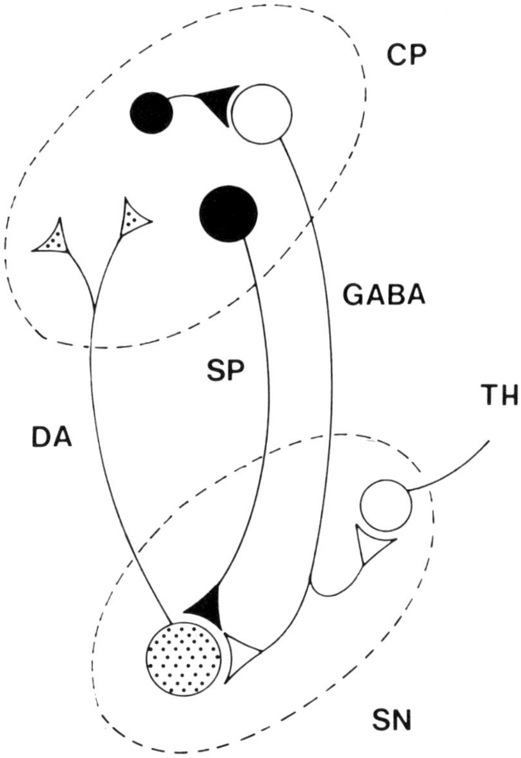


**Fig. 6.** *Schematic diagram of some neuronal pathways connecting caudatoputamen (CP) and substantia nigra (SN): Shown are the nigrostriatal dopamine pathway (DA), the striatonigral substance P pathway (SP), and the striatonigral GABAergic pathway (GABA). TH = thalamus.*

Electric stimulation of the caudate nucleus evokes inhibitory postsynaptic potentials in neurons recorded in the pars compacta of the substantia nigra [Yoshida and Precht, 1971]. The evoked mass inhibitory postsynaptic potentials were found to be dose-dependently depressed by bicuculline and picrotoxin confirming the involvement of GABA. While various benzodiazepines had no effect by themselves on this inhibitory striatonigral evoked response, they dose-dependently reversed the effect of bicuculline [Haefely et al. 1975]. The lack of a direct effect of benzodiazepines on the response might be due to a high safety of transmission (e.g., high GABA receptor reserve) in these synapses. Benzodiazepines would enhance transmission in these synapses only when the receptor reserve is reduced or eliminated by the GABA antagonist bicuculline. Similar results as on the striatonigral pathway have been obtained on the GABAergic pathway projecting from the nucleus accumbens septi to the substantia nigra zona reticulata [Strahlendorf and Barnes 1983].

The spontaneous firing rate of neurons both in the pars reticulata and pars compacta of the substantia nigra is reduced by various benzodiazepine receptor agonists in anesthetized and nonanesthetized animals [Ross et al., 1982; Laurent et al., 1983. Mereu et al., 1983; Trulson 1984]. This effect is antagonized by Ro 15-1788, bicuculline, picrotoxin, and caffeine. β-CCM, β-CCE, and DMCM increased the firing of reticulata neurons, an effect which was reversed by diazepam as well as by Ro 15-1788 [Mereu and Biggio, 1983., Mereu et al., 1983; Weissman et al. 1983]. The peripheral benzodiazepine binding site ligand Ro 5-4864 was reported to have a slight excitatory effect on the firing of reticulata neurons, which was not reversed by Ro 15-1788 [Weissman et al., 1983]. This is further evidence that Ro 5-4864 may affect GABA receptor function, however, not through the associated benzodiazepine receptor but via another site on this complex, presumably the channel domain.

The depressant effect of exogenous GABA (and muscimol) on substantia nigra neurons was enhanced by diazepam [Waszczak, 1983].

The available evidence convincingly shows that benzodiazepine receptor agonists enhance GABAergic inhibition of both dopaminergic and GABAergic output neurons of the substantia nigra. This is consistent with biochemical and behavioral data pointing to a reduction of the activity of nigral dopamine neurons [Keller et al., 1976]. The substantia nigra has been looked at for two decades almost exclusively as a source of dopaminergic neurons. It is increasingly evident that the nondopaminergic neurons of the substantia nigra may be functionally as important in motor functions as the dopamine cells and that the substantia nigra is a very important area in the control of generalization of paroxysmal epileptic activity throughout the CNS [Iadarola and Gale, 1982; McNamara et al., 1984]. Depression of the excitability of nigral reticulata neurons by GABAergic mechanisms may be relevant for the anticonvulsant activity of benzodiazepine receptor agonists (in addition to other sites).

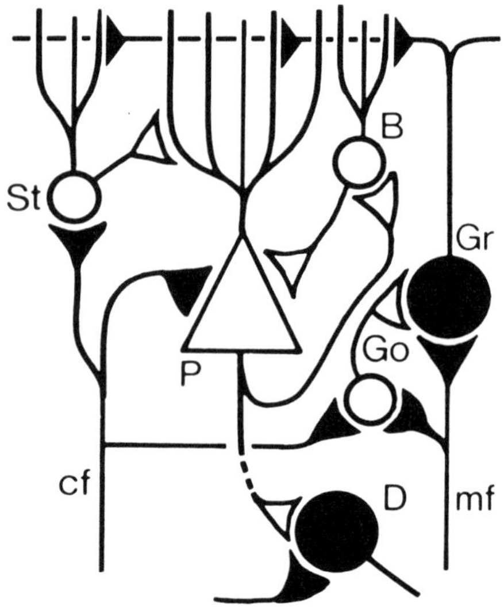


**Fig. 7.** *Schematic diagram of neuronal pathways in the cerebellar cortex and in the lateral vestibular nucleus. (Inhibitory neurons, white; excitatory neurons, black). B = basket cell; cf = climbing fiber; D = Deiters' neuron; Go = Golgi cell; Gr = granule cell; mf = mossy fiber; P = Purkinje cell; St = stellate cell.*

### Cerebellar Cortex

In the cerebellar cortex, one type of intrinsic neuron, the granule cell, is excitatory. The other three types of inhibitory interneurons, the basket, stellate, and Golgi neurons, are probably GABAergic, as is the only output neuron of the cerebellar cortex, the Purkinje cell (Fig. 7). Since the cerebellar cortex has a high density of benzodiazepine and channel agent–binding sites, benzodiazepine ligands and barbiturates are very likely to affect GABA effects in this area.

A relatively simple way in which GABAergic synaptic inhibition of Purkinje cells can be studied is to record the activity of single Purkinje cells after single electric stimuli are applied to parallel fibers, the axons of granule cells running in dense bundles on the cortical surface (parallel fiber stimulation, surface stimulation). Activation of the parallel fibers first activates Purkinje cell by monosynaptic excitation and, after a short latency, depresses their firing rate indirectly by activating the (inhibitory) GABAergic stellate and basket cell pathways.

Various benzodiazepine receptor agonists (diazepam, flurazepam, flunitrazepam, alprazolam, RU 32007, the thienodiazepine Ro 11-7800, the triazolopyridazine CL 218,872) were reported to enhance synaptic inhibition of Purkinje cells by stellate and basket cells after both systemic, local (on the cerebellar surface), and microionophoretic application [Curtis et al., 1976; Geller et al., 1978; Lippa et al., 1979; Okamoto and Sakai, 1979; Montarolo et al., 1979; Gardner, 1984]. In some of these studies as well as in others [Pieri and Haefely, 1976; Mariani and Delhaye-Bouchaud, 1978; Sinclair et al., 1982; Sanghera and German, 1983; Sorensen and Freedman, 1983; Waterhouse et al 1984; Gardner, 1984] benzodiazepine receptor agonists were found to depress the spontaneous firing rate of Purkinje cells as one would expect by virtue of their enhancement of the efficiency of the GABAergic input. Gardner [1984] has shown that there is an exponential relationship between the duration of poststimulus inhibition and the firing rate of Purkinje cells in the absence as well as in the presence of benzodiazepine receptor agonists. In contrast, β-CCE was observed to increase the spontaneous activity of Purkinje cells and to decrease the inhibitory effect of parallel fiber stimulation [Gardner, 1984]. Ro 15-1788 had no effect by itself, but blocked the effect of β-CCE and of agonists [Gardner, 1984].

The inhibitory effect of microionophoresed GABA on Purkinje cells was observed to be augmented by various benzodiazepine receptor agonists [Assumpçao et al., 1979; Okamoto and Sakai, 1979; Jiang, 1981; Sanghera and German, 1983; Waterhouse et al., 1984]. In some of these studies the potentiating effect was found to be specific for GABA and absent for glycine and norepinephrine. No effect of benzodiazepines on the excitatory effect of glutamate, aspartate, and acetylcholine was observed. Ro 15-1788 antagonized the GABA–enhancing action of benzodiazepines without by itself affecting GABA responses [Bold et al., 1982]. In rats chronically treated with flurazepam, the GABA–potentiating action of this benzodiazepine applied by microionophoresis was reduced [Waterhouse et al., 1984].

One of the major functions of the cerebellum is the control of muscle tone and coordination. It is, therefore, tempting to associate the muscle relaxant (antispastic) as well as the ataxic effects of benzodiazepine receptor agonists with GABAergic synapses in the cerebellum. Other known or supposed functions of the cerebellum, e.g. control of the cardiovascular system, may be modified by benzodiazepine receptor ligands. Because of the very complicated neuronal wiring diagram of the intracerebellar pathways, which contain GABAergic neurons arranged in series (thereby permitting disinhibition of some neurons to result from a potentiation of GABAergic inhibitory transmission), it is presently impossible to understand the contribution of the various intracerebellar neuronal elements to the effect of GABA-modulating drugs on the cerebellar output.

## Hypothalamus

The various hypothalamic nuclei differ in their density of GABA and benzodiazepine receptors. Hypothalamic GABAergic neurons seem to innervate the pituitary gland. Our knowledge of the organization of GABAergic neurons in hypothalamic neuronal circuits is very limited. From studies on neurons of the tuberal part of the hypothalamus in tissue culture, GABAergic interneurons are known to mediate recurrent inhibition of principal neurons with unidentified function [Geller and Woodward, 1979].

Diazepam and flurazepam enhance this recurrent inhibition; this effect is reversed by picrotoxin [Geller et al., 1978; Geller, 1979]. The two benzodiazepines also potentiate the depressant effect of GABA on the spontaneous firing rate, but not that of glycine. The benzodiazepine receptor antagonist Ro 15-1788 and its close congener, Ro 14-7437, blocked the GABA-potentiating effect of flurazepam [Krespan et al., 1984].

Three different barbiturates were reported to increase postsynaptic inhibition of single neurons in the preoptic–anterior region of the hypothalamus (ovulatory trigger region) evoked by stimulation of the arcuate–ventromedial hypothalamic area. [Mayer, 1981]. These drugs also selectively enhanced the depressant effect of GABA.

Benzodiazepines and barbiturates are well known to reduce a number of behavioral, autonomic, and endocrine responses to emotional stimuli [Haefely et al., 1981]. It is reasonable to assume that some of these effects are due to an action on GABAergic synapses in the hypothalamus, but no precise information is available yet.

## Hippocampus

The hippocampal formation (hippocampus proper and dentate gyrus) is characterized by a simple laminar structure [Teyler and Di Scenna, 1984]. The output neurons (hippocampal pyramidal and dentate granule cells) are arranged in a thin layer with their axons and basal dendrites on the one side and the apical dendrites on the opposite side of the layer. Separate afferent systems cross the two dendritic layers. Interneurons, the (presumably prevalent GABAergic) basket cells, are excited both by recurrent collaterals of the pyramidal cell axons (feedback inhibition) and by collaterals of incoming afferents (feedforward inhibition), as schematically depicted in Fig. 8. Hippocampal pyramidal cells have the characteristic tendency to fire repetitively in bursts, due to particular inherent membrane properties. This tendency is greatly reduced under normal conditions by the strong GABAergic synaptic input; indeed, when GABAergic inhibition is depressed experimentally, these cells are easily driven to paroxysmal bursting discharges. This paradigm of epileptogenesis is being studied intensively in hippocampal slice preparations.

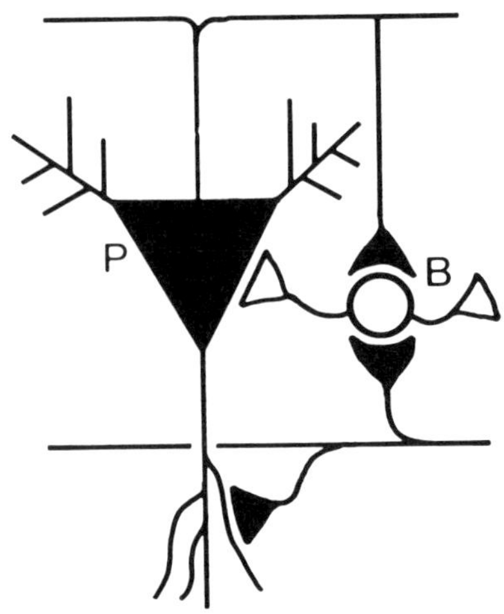


**Fig. 8.** *Schematic diagram of the recurrent and the feedforward inhibitory pathways in the hippocampus. B = basket cell; P = pyramidal cell. Only afferents in the stratum radiatum ending on apical pyramidal cell dendrites and on basket cells are shown.*

*Benzodiazepine receptor agonists* are consistently found to enhance recurrent inhibition as measured by a prolongation of the postexcitatory synaptic depression or by the augmentation of inhibitory postsynaptic potentials [Wolf and Haas, 1977; Tsuchiya and Fukushima, 1978; Lee et al., 1979; Adamec et al., 1981, Alger and Nicoll, 1982a,b; Polc et al., 1982; Krespan et al., 1984; Segal and Barker, 1984]. This enhancement of feedback and probably also of feedforward inhibition [Alger and Nicoll, 1982a,b], results in a reduction of the spontaneous firing rate [Steiner and Hummel, 1968; Olds and Olds, 1969; Umemoto and Olds, 1975; Matthews and Connor, 1977; Wolf and Haas, 1977; Laurent et al., 1983]. The facilitation of pyramidal cell discharge by repetitive frequent stimulation, which might be due to a failure of GABAergic inhibition by desensitization [Ben-Ari et al., 1979], is reduced by diazepam and

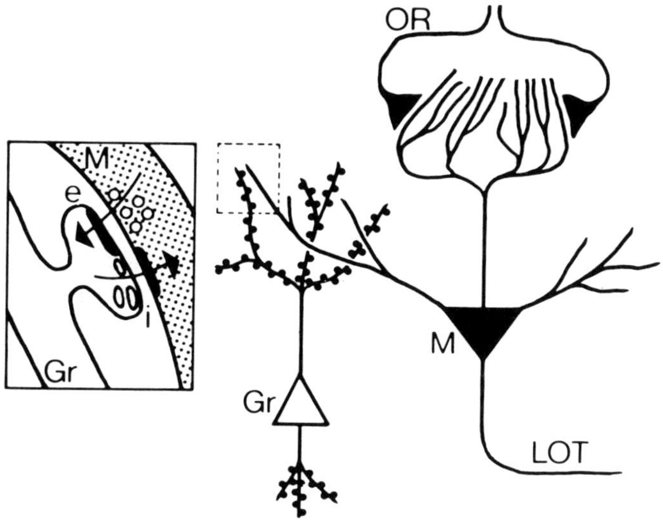


**Fig. 9.** *Schematic diagram of the neuronal pathways in the olfactory bulb. G = GABAergic granule cell; LOT = lateral olfactory tract; M = mitral cell; OR = olfactory receptor neuron. The inset shows the reciprocal mitral cell/granule cell dendritic synapse with its excitatory (e) and inhibitory (i) parts.*

midazolam [Polc et al., 1981a; Brown et al., 1983]. The benzodiazepine receptor blocker Ro 15-1788 prevents the effect of benzodiazepines on hippocampus [Polc et al., 1981b]. The inverse agonists β-CCM and β-CCE produce the opposite effect of agonists on spontaneous firing [Paterson and Roberts, 1983] and frequency facilitation [Polc et al. 1981a].

The hyperpolarizing and depolarizing effect of GABA mimetics on hippocampal pyramidal cells was found to be enhanced by benzodiazepine receptor agonists [Jahnsen and Laursen, 1981; Alger and Nicoll, 1982b; Carlen et al., 1983; Biscoe et al., 1983, Krespan et al., 1984; Segal and Barker, 1984]. β-CCM and β-CCE attenuate the depressant effect of GABA on the spontaneous activity of hippocampal pyramidal cells in vivo [Paterson and Roberts, 1983].

*Barbiturates* also enhance GABA-mediated recurrent and feedforward inhibition of hippocampal pyramidal cells [Wolf and Haas, 1977; Tsuchiya and Fukushima, 1978; Alger and Nicoll, 1982a, b; Collingridge et al., 1984]. Accordingly, the spontaneous activity of these cells is depressed by barbiturates [Steiner and Hummel, 1968; Olds and Olds, 1969; Wolf and Haas, 1977]. The chloride conductance increase induced by exogenous GABA is enhanceed by barbiturates [Segal and Barker, 1984].

The marked effects of benzodiazepine receptor agonists and barbiturates in the hippocampus are very likely to be mechanisms that contribute to a relevant degree to the anticonvulsant and anxiolytic activity, and, perhaps the induction of anterograde amnesia.

### Olfactory Bulb

Axon terminals of olfactory receptor neurons, whose cell bodies lie in the nasal mucosa, terminate on dendrites of the two classes of output neurons of the bulb (second-order olfactory neurons), the mitral cells and the tufted cells. In addition to these principal cells, a variety of interneurons are present in a large number, the greater part of them being GABAergic [Mugnaini et al., 1984]. This suggests an important role of GABA in signal processing in this relay structure. A unique feature of the olfactory bulb (Fig. 9) is the presence of reciprocal dendro-dendritic synapses between (excitatory) mitral cells and GABAergic (inhibitory ) granule cells [ Shepherd, 1983]. Depolarization of a mitral cell dendrite results in the local release of the excitatory neurotransmitter, which excites a granule cell dendrite. This local depolarization of a granule cell dendrite, in turn, provokes the local release of GABA, which acts back on the adjacent mitral cell dendrite to produce hyperpolarization.

Barbiturates were shown by Nicoll [1972, 1978, 1980] to enhance in the rabbit the GABA-mediated depression of mitral cell excitability that follows an antidromic mitral cell action potential induced by stimulation of its axon in the lateral olfactory tract. At anesthetic concentrations at which GABAergic dendro-dendritic inhibition was enhanced, barbiturates did hardly affect the mitral cell–granule cell excitatory synapse.

The effect of benzodiazepine receptor ligands in the olfactory bulb has not been studied. Changes in olfactory perception do not seem to have been observed with barbiturates or benzodiazepines in humans.

### Neocortex

The neocortex has a very high density of GABAergic neurons, GABA receptors and benzodiazepine receptors. Because of the complexity of the cortical neuronal circuits and the heterogeneity of the GABAergic interneuron population [Somogyi et al., 1984], it is the experimentally accessible recurrent inhibitory synapse on cortical pyramidal cells (similar to recurrent inhibition in hippocampus and hypothalamus) for which most pharmacological information is available.

Diazepam was found by Raabe and Gumnit [1977] to enhance recurrent inhibition of pyramidal tract cells in the cat cerebral cortex. Consistent with a potentiation of GABAergic postsynaptic inhibition is the finding by White et al., [1981] of an increase in the amplitude of bicuculline-sensitive inhibitory postsynaptic potentials in cultured rat cortical neurons in the presence of diazepam. Recurrent inhibition was also enhanced by diazepam, clonazepam, flurazepam, and chlordiazepoxide in slices of the guinea-pig cortex [Scholfield, 1983; Riley and Scholfield, 1983] and in the rat olfactory cortex slice [Collins 1981]. Ro 15-1788 reversed this effect of benzodiazepines. Enhancement of GABAergic transmission in recurrent and other local inhibitory circuits might underlie the facilitation by diazepam of the postexcitatory depression of the local cortical response and of the inhibitory period in the peripherally evoked neuronal activity observed by Zakusov et al. [1975, 1977]. The absence of a depressant effect of benzodiazepines, even in very high doses (50 mg/kg i.v.), on excitatory synaptic transmission might be inferred from their failure to affect short-latency cortical responses to stimulation of the forepaw in rats [Angel and Gratton, 1982], quite in contrast to barbiturates.

Benzodiazepines have also been found in several studies to enhance the depressant effect of ionophoretic GABA or muscimol on the firing rate of cortical neurons [Kozhechkin and Ostrovskaya, 1977; Zakusov et al., 1977; Nestoros and Nistri, 1978, 1979; Schmidt et al., 1978; Nestoros 1982; Jiang, 1981, White et al., 1981; Tancredi et al., 1983].

Various *barbiturates* were shown to potentiate inhibitory phenomena in the cerebral cortex [Pickles and Simmonds, 1978; Scholfield, 1978]. At high concentrations pentobarbital increased membrane conductance of cortical neurons; this effect was reversed by picrotoxin [Scholfield, 1978].

Although benzodiazepines and barbiturates markedly facilitate inhibitory synaptic mechanisms in the cerebral cortex, the complexity of this structure makes any association between enhanced GABAergic transmission and specific pharmacological effects impossible, except for their attenuation of seizure propagation.

### Sympathetic Ganglion Cells

Although sympathetic ganglia do not contain GABAergic neurons (with the possible exception of abdominal ganglia), the ganglionic cells have GABA receptors and respond to GABA-mimetics with a bicuculline-sensitive depolarization [Adams and Brown, 1975], due to a higher intracellular chloride concentration than exists in most central neurons. GABA also depolarizes preganglionic nerve endings [Brown and Higgins, 1979].

Chlordiazepoxide was found in an early study to partially reverse the GABA antagonistic effect of bicuculline in the isolated rat superior cervical ganglion without affecting consistently the depolarizing effect of GABA in the absence of a GABA antagonist [Bowery and Dray, 1978]. This suggested that the benzodiazepine interacts with the GABA antagonist rather than with GABA itself on the GABA receptors. However, Brown and Constanti [1978] observed a potentiation of the GABA-induced conductance increase in single cells of the rat superior cervical ganglion by chlordiazepoxide. Recently, Little [1984] showed that chlordiazepoxide in low micromolar concentrations shifted to the left the dose–response curve for the depolarizing action of GABA in the absence of bicuculline. Smaller and more variable effects were seen with flurazepam. The GABA potentiating action of the benzodiazepines was abolished by Ro 15-1788. The inverse agonist β-CCE produced a shift to the right of the GABA dose–response curve at low, but not at higher concentrations. Ro 15-1788 also reversed the effect of β-CCE. A more marked depression of GABA effects was obtained with β-CCM. Ro15-1788 at concentrations reversing the effect of the β-carbolines had no intrinsic effect on GABA-induced depolarization. In submillimolar concentrations Ro 15-1788 decreased the effect of GABA. The similar depolarization induced by a high dose of GABA in the presence of bicuculline was increased by Ro 15-1788. The results of Little [1984] indicate that the GABA receptors in the rat superior cervical ganglion contain functional benzodiazepine receptors similar to those in the CNS. The density of these benzodiazepine receptors in the ganglion seems to be very low, as it has not yet been possible to visualize them with the autoradiographic technique using radioligand benzodiazepines [Richards, unpublished results].

Barbiturates were also reported to reverse the GABA–antagonistic action of bicuculline [Bowery and Dray, 1976]. It was later found that these drugs consistently enhance the effect of GABA in the absence of a GABA antagonist [Brown and Constanti, 1978; Nicoll and Iwamoto, 1978, Little, 1984]. Pentobarbital inconsistently induced a direct conductance increase and depolarization [Bowery and Dray, 1978; Evans, 1979; Brown and Constanti, 1978; Nicoll and Iwamoto 1978; Schlosser and Franco, 1979].

In view of the absence of GABAergic synapses in autonomic ganglia, the GABA–potentiating action of benzodiazepine receptor agonists and barbiturates is therapeutically irrelevant. The intrinsic chloride conductance–inducing effect of barbiturates may result in depression of ganglionic transmission at anesthetic concentrations.

### Retina

Horizontal cells and a subpopulation of amacrine cells are GABAergic. The functional role of GABAergic synapses in the transmission between photoreceptors and bipolar and ganglion cells is still largely unknown.

A subpopulation of amacrine cells is dopaminergic, and it has been shown that illumination of the retina increases the turnover of dopamine in the retina. GABA receptor stimulation by muscimol reduces the light-induced rise in dopamine turnover, an effect that is blocked by picrotoxin [Morgan and Kamp, 1980]. Bicuculline increases dopamine turnover in the darkness [Kamp and Morgan, 1980a, b]. Consistent with the presence of benzodiazepine receptors in the retina, benzodiazepines were found to attenuate the light-induced increase of dopamine turnover [Kamp and Morgan, 1982]. It is not yet known whether therapeutic concentrations of benzodiazepines affect retinal function.

### GABA –Receptor Complex of Chicken Brain Expressed in Frog Oocyte Membrane

The injection of a mixture of messenger

(m) RNA from embryonic chicken or rat brain into frog oocytes, which normally do not contain GABA receptors, results in the responsiveness of these cells to GABA applied on their surface. The response is a depolarization produced by an increase of the membrane permeability for chloride; it is blocked by bicuculline and picrotoxin and enhanced by chlorazepate and pentobarbital [Smart et al., 1983; Constanti et al., 1984; Houamed et al., 1984]. Thus, the mRNA encoding the receptor–ionophore complex in the chicken CNS led to the synthesis of the complex by the oocyte and its incorporation into the membrane in a form functioning basically like the complexes studied in the various neurons of the mammalian CNS.

## CONCLUSIONS

Benzodiazepine receptor ligands and barbiturates affect the chloride conductance gating function of the $GABA_A$ receptor through allosteric modulatory sites on the receptor–ionophore complex.

The uniqueness of the benzodiazepine receptor is that it reacts to ligands which affect the gating function of the GABA receptor in two opposite ways: benzodiazepine receptor agonists improve this gating function and thereby enhance the GABA effect. Their action is an increase of the probability of channel openings in the presence of GABA. These agonists are said to have positive intrinsic efficacy. This intrinsic efficacy is measured by the amplitude of the left-hand shift of the dose–response curve for the conductance increase of GABA. No direct relationship between affinity for the benzodiazepine receptor (potency of the GABA–enhancing effect) and the intrinsic efficacy has been found. Inverse agonists of benzodiazepines are ligands that, in contrast, decrease the gain of the GABA receptor/ionophore function; they are said to have negative intrinsic efficacy. Benzodiazepine receptor blockers (antagonists) are ligands with negligible intrinsic efficacy; they do not affect the gain of the GABA receptor/ionophore function to any significant extent, but block the effect of both agonists and inverse agonists. Ligands covering the whole spectrum of intrinsic efficacy exist, i.e., partial agonists and partial inverse agonists. It is proposed that submaximal intrinsic efficacy together with a varying GABA receptor reserve on various neurons may explain mixed agonist–antagonistic profiles of activity without the need for postulating benzodiazepine receptor multiplicity. The effects of various benzodiazepine receptor ligands on GABAergic synaptic transmission in various regions of the CNS are reported and their possible role for the behavioral and somatic effects of benzodiazepine receptor ligands are discused.

Barbiturates increase the effect of GABA by an interaction with a binding site on the chloride ionophore. In a lower concentration range, they increase the channel lifetime; in a higher concentration range, some of them result in a GABA-independent opening of the chloride ionophore. Accordingly, the effect of barbiturates on the dose–response curve for the conductance–inducing effect of GABA is more complex than that of benzodiazepine receptor agonists. Although some barbiturates also have excitatory effects, the available evidence suggests that these are not due to an inverse agonistic action.

Benzodiazepine receptor ligands and barbiturates (in part) are the first psychotropic drugs that have been shown to produce their pharmacological and therapeutic effects by allosterically modulating the function of a neurotransmitter receptor.

## ACKNOWLEDGMENT

We thank Dr. R. Schaffner for careful reading of the manuscript.

## REFERENCES

Adamec RE, McNaughton B, Racine R, Livingston KE (1981): Effect of diazepam on hippocampal excitability in the rat: Action in the dentate area. Epilepsia 22:205–215.

Adams PR, Brown DA (1975): Actions of GABA on sympathetic ganglion cells. J Physiol (Lond.) 250:85–120.

Alger BE, Nicoll RA (1982a): Feed-forward dendritic inhibition in rat hippocampal pyramidal cells studied in vitro. J. Physiol 328:105–123.

Alger BE, Nicoll RA (1982b): Pharmacological evidence for two kinds of GABA receptor in rat hippocampal pyramidal cells studied. J Physiol 328:125–141.

Angel A, Gratton DA (1982): The effect of anaesthetic agents on cerebral cortical responses in the rat. Br J Pharmacol 76:541–549.

Assumpção JA, Bernardi N, Brown J, Stone TW (1979): Selective antagonism by benzodiazepines of neuronal responses to excitatory amino acids in the cerebral cortex. Br J Pharmacol 67:563–568.

Banna NR, Jabbur SJ (1969): Pharmacological studies on inhibition in the cuneate nucleus of the cat. Int J Neuropharmacol 8:299–307.

Barker JL, Gratz D, Owen DG, Study RE (1984): Pharmacological effects of clinically important drugs on the excitability of cultured mouse spinal neurons. In Bowery NG (ed): Actions and Interactions of GABA and Benzodiazepines. New York: Raven, pp 203–216.

Barker JL, Mathers DA (1981): GABA receptors and the depressant action of pentobarbital. Trends Neurosci 4:10–13.

Barker JL, Nicoll RA (1972): Gamma-aminobutyric acid: role in primary afferent depolarization. Science 176:1043–1045.

Barker JL, Ransom BR (1978): Pentobarbitone pharmacology of mammalian central neurons in tissue culture. J. Physiol (Lond.) 280:355–372.

Barmack NH, Pettorossi VE (1980): The influence of diazepam on the activity of secondary vestibular neurons in the rabbit. Neurosci Lett 16:339 344.

Belin MF, Aguera M, Tappaz M, McRae-Degneuree A, Bobillier P, Pujol JF (1979): GABA-accumulating neurons in the nucleus raphé dorsalis and periaqueductal gray in the rat: a biochemical and radioautographic study. Brain Res 170:279–297.

Bell JA, Anderson EG (1972): The influence of semicarbazide-induced depletion of γ-aminobutyric acid on presynaptic inhibition. Brain Res 43:161–169.

Ben-Ari Y, Krnjević K, Reinhardt W (1979): Hippocampal seizures and failure of inhibition. Can J Physiol Pharmacol 57:1462–1466.

Biscoe TJ, Duchen MR, Pascoe JE (1983): GABA/benzodiazepine interactions in the mouse hippocampal slice. J Physiol 341:8P–9P.

Boireau A, Stutzmann J-M, Ganet C, Julou L, Blanchard J-C (1984): Suriclone, a new anxiolytic of the cyclopyrrolone family: evidence for possible interference with GABAergic systems. Eur J Pharmacol 104:139–144.

Bold JM, Gardner CR, Walker RJ (1982): An iontophoretic study of Ro 15-1788, a selective benzodiazepine antagonist, on rat cerebellar neurones. Br J Pharmacol 76:241 P.

Bormann J, Sakmann B (1984): Properties of inhibitory chloride channels and the action of diazepam. Abstracts 9th Int Congr Pharmacol London, 1984:13–14.

Bowery NG, Dray A (1978): Reversal of the action of amino acid antagonists by barbiturates and other hypnotic drugs. Br J Pharmacol 63:197–215.

Braestrup C, Honoré T, Nielsen M, Petersen EN, Jensen LH (1983): Benzodiazepine receptor ligands with negative efficacy: chloride channel coupling. In Biggio G and Costa E (eds): "Benzodiazepine recognition Site Ligands: Biochemistry and Pharmacology." New York: Raven pp 29–36.

Braestrup C, Honoré T, Nielsen M, Petersen EN, Jensen LH (1984): Ligands for benzodiazepine receptors with positive and negative efficacy. Biochem Pharmacol 33:859–862.

Brown DA, Constanti A (1978): Interaction of pentobarbitone and γ-aminobutyric acid on mammalian sympathetic ganglion cells. Br J Pharmacol 63:217–224.

Brown DA, Higgins AJ (1979): Presynaptic effect of γ-aminobutyric acid in isolated rat superior cervical ganglia. Br J Pharmacol 66:108P–109P.

Brown MW, Rose D, Ahlquist J (1983): Amnesia-producing drugs affecting hippocampal frequency potentiation. Neuroscience 10:697–706.

Calvillo O, Madrid J, Rudomin P (1982): Presynaptic depolarization of unmyelinated primary afferent fibres in the spinal cord of the cat. Neuroscience 7:1389–1400.

Carlen PL, Gurevich N, Polc P (1983): Low-dose benzodiazepine neuronal inhibition: enhanced $Ca^{2+}$–mediated $K^{+}$–conductance. Brain Res 217:358–364.

Chan CY, Farb DH (1985): Modulation of neurotransmitter action: Control of the GABA response through the benzodiazepine receptor. J Neurosci (in press).

Choi DW, Farb DH, Fischbach GD (1981a): Chlordiazepoxide selectively potentiates GABA conductance of spinal cord and sensory neurons in cell culture. J Neurophysiol 45:621–631.

Choi DW, Farb DH, Fischbach GD (1981b): GABA-mediated synaptic potentials in chick spinal cord and sensory neurons. J Neurophysiol 45:632–643.

Collins GGS (1981): The effects of chlordiazepoxide on synaptic transmission and amino acid neurotransmitter release in slices of rat olfactory cortex. Brain Res 224:389–404.

Collingridge GL, Gage OW, Robertson B (1984): Inhibitory postsynaptic currents in rat hippocampal $CA_1$ neurons. J Physiol (Lond.) 356:551–564.

Connors BW (1981): A comparison of the effects of

pentobarbital and diphenylhydantoin on the GABA sensitivity and excitability of adult sensory ganglion cells. Brain Res 207:357–369.

Constanti A, Houamed KM, Smart TG, Bilbe G, Brown DA, Barnard EA (1984): Functional brain GABA and glycine receptors in mRNA-injected Xenopus oocytes. Neuropharmacology 23:817–818.

Costa E (1983): The supramolecular organization of receptors for gamma-aminobutyric acid (GABA). In Biggio G, Costa E, Gessa GL, Spano PF (eds): "Receptors as Supramolecular Entities." New York: Pergamon Press, pp 213–235.

Costa E, Guidotti A, Toffano G (1978): Molecular mechanisms mediating the action of diazepam on GABA receptors. Br J Pharmacol 133:239–248.

Cullheim S, Kellerth J.-O. (1981): Two kinds of recurrent inhibition of cat spinal α-motoneurones as differentiated pharmacologically. J Physiol (Lond.) 312:209–224.

Curtis DR, Duggan AW, Felix D, Johnston GAR (1971): Bicuculline, an antagonist of GABA and synaptic inhibition in the spinal cord. Brain Res 32:69–96.

Curtis DR, Johnston GAR (1974): Amino acid transmitters in the mammalian central nervous system. Ergebn Physiol 69:97–188.

Curtis DR, Lodge D (1977): Effect of pentobarbitone on inhibition of spinal interneurones in the cat by glycine and GABA. J. Physiol (Lond.) 272:48P.

Curtis DR, Lodge D, Johnston GAR, Brand SJ (1976): Central actions of benzodiazepines. Brain Res 118:344–347.

Davidoff RA (1972): Gamma-aminobutyric acid antagonism and presynaptic inhibition in the frog spinal cord. Science 175:331–333.

Désarménien M, Lamour Y, Feltz P (1980): Effects of diazepam on GABA-evoked depolarization in rat dorsal root ganglia in vivo. Prog Neuropsychopharmacol 4:31–36.

Dray A, Straughan DW (1976): Benzodiazepines: GABA and glycine receptors on single neurons in the rat medulla. J Pharm Pharmacol 28:314–315.

Eccles JC (1964): "Physiology of Synapses." Berlin: Springer.

Eccles JC, Malcolm JL (1946): Dorsal root potentials of the spinal cord. J Neurophysiol 9:139–160.

Evans RH (1979): Potentiation of the effects of GABA by pentobarbitone. Brain Res 171:113–120.

Faingold CL (1978): Pentylenetetrazol-induced enhancement of responses of mesencephalic reticular neurons to visual stimuli. Brain Res 150:418–423.

Faingold CL (1980): Enhancement of mesencephalic reticular neuronal responses to sensory stimuli with pentylenetetrazol. Neuropharmacology 19:53–62.

Faingold CL, Hoffmann WE, Caspary DM (1983): Bicuculline-induced enhancement of sensory responses and cross-correlations between reticular formation and cortical neurons. EEG and Clin Neurophysiol 55:301–313.

Farb DH, Borden LA, Chan CY, Czajkowski CM, Gibsbs TT, Schiller GD (1984): Modulation of neuronal function through benzodiazepine receptors: biochemical and electrophysiological studies of neurons in primary monolayer cell culture. Ann NY Acad Sci 435:1–31.

File SE, Simmonds MA (1984): Interactions of two phenylquinolines with picrotoxin and benzodiazepines in vivo and in vitro. Eur J Pharmacol 97:295–300.

Furchgott RF (1966): The use of β-haloalkylamines in the differentiation of receptors and in the determination of dissociation constants of receptor-agonist complexes. Advances in Drug Research 3:21–55.

Galindo A (1969): Effects of procaine, pentobarbital and halothane on synaptic transmission in the central nervous system. J Pharmacol Exp Ther 169:185–195.

Gallager DW (1978): Benzodiazepines: Potentiation of a GABA inhibitory response in the dorsal raphé nucleus. Eur J Pharmacol 49:133–143.

Gallager DW, Lakoski JM, Gonsalves SF, Rauch SL (1984): Chronic benzodiazepine treatment decreases postsynaptic GABA sensitivity. Nature 308:74–76.

Gallagher JP, Inokuchi H, Nakamura J, Shinnick-Gallagher P (1981): Effects of anticonvulsants on excitability and GABA sensitivity of cat dorsal root ganglion cells. Neuropharmacology 20:427–433.

Gardner CR (1984): Agonist and inverse agonist effects of CL 218872 on benzodiazepine receptors: Neurosci Lett 51:1–6.

Geller HM (1979): Water soluble benzodiazepines with agonistic and antagonistic actions on GABA-induced inhibition in cultured hypothalamus. Neurosci Lett 15:313–318.

Geller HM, Taylor DH, Hoffer BJ (1978): Benzodiazepines and central inhibitory mechanisms. Naunyn Schmiedebergs Arch Pharmacol 304:81–88.

Geller HM, Woodward DJ (1979): Synaptic organization of tuberal hypothalamus in tissue culture: Effects of electrical stimulation and blockers of synaptic transmission. Exp Neurol 64:535–552.

Gray JA (1982): "The Neuropsychology of Anxiety: An Enquiry into the Function of the Septo-Hippocampal System." Oxford: Oxford University Press.

Haefely W (1984): Pharmacological profile of two benzodiazepine partial agonists: Ro 16-6028 and Ro 17-1812. Clin Neuropharmacol 7 (Suppl. 1):S363–364.

Haefely W, Polc P (1983): Electrophysiological studies on the interaction of anxiolytic drugs with GABAergic mechanisms. In Malick JB, Enna SJ, Yamamura HI (eds): "Anxiolytics: Neurochemical, Behavioral and Clinical Perspectives." New York: Raven, pp 113–145.

Haefely W, Keller HH, Pieri L, Polc P Schaffner R, Zihlmann R (1978): Interaction of minor tranquillizers with synaptic processes mediated by GABA. In Deniker P, Radouco-Thomas C, Villeneuve A (eds):

"Neuro-Psychopharmacology." Oxford: Pergamon, pp 907–916.
Haefely W, Kulcsár A, Möhler H, Pieri L, Polc P, Schaffner R (1975): Possible involvement of GABA in the central actions of benzodiazepines. In Costa E, Greengard P (eds): "Mechanism of Action of Benzodiazepines." New York: Raven, pp 131–151.
Haefely W, Pieri L, Polc P, Schaffner R (1981): General pharmacology and neuropharmacology of benzodiazepine derivatives. In Hoffmeister F, Stille G (eds): "Handbook of Experimental Pharmacology," Vol. 55/II. Berlin: Springer, pp 13–262.
Haefely W, Kyburz E, Gerecke M, Möhler H (1985): Recent advances in the molecular pharmacology of benzodiazepine receptors and in the structure activity relationships of their agonists and antagonists. Advances in Pharmacology 14:165–322.
Hamill OP, Bormann J, Sakmann B (1983): Activation of multiple-conductance state chloride channels in spinal neurones by glycine and GABA. Nature 305:805–808.
Headley PM, Désarménien M, Santangelo F, Feltz P (1981): Direct action of pentobarbitone in potentiating the response to GABA of rat dorsal root ganglion neurons in vitro. Neurosci Lett 24:273–278.
Higashi H, Nishi S (1982): Effect of barbiturates on the GABA receptor of cat primary afferent neurones. J Physiol 332:299–314.
Houamed KM, Bilbe G, Smart TG, Constanti A, Brown DA, Barnard EA (1984): Expression of functional GABA, glycine and glutamate receptors in Xenopus oocytes injected with rat brain mRNA. Nature 310:318–321.
Huang L-YM, Barker JL (1980): Pentobarbital: Stereospecific action of (+) and (−) isomers revealed on cultured mammalian neurons. Science 207:195–197.
Iodarola MJ, Gale K (1982): Substantia nigra: Site of anticonvulsant activity mediated by γ-aminobutyric acid. Science 218:1237–1240.
Jahnsen H, Laursen AM (1981): The effects of a benzodiazepine on the hyperpolarizing and the depolarizing responses of hippocampal cells to GABA. Brain Res 207:214–217.
Jensen MS, Lambert JDC (1983): The interaction of the β-carboline derivative DMCM with inhibitory amino acid responses on cultured mouse neurones. Neurosci Lett 40:175–179.
Jiang ZG (1981): A comparison of the effects of flurazepam on γ-aminobutyric acid mediated depression of cerebellar and cerebral cortical neurons. Can J Physiol Pharmacol 59:595–598.
Johnston GAR, Skerritt JH (1983): GABARINS and the nexus between GABA and benzodiazepine receptors. In Bowery NG (ed): "Actions and Interactions of GABA and Benzodiazepines." New York: Raven, pp 179–189.
Kamp CW, Morgan WW (1980a): GABA antagonists enhance dopamine turnover in the rat retina in vivo. Eur J Pharmacol 69:273–279.
Kamp CW, Morgan WW (1980b): Some barbiturates enhance the effect of muscimol on dopamine turnover in the rat retina. J Pharmacol Exp Ther 213:332–336.
Kamp CW, Morgan WW (1982): Benzodiazepines suppress the light response of retinal dopaminergic neurons in vivo. Eur J Pharmacol 77:343–346.
Keller HH, Schaffner R, Haefely W (1976): Interaction of benzodiazepines with neuroleptics at central dopamine neurons. Arch Pharmacol 294:1–7.
Kenakin TP (1984): The classification of drugs and drug receptors in isolated tissues. Pharmacol Revs 36:165–222.
Kirsten EB, Schoener EP (1972): Antagonism of pentylenetetrazol excitation by anticonvulsants on single brain stem neurons. Neuropharmacology 11:591–599.
Koob GF, Thatcher-Britton K, Britton DR, Roberts DCS, Bloom FE (1984): Destruction of the locus coeruleus and the dorsal NE bundle does not alter the release of punished responding by ethanol and chlordiazepoxide. Physiology & Behavior 33: 479–486.
Kozhechkin SN, Ostrovskaya RU (1977): Are benzodiazepines GABA antagonists? Nature 269:72–73.
Krespan B, Springfield SA, Haas H, Geller HM (1984): Electrophysiological studies on benzodiazepine antagonists. Brain Res 295:265–274.
Lamour Y (1979): Les propriétés dépolarisantes du GABA au niveau des neurones du ganglion rachidien du rat: Application à la pharmacologie de l'inhibition présynaptique. Ph.D. Thesis, Université Pierre-et-Marie-Curie, Paris.
Larson MD, Major MA (1970): The effect of hexobarbitone on the duration of the recurrent IPSP in cat motoneurons. Brain Res 21:309–311.
Laurent JP, Mangold M, Humbel U, Haefely W (1983): Reduction by two benzodiazepines and pentobarbitone of the multiunit activity in substantia nigra, hippocampus, nucleus locus coeruleus and nucleus raphé dorsalis of encéphale isolé rats. Neuropharmacology 22:501–511.
Leah JD, Malik R, Curtis DR (1983): Actions of midazolam in the spinal cord of the cat. Neuropharmacology 22:1349–1356.
Lee HK, Dunwiddie TW, Hoffer BJ (1979): Interaction of diazepam with synaptic transmission in the in vitro rat hippocampus. Naunyn Schmiedebergs Arch Pharmacol 309:131–136.
Le Fur G, Mizoule J, Burgevin MC, Fenis O, Heanlure M, Gauthier A, Guérémy C, Uzan A (1981): Multiple benzodiazepine receptors: evidence of a dissociation between anticonflict and anticonvulsant properties by PK 8165 and PK 9084 (two quinoline derivatives). Life Sci 28:1439–1448.
Levy RA (1980): Presynaptic control of input to the central nervous system. Can J Physiol Pharmacol 58:751–766.
Lippa AS, Critchett D, Sano MC, Klepner CA, Green-

blatt EN, Coupet JM, Beer B (1979): Benzodiazepine receptors: Cellular and behavioral characteristics. Pharmacol Biochem Behav 10:831–843.

Little HJ (1984): The effects of benzodiazepine agonists, inverse agonists and Ro 15-1788 on the response of the superior cervical ganglion to GABA in vitro. Br J Pharmacol 83:57–68.

Lodge D, Anis NA (1984): Effects of ketamine and three other anaesthetics on spinal reflexes and inhibitions in the rat. Br J Anesthesiol 56:1143–1151.

Lodge D, Curtis DR (1978): Time course of GABA and glycine actions on cat spinal neurones: Effect of pentobarbitone. Neurosci Lett 8:125–129.

Macdonald RL, Barker JL (1978): Different actions of anticonvulsant and anesthetic barbiturates revealed by use of cultured mammalian neurons. Science 200:775–777.

Mugnaini E, Wonterlood FG, Dahl AL, Oertel WH (1984): Immunocytochemical identification of GABAergic neurons in the main olfactory bulb of the rat. Arch Ital Biol 122:83–113.

Nanopoulos D, Belin MF, Maitre M, Vincendon G, Pujol JF (1980): Immunocytochimie de la glutamate décarboxylase: en évidence d'éléments neuronaux GABAergiques dans le noyau raphé dorsalis du rat. Comptes Rendus de l'Académie des Sciences (Paris) 290:1153–1156.

Nestoros JN (1982): Benzodiazepine and GABA receptors are functionally related: further electrophysiological evidence in vivo. Progr Neuropsychopharmacol Biol Psychiat 6:417–420.

Nestoros JN, Nistri A (1978): A presynaptic component of the action of iontophoretically applied flurazepam on feline cortical neurones. Can J Physiol Pharmacol 56:889–892.

Nestoros JN, Nistri A (1979): Effects of microiontophoretically applied flurazepam on responses of cerebral cortical neurones to putative neurotransmitters. Can J Physiol Pharmacol 57:1324–1329.

Nicoll RA (1972): The effects of anaesthetics on synaptic excitation and inhibition in the olfactory bulb. J Physiol (Lond.) 223:803–814.

Nicoll RA (1975a): Presynaptic action of barbiturates in the frog spinal cord. Proc Natl Acad Sci USA 72:1460–1463.

Nicoll RA (1975b): Pentobarbital: Action on frog motoneuron. Brain Res 96:119–123.

Nicoll RA (1978): Selective actions of barbiturates on synaptic transmission. In Lipton MA, DiMascio A, Killam KF (ed): "Psychopharmacology: A Generation of Progress." New York: Raven, pp 1337–1348.

Nicoll RA (1980): Sedative-hypnotics: animal pharmacology. In Iversen LL, Iversen SD, Snyder SH (eds): "Handbook of psychopharmacology." New York: Plenum Press, pp 187–234.

Nicoll RA, Alger BE (1979): Presynaptic inhibition: transmitter and ionic mechanisms. Int Rev Neurobiol 21:217–258.

Nicoll RA, Eccles JC, Oshima T (1975): Prolongation of hippocampal inhibitory postsynaptic potentials by barbiturates. Nature 258:625–627.

Nicoll RA, Iwamoto ET (1978): Action of pentobarbital on sympathetic ganglion cells. J. Neurophysiol 41:977–986.

Nicoll RA, Madison DV (1982): General anesthetics hyperpolarize neurons in the vertebrate central nervous system. Science 217:1055–1057.

Nicoll RA, Wojtowicz JM (1980): The effects of pentobarbital and related compounds on frog motoneurons. Brain Res 191:225–237.

Nishikawa T, Scatton B (1983): Evidence for a GABAergic inhibitory influence on serotonergic neurons originating from the dorsal raphe. Brain Res 279:325–329.

Nistri A, Berti C (1983): Potentiating action of midazolam on GABA- mediated responses and its antagonism by Ro 14-7437 in the frog spinal cord. Neurosci Lett 39:199–204.

Nistri A, Berti C (1984a): Influence of benzodiazepine on GABA- evoked responses of amphibian brain and spinal neurones in vitro. Neuropharmacology 23:851–852.

Nistri A, Berti C (1984b): GABA-induced depolarizing responses of the frog spinal cord can be either enhanced or antagonized by the benzodiazepine midazolam and the methylxanthine caffeine. Neurosci Lett 47:277–281.

Nistri A. Constanti A (1978): Effects of flurazepam on amino acid- evoked responses recorded from the lobster muscle and the frog spinal cord. Neuropharmacology 17:127–135.

Nistri A, Constanti A, Krnjevič K (1980): Electrophysiological studies of the mode of action of GABA on vertebrate central neurons. In Pepeu G, Kuhar MJ, Enna SJ (eds): "Receptors of Neurotransmitters and Peptide Hormones." New York: Raven, pp 81–90.

Okamoto K, Sakai Y (1979): Augmentation by chlordiazepoxide of the inhibitory effects of taurine, beta-alanine and GABA on spike discharges in guinea-pig cerebellar slices. Br J Pharmacol 65:227–285.

Olds ME, Olds J (1969): Effects of anxiety-relieving drugs on unit discharges in hippocampus reticular midbrain, and preoptic area in the freely moving rat. Int J Neuropharmacol 8:87–103.

Olsen RW (1982): Drug interactions at the GABA receptor-ionophore complex. Ann Rev Pharmacol Toxicol 22:245–277.

Ongini E (1983): Behavioral and EEG effects of benzodiazepines and their antagonists in the cat. In Biggio G, Costa E (eds): "Benzodiazepine Recognition Site Ligands: Biochemistry and Pharmacology." New York: Raven, pp 211–225.

Paterson IA, Roberts MHT (1983): Methyl-$\beta$-carboline-3-carboxylate attenuates GABA effects in rat brain hippocampus in vivo. Brain Res 278:274–278.

Pellow S, File SE, Simmonds MA (1984): The benzodiazepine Ro 5-4864 antagonizes the electrophysiological effects of flurazepam and has convulsant and anxiogenic properties. 13th CINP Congress, Florence, P-226.

Pickles HG, Simmonds MA (1978): Field potentials, inhibition and the effect of pentobarbitone in the rat olfactory cortex slice. J Physiol (Lond.), 275:135–148.

Pieri L, Haefely W (1976): The effect of diphenylhydantoin, diazepam and clonazepam on the activity of Purkinje cells in the rat cerebellum. Arch Pharmacol 296:1–4.

Polc P (1978): Intravenous GABA, muscimol and diazepam: GABAergic transmission in the spinal cord and cuneate nucleus of cats. In: "Abstracts 11th C.I.N.P. Congress, Vienna," p 40.

Polc P (1981): Effects of the selective benzodiazepine antagonist Ro 15-1788 on the cat spinal cord. Experientia 37:674.

Polc P, Haefely W (1976): Effects of two benzodiazepines, phenobarbitone and baclofen on synaptic transmission in the cat cuneate nucleus. Naunyn Schmiedebergs Arch Pharmacol 294:121–131.

Polc P, Haefely W (1982): Benzodiazepines enhance the bicuculline-sensitive part of recurrent Renshaw inhibition in the cat spinal cord. Neurosci Lett 28:193–197.

Polc P, Schaffner R (1983): Electrophysiological effects of the peripheral-type benzodiazepine receptor ligand Ro 5-4864 in cat spinal cord and rat hippocampus. Neurosci Lett (Suppl) 14:S 288.

Polc P, Möhler H, Haefely W (1974): The effect of diazepam on spinal cord activities: possible sites and mechanisms of action. Naunyn Schmiedebergs Arch Pharmacol 284:319–337.

Polc P, Ropert N, Wright PM (1981a): Ethyl $\beta$-carboline-3-carboxylate antagonizes the action of GABA and benzodiazepines in the hippocampus. Brain Res. 217:216–220.

Polc P, Laurent J-P, Scherschlicht R, Haefely W (1981b): Electrophysiological studies on the specific benzodiazepine antagonist Ro 15-1788. Naunyn Schmiedebergs Arch Pharmacol 316:317–325.

Polc P, Bonetti EP, Schaffner R, Haefely W (1982): A three-state model of the benzodiazepine receptor explains the interactions between the benzodiazepine antagonist Ro 15-1788, benzodiazepine tranquilizers, $\beta$-carbolines, and phenobarbitone. Naunyn Schmiedebergs Arch Pharmacol 321:260–264.

Polzin RL, Barnes CD (1976): The effect of diazepam and picrotoxin on brainstem evoked dorsal root potentials. Neuropharmacology 15:133–137.

Raabe W, Gumnit RJ (1977): Anticonvulsant action of diazepam: Increase of cortical postsynaptic inhibition. Epilepsia 18:117–120.

Ransom BR, Barker JL (1975): Pentobarbital modulates transmitter effects on mouse spinal neurones grown in tissue culture. Nature 254:703–705.

Ransom BR, Barker JL (1976): Pentobarbital selectively enhances GABA-mediated post-synaptic inhibition in tissue cultured mouse spinal neurons. Brain Res 114:530–535.

Richards JG, Möhler H, Haefely W (1982): Benzodiazepine binding sites: receptors or acceptors? Trends in Pharmacological Sciences 3:233–235.

Richards JG, Möhler H, Haefely W (1985): Benzodiazepine receptors and their ligands. In Woodruff GN (ed): "Mechanisms of Drug Action" (in press).

Riley M, Scholfield CN (1983): Diazepam increases GABA-mediated inhibition in the olfactory cortex slice. Pflügers Arch 397:312–318.

Roberts E (1979): New directions in GABA research. 1. Immunocytochemical studies of GABA neurons. In Krogsgaard-Larsen P, Scheel-Krüger J, Kofod H (eds): "GABA-Neurotransmitters." Copenhagen: Munksgaard, pp 28–45.

Ross RJ, Waszczak BL, Lee EK, Walters JR (1982): Effects of benzodiazepines on single unit activity in the substantia nigra pars reticulata. Life Sci 31:1025–1035.

Ruffolo RR (1982): Important concepts for receptor theory. J Auton Pharmacol 2:277–295.

Sanghera MK, German DC (1983): The effects of benzodiazepine and non-benzodiazepine anxiolytics on locus coeruleus unit activity. J Neural Transmission 57:267–279.

Schlosser W, Franco S (1979): Modification of GABA-mediated depolarization of the cat ganglion by pentobarbital and two benzodiazepines. Neuropharmacology 18:377–381.

Schmidt RF (1963): Pharmacological studies on the primary afferent depolarization of the toad spinal cord. Pflugers Arch 277:325–346.

Schmidt RF, Vogel ME, Zimmermann M (1967): Die Wirkung von Diazepam auf die präsynaptische Hemmung und andere Rückenmarksreflexe. Naunyn Schmiedebergs Arch Pharmacol 258:69–82.

Schmidt I, Eler I, Fisenko VP (1978): Effect of clonazepam on neuronal activity in the sensomotor cortex in rabbit's brain. Farmakol Toksikol 41:231–234.

Schoch P, Richards JG, Häring P, Takacs B, Stähli C, Staehelin T, Haefely W, Möhler H (1985): Co-localisation of $GABA_A$ receptors and benzodiazepine receptors in the brain shown by monoclonal antibodies. Nature 314:168–171.

Scholfield CN (1978): A barbiturate induced intensification of the inhibitory potential in slices of guinea-pig olfactory cortex. J Physiol (Lond.) 275:559–566.

Scholfield CN (1983): Ro 15-1788 is a potent antagonist of benzodiazepines in the olfactory cortex slice. Pflügers Arch 396:292–296.

Schulz DW, Macdonald RL (1981): Barbiturate enhancement of GABA-mediated inhibition and activation of chloride ion conductance: Correlation with anticonvulsant and anesthetic actions. Brain Res

209:177–188.
Segal M, Barker JL (1984): Rat hippocampal neurons in culture: voltage-clamp analysis of inhibitory synaptic connections. J Neurophysiol 52:469–487.
Shepherd GM (1983): "The Synaptic Organization of the Brain." Oxford: Oxford University Press.
Sher PK, Study RE, Mazzetta J, Barker JL, Nelson PG, (1983): Depression of benzodiazepine binding and diazepam potentiation of GABA-mediated inhibition after chronic exposure of spinal cord cultures to diazepam. Brain Res 268:171–176.
Simmonds MA (1981): Distinction between the effects of barbiturates, benzodiazepines, and phenytoin on responses to $\gamma$-aminobutyric acid receptor activation and antagonism by bicuculline and picrotoxin. Br J Pharmacol 73:739–747.
Simmonds MA (1984): Physiological and pharmacological characterization of the actions of GABA. In Bowery NG (ed): "Actions and Interactions of GABA and Benzodiazepines." New York: Raven: pp 27–41.
Sinclair JG, Lo GF, Harris DP (1982): Flurazepam effects on rat cerebellar Purkinje cells. Gen Pharmacol 13:453–456.
Skerritt JH, Macdonald RL (1984a): Benzodiazepine receptor ligand actions on GABA responses. Benzodiazepines, CL 218872, zopiclone. Eur J Pharmacol 101:127–134.
Skerritt JH, Macdonald RL (1984b): Benzodiazepine receptor ligand actions on GABA responses. $\beta$-Carbolines, purines. Eur J Pharmacol 101:135–141.
Skerritt JH, Werz MA, McLean MJ, Macdonald RL (1984): Diazepam and its anomalous p-chloro-derivative Ro 5-4864: comparative effects on mouse neurons in cell culture. Brain Res 310:99–105.
Skerritt JH, Willow M, Johnston AR (1982): Diazepam enhancement of low affinity GABA binding to rat brain membranes. Neurosci Lett 29:63–66.
Smart TG, Constanti A, Bilbe G, Brown DA, Barnard EA (1983): Synthesis of functional chick brain GABA-benzodiazepine-barbiturate receptor complex in mRNA-injected xenopus oocytes. Neurosci Lett 40:55–59.
Somogyi P, Hodgson AJ, Smith AD, Nunzi MG, Gorio A, Wu JY (1984): Different populations of GABAergic neurons in the visual cortex and hippocampus of cat contain somatostatin — or cholecystokinin-immunoreactive material. J Neurosci 4:2590–2603.
Sorensen S, Freedman R (1983): Effects of alprazolam on the activity of rat cerebellar Purkinje neurons: evidence for mediation by norepinephrine. Drug Development Research 3:550–560.
Speth RC, Wastek GJ, Yamamura HI (1979): Benzodiazepine receptors: Temperature dependence of [$^3$H]flunitrazepam binding. Life Sci 24:351–358.
Squires RF, Braestrup C (1977): Benzodiazepine receptors in rat brain. Nature 266:732–734.
Strahlendorf HK, Barnes CD (1983): Control of substantia nigra pars reticulata neurons by the nucleus accumbens. Brain Res Bull 11:259–263.
Steiner FA, Hummel P (1968): Effects of nitrazepam and phenobarbital on hippocampal and lateral geniculate neurones in the cat. Int J Neuropharmacol 7:61–69.
Study RE, Barker JL (1981): Diazepam and (−) pentobarbital: Fluctuation analysis reveals different mechanisms for potentiation of GABA responses in cultured central neurons. Proc Natl Acad Sci USA 78:7180–7184.
Tallman JF, Paul SM, Skolnick P, Gallager DW (1980): Receptors for the age of anxiety: pharmacology of the benzodiazepines. Science 207:274–281.
Tancredi V, Frank C, Brancati A, Avoli M, White P (1983): Interactions between amino acid neurotransmitters and flurazepam in the neocortex of unanesthetized rats. J Neurosci Res 9:159–164.
Teyler TJ, DiScenna P (1984): The topological anatomy of the hippocampus: a clue to its function. Brain Res Bull 12:711–719.
Thiébot M-H, Hamon M, Soubrié P (1982): Attenuation of induced-anxiety in rats by chlordiazepoxide: role of raphe dorsalis benzodiazepine binding sites and serotoninergic neurons. Neuroscience 7:2287–2294.
Thiébot M-H, Soubrié P, Hamon M, Simon P (1984): Evidence against the involvement of serotonergic neurons in the antipunishment activity of diazepam in the rat. Psychopharmacology 82:355–359.
Trulson ME (1984): Effects of diazepam on behavior and dopamine-containing substantia nigra units in freely moving cats. Psychopharmacology 84:91–95.
Trulson ME, Trulson VM (1983): Chloral hydrate anesthesia alters the responsiveness of dorsal raphé neurones to psychoactive drugs. Life Sci 32:949–956.
Trulson ME, Preussler DW, Howell GA, Frederickson CF (1982): Raphe unit activity in freely moving cats: effects of benzodiazepines. Neuropharmacology 21:1045–1050.
Tsuchiya T, Fukushima H (1978): Effects of benzodiazepines and pentobarbitone on the GABAergic recurrent inhibition of hippocampal neurons. Eur J Pharmacol 48:421–424.
Umemoto M, Olds ME (1975): Effects of chlordiazepoxide, diazepam and chlorpromazine on conditioned emotional behavior and conditioned neuronal activity in limbic, hypothalamic and geniculate regions. Neuropharmacology 14:413–425.
Waszczak BL (1983): Diazepam potentiates GABA-, but not adenosine-mediated, inhibition of neurons of the nigral pars reticulata. Neuropharmacology 22:953–959.
Waterhouse BD, Moises HC, Yeti HH, Geller HM, Woodward DJ (1984): Comparison of norepinephrine- and benzodiazepine-induced augmentation of Purkinje cell responses to $\gamma$-aminobutyric acid

(GABA). J Pharmacol Exp Ther 228:257–267.
Weakey JN, Esplin DW, Zablocka B (1960): Criteria for assessing effects of drugs on postsynaptic inhibition. Arch Int Pharmacodyn Ther 171:385–393.
Weissmann BA, Cott J, Paul SM, Skolnick P (1983): Ro 5-4864: a potent benzodiazepine convulsant. Eur J Pharmacol 90:149–150.
White WF, Dichter MA, Snodgrass SR (1981): Benzodiazepine binding and interactions with the GABA receptor complex in living cultures of rat cerebral cortex. Brain Res 215:162–176.
Wolf P, Haas HL (1977): Effects of diazepines and barbiturates on hippocampal recurrent inhibition. Arch Pharmacol 299:211–218.
Yoshida M, Precht W (1971): Monosynaptic inhibition of neurons on the substantia nigra by caudato-nigral fibres. Brain Res 32:225–228.
Zakusov VV, Ostrovskaya RU, Kozhechkin SN, Markovitch VV, Molodavkin GM, Voronina TA (1977): Further evidence for GABAergic mechanisms in the action of benzodiazepines. Arch Int Pharmacodyn Ther 229:313–326.
Zakusov VV, Ostrovskaya RU, Markovitch, Molodavkin GM, Bulayev VM (1975): Electrophysiological evidence for an inhibitory action of diazepam upon cat brain cortex. Arch Int Pharmacodyn Ther 214: 188–205.

**Benzodiazepine/GABA Receptors and Chloride Channels: Structural and Functional Properties, pages 135–165**


# 6

# Electrophysiological Pharmacology of GABA and Diazepam in Cultured CNS Neurons

**Jeffery L. Barker and David G. Owen**

*Laboratory of Neurophysiology, National Institute of Neurological and Communicative Disorders and Stroke, National Institutes of Health, Bethesda, Maryland 20892*

As described in other chapters in this volume, there is abundant immunohistochemical and biochemical evidence for the widespread distribution of GABAergic neurons throughout the mammalian CNS. Pharmacological applications of GABA onto individual central neurons in vivo and in vitro have almost invariably led to depression of spontaneous or evoked electrical excitability, so that GABA is now considered primarily to be an "inhibitory" neurotransmitter. The transmitter signal is conceptualized as "inhibitory" because it serves to decrease the probability that a neuron will generate rapid and brief, all-or-none changes in the existent voltage through activation of $Na^+$ conductance mechanisms in the membrane (commonly known as "action potentials"). The molecular mechanisms underlying the $Na^+$ conductance are under intensive study throughout the world. This conductance assumes special importance because many, if not all, CNS functions are derived in some way from such "electrical" excitability. Electrically excitable membrane properties help to dictate modes of intercellular communication and patterns of synaptic and extrasynaptic transmission. The molecular mechanisms underlying "inhibitory" transmitter signals generated in CNS tissue have been studied in most detail using embryonic neurons cultured from different regions of the CNS, mainly because such monolayers are readily accessible to electrophysiological assays of cell biological properties at relatively quantitative levels. The results obtained from these and other in vitro preparations of invertebrate and vertebrate CNS tissue show that the microscopic mechanisms underlying both electrical ("action potential"–like) and chemical (transmitter-receptor-regulated) forms of excitability are, in principle, quite similar in that they both involve conductance of ions through momentary "openings" or "channels" in the outer membrane.

In vitro preparations of CNS tissue have also been used to examine the pharmacologic effects of benzodiazepines on electrically and chemically excitable membrane properties. Several distinctly different actions have been observed, all of which likely contribute to the therapeutic effects of the drugs. In this chapter we review some recent electrophysiological studies on the electrophysiological pharmacology of GABA and benzodiazepines studied using neurons cultured from the embryonic CNS. Most, if not all of the results reported here have also been recorded with normally organized CNS tissue examined in vivo or in slices [reviewed in Haefely and Polc, this volume].

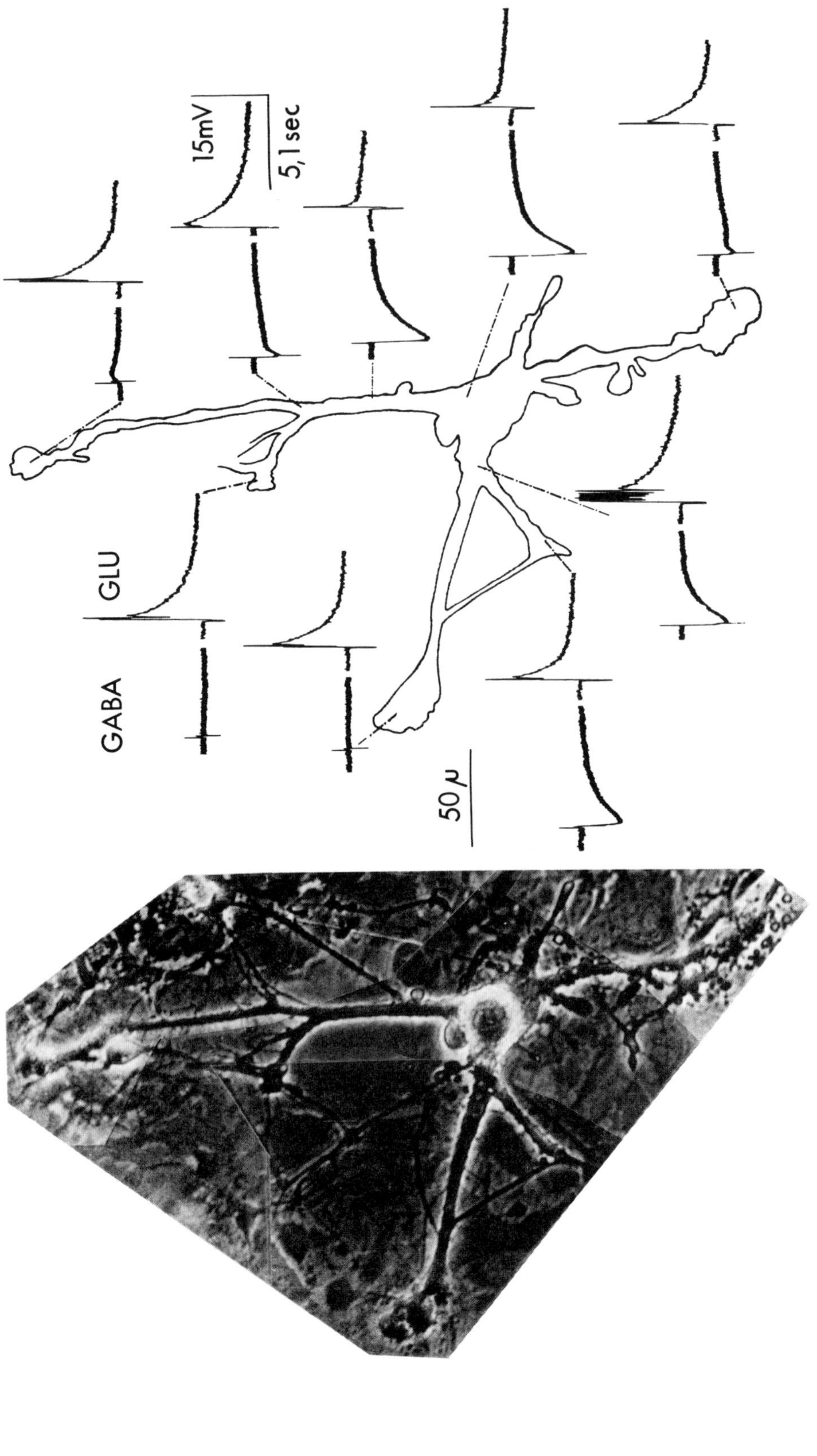
GABA
GLU
15mV
5,1 sec
50 μ

## $Cl^-$ ION CONDUCTANCE

We and others have studied the pharmacological actions of GABA in CNS cells cultured from the embryonic CNS. The most commonly observed pharmacological action involves a rapid hyperpolarization or increase in the membrane potential, which effectively reduces the probability that a cell will trigger an action potential (Fig. 1). In contrast, the acidic amino amino acid glutamate invariably evokes an "excitatory" depolarizing response sufficient to evoke actiion potentials in the same cell (Fig. 1). Responses evoked by GABA on neuronal processes are often more complex, consisting of both depolarizing and hyperpolarizing components (Fig. 2). The depolarizing phase desensitizes rapidly but has not yet been examined in further detail.

The amplitude and polarity of the hyperpolarizing response to GABA are quite sensitive to the immediate "history" of the cell in terms of its "resting" membrane potential. This reflects lability in the principal ion gradient associated with the pharmacologic action. Extrinsic current injected across the cell's membrane to hyperpolarize or depolarize the cell relative to its resting level results in time-dependent changes in the voltage responses evoked by repetitive applications of GABA (or the structurally similar amino acid glycine) (Fig. 3A,C). Sustained conditioning of the cell at hyperpolarized levels invariably shifts the "reversal" potential ($E_{rev}$) of the response in a hyperpolarizing direction (and vice versa) (Fig. 3B,D). $E_{rev}$ is the potential at which electrical and ionic gradients are equal and opposite so that although membrane conductance is activated by GABA, ion current does not flow into or out of the cell. For current to flow there must be "driving force," or an imbalance in the ionic and potential gradients. Electrical polarization alters the voltage gradient and ion current flows according to the newly established driving force acting on the GABA-activated conductance. The potential- and time-dependencies of $E_{rev}$ reflect the subsequent shift in the ion gradient, which reaches a new steady level in less than a minute, as the involved ions concentrate or dispel with continued activations (Fig. 3C).

Quick changes in $[Cl^-]_0$ induce time-dependent changes in the polarity of the GABA-evoked voltage response, while having no detectable effects on glutamate-elicited excitation (Fig. 4). Such shifts in the $Cl^-$ion gradient also have relatively little effect on the "passive" electrical properties of many cultured neurons. The peak conductance evoked by GABA is an order of magnitude or more greater in the presence of the full complement of $Cl^-$ions. Altering the $Cl^-$ gradient by actively injecting $Cl^-$ions into the cell shifts $E_{rev}$ for GABA-evoked responses in the depolarizing direction by about 40mV as more $Cl^-$ions concentrate in the cell and effectively change the $Cl^-$ion gradient (Fig. 4, bottom panel). All of these results, as well as those reported by other investigators using other in vitro preparations of central neurons, indicate that the hyperpolarizing response evoked by GABA in embryonic neurons cultured from spinal, hippocampal, and cortical regions involves primarily, if not exclusively $Cl^-$ion conductance mechanisms.

When the membrane potential and thus the "driving force" is held constant with the "voltage-clamp" feedback circuit, pharmacological applications of GABA trigger

**Fig. 1.** *Amino acid pharmacology recorded in a spinal cord neuron cultured 4 weeks from an E13 mouse embryo. The recording has been made at 23°C with a 4M $K^+$acetate microelectrode in a balanced saline containing 10 mM $MgCl_2$ and 1 mM $CaCl_2$. This divalent cation ratio isolates the cell from spontaneous synaptic activity and raises action potential threshold so that there is a relatively quiet and passive baseline upon which to evoke pharmacological responses. GABA and glutamate have been iontophoresed with identical pulses of current passed through the pipettes at different sites on the neuron (illustrated with phase-contrast optics on the left and schematized on the right). Clear and convincing hyperpolarizing responses to GABA are detectable in and about the cell body but become less clear or undetectable at the terminals. [Modified from Barker and Ransom, 1978a.]*

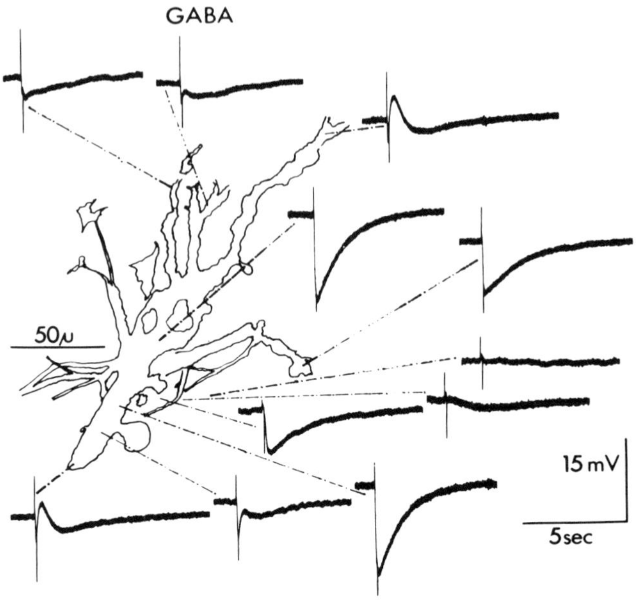


**Fig. 2.** *GABA-activated responses are simple at the cell body, complex at processes. $K^+$ acetate recording at room temperature from a cultured mouse spinal cord neuron. GABA was iontophoresed with identical brief charge pulses at different sites on the cell, which is drawn in outline form from a phase-contrast picture. The voltage responses are uniformly hyperpolarizing at the level of the cell body, but exhibit more complex behavior that includes a rapid depolarizing component superimposed on the hyperpolarization (e.g., responses evoked at the tips of the two processes arranged along the long axis of the cell). The effective pharmacological spread of the iontophoretic pulses is about 25 micrometers, as judged by the step-wise disappearance of the response as the drug pipette is removed from the cell body to the plate (set of traces at about 3 o'clock with respect to the cell body). The three traces at the bottom of the illustration, obtained over a 50 micrometer portion of one process, reveal the transition from complex de- and hyperpolarizing to simple, hyperpolarizing response characteristics. [Modified from Barker and Ransom, 1978a].*

$Cl^-$ ion current (rather than potential) responses. This occurs because the conductance is related to both the current and the driving force so that with the latter held constant by "voltage-clamping" the cell the underlying conductance response will be directly proportional to the current flowing. Brief "iontophoretic" applications of GABA using increasing iontophoretic current intensities elicit increasingly more membrane current response as more conductance is activated (Fig. 5, top). The relationship between applied iontophoretic charge and membrane current response is nonlinear (Fig. 5, left). Thus, there is disproportionately more current (reflecting $Cl^-$ conductance) over the lower concentration range of iontophoretically applied transmitter. Similar nonlinear dose–response curves have been developed using pressure applications of GABA. This result is true also for glycine, another putative "inhibitory" amino acid that activates $Cl^-$ conductance, but not for glutamate whose "dose–response" curves routinely have slopes about 1.0 [Barker and Ransom, 1978a]. One interpretation of this positive cooperativity in the case of GABA (and glycine) responses has been that there are two or more sites for agonist binding, which then leads to activation of the conductance.

## FLUCTUATION ANALYSIS OF Cl – ION CURRENTS

The microscopic mechanism underlying the macroscopic conductance (and current) re-

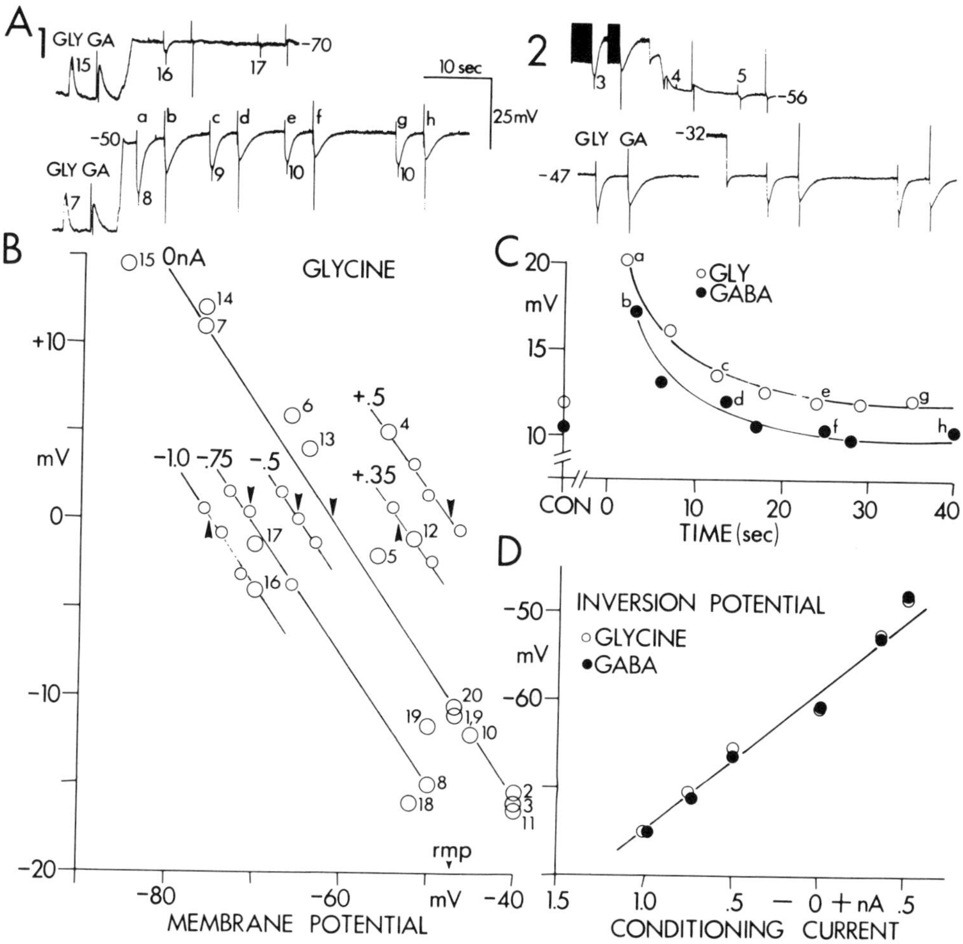


**Fig. 3.** *Reversal of amino acid–evoked responses is sensitive to conditioning-potential in a $K^+$ acetate recording from a brainstem neuron cultured 4 weeks. Panel A: Brief pulses of glycine (18 nA for 200 msec) and GABA (30 nA for 200 msec) produce responses that vary as a function of membrane potential and time. Individual glycine responses are numbered and lettered, while GABA responses are lettered. (1) In the upper pen-recorder trace, depolarizing responses to both glycine (numbered 15) and GABA evoked at a conditioning potential of −90 mV become hyperpolarizing (e.g., 16), then null (e.g., 17) as the cell is shifted to −70 mV. In the lower trace, the responses are depolarizing at about −75 mV (e.g., 7) and then invert at −50 mV to become hyperpolarizing (e.g., 8). (2) Upper trace. With the cell depolarized sufficiently to generate spontaneous action potential activity (responses look like "black rectangles" because of their high frequency and attenuation by the frequency response of the pen-recorder), the amino acid applications are clearly "inhibitory" to excitability. Rapid membrane hyperpolarization to −50 mV transiently inverts the responses (e.g., 4), which later revert to hyperpolarizing (e.g., 5). Lower trace. With the cell at rest (−47 mV), glycine and GABA elicit about 10 mV hyperpolarizing responses. Notice that in this and in all other pairs of responses those evoked by glycine are consistently shorter-lasting than those triggered by GABA. Panel B: Response amplitudes evoked by glycine when the cell is conditioned at various membrane potentials with sustained current injections are plotted against absolute membrane potentials. Arrowheads indicate measured or extrapolated inversion potentials. When inversion potential is plotted as a function of injected current, there is a direct, linear relationship between current level and absolute value of the inversion potential (Panel D). Panel C shows that the time courses of changes in pharmacological response amplitude after returning to resting levels following conditioning hyperpolarization are similar for both GABA and glycine and require 10–20 seconds. [Barker and Ransom, unpublished observations].*

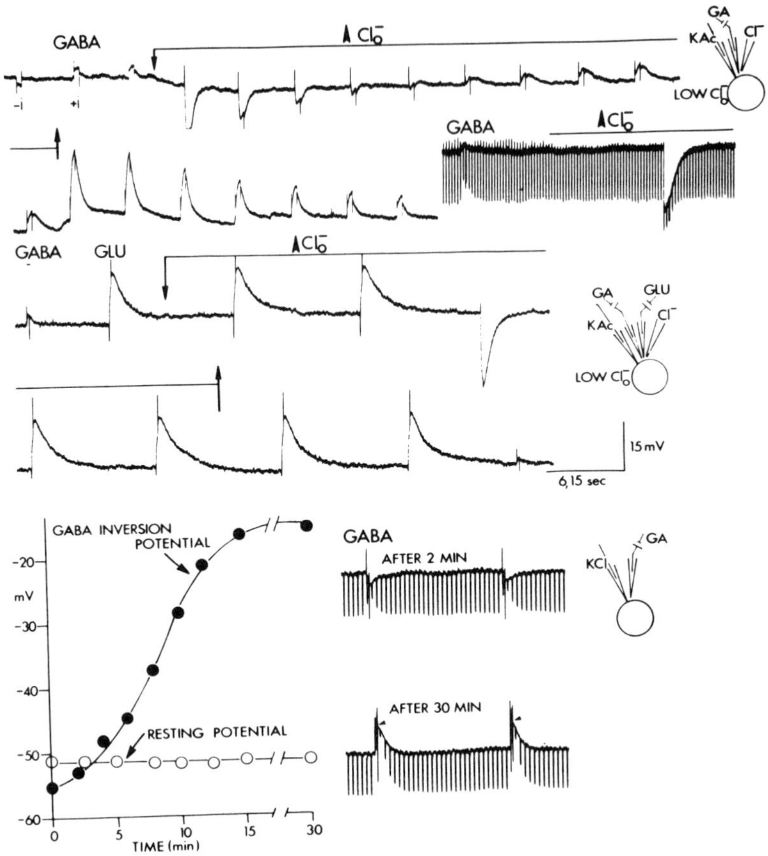


**Fig. 4.** *$Cl^-$ ion dependency of GABA-activated voltage and conductance resposonses. The intracellular recording has been made from a 4-week-old cultured mouse spinal cord neuron using a $K^+$ acetate microelectrode with cell bathed in low $Cl_0^-$ (mM; substitute anion: sulfate). When GABA is iontophoresed with 10 nA current pulses of either polarity ($-i$, $+i$) in the low $Cl_0^-$, there is little voltage response and much of what is registered is a symmetrical electrical coupling artifact between the current-passing and voltage-recording electrodes. Topmost trace: Beginning at the downward arrowhead, pressure is applied to a micropipette containing 159 mM $Cl^-$, thereby rapidly increasing the $[Cl^-]_0$ in the vicinity of the cell (experimental arrangements schematized in upper-right-hand corner). A slight hyperpolarization of the cell also suggests steady-state $Cl^-$ conductance occurs, which is undetectable during the other $Cl^-$ ion applications illustrated. Positive current pulses applied to the iontophoretic pipette now elicit a hyperpolarizing voltage response so large that the pen-recorder initially saturates. Repeated pulses quickly lead to apparent loss, then clear inversion of the voltage response. Termination of the $Cl^-$ ion application after the first response on the second trace (indicated by upward arrowhead) leads to a discernible, transient 3 mV depolarization coincident with a transient and marked increase in the initial responses. By the end of the trace (about 30 seconds) there is little biological effect of GABA left. Right-hand trace: brief constant-current hyperpolarizing pulses have been injected to detect membrane conductance responses to GABA. In low $Cl_0^-$ there is a measurable conductance but little voltage response. Application of $Cl_0^-$ has little, if any effect on resting properties but leads to a 20 mV hyperpolarizing response to GABA associated with a marked increase in conductance. In the next set of traces the excitatory amino acid glutamate has also been iontophoresed (schematized at right). A rapid increase in $Cl_0^-$ does not change the voltage responses to glutamate, but induces a 20 mV hyperpolarization to GABA, which disappears within 30 seconds of terminating the $Cl_0^-$ application. In the bottom panel of the figure, the recording has been made with a KCl-filled microelectrode and the resting potential of the cell and the inversion potential of the GABA response have been measured. Neither resting potential nor input resistance are affected. The GABA response is initially hyperpolarizing, inverting at $-55$ mV, then gradually shifts in a depolarizing direction over the next 15 minutes and stabilizes at about $-15$ mV. The conductance associated with the voltage change increases slightly. [Modified from Barker and Ransom, 1978a].*

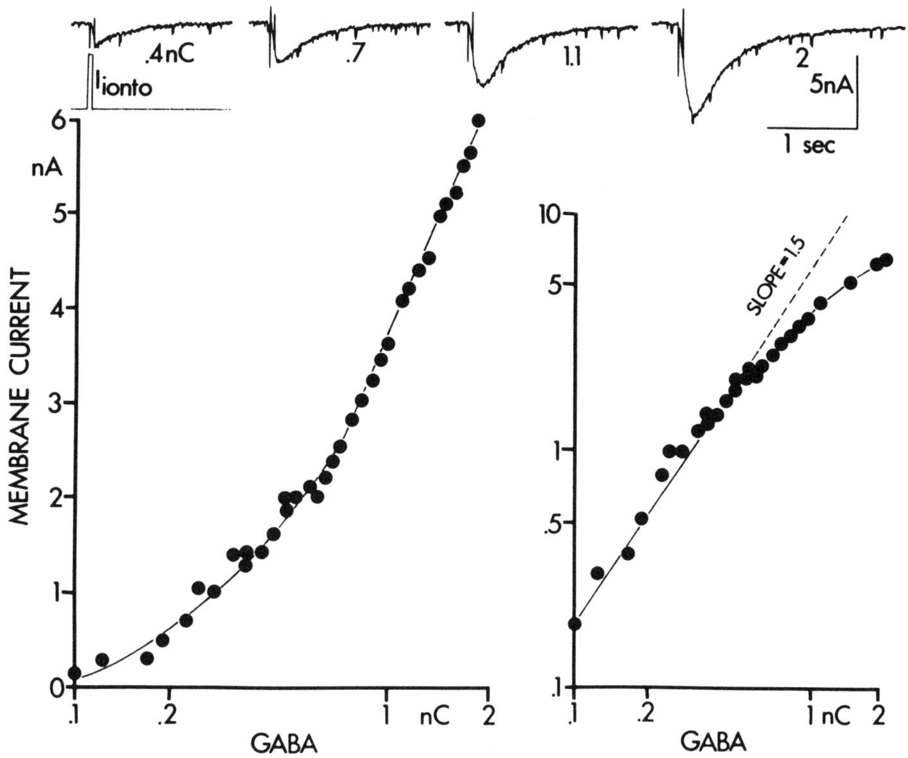


**Fig. 5.** *Iontophoretic charge-membrane current response curves for GABA. A mouse spinal cord neuron has been voltage-clamped at −70 mV with two KCl microelectrodes and increasing intensities of positive current pulsed through an iontophoretic pipette containing 1 M GABA. The iontophoretic charges are shown below some of the GABA-evoked membrane current responses in nanoCoulombs. The log–log transformation plotted on the right of the semi-log data illustrated on the left reveals a slope of 1.5 over the 0.1–1.0 nC range. [Barker and Macdonald, unpublished observations].*

sponses evoked by GAB has been studied in two ways. The first involves "noise" analysis, or spectral analysis of the fluctuations in membrane current so readily apparent during many responses evoked pharmacologically on excitable membranes by transmitters. Katz and Miledi [1972] were prescient to realize that the fluctuations might be a complex biological signal rather than instrument artifact. Since then, fluctuation analysis has been carried out in many systems with a number of cholinergic and amino acid transmitter substances.

Briefly, the strategy is based on the notion that the macroscopic current fluctuations represent moment-to-moment fluctuations in the number of microscopic, unitary-sized conductance events. The latter are conceptualized as all-or-none, picoSiemen-sized transitions in membrane conductance that occur randomly and independently and give rise to microscopic currents passing across the membrane through ion channels (Fig. 6A). The randomized conducting periods (and resulting elementary currents) arising from successful agonist–receptor interactions sum together to yield a total level of conductance (and corresponding current) that fluctuates in time. This conductance (and current) signal can be analyzed statistically for its mean level and average duration. Schematic representation of unitary activity in 10 hypothetical channels is illustrated in Figure 6A. By extension, the macroscopic current response to agonist and

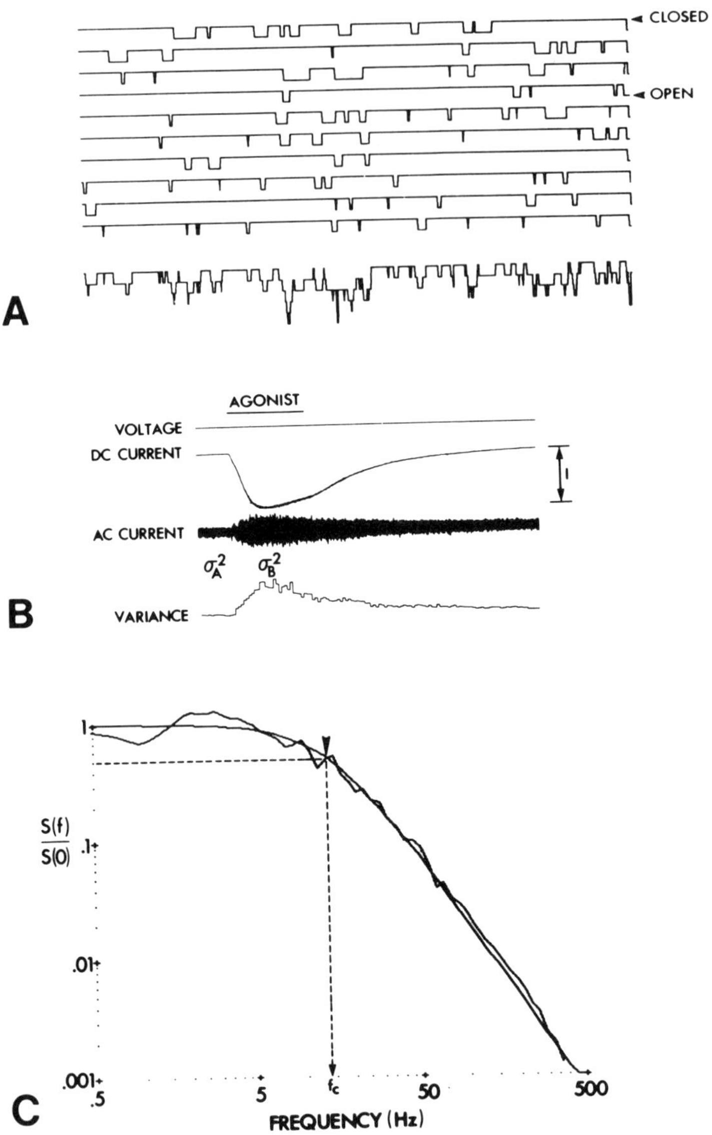


**Fig. 6.** *Relationship between microscopic "ion channels" and macroscopic "noise." A. Computer-simulated two-state (open–closed) ion channel activities. The random activities of the independent ion channels are illustrated above a trace of their summed activities. The majority of their amplitudes are similar so that the mean number of channels open, and hence the average level of summed single-channel current fluctuates. B. Two-electrode voltage-clamp recording of macroscopic $Cl^-$ ion current (I) induced in the whole cell by a GABAmimetic agonist. The current has been band-pass filtered and amplified 10× to reveal the clear increase in fluctuations occurring during the response in direct proportion to the response magnitude. One-second integrations of the membrane current variance show that second-by-second $\sigma^2$ values correspond precisely to the macroscopic response and associated fluctuations. C. A normalized power spectral density plot calculated from membrane current fluctuations induced by an agonist. The power at each frequency (S(f)) over the 0.5–500 Hz range has been divided by the zero frequency asymptote power (S(0)). The "corner" frequency ($F_c$), defined as that frequency at which the power is 50% of S(0), is indicated by the arrowhead and dashed lines. The uneven line reflects data points, while the smooth curve describes an equation that implies single exponential kinetics for the underlying process. In this example $F_c$ = 12 Hz, which corresponds to an estimated average channel duration, τ, of 13.2 msec according to the relationship $\tau = (2\pi f_c)^{-1}$.*

its associated "noise" or variance (Fig. 6B) are thought to arise from the simultaneous activities of several thousand ion channels.

Statistical and spectral analysis of the complex, fluctuating signal can thus be utilized for *estimating* the average elementary electrical dimensions of the underlying ion channel activity. [For details, the reader should consult Mathers and Barker, 1982]. If the ion channel activity is random in nature, channel open times will be exponentially distributed and the calculated power density spectrum will be described by a Lorentzian function, where the corner frequency is related to the closing rate constant (Fig. 6C). Spectral analysis of the fluctuating current response therefore allows resolution of the estimated channel duration, $\tau$. If the spectral analysis reflects the exponential kinetics of essentially all-or-none ion channel activities, then the average level of membrane current variance divided by the mean level of membrane current gives the estimated amplitude of the all-or-none microscopic current step. Microscopic current amplitude can then be related to the elementary conductance by knowing the "driving force" associated with the current. Fluctuation analysis of macroscopic current responses can thus yield estimates of microscopic ion channel conductance and duration.

## PATCH-CLAMP ANALYSIS OF $Cl^-$ ION CHANNELS

The second manner in which the microscopic mechanism has been examined involves direct electrical recording of the individual transitions in membrane conductance in micron-sized patches of the cell (Fig. 7). By forming a physically tight seal with the outer plasma membrane, thereby effectively insulating the patch electrically from the rest of the cell, it is possible to detect openings of individual ion channels with fine temporal resolution. All-or-none transitions can be evaluated for their amplitude and duration, and current-amplitude and current-time histograms derived.

Initially, resolution of such ion channel activities revealed, as predicted from fluctuation analysis, a unimodal, Gaussian distribution of amplitudes of single channel currents and single, exponential distributions in nonconducting and conducting periods. However, more recent experiments and improved experimental resolution have revealed more complex kinetics and multiple conductance states for a number of transmitter–receptor interactions. *In the case of single-channel currents activated by GABA, most (>90%) of the actual $Cl^-$ ion charge carried across the membrane occurs at one primary conducting level with kinetics approximated by the slower of two exponential distributions in channel conducting times observed in current-time histograms.* Hence, patch-clamp resolution of molecular pharmacological transmitter actions complements whole-cell strategies by providing direct recordings of individual elementary current properties under equilibrium conditions. There are still some uncertainties regarding the excess number of exponentially distributed fast channel events ($\tau$ = 2-5 msec at 23° and −60mV), and it is noteworthy that spontaneous, apparently agonist-independent, $Cl^-$ channel activities have been recorded at this potential [J.F. MacDonald, D.G. Owens, and J.L. Barker, 1985]. These $Cl^-$ channels have open times on the order of 1 msec at this potential and two exponentially distinct open times at +60mV that are similar in value to those activated by GABA at −60mV.

## MICROSCOPIC AND MACROSCOPIC PROPERTIES COMPARED

There is recognizable agreement between estimates from fluctuation analysis of whole-cell responses involving populations of ion channels activated by GABA and patch-clamp observations of GABA-evoked microcurrents (Fig. 8). The short-lived channel kinetics recorded in patches [Jackson et al., 1982; Sakmann et al., 1983] are not clearly and consistently recorded in fluctuation analysis of whole-cell recordings [but see Ozawa and Yuzaki, 1984]. However, the long open-time

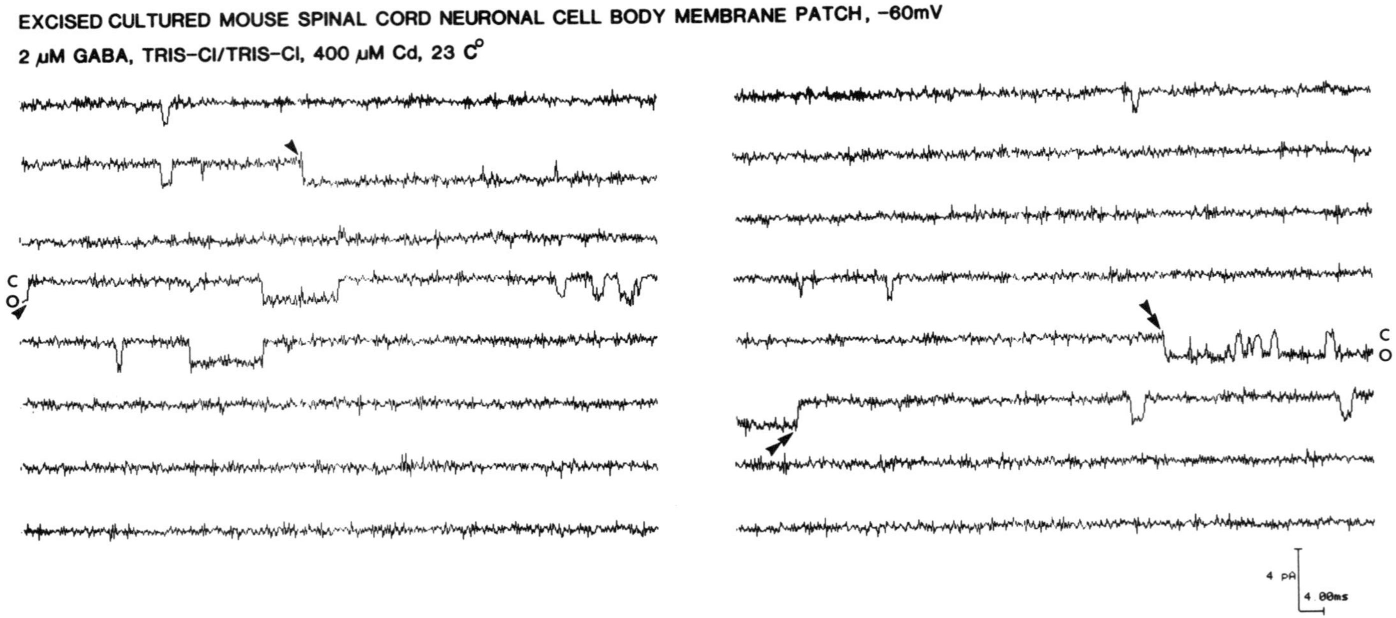


**Fig. 7.** *GABA-activated single channel currents. The microscopic current transitions have been recorded at 1 KHz bandwidth from a patch of cell body membrane excised from a 2-week-old cultured mouse spinal cord neuron. Symmetrical TRIS-Cl/TRIS-Cl solutions containing 0.4 mM $Cd^{2+}$ions and 1 μM TTX have been used to eliminate cation currents. No $K^{+}$ions were added in either bath or pipette solutions. Two-micromolar GABA has been included in the patch pipette, and elementary transitions in $Cl^{-}$ conductance have been revealed by holding the patch at −60 mV, thus providing a driving force of about 60 mV, since the currents invert at 0 mV when identical [$Cl^{-}$]s are used in solutions bathing both faces of the membrane patch. About 700 msec of continuous recording are shown. The single downward arrowhead in the set of left-hand traces marks the opening of a channel that remains conducting for well over 100 msec. The upward arrowhead in the fourth trace marks the transition from the open (O) to the closed state (C). A series of rapidly occurring transitions between closed and open states comprising "burst"-like activity are indicated in the right-hand set by double-arrowheads. [Redmann, Lecar, Barker, unpublished observations.]*

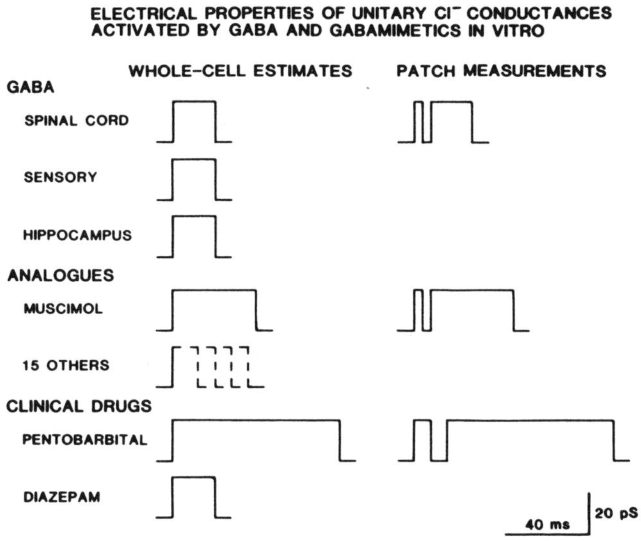


**Fig. 8.** *Comparison of estimated and measured molecular $Cl^-$ conductances in cultured CNS neurons. Indirect estimates have been made using fluctuation analysis of 1–3 nA membrane current responses generated in the whole cell with agonists being applied at the 10–200 micromolar concentration range. Direct measurements have been carried out in somal patches at 0.5–20 micromolar agonist concentration. The estimated elementary electrical properties activated by GABA appear to be similar for neurons cultured from different regions and correspond to the longer-lasting of the two exponentially distributed, single-channel populations recorded in patches. Muscimol-activated channels are about twice as long as those triggered by GABA but, like many analogues, have about the same conductance. Two clinically important drugs with agonist actions at $Cl^-$ conductance mechanisms [Barker and Ransom, 1978b; MacDonald and Barker, 1982] also activate two-state ion channels whose conductance approximates that of GABA. [Results reported in Mathers and Barker, 1980; Barker and Mathers, 1981; Barker, MacDonald and McBurney, 1982; Segal and Barker, 1984a; and Owen, Barker, Segal, Study, 1985].*

component derived from single-channel analysis corresponds closely to the estimates of average ion channel duration obtained in fluctuation analysis of membrane currents in recordings from whole cells, as do observed and estimated primary-level conductance values. *Since the long-lived open channel state accounts for more than 90% of the charge transfer, fluctuation analysis allows reasonable first approximation of the elementary electrical details of receptor-coupled $Cl^-$ ion channel activity.*

## CHANNEL DURATION CORRELATES WITH MOLAR POTENCY

Estimates of GABA-activated $Cl^-$ ion channels using fluctuation analysis in neurons cultured from different regions of the CNS show that the unitary event has relatively constant dimensions in functionally distinct neuronal phenotypes grown in culture independent of evidence for GABAergic innervation (Fig. 8). This is different from cholinergically operated channels at neuromuscular junctions where differences in channel properties can consistently be detected at synaptic and extrasynaptic sites in muscle membranes [Fischbach and Schuetze, 1980; Brenner and Sakmann, 1983; Schuetze and Vicini, 1984].

Muscimol, which is twice as potent on a molar basis as GABA in activating $Cl^-$ conductance, opens channels that are, on average, twice as long as those activated by GABA, suggesting that *channel duration* is a molecular correlate of *molar potency* [Barker

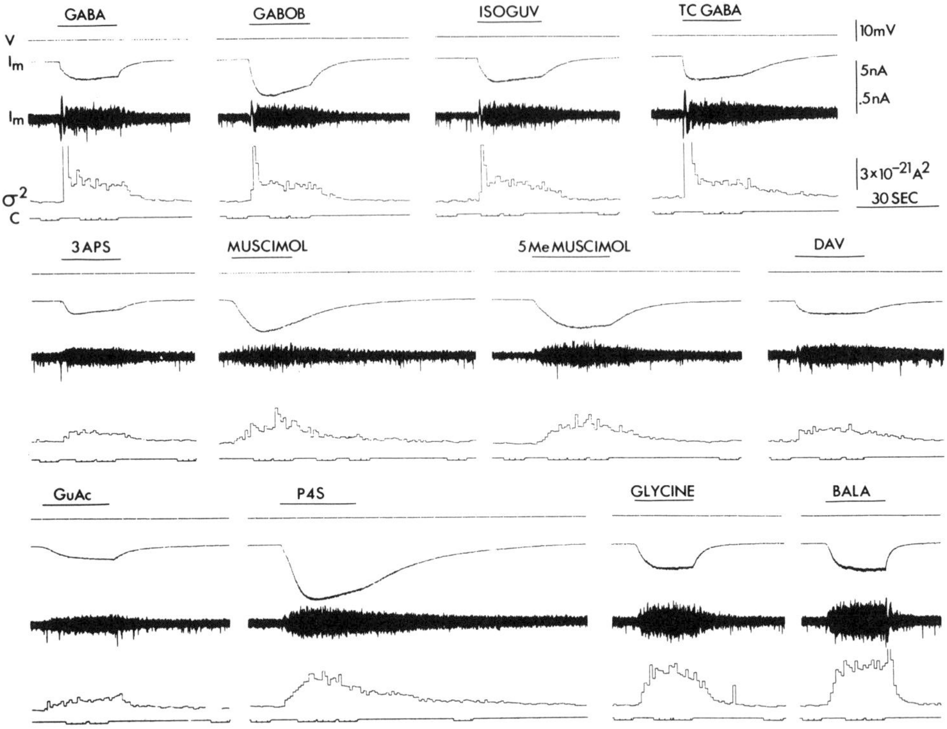


**Fig. 9.** *GABAmimetic analogues activate membrane current responses like GABA in a cultured spinal neuron. The cultured mouse spinal cord neuron has been voltage-clamped at −70 mV (trace marked V) with two KCl-filled microelectrodes and various analogues of GABA (at 10-500 micromolar concentrations) have been applied by pressure (marked by bars above traces). All of the agonists induce macroscopic current responses (upper traced labeled $I_m$) associated with coincident fluctuations (lower trace labeled $I_m$) and variance ($\sigma^2$) integrated each second. C marks the trace where data has been acquired by computer for analysis. Although both of the amino acid transmitters glycine and β-alanine evoke current responses comparable in magnitude to those triggered by several analogues (e.g., isoguvacine, transcyclopropane-GABA, and 5-methylmuscimol), measurably greater fluctuations and more variance can be detected for each. [Barker and Mathers, unpublished observations.]*

and Mathers, 1981]. Many other analogues that compete with labeled GABA in equilibrium binding assays conducted with membranes isolated from the CNS activate $Cl^-$ conductance in cultured CNS neurons (Fig. 9). Fluctuation analysis of these responses shows that each exhibits a "Lorentzian" power spectrum whose corner frequency is different (Fig. 10). Thus, on average each agonist opens $Cl^-$ ion channels for different periods; however, the elementary conductance estimated for each is similar, if not identical (Fig. 11). Coincident conclusions regarding cholinergic analogues and cation channel properties in muscle membranes have recently been reached [Gardner et al., 1984].

The estimated durations of the agonist-induced channels are highly and significantly correlated with the concentrations effective at displacing 50% of specific GABA binding in membrane preparations (Fig. 12). Thus, for a variety of structural analogues the molar potency of an agonist in competitively displacing GABA predicts its relative molecular efficacy in stabilizing $Cl^-$ conductance. Unfortunately, the degrees of freedom in molec-

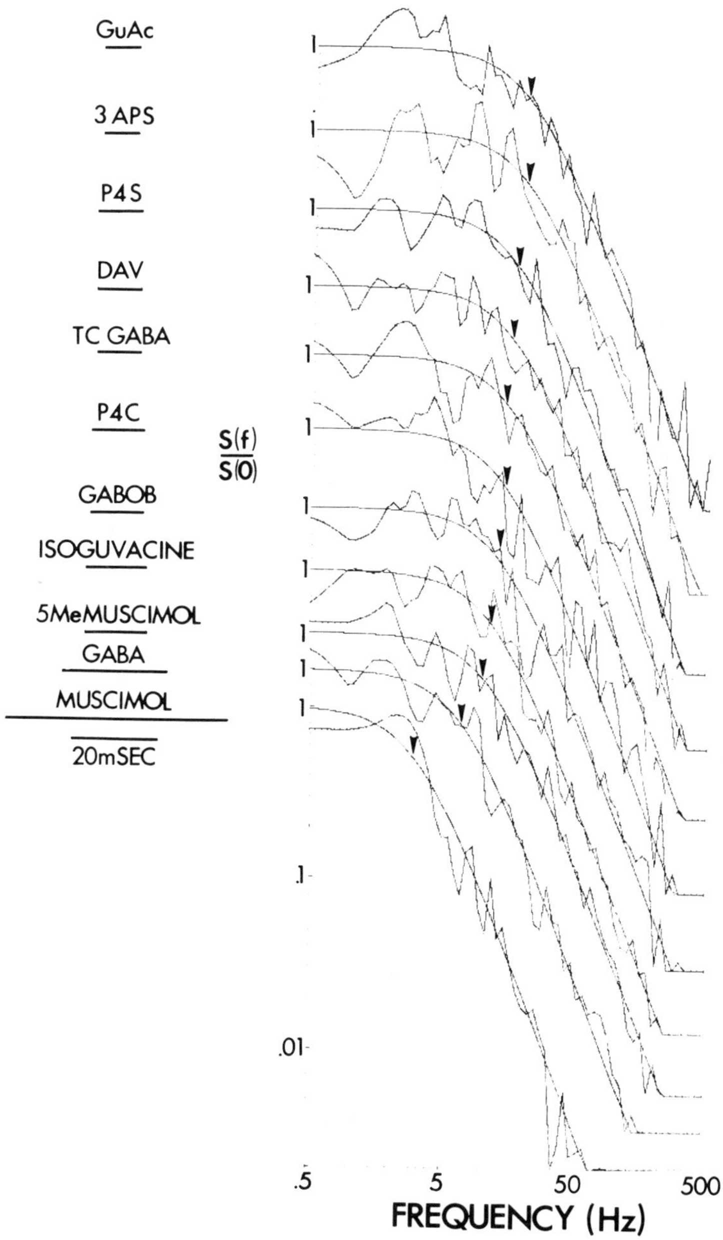


**Fig. 10.** *Normalized spectral density plots of agonist-evoked current fluctuations from the same experiment illustrated in Fig. 9. The plots have been aligned to demonstrate the spectrum of "corner" frequencies (marked by arrowheads) and the corresponding estimated channel durations (bars at left) for each of the GABAmimetic agonists (identified above bars). [Barker and Mathers, unpublished observations.]*

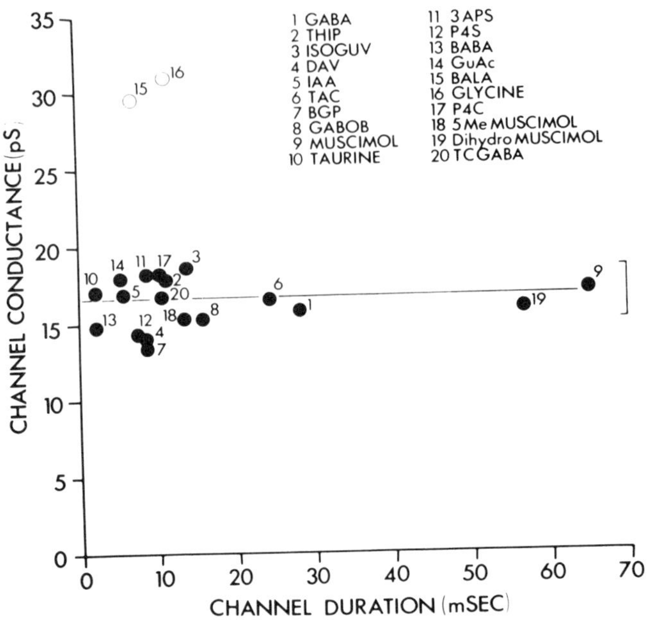


**Fig. 11.** *Constant-conductance for Cl⁻ channels activated by GABAmimetics. All of the data are derived from fluctuation analysis of membrane current responses evoked by GABA and 17 analogues that have effects in competitive binding assays on membrane preparations. Estimated values of channel conductance have been plotted as a function of estimated values of channel duration. The average value for channel conductance activated by GABA and its analogues under these conditions is 16.5 pS (indicated by horizontal bar) with a standard deviation of ±1.5 pS (marked by vertical bar at right). Almost all of the conductance values lie within a standard deviation of the mean. There is a 20-fold range in channel duration and only muscimol and dihydromuscimol activate channels with durations longer than that opened by GABA. β-Alanine (15) and glycine (16), which are ineffective in competitive binding assays with GABA, activate relatively short channels of significantly more conductance. [Modified from Barker and Mathers, 1981, and Mathers and Barker, 1982.]*

ular structure are too great to determine the precise structural requirements for receptor binding and subsequent activation of $Cl^-$ conductance.

## DRUGS ACTIVATE $Cl^-$ CHANNELS

Pentobarbital and diazepam, which do *not* competitively displace GABA in binding assays, yet independently induce $Cl^-$ ion conductance, activate ion channels whose estimated and observed elementary conductance is similar, if not identical to that activated by GABA [Mathers and Barker, 1981; Jackson et al., 1982]. From the experiments done thus far, the average duration of the elementary event activated by diazepam is also close to, if not identical to, that evoked in the same cell by GABA, while the average lifetime of pentobarbital-triggered channels is typically longer (Fig. 8), although durations identical to that estimated for GABA have also been recorded [R.E. Study and J.L. Barker, unpublished observations]. Furthermore, the conductance activated by diazepam is blocked by bicuculline, an antagonist of GABA binding and function, and *occludes,* rather than potentiates, that generated by GABA. That induced by pentobarbital, which, like GABA's interaction, is potentiated by diazepam and attenuated by both bicuculline and picrotoxin, invariably *cooperates* with, rather than

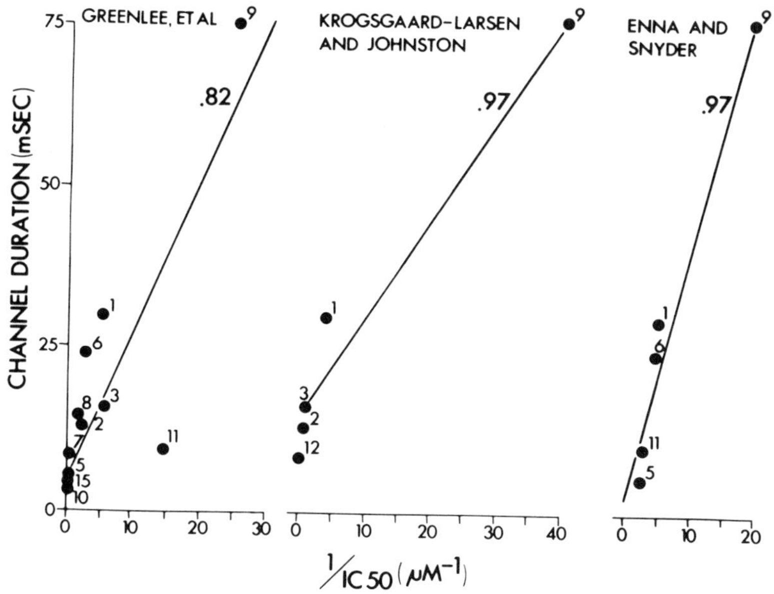


*Fig. 12.* *Channel duration in cultured spinal neurons correlates with $IC_{50}$ in competitive binding studies. Channel duration data is derived from fluctuation analysis of agonist-induced responses at 23°C in cultured spinal neurons. $IC_{50}$ is the agonist concentration blocking 50% of "specific" GABA binding at 4°C in membranes prepared from adult CNS tissue. In each study there is a high correlation between channel duration and $IC_{50}$ values. The latter values have arbitrarily been inverted in the figure to suggest that molecular efficacy (channel duration) corresponds functionally to molar potency ($1/IC_{50}$). The numbers refer to agonists identified in Fig. 11. [References: Greenlee, VanNess, Olsen, 1978; Krogsgaard-Larsen and Johnston, 1978; Enna and Snyder, 1977.]*

occludes that evoked by GABA. These pharmacological and biophysical results indicate that the drugs closely mimic transmitter actions at a high level of resolution. They may utilize the same $Cl^-$ conductance mechanism as that coupled to GABA receptors.

## $Cl^-$ ION CHANNEL PROPERTIES ARE TRANSMITTER-SPECIFIC

Fluctuation analysis of current responses evoked by two other neutral amino acid transmitters (β-alanine and glycine, Fig. 9) predict that these transmitters activate channels whose conductance is, on average, greater but significantly shorter-lasting than that associated with GABA [Barker and McBurney, 1979a; Barker et al., 1982; Fig. 11]. Different electrical properties have also been reported for GABA- and glycine-activated $Cl^-$ conductance generated in normally developed CNS tissue [Gold and Martin, 1983a,b], indicating that these differences in estimated ion channel properties are not simply a reflection of development in vitro.

Recent results derived from patch-clamp study of glycine-activated channels have confirmed and extended the whole-cell level of observations in central neurons [Hamill et al., 1983]. In addition, certain aspects of the elementary channel properties resolved by the high fidelity of patch-clamp analysis are potentially illuminating.

1. The single-channel level of sensitivity of the electrical assay reveals consistent microscopic current activity at 500 nM GABA

[Jackson et al., 1982], about an order of magnitude lower than that necessary to evoke detectable whole-cell responses.

2. A variable number of the somal patches respond to *both* transmitters and both *agonists activate $Cl^-$ ion channels with multiple levels of conductance, about 90% of the channel activity occurring at the principal levels of conductance, which, as mentioned above, are close to those predicted from fluctuation analysis.*

3. While each agonist stabilizes the membrane at different primary conducting levels, the secondary level activated occasionally by glycine corresponds to the primary GABA-regulated level, and the tertiary level activated (even less often) by glycine corresponds to the secondary level occasionally observed with GABA. Coincidental conductance states suggest similarities in the transmitter–receptor interactions that culminate in $Cl^-$ ion movement.

It is even possible that the two receptors physically couple to the same channel mechanism, since 1) channel activation can occur in the same micrometer patch of membrane; 2) macroscopic conductance responses occlude [Barker and McBurney, 1979a], and 3) there is a high and significant correlation between amplitudes of responses evoked by the two agonists at different sites on the same cell [Study and Barker, in preparation].

Thus, the results from pharmacological experiments conducted at macro- and microscopic levels demonstrate that, although the amino acids all utilize the existing $Cl^-$ ion gradient to "inhibit" excitability, *each transmitter generates a molecular signal with unique electrical dimensions.* If these pharmacological results are relevant to physiologically generated signals, as they are when other transmitter-evoked responses have been compared with physiologically elaborated synaptic potentials in peripheral vertebrate and central invertebrate cells, then we might expect that *in the mammalian CNS each amino acid transmitter mediates an inhibitory postsynaptic signal with unique kinetics* (compare time courses of inhibitory signals illustrated in Figs. 19, 21).

## DRUG-MODULATION OF GABA-MEDIATED SIGNALS

Clinically important drugs of the benzodiazepine and barbiturate classes modulate $Cl^-$ conductance responses activated by GABA in cultured neurons at concentrations below those required for the agonist actions of the drugs described above [Barker and Ransom, 1978b; Huang and Barker, 1980; Study and Barker, 1981]. The most commonly observed modulatory effect involves a dose-dependent increase in the amplitude and/or time-course of the pharmacologic response. The diazepam effect is attenuated by an antagonist that has little or no detectable effects of its own (Fig. 13). Several other classes of CNS depressants do not modualte GABA-evoked responses at clinically relevant concentrations nor do they noticeably affect diazepam potentiation (Fig. 13). Micromolar concentrations of the drugs are usually necessary to enhance responses, although some benzodiazepine drugs like flurazepam have been observed to be 1000-fold more potent on a molar basis [MacDonald and Barker, 1982]. Dose–response curves obtained in the presence and absence of either of the modulatory drugs show a parallel, leftward shift (Fig. 14) without change in the maximal response obtained [Choi et al., 1981].

Under voltage-clamp the drugs increase the amplitude and/or time-course of the GABA-evoked current responses. In the case of the benzodiazepine class the peak and time course of the pharmacological response are both increased. For example, in Figure 15 is illustrated a pair of membrane current responses to 100 msec applications of 20 $\mu$M GABA evoked at a nominal resting potential of $-70$mV before and during coincident application of 5 $\mu$M diazepam. The drug increases the amplitude by about three-fold without measurably affecting time-to-peak. There is also a modest increase in the time course of

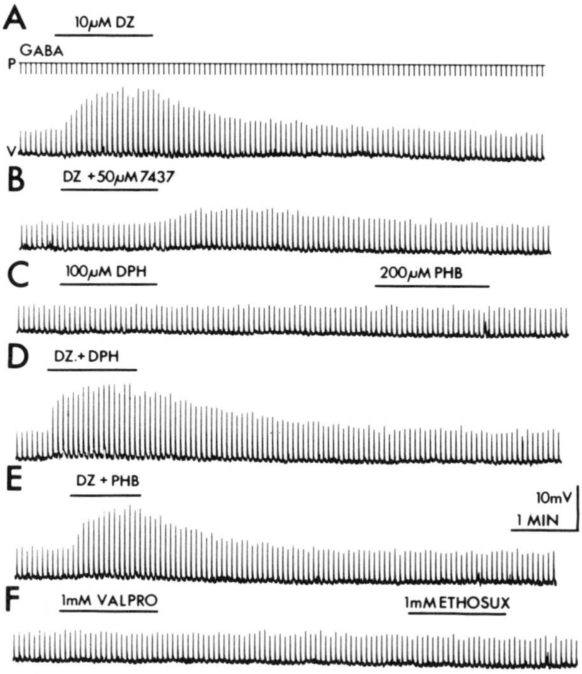


**Fig. 13.** *Diazepam, but not other anticonvulsant drugs potentiate GABA-evoked responses. All of the traces were obtained on the same cell held at −60 mV with a KCl-filled microelectrode (so as to amplify the "driving force" acting on the conductance, thereby enlarging the $Cl^-$-dependent voltage signal). Fifteen micromolar GABA was applied by low pressure in 100 msec pulses at the times indicated by the trace marked "P," while the rest of the drugs were delivered for the times indicated by bars above the various voltage responses (V). GABA responses are quite stable in amplitude and do not desensitize. Panel A: 10 micromolar diazepam (DZ) gradually enhances pharmacologic responses to GABA by slightly more than two-fold over a period of more than a minute. Recovery from the equilibrium state takes minutes. Panel B: Inclusion of a competitive antagonist (50 micromolar (R014-7437) delays the onset of a detectable but markedly attenuated potentiation until after the coincident application period is over. Panel C: Clinical and supraclinical concentrations of diphenylhydantoin (DPH) and phenobarbital (PhB) do not augment GABA-evoked responses nor significantly change potentiations by diazepam (Panels D,E); Panel F: Supraclinical concentrations of valproate and ethosuximate are also ineffective [Barker and Owen, unpublished observations].*

the decay of the response (less than 20%). In contrast, pentobarbital increases the peak amplitude and decay both several-fold.

## FLUCTUATION ANALYSIS OF MODULATED GABA-EVOKED CURRENTS

Fluctuation analysis of the drug-modulated current responses evoked by GABA indicates that the drugs have distinctly different effects at the single channel level. Spectral analysis of the potentiated responses shows that, as with control responses, many spectra can be fit adequately by a "Lorenztian" curve, indicating exponential channel kinetics. There is a definite shift to lower frequencies in the corner frequency of the spectrum derived from pentobarbital-potentiated responses, while diazepam-modulated responses often show little, if any, detectable change in this parameter (Fig. 16). When estimated channel duration is plotted as a function of the potentiated current response amplitude, it is clear that increasing duration of the ion channel

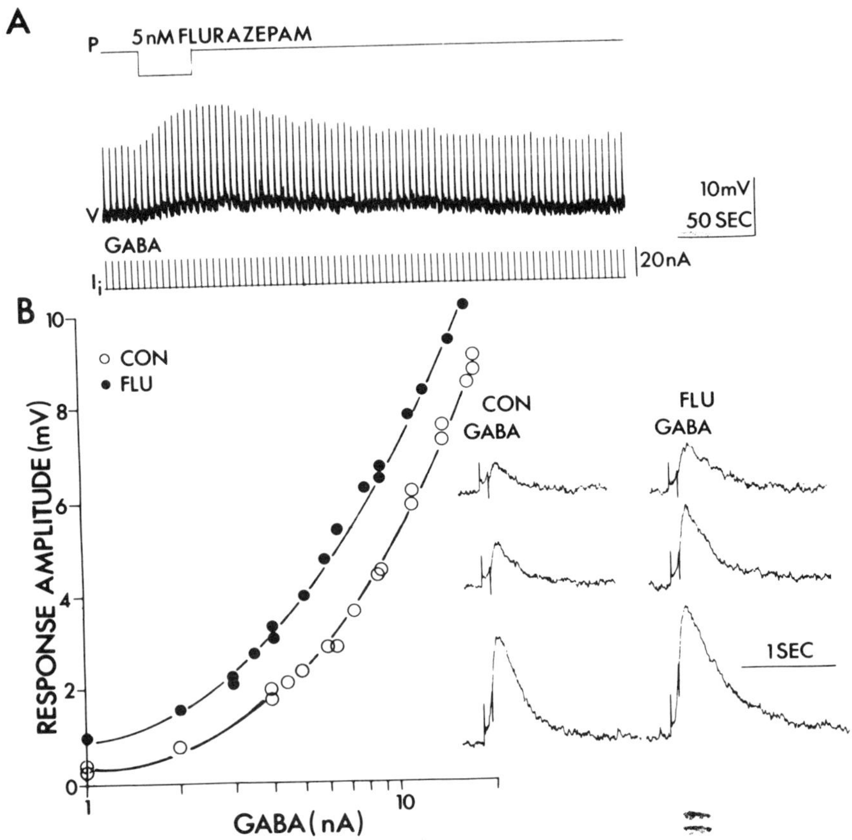


**Fig. 14.** *Flurazepam shifts the dose–response curve leftward in a parallel manner. A: The cell was held at −60 mV and GABA applied with 18 nA–50 msec iontophoretic current pulses ($I_i$). Five nanomolar flurazepam, applied by pressure (P) for about 50 sec, augments GABA-activated responses in a detectable manner. B: Dose–response curves generated under control conditions (CON) and in the presence of the drug (FLU). Samples of raw data are shown to the right. [Modified from MacDonald and Barker, 1982].*

conductance is more than adequate to account for the degree of pentobarbital-induced potentiation of macroscopic responses (Fig. 17). Pentobarbital amplifies GABA-evoked responses primarily by stabilizing ion channels in the open, conducting state for longer periods so that there is an almost stoichiometric relationship between channel duration and potentiated current amplitude. Thus, approximately the same number of channels is activated in the presence of pentobarbital, but on average, each passes more $Cl^-$ ion charge by conducting for a longer period.

In contrast, although channel duration ($\tau$) estimated for responses potentiated by diazepam also increases, the relative change in $\tau$ and the macroscopic current (I) has a stoichiometry of considerably less than 1 (Fig. 17). Furthermore, the degree of diazepam-induced potentiation, as defined by peak current amplitude, consistently asymptotes at between two- and three-fold, considerably less than that observed with pentobarbital-mediated modulation. Fluctuation analysis of GABA-evoked responses potentiated by increasing doses of diazepam reveals that, as well as

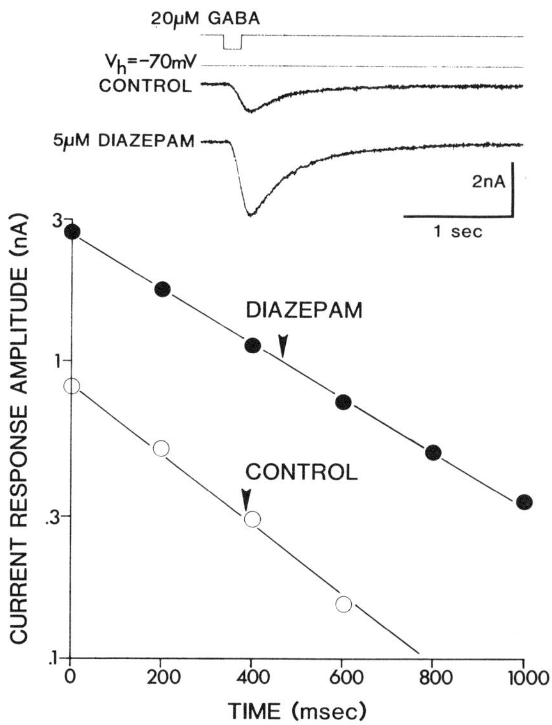


**Fig. 15.** *Diazepam augments GABA-activated membrane current response with little change in time course. The cell was voltage-clamped at −70 mV with two KCl electrodes and GABA applied for the period indicated under control conditions and in the presence of diazepam. Although the membrane response to GABA increases more than three-fold in peak amplitude in the presence of diazepam, there is relatively little, if any change in the time-to-peak and at most a 15% increase in the macroscopic time constant of decay (arrowheads). [Modified from Study and Barker, 1982].*

prolonging channel lifetime, the drug markedly increases the rate at which channels open (Fig. 18). Thus, using fluctuation analysis, it is possible to account in a relatively quantitative manner for the drug modulation in terms of estimated channel conductance and duration and frequency, or rate of occurrence. Theoretically, the measured macroscopic current (I) is the product of the microscopic current (i), the channel lifetime ($\tau$) and the channel opening frequency ($\nu$) (i.e., $I = i.\tau.\nu$). In the experiment illustrated in Figure 18, for example, the average amplitude of the macroscopic current response evoked by GABA under control conditions is 0.65 nA, the estimated microcurrent amplitude is 0.9 pA, the estimated average duratiion is 20 msec, and the estimated channel frequency (derived from the variance, mean macroscopic current measurement, and corner frequency in the spectrum) is $3.6 \times 10^4$ Hz.

A variety of clinically important drugs modulate pharmacologic responses to GABA and fluctuation analysis of the modulated responses indicates that in each pharmacologic interaction the modulation can be accounted for primarily, if not exclusively, in terms of changes in channel kinetics, rather than channel conductance (summarized in Table I). Thus, changes in channel opening rate (frequency) and/or channel closing rate (open-time) rather than ion flux rate through the open channel (conductance) lead to corresponding changes in the macroscopic current (and conductance) responses.

## SYNAPTICALLY ACTIVATED $Cl^-$ CONDUCTANCE

Neurons maintained in monolayer culture exhibit spontaneous and evoked forms of synaptic activity. Intracellular recordings from both mouse spinal cord and rat hippocampal neurons reveal spontaneous occurring, rapid (< 5 msec) depolarizing excitatory signals, and several, slower types of hyperpolarizing inhibitiory signals, one of which reverses polarity at $E_{rev}$ for GABA-activated, $Cl^-$-dependent responses [Ransom et al., 1977; Barker and McBurnery, 1979b; Segal and Barker, 1984b; Fig. 19]. The available evidence suggests that these inhibitory signals probably result from $Cl^-$ conductance responses generated at subsynaptic membranes by GABA (see below).

## DRUGS MODULATE SYNAPTIC SIGNALS

Under voltage clamp these inhibitory postsynaptic signals appear as $Cl^-$ currents that

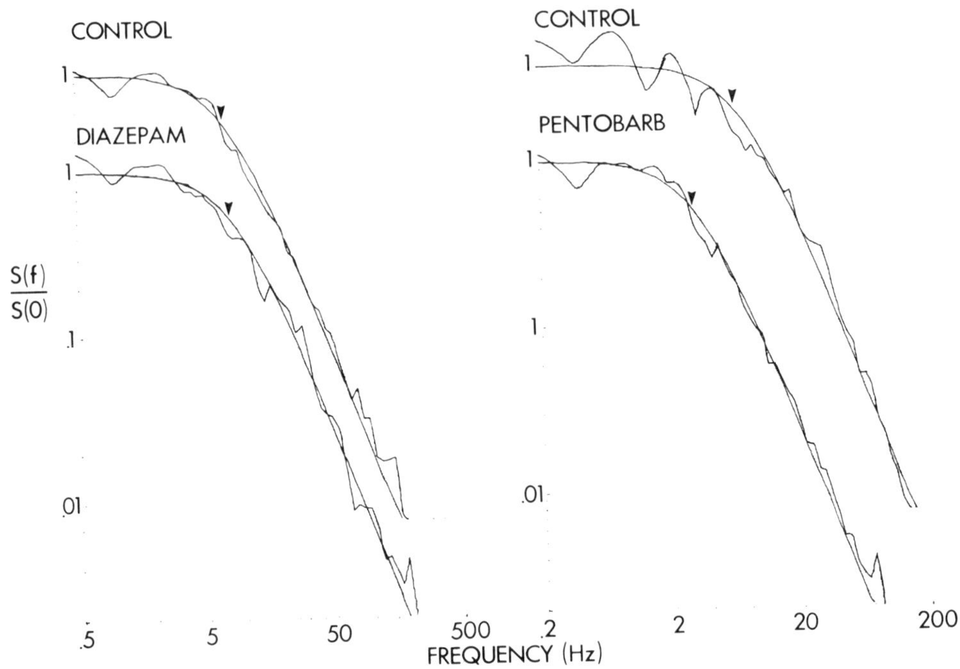


**Fig. 16.** *Spectral comparison of drug-potentiated and control responses to GABA. The data was collected from cultured mouse spinal neurons voltage clamped at −70 mV with two KCl microelectrodes. Jagged, uneven lines represent data, and smooth lines are best-fits of Lorentzian model to the data. Control spectrum is displayed above the "drug-potentiated" spectrum to emphasize corner-frequencies (arrowheads). In this experiment the spectrum calculated for a diazepam-potentiated GABA response has a corner-frequency slightly higher than control. Thus, channels underlying the potentiation are slightly shorter. The pentobarbital-potentiated spectrum is significantly shifted to lower frequencies, indicating channels associated with augmented responses are significantly longer. [Modified from Mathers and Barker, 1982].*

decay with *single exponential kinetics* and decay time constants ($\tau_{IPSC}$) consistently greater than 5 msec (Fig. 20). There is little, if any, convincing evidence for an initial, rapidly decaying exponential phase. The exponential decay is sensitive to membrane voltage, invariably becoming *longer* at more depolarized levels. The amplitude and/or time course are affected by a variety of clinically important drugs, including diazepam (Fig. 20), picrotoxin, and pentobarbital [Segal and Barker, 1984b]. Diazepam consistently increases both the average amplitude and $\tau_{IPSC}$. For example, in Figure 20 the inhibitory postsynaptic current amplitude (IPSC) increases from 1.7 nA in control to 2.6 nA in the presence of diazepam and $\tau_{IPSC}$ increases from about 9 msec to about 18 msec. On average, $\tau_{IPSC}$ increases by about 50% and the amplitude of the current increases by about 40% in the presence of 10$\mu$M diazepam (Table II). (−)Pentobarbital, at concentrations subthreshold for direct activation of $Cl^-$ conductance, increases $\tau_{IPSC}$ three-fold (Table II), while having little, if any, detectable effect on current amplitude [Segal and Barker, 1984b]. Thus, both drugs effectively facilitate $Cl^-$ion charge transfer during the average-sized inhibitory synaptic signal.

Picrotoxin, which attenuates GABA-induced $Cl^-$ conductance, apparently by depressing channel opening frequency without altering channel conductance [Barker et al.,

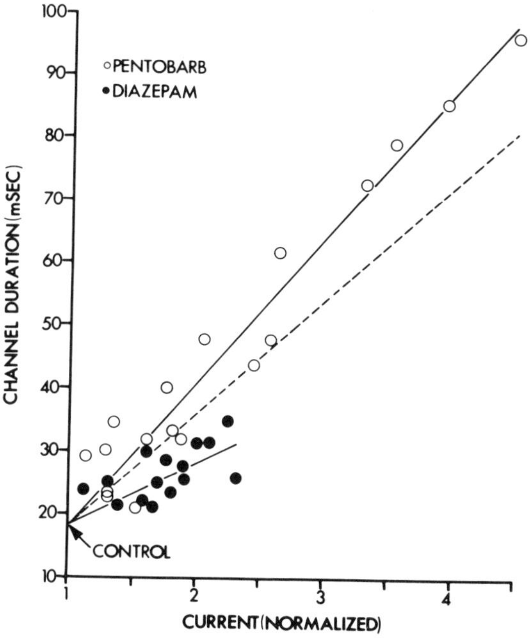


**Fig. 17.** *Stoichiometries between drug-induced changes in estimated microscopic and measured parameters. Spectral analyses were carried out on control current responses to GABA and on responses potentiated to different degrees by 10 micromolar and 50 micromolar (−)pentobarbital. The dashed line directly relates changes in estimated channel duration to changes measured in macroscopic membrane current responses in a linear one-for-one manner. Drug-induced changes in micro- and macroscopic parameters are also linearly related, but for each the stoichiometric relationship is different with the barbiturate-inducing channels slightly longer than expected if duration were to account for all of the macroscopic-level potentiation. The benzodiazepine, which potentiates the macroscopic response only 2–2.5 fold, increases channel duration in a less than one-for-one manner, indicating that simple increases in channel durations cannot account for the potentiated amplitude of the macroscopic current response. [Modified from Study and Barker, 1981.]*

1983; Segal and Barker, 1984b], depresses IPSC amplitude without changing $\tau_{IPSC}$ [Table II; Segal and Barker, 1984b]. If channels are opening they must be blocked so fast that they do not register at the whole-cell level.

These drug effects on the synaptic signals could result from pre- and/or postsynaptic actions, especially since these drugs have been observed to affect voltage-dependent properties important in neurotransmitter secretion in CNS neurons [MacDonald and Barker, 1982; Macdonald et al., 1982]. One of these effects—attenuation of fast voltage-dependent Ca spikes by subanesthetic concentrations of (−)pentobarbital [Owen et al., 1985]—would be expected to depress transmitter secretion. Another action seen with both benzodiazepines and barbiturates, involving reduction in cellular spike generation during sustained injections of depolarizing current (see below), would likewise attenuate excitability and reduce transmitter release during trains of action potentials.

However, $\tau_{IPSC}$ for the synaptic current is similar, if not identical, to estimates of GABA-activated $Cl^-$ ion channel duration ($\tau_{GABA}$) under different experimental and pharmacological conditions (Table II), indicating that the IPSCs recorded in cultured neurons are likely to be mediated by GABA and that the *decay of the IPSC is almost entirely a postsynaptic characteristic*. Thus, while drug-induced changes in IPSC amplitude could conceivably result from presynaptic effects of the drugs, changes in decay appear to reflect modulation of the postsynaptic receptor–channel coupling processes.

If GABA does mediate these events, then we can estimate the number of ion channels that, on average, comprise the IPSC peak from the estimated elementary current amplitude derived from fluctuation analysis of currents evoked by exogenous GABA in the same cell at the same potential. Thus, the peak of the synaptic signal reflects simultaneous activation of about 1700 channels. The single exponential time course of IPSC decay closely approximates a major component reported in "current-time" histograms, which describe durations of channel openings accumulated over time in patch-clamp assays of single-channel behavior. The exponential time course of signal decay also corresponds closely to $\tau_{GABA}$ estimated from fluctuation analysis of many channels activated during a

**TABLE I. Drugs Modulate Macroscopic $Cl^-$ Ion Conductance Responses Evoked by GABA in Cultured CNS Neurons by Acting on Microscopic Channel Kinetics***

| Drug | Modulation | Channel kinetics | References[2] |
|---|---|---|---|
| Bicuculline | ↓ | ↓frequency | A |
| Picrotoxin | ↓ | ↓frequency | A, B |
| Metrazol | ↓ | ↓frequency (↓life-time) | A |
| Methyl- β-carboline-3-carboxylate | ↓ | ↓frequency | C |
| Dimethoxy-β-carboline (DMCM) | ↓ | ↓frequency | C |
| Butyl-β-carboline-3 carboxylate | ↑ | ↑frequency (↑open-time) | C |
| Lorazepam | ↑ | ↑frequency | C |
| Irazepine | ↑ | ↑frequency | C |
| Diazepam | ↑ | ↑frequency (↑open-time) | B, D |
| (−)Pentobarbital | ↑ | ↑open-time (↓frequency) | B, D |

*All data are from two-electrode (KCl, KCl) voltage-clamp recordings of cells held at −70mV and 22–24°C in a physiological saline containing 10mM Mg $Cl_2$. The "channel mechanism" data is derived from fluctuation analysis of membrane current responses evoked pharmacologically with 10–20 μM GABA under control conditions and in the presence of 10–50 μM drug. Effective instrument bandwidth is about 200 Hz. Drug-induced changes resulting in faster channel kinetics will not be detected in a reliable manner. The secondary channel parameter that apparently changes during the modulatory action is shown in parentheses. Frequency refers to rate of channel opening. Open-time refers to channel duration (inverse of channel closing rate).

[2]*References:*
A. Barker, McBurney and Mathers [1983]; B. Segal and Barker [1984b]; C. Owen, Study, Gratz, and Barker [1982]; D. Study and Barker [1981].

pharmacological response to GABA at the whole-cell level of assay. In vitro GABA molecules, whether released physiologically from presynaptic stores or applied pharmacologically at concentrations ranging from 500 nM–20 μM at room temperature, induce $Cl^-$ conductance mechanisms with quite similar kinetics.

From these observations we can tentatively conclude that the exponential time course of IPSC decay, $\tau_{IPSC}$, reflects the distribution in the durations of the 1700 channels comprising the peak, with most channels closing quite quickly and successively fewer and fewer remaining open for longer and longer times. The parallel changes in $\tau_{IPSC}$ and $\tau_{GABA}$ induced by the drugs strongly suggest that the changes in synaptically activated $Cl^-$ conductance result from drug modulation of channel kinetics. In the presence of diazepam, more channels open more-or-less simultaneously and the exponential decay, though still monoexponential, is clearly longer (by about 50%). Presumably, the drug promotes the tendency of channels to enter a conducting state and also reduces the tendency to leave that state. Alternatively, diazepam may facilitate channel reopening at a rate that is too fast to resolve at the whole-cell level of recordings. *Since the drug can have full-agonist properties at $Cl^-$ conductance mechanisms in these cells* (see above), *it is conceivable that some of the changes in synaptic signal result from the drug's potential agonist properties.* In fact, micromolar concentrations of GABA, and a variety of other substances that activate $Cl^-$ ion channels having the same estimated elementary conductance as that opened by GABA, and both GABA analogues and barbiturate-like drugs *promote the binding of diazepam to specific membrane sites,* especially in the presence of a full complement of $Cl^-_0$

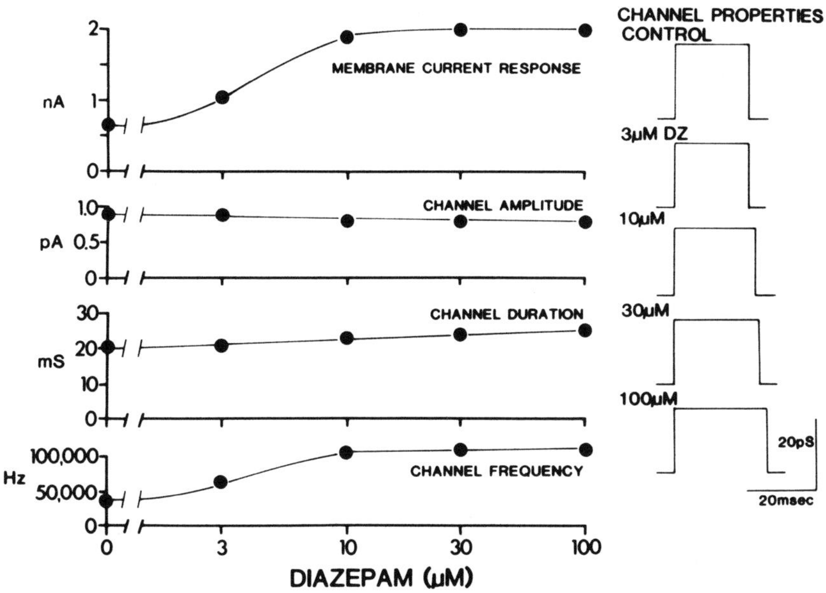


**Fig. 18.** *Changes in elementary kinetics, not conductance account for diazepam-induced potentiation of GABA-evoked membrane current response. The experimental arrangements were described in Fig. 15. Increasing concentrations of diazepam were applied by low pressure and augmented current responses generated in the whole cell were studied with the fluctuation analysis protocol described previously. Macroscopic membrane current and the parameters estimated from fluctuation analysis have been plotted as a function of diazepam concentration. Most of the potentiation recorded at the macroscopic level occurs over the 3–10 micromolar concentration of diazepam. Likewise, channel frequency increases from 40,000 openings/second to 100,000 openings/second over the same range. Channel amplitude, reflecting conductance, does not change, while channel duration increases by about 40%. Thus, the combination of increased channel opening rate and channel duration accounts for the potentiation. [Barker and Study, unpublished observations.]*

**TABLE II. Comparison of $\tau_{IPSC}$ and $\tau_{GABA}$ in Rat Hippocampal Neurons***

| | $\tau_{IPSC}$(msec) | $\tau_{GABA}$(msec) | Ratio | Reference[2] |
|---|---|---|---|---|
| Culture | | | | |
| Picrotoxin | 18 | 21 | 0.8 | A |
| Pentobarbital | 70 | 50 | 1.4 | A |
| Diazepam | 30 | 26 | 1.1 | A |
| −60 mV | 20 | 20 | 1.0 | A |
| +10 mV | 30 | 30 | 1.0 | A |
| Slice | | | | |
| −60 mV | 12 | — | — | B |
| +10 mV | 20 | — | — | B |
| Pentobarbital | 43 | — | — | B |

*Data obtained from two-electrode (culture) and single-electrode (slice) voltage-clamp recordings made with KCl-filled microelectrodes at 23°C in physiological salines. Ratio refers to ratio of $\tau_{IPSC}/\ \tau_{GABA}$.
[2]*References:*
A. Segal and Barker [1984b]; B. Collingridge, Gage, Robertson [1984].

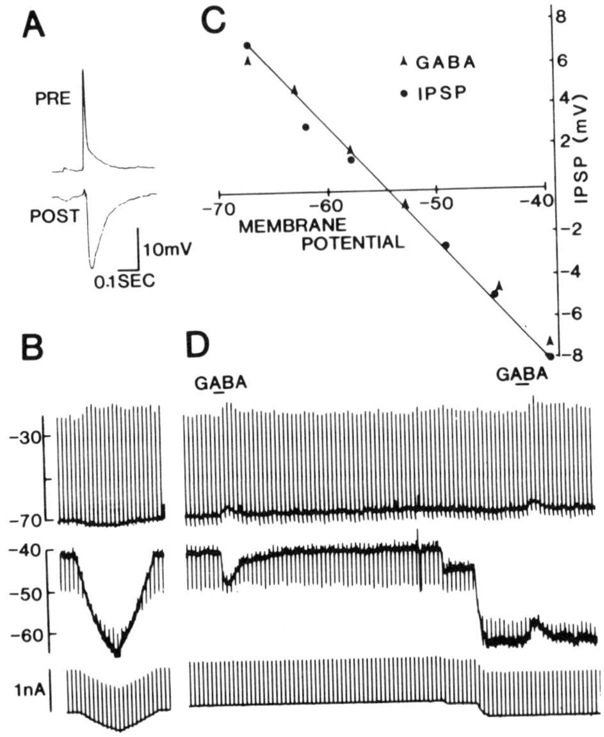


**Fig. 19.** *Inhibitory synaptic potentials invert at the same membrane potential as pharmacological responses evoked by GABA. Simultaneous intracellular recordings were made in two neurons cultured from the embryonic rat. A: An action potential triggered in a presynaptic cell evokes a long-lasting hyperpolarizing synaptic potential in the postsynaptic neuron. B: Pen-recorder traces of pre- and postsynaptic activity show that the downward-going evoked synaptic potentials invert in polarity as the postsynaptic membrane is polarized from about −40 mV to beyond −60 mV with DC current injection. C: The inhibitory postsynaptic potential (IPSP) inverts at about −54 mV, which is identical to the inversion potential of GABA-activated voltage responses (shown in panel D). [Modified from Segal and Barker, 1984b.]*

[see Ticku; Karobath and Supavilai; Squires; and Fischer and Olsen, this volume]. There is a high and significant correlation between agonist concentration required to increase diazepam binding to the same degree and $\tau_{agonist}$, the estimated mean channel duration for different GABAmimetic agonists (see above). Pharmacologically activated channels that conduct $Cl^-$ ions promote diazepam binding, and this occurs in direct proportion to the molecular efficacy of the agonist in promoting charge transfer. Hence, when $Cl^-$ channels are activated physiologically by GABA, it is possible that the stabilized open-channel complex might itself promote binding of diazepam as a ligand cooperating in channel opening. Indeed, the duration that brief GABA-evoked pharmacological responses remain potentiated by diazepam is dependent on the rate at which the responses are elicited. The more rapidly the responses are evoked, the longer-lasting is the period of potentiation, as if channel activation per se increased the binding and subsequent "modulatory" actions of diazepam [Barker et al., 1982]. Thus, although changes in the IPSC amplitude and time course, like the above-described modulation of GABA-activated $Cl^-$ conductance,

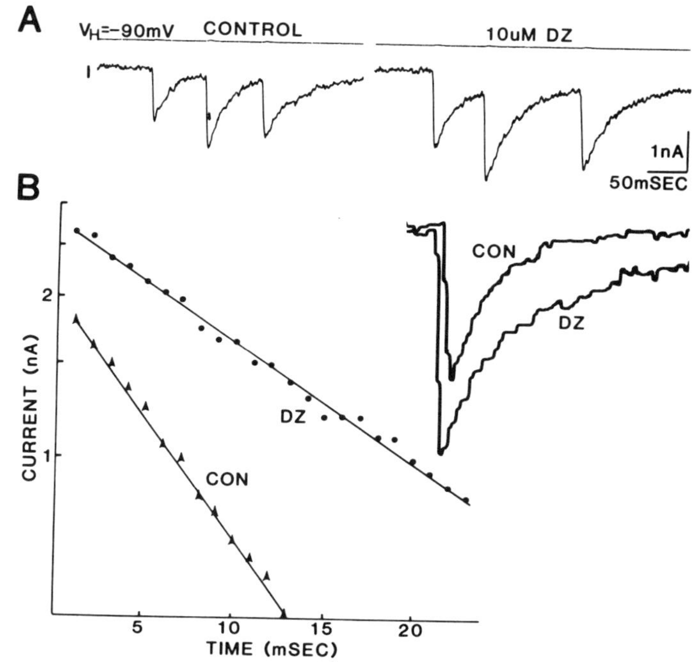


**Fig. 20.** *Diazepam increases inhibitory synaptic current amplitude and time course of decay. The cultured rat hippocampal neuron was clamped at −90 mV and inhibitory postsynaptic currents (IPSCs) triggered in the postsynaptic membrane. Pressure application of diazepam increases both the IPSC amplitude (from 1.9 to 2.7 nA) and the monoexponential time constant of signal decay (from about 9 msec to about 18 msec). [Modified from Segal and Barker, 1984b.]*

can be ascribed to changes in ion channel kinetics, it is not clear precisely how the receptor-coupled $Cl^-$ ion channels activate in the presence of both transmitter and drug. A unique and suitable model relating closed- and open-channel states with agonist(s) binding has yet to be developed. A similar conclusion might also be drawn from barbituarate "modulation" of $\tau_{IPSC}$ and $\tau_{GABA}$, since barbiturates, and (−)pentobarbital in particular, have full-agonist properties at concentrations higher than those necessary for modulatory actions (vide supra).

It should be pointed out that there is relative specificity in the modulation of synaptically activated $Cl^-$ conductance. Barbiturates and benzodiazepines do not alter strychnine-sensitive $Cl^-$ conductance evoked by glycine [Barker and Ransom, 1978b; Macdonald and Barker, 1978], nor $Cl^-$-dependent synaptic signals that decay in less than 10 msec and are blocked by relatively low concentrations of strychnine [Choi et al., 1981; Fig. 21]. Since there is increasing evidence that GABA and glycine receptors may share a common mechanism to open $Cl^-$ ion channels [Barker and McBurney, reviewed in Barker, 1985], the specificity of the drug modulation suggests that functionally the effect occurs at a site other than the $Cl^-$ ion channel site, perhaps nearer the receptor function.

## BENZODIAZEPINES BLOCK CONVULSANT ACTIVITY

Diazepam and other CNS depressants, including phenobarbital and diphenylhydantoin, can attentuate or block the paroxysmal depolarizing event (PDEs) triggered in the presence of agents that evoke lethal convulsions

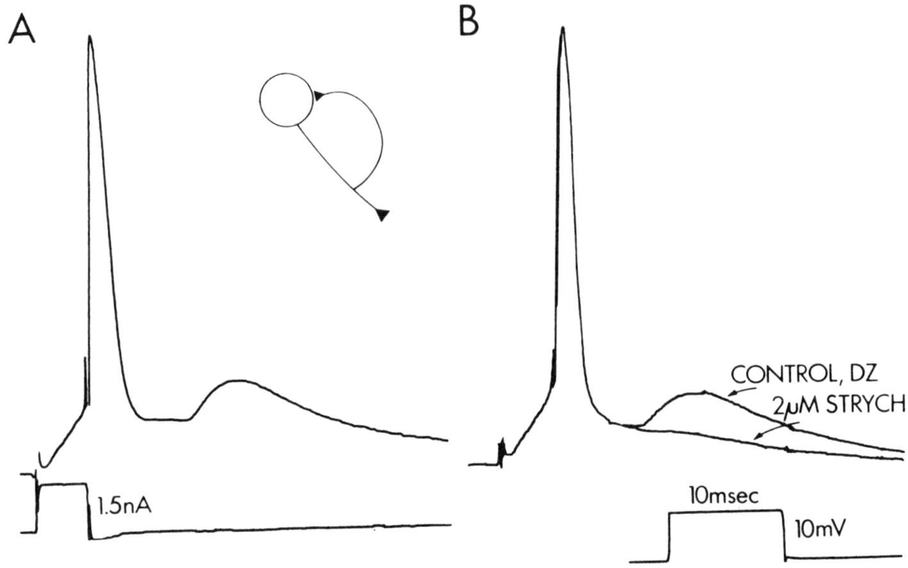


**Fig. 21.** *Diazepam does not alter strychnine-sensitive short synaptic potential recorded in a cultured mouse spinal neuron. A: The cell was held at −60 mV with a KCl microelectrode and a 5 msec–1.5nA current pulse injected to trigger an action potential. The latter gives rise to a depolarizing synaptic potential of short duration (< 10 msec). The inhibitory synaptic potential is depolarizing under these recording conditions and may be a recurrent signal (see insert) since the event faithfully follows high frequency stimulation. B: Three traces are superimposed. Two micromolar strychnine eliminates the synaptic signal with little, if any change in the action potential. Diazepam (10 micromolar) has no effect on this synaptic signal. [Barker and Rogawski, unpublished observations.]*

in vivo, at clinically relevant concentrations [Fig. 22—panels 1a,b; MacDonald and Barker, 1982; Macdonald et al., 1985]. Although the time course and appearance of such electrical events are quite variable, they can consistently be inverted in polarity near O mV, suggesting an underlying ionic mechanism whose principal driving force is close to O mV [Barker and MacDonald, 1980]. The PDEs are blocked by tetrodotoxin or solutions nominally free of $Ca^{2+}$ions. They can be voltage-clamped to reveal bursts of "synaptic-like" currents superimposed on a much larger envelope of current. These results suggest that the PDEs in these cultured spinal neurons originate at synaptic membranes, perhaps the result of "irritable" presynaptic boutons discharging erratic amounts of excitatory transmitter substances. It should also be noted that *abbreviated versions of PDEs that include inhibitory synaptic potentials can be recorded in cultured neurons in the absence of convulsant drugs,* suggesting that such episodic, though sometimes highly organized, even pacemaker-like forms of intercellular signaling may occur naturally. However, since the mechanisms underlying the presumably postsynaptic reflection of presynaptic release have not been elucidated in a convincing manner, our understanding of the anticonvulsant actions of the drugs is somewhat limited. The anticonvulsant actions of all three drug classes do *not* result from drug-induced shunting of the postsynaptic membrane, which would effectively act to reduce the amplitude of the PDEs [Barker et al., 1980; MacDonald and Barker, 1982; Fig. 22, panel 2].

## BENZODIAZEPINES MODULATE ELECTRICAL EXCITABILITY

In addition to their well-established actions on GABA-mediated inhibitory synaptic sig-

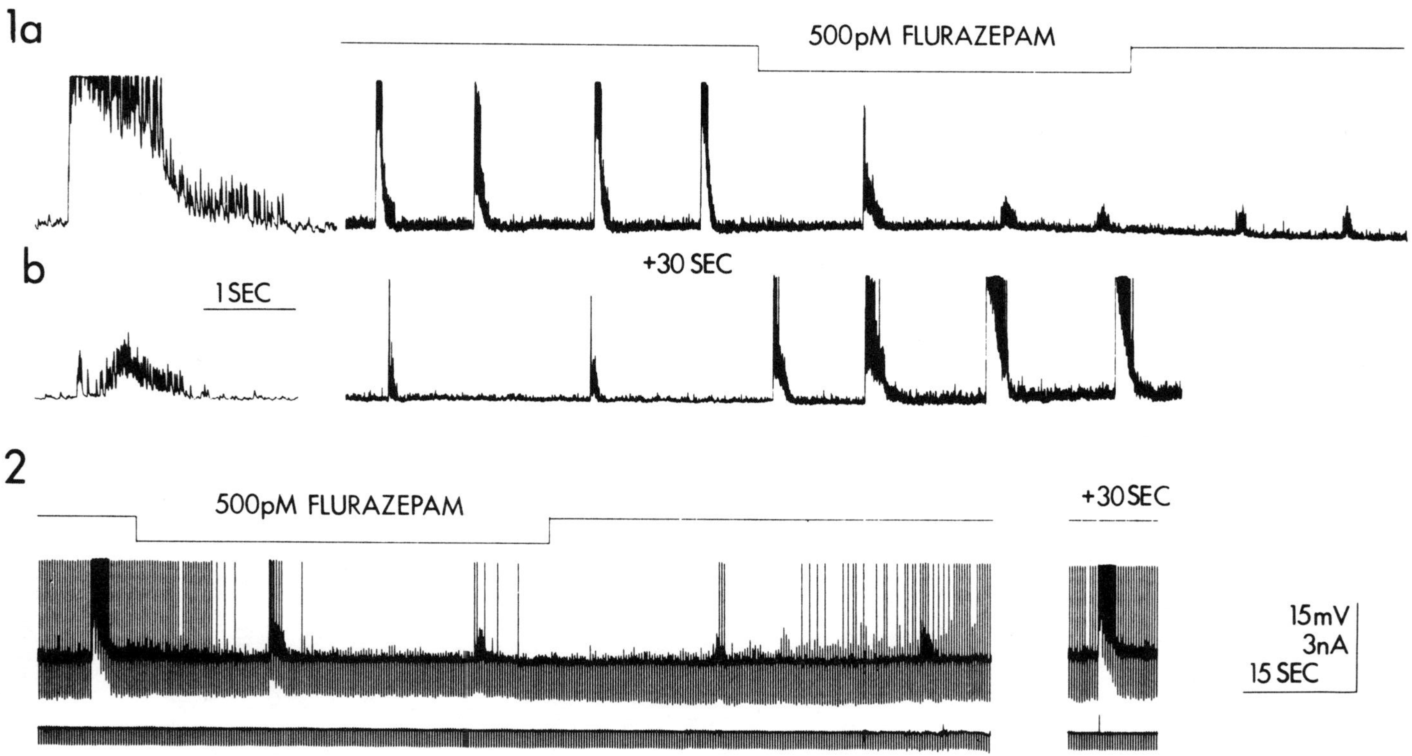


**Fig. 22.** *Flurazepam attenuates convulsive activity induced by picrotoxin. $K^+$ acetate recording from a cultured mouse spinal neuron in bathing medium containing 50 micromolar picrotoxin. Panels 1a,b: Pen-recorder traces from a cell polarized to −65 mV show paroxysmal depolarizing shifts 25 mV or 50 in amplitude which saturate the pen-recorder. At left is an expanded example of a large-amplitude depolarization showing the superimposed fast depolarizing activity. 500 Picomolar flurazepam applied in the vicinity of the cell (during indicated period) attenuates the convulsive activity in a reversible manner. Panel 2: Flurazepam attenuates the depolarizing shifts without significantly changing the passive membrane properties of the cell, as evidenced by the lack of change in the voltage responses to brief constant-current hyperpolarizing stimuli [Modified from MacDonald and Barker, 1982].*

nals, benzodiazepines (and several other CNS depressants) at a concentration that is effectively anticonvulsant in the same cell have been found to alter electrically excitable membrane properties in such a way as to render cells less able to generate action potentials [Barker et al., 1980; MacDonald and Barker, 1982; Macdonald et al., 1985]. Such an action is illustrated in Fig. 23 from a recording of a cell in relative synaptic isolation. Depolarizing constant-current pulses just sufficient to trigger action potentials rapidly become ineffectual in the presence of the benzodiazepine flurazepam (Fig. 23, panel A1). The voltage responses evoked by depolarizing currents clearly decrease in amplitude, while those elicited in response to hyperpolarizing currents do not (Fig. 23, panel A2). Thus, the attenuation of electrically triggered action potentials does not result from a shunting of the passive membrane, rather the modulation occurs through some effect on voltage-dependent conductance mechanisms. In fact, a family of increasingly intense depolarizing currents sufficient to trigger an action potential before, during and after drug application shows that the drug apparently *elevates action potential threshold* (Fig. 23, panels A3,B2) and attenuates repetitive action potential activity. Plots of the current-voltage (I–V) relations over the range of membrane potential between resting level and threshold indicate that the drug markedly changes the slope of the I–V curve so that more depolarizing current must be injected to induce the same voltage response (Fig. 23, panel B1). The ionic and molecular mechanisms of this action have not been elucidated, although several putative conductances that act to regulate action potential activity have been characterized in CNS cells and are obvious targets for experimental attention.

## Conclusion

Electrophysiological analysis of the pharmacological effects of benzodiazepines on embryonic spinal and hippocampal neurons indicates that the drugs interact with both chemically and electrically excitable membrane properties in ways that would reduce interneuronal communication and tend to isolate cells electrically from each other in vivo. With the exception of a series of observations regarding flurazepam, which is active in the pM–nM range of concentration, pharmacological actions of benzodiazepines like diazepam and chlordiazepoxide require $\mu$M levels for detectable activity. Thus, much of the electropharmacology reviewed here necessitates 1000-fold greater concentration than is consistently reported for benzodiazepine binding kinetics from isolated membranes (see other chapters, this volume) or for intact cultured cells [Sher et al., 1983] where nM–10nM affinities are typical. Some of this difference may be due to the fact that 20°C differences exist between the biochemical and electrophysiological assays and that ligand binding likely is transformed into electrical signal at less than 100% efficiency. Chronic exposure of cultured cells to diazepam markedly and reversibly reduces intact-cell binding of diazepam and potentiation of GABA-evoked $Cl^-$ conductance without evidence for chronic alteration of GABA-activated $Cl^-$ conductance per se [Sher et al., 1983]. The parallelism suggests that the biochemical and electrical signals may well be related in some way. If this is true, it may be possible to account more precisely for the selective receptor-coupled modulation of GABA-activated $Cl^-$ion channels by benzodiazepines and the biological mechanism of pharmacological regulation of drug binding and function.

It should not be surprising that in addition to modulatory and agonist actions at GABA receptor-coupled $Cl^-$ion channels, benzodiazepines have "modulatory" effects on electrically excitable membrane properties. Putative peptide transmitters exhibit the same multiplicity of electrophysiological actions in these central neurons [Barker et al., 1980]. These other actions may involve a binding reaction that is detectably different in bio-

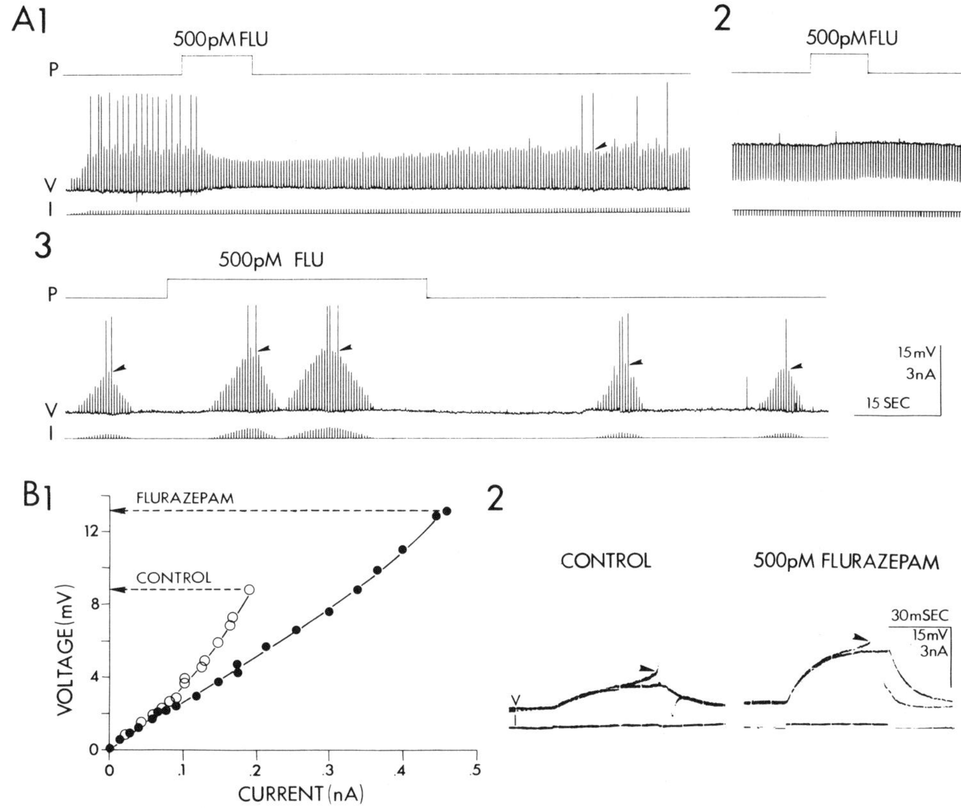


**Fig. 23.** *Flurazepam elevates threshold for action potential generation without changing resting membrane properties of a cultured spinal nueron. Panel A1: Pen-recorder traces of a KCl recording showing membrane potential responses (V) to 50 msec threshold-straddling depolarizing current stimuli (I) before, during, and after pressure application of flurazepam from a pipette containing 500 picomolar drug positioned close to the cell. Arrowhead delineates full-amplitude spike from subthreshold voltage responses. Flurazepam depresses action potential generation for more than a minute after termination of the application. A2: The drug does not alter passive membrane properties, as evidenced by the lack of change in resting potential or conductance. A3: Flurazepam elevates threshold for action potential generation in a reversible manner, as manifested by the shift in the arrowheads to more depolarizing levels. B1: Current-voltage relations plotted under control conditions and during flurazepam application reveals that the elevation in threshold is associated with an apparent increase in slope conductance. B2: Oscillographic traces for superimposed current stimuli revealing the increase in spike threshold. Resting potential: −56 mV. [Modified from MacDonald and Barker, 1982.]*

chemical or morphological studies from that underlying the $Cl^-$ conductance-related actions, but systematic structure–activity studies with highly specific and highly selective chemical or immunological probes have not yet been conducted.

Another issue worth considering involves elucidating the transmitter/hormone neurochemistry that might physiologically mimick this biochemical and electrical pharmacology. There is accumulating evidence from biochemical fractionation experiments that substances can be isolated from the CNS that have some level of pharmacological activity in biochemical, behavioral, and electrophysiological assays similar to those of benzodiazepine actions. There may be a "cascade" of specific chemical signals and electrical patterns that corresponds to the emotional states of anxiety and fear; this cascade could involve the drug-sensitive chemical and electrical excitability mechanisms discussed in this chapter. Characterizing the cellular mechanisms underlying such a fundamental and teleologically important behavior common to many, if not most, nervous systems studied thus far is a focus for future research.

## References

Barker JL (1985): "GABA and Glycine Channel Mechanisms." In Rogawski MA and Barker JL (ed): Neurotransmitter Actions in the Central Nervous System. New York: Plenum.

Barker JL and MacDonald JF (1980): Picrotoxin convulsions involve synaptic and non-synaptic mechanisms on cultured mouse spinal neurons. Science 208:1054–1056.

Barker JL, Mathers DA (1981): GABA analogues activate channels of different duration on cultured mouse spinal neurons. Science 212:358–361.

Barker JL, McBurney RN (1979a): GABA and glycine may share the same conductance channel on cultured mammalian neurones. Nature 277:234–236.

Barker JL, McBurney RN (1979b): Phenobarbitone modulation of post-synaptic GABA receptor function on cultured mammalian neurons. Proc Roy Soc Lond Ser B 206:318–326.

Barker JL, Ransom BR (1978): Amino acid pharmacology of mammalian central neurones grown in tissue culture. J Physiol 280:331–354.

Barker JL, McBurney RN, MacDonald JF (1982): Fluctuation analysis of neutral amino acid responses in cultured mouse spinal neurons. J Physiol (Lond) 322:365–387.

Barker JL, McBurney RN, Mathers DA (1983): Convulsant-induced depression of amino acid responses in cultured mouse spinal neurons studied under voltage clamp. Br J Pharm 80:619–629.

Barker JL, Gratz E, Owen DG, Study RE (1984): "Pharmacological effects of clinically important drugs on the excitability of cultured mouse spinal neurons." In NG Bowery (ed): Actions and Interactions of GABA and Benzodiazepines. New York: Raven Press, pp 203–216.

Barker JL, Gruol DL, Huang LM, MacDonald JF, Smith TG (1980): Peptides: Three forms of chemical excitability on cultured mouse spinal neurons. Neuropeptides 1:63–82.

Barker JL, MacDonald JF, Mathers DA, McBurney RN, Oertel W (1980): "GABA receptor functions in cultured mouse spinal neurons." In FV DeFeudis, P Mandel (eds): Amino Acid Neurotransmitters. New York: Raven Press, pp 281–293.

Brenner, Sakmann B (1983) Neurotrophic control of channel properties at neuromuscular synapses of rat muscle. J Physiol (Lond) 337:159–171.

Choi DW, Farb DH, Fischbach GD (1981): Chlordiazepoxide selectively potentiates GABA conductance of spinal cord and sensory neurons in cell culture. J Neurophysiol 45:621–631.

Collingridge GL, Gage PW, Robertson B (1984): Inhibitory post-synaptic currents in rat hippocampal CA1 neurones. J Physiol 356:551–564.

Enna S, Snyder SH (1977): Influences of ions, enzymes, and detergents on $\gamma$-aminobutyric acid receptor binding in synaptic membranes of rat brain. Mol Pharmacol 13:442–453.

Fischbach GD, Schuetze S (1980): A postnatal decrease in acetylcholine channel open time at rat endplates. J Physiol (Lond) 303:125–137.

Gardner P, Ogden DC, Colquhoun D (1984): Conductances of single channels opened by nicotinic agonists are indistinguishable. Nature 309:160–162.

Gold MR, Martin AR (1983a): Analysis of glycine-activated inhibitory post-synaptic channels in brainstem neurons of the lamprey. J Physiol 342:88–98.

Gold MR, Martin AR (1983b): Characteristics of inhibitory postsynaptic currents in brainstem neurons of the lamprey. J Physiol 342:99–117.

Greenlee DV, Van Ness PC, Olsen RW (1978): Gamma-aminobutyric acid binding in mammalian brain: receptor-like specificity of sodium-independent sites. J Neurochem 31:933–938.

Hamill OP, Bormann J, Sakmann B (1983): Activation of multiple-conductance state chloride channels in spinal neurons by glycine and GABA. Nature 305:805–808.

Huang L-Y M, Barker JL (1980): Pentobarbital: Stereospecific actions of (+) and (−) isomers revealed on cultured mammalian neurons. Science 207:795–797.

Jackson MB, Lecar H, Mathers DA, Barker JL (1982): Single channel currents activated by GABA, muscimol, and (−)pentobarbital in cultured mouse spinal neurons. J Neurosci 2:889–894.

Katz B and Miledi R (1972): The statistical nature of the acetycholine potential and its molecular components. J Physiol 224:665–699.

Krogsgaard-Larsen P, Johnston GAR (1978): Structure–activity studies on the inhibition of GABA binding to rat brain membranes by muscimol and related compounds. J Neurochem 30:1377–1382.

MacDonald JF and Barker JL (1982): Multiple actions of picomolar concentrations of flurazepam on the excitability of cultured mouse spinal neurons. Brain Res 246:257–264.

MacDonald JF, Owen DG, Barker JL (1985): Voltage-sensitive, spontaneously occuring $Cl^-$channels in

cultured spinal neurons. Soc Neurosci Abst (in press).
Macdonald RL and Barker JL (1978): Benzodiazepines specifically modulate GABA-mediated postsynaptic inhibition in cultured mammalian neurons. Nature 271:563–564.
Macdonald RL, McLean MJ, Skerritt JH (1985): Anticonvulsant drug mechanisms of action. Fed Proc 44:2634–2639.
Mathers DA, Barker JL (1980): Pentobarbital opens channels of long duration on cultured mouse spinal neurons. Science 209:507–509.
Mathers DA, Barker JL (1982): Chemically-induced ion channels in nerve cell membranes. Int Rev Neurobiol 23:1–34.
Owen DG, Study, RE, Gratz, E, Barker JL (1982): Pharmacological modulation of GABA responses in cultured mouse spinal neurones. 12th Ann Meeting Soc Neurosci (Minneapolis), Abstract 239.5.
Ozawa S, Yuzaki M (1984): Patch-clamp studies of chloride channels activated by Gamma-aminobutyric acid in cultured hippocampal neurons of the rat. Neuroscience Research 1:275–293.
Sakmann B, Hamill OP, Bormann J (1983): Patch-clamp measurements of elementary chloride currents activated by the putative inhibitory transmitters GABA and glycine in mammalian spinal neurons. J Neural Transmission 18:83–95.
Schuetze S, Vicini S (1984): Neonatal denervation inhibits the normal postnatal decrease in endplate channel open time. J Neurosci 4:2297–2302.
Sher P, Study RE, Mazzetta J, Barker JL, Nelson PG (1983): Depression of benzodiazepine binding and diazepam potentiation of GABA-mediated inhibition after chronic exposure of spinal cord cultures to diazepam. Brain Res 268:171–176.
Study RE, Barker JL (1981): Diazepam and (−)pentobarbital: fluctuation analysis reveals different mechanisms for potentiation of GABA responses in cultured central neurons. Brain Research 268:171–176.

Benzodiazepine/GABA Receptors and Chloride Channels:
Structural and Functional Properties, pages 167–184


# 7

# Benzodiazepine Receptor Binding In Vivo and Efficacy

**Claus Braestrup and Mogens Nielsen**

*Novo Industri A/S, Pharmaceuticals R&D, DK-2880 Bagsvaerd (C.B.), and Psychopharmacological Research Laboratory, Sct. Hans Mental Hospital, DK-4000 Roskilde (M.N.), Denmark*

Specific binding sites for the benzodiazepine type of minor tranquillizers have been thoroughly demonstrated in the central nervous system (CNS) of all higher vertebrates, including man [Squires and Braestrup, 1977; Möhler and Okada, 1977; for review see Braestrup and Nielsen, 1983]. Extensive evidence suggests that these binding sites are the pharmacological receptors for benzodiazepines; benzodiazepines exert their pharmacological and clinical actions by combining with these receptors. The physiological function of the benzodiazepine receptor is still unknown; but since it is clearly demonstrated that the benzodiazepine receptor is ultimately coupled to the GABA receptor, it is conceivable that benzodiazepine receptors function to regulate the "gain" in the GABA-ergic neurotransmission system in such a way that the benzodiazepine receptor, within certain limits, determines the magnitude of effects elicited by a given concentration of GABA. When benzodiazepines are present at the benzodiazepine receptor, they indirectly facilitate the opening of chloride channels by GABA. Benzodiazepines do not open chloride channels by themselves; they demand the presence of GABA to exert their effects.

A few years ago new types of benzodiazepine receptor ligands were discovered which bound tightly to the receptors and yet do not elicit the well-known pharmacological effects of benzodiazepines. The first representative of this kind of ligands was ethyl β-carboline-3-carboxylate (β-CCE) [Braestrup et al., 1980; Mitchell and Martin, 1980; Tenen and Hirsch, 1980; Cowen et al., 1981; Polc et al., 1981], rapidly followed by several other ligands, all to be known as benzodiazepine receptor antagonists; Ro 15-1788 [Hunkeler et al., 1981], CGS 8216 [Czernik et al., 1982], ZK 93423 [Jensen et al., 1984], CGS 9895 [Brown and Martin, 1984].

A common feature of all these benzodiazepine receptor antagonist ligands was that they antagonized all pharmacological, electrophysiological, and other effects of benzodiazepines and other ligands acting directly on the benzodiazepine receptor. A group of ligands which are intermediates between benzodiazepines and benzodiazepine receptor antagonists has been described in terms of partial agonists ZK 91296 [Petersen et al., 1984] and CGS 9896 [Yokoyama et al., 1982; Bennett and Petrack, 1984; see further below].

Subsequent to the discovery of benzodiazepine receptor antagonists yet a third group of benzodiazepine receptor ligands, convulsive ligands, were identified [Braestrup et al., 1982a] inspired by studies showing that some of the β-carboline ligands for benzodiazepine receptors in some paradigms exerted definite but weak effects that pointed in the opposite direction of those of benzodiazepines [Braestrup and Nielsen, 1981; Jones and Oakley, 1981; Cowen et al., 1981; File et al., 1982]. The identification of DMCM as a highly convulsant benzodiazepine receptor ligand, and one which furthermore exhibited a GABA receptor coupling which was "opposite" to that of the benzodiazepines, created a firm basis for the idea that one single receptor, the benzodiazepine receptor, could be affected in two opposite directions, depending on the nature of the ligand [Braestrup et al., 1982a]. Nutt et al. [1982] arrived at a similar conclusion using β-CCE to lower the seizure threshold for pentylenetetrazol and to reduce GABA depolarization in a ganglionic preparation. The extensive literature showing that inverse agonists produce the opposite pharmacological effects of benzodiazepines shall not be reviewed here.

In the following, we will allude first to the techniques used to demonstrate benzodiazepine receptor interactions and occupancy in the living brain, then to a discussion of how a comparison of benzodiazepine receptor occupancy with pharmacological effects identifies a continuum of ligands for benzodiazepine receptors, spanning from ligands with positive efficacy (the benzodiazepines) over ligands with nil efficacy (receptor antagonists) to ligands with negative efficacy (inverse agonists). Finally, we will discuss biochemical in vitro measures which to varying degrees reveal benzodiazepine receptor ligand efficacy.

## IN VIVO DETERMINATION OF BENZODIAZEPINE RECEPTOR OCCUPANCY

In the early benzodiazepine receptor studies great effort was laid in establishing correlations between the in vitro affinity of benzodiazepines for benzodiazepine receptors on the one hand and a number of functional effects of benzodiazepines on the other. The functional effects were all behavioral (e.g., inhibition of convulsions, anticonflict effect); no functional studies were done using isolated organs or any other system where the concentration of drug in the relevant biophase could be reliably assessed. These early studies were therefore compromised by the lack of knowledge of the actual drug concentration at the receptors in the brain after systemic treatment of experimental animals. For example, two benzodiazepines might have exactly the same affinity for the receptors and yet, if one of them had low bioavailability, high plasma protein binding, low penetration over the blood brain barrier, fast metabolism, etc., it would show low potency in whole animal pharmacological investigations.

Two different approaches to obtaining more meaningful correlations of receptor interaction and effect have been developed. Methods have been developed to estimate the occupation of benzodiazepine receptor ligands in vivo; secondly, functional in vitro methods have been developed where in vitro affinity can be directly correlated to in vitro function, examples of this will be voltage-clamped neurons in culture to be described later in this chapter under "Biochemical and Electrophysiological Correlations to Efficacy of Benzodiazepine Receptor Ligands."

Several methods have been published for determining benzodiazepine receptor binding in the brain of living rats or mice [Chang and Snyder, 1978; Tallman et al., 1979; Blanchard et al., 1979; Duka et al., 1979; Hirsch and Lydigsen, 1981; Braestrup et al., 1982b; Fehske and Müller, 1982; Minchin and Nutt, 1983; see also discussion in Mennini and Garattini, 1982]. In principle these methods are all very similar. A tritium-labeled benzodiazepine receptor ligand ($^3$H-diazepam, $^3$H-flunitrazepam, $^3$H-Ro 15-1788, $^3$H-β-CCE, or $^3$H-lormetazepam) is administered via the tail vein. After a certain time interval (usually from 1–20 min), in which the radioactive ligand reaches the brain and combines with the receptors, the animals are decapitated, and the brains rapidly removed, homogenized, and filtered under vacuum through glass fiber filters. Radioactivity bound to the filters expresses total bound radioligand. These manipulations are all done quickly (60–100 seconds) to minimize the degree of radioligand redistribution after death. The aim is to measure the amount of binding present in the living brain at the time of decapitation—not higher binding values, which might result from lowering the temperature from body temperature to room or ice-bath temperature, and not lower binding values, which will occur if the tissue is homogenized in a large volume or left to stand before filtration. In this latter case the radioligand will dissociate from the receptors and a new equilibrium will be reached depending on the volume, the amount of radioligand present, and the temperature.

In vivo binding determined as described above represents total binding. To obtain values for specific binding, nonspecific binding is subtracted from total binding. Nonspecific binding is obtained by occupying all brain benzodiazepine receptors prior to administration of the radioligand—for example, by administration of the brain-specific benzodiazepine clonazepam in a high dose, 10–25 mg/kg i.p. After this dose of clonazepam, $^3$H-flunitrazepam will bind only to nonbenzodiazepine receptor–related sites. Nonspecific binding typically represents 10% of total binding when $^3$H-flunitrazepam is used as radioligand, more when $^3$H-diazepam or $^3$H-CGS 8216 are used, and slightly less when $^3$H-Ro 15-1788 is used.

Alternative methods to those referred above have been proposed [Williamson et al., 1978; Mennini and Garattini, 1982; Mennini et al., 1982a,b]. In these methods nonspecific binding is determined by adding excess benzodiazepine to the brain homogenate after decapitation, and a certain time period is allowed for the radioligand to redistribute after homogenization. These methods essentially measure the active drug concentration in the brain by a radioreceptor assay and not the occupancy. The drug concentration may not always be directly related to the occupancy; discrepancies between the occupancy method and the radioreceptor assay method have been demonstrated only in a few cases. For example, using the occupancy method we found that estazolam inhibited $^3$H-flunitrazepam binding with an $ED_{50}$ of 2.2 mg/kg i.p. (35 minutes), while Mennini et al. [1984] using the radioreceptor assay method found that estazolam enhances $^3$H-diazepam binding.

The method described above for in vivo determination of benzodiazepine receptor binding can probably be used to obtain an estimate of the occupancy of a given ligand for benzodiazepine receptors after a given dose. Administration of increasing doses of diazepam to mice (NMRI strain) prior to the administration of $^3$H-flunitrazepam will inhibit specific $^3$H-flunitrazepam binding in a dose-dependent manner (see Fig. 1). Inhibition of $^3$H-flunitrazepam binding by 50% will correspond to an occupancy by diazepam of brain benzodiazepine receptors also by 50%, provided that at least three conditions are fulfilled:

1) $^3$H-Flunitrazepam is present only in tracer amount. This is fulfilled because $^3$H-flunitrazepam labels only 0.6% of the total receptor population in experiments as described in Figure 1.

2) There is no postmortem redistribution of radioactivity that will affect the results. This precondition is difficult to verify. $^3$H-Flunitrazepam will undoubtedly associate with and dissociate from the receptors to some degree during homogenization. However, when excess unlabeled diazepam is present in the homogenization buffer [Tallman et al., 1979; Braestrup, unpublished] no measurable effect is observed on the amount of total binding when the process is rapidly conducted at 0°C. Redistribution in the period of 30–40 sec it takes from killing until the brain is excised cannot be measured, but since the temperature remains close to body temperature, and since there is no net flow of radioactivity in or out of the brain, there is no reason to believe that cessation of blood flow will change the equilibrium of $^3$H-flunitrazepam

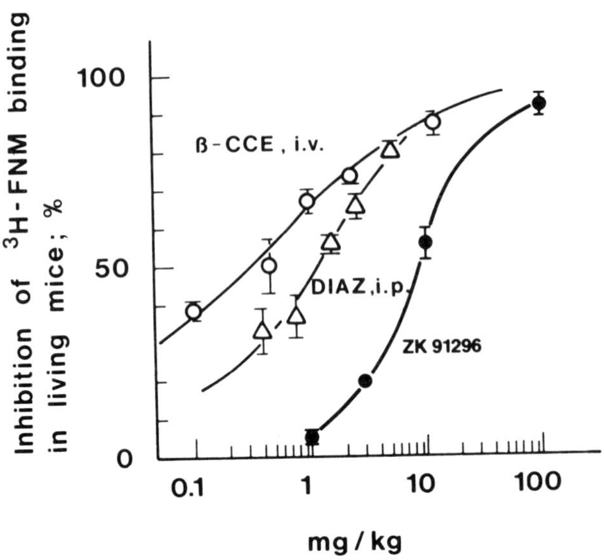


**Fig. 1.** *Inhibition of specific $^3$H-flunitrazepam ($^3$H-FNM) binding in the forebrain of the mouse by diazepam, ZK 91296 and β-CCE. Increasing doses of diazepam or ZK 91296 were administered by the intraperitoneal route 35 min before decapitation. β-CCE was administered by the intravenous route 7 min before decapitation. $^3$H-FNM was given intravenously (4 μCi per mouse, 19-21 g NMRI strain) mainly according to Chang and Snyder [1978]. The mice were decapitated 20 min after $^3$H-FNM administration (except in the experiment with β-CCE where 7 min elapsed), whole forebrains (about 285 mg) were rapidly excised and homogenized in 12.0 ml of ice-cold 25 mM $K_2HPO_4$–$KH_2PO_4$ buffer, pH 7.1. One milliliter aliquots of this crude homogenate were immediately filtered through Whatman GF/C glass fiber filters and washed with 2 × 5 ml ice-cold buffer. Tritium on the filters which represents total membrane bound $^3$H-FNM (about 50% of total radioactivity in the homogenate) was determined by conventional scintillation counting. Specific binding is total membrane bound $^3$H-FNM minus non-specific binding, which was obtained in separate groups of mice receiving a systemic administration of a supramaximal dose of clonazepam (25 mg/kg i.p.) 30 min before $^3$H-FNM administration. Non-specific binding represents approximately 10% of total membrane bound $^3$H-FNM. Specific binding was 622 ± 19 cpm/assay corresponding to 24 mg of tissue (mean ± sem of 20 values from 20 independent experiments).*

on and in the vicinity of the receptors, again provided that the handling is not delayed. In conclusion, redistribution does not seem to affect the results to any great extent. Calculations based on the on- and off-rates for $^3$H-flunitrazepam would not have eliminated the time factor as an important variable; in vitro the off-rate for $^3$H-flunitrazepam at 0°C and 30°C is $k = 0.013\ min^{-1}$ and $k = 0.6\ min^{-1}$, respectively, which indicates that 45% of bound $^3$H-flunitrazepam will dissociate from the receptor in one minute at 30°C.

3) If $^3$H-flunitrazepam preferentially labeled a subpopulation of benzodiazepine receptors, a population which for example only represented 30–50% of the whole receptor population, there would obviously not be any correlation between inhibition of $^3$H-flunitrazepam binding and occupancy of benzodiazepine receptors in general. To our knowledge data have previously not been presented to show that $^3$H-flunitrazepam in fact does bind to the total population of benzodiazepine receptors in the living mouse forebrain, even though $^3$H-flunitrazepam would be expected to do so because of its lack of receptor selectivity. We therefore initiated experiments to show that $^3$H-flunitrazepam binds to all known classes of benzodiazepine receptors also in in vivo experiments. This was done as a saturation experiment in the living mouse brain. The data were subsequently transformed by

Scatchard analysis, as shown in Figure 2. As expected from previous studies, we found that the apparent affinity constant, $K_D$, for $^3$H-flunitrazepam binding in vivo, 148 nM (Fig. 2), is much higher than the corresponding in vitro value [Chang and Snyder, 1978; Braestrup et al., 1982b; Fig. 2]. Furthermore we observed that the $B_{max}$-value determined in vivo was almost exactly identical to the $B_{max}$-value determined in vitro. This shows that $^3$H-flunitrazepam labels all benzodiazepine receptor subclasses in vivo.

We conclude that inhibition of $^3$H-flunitrazepam binding in the forebrain of living mice by a benzodiazepine receptor ligand confers a measure of the actual occupancy of benzodiazepine receptors ($BZ_1$ plus $BZ_2$) by this ligand plus its active metabolites (if any). Hill analyses are not available for a great number of substances. Hill analysis of diazepam gives a slope of 1.03 (not shown). Thus for diazepam the occupancy follows the law of mass action and can be calculated for a desired dose level by using the Hill equation and occupancy measures obtained at another dose level. It is important to realize, however, that the mass action law may not be applicable to all benzodiazepine receptor ligands, neither for example to agents with dose-dependent pharmacokinetics, nor for agents which exert benzodiazepine receptor subclass selectivity. For this reason it is advisable to use calculations of occupancy values only within narrow ranges of experimentally determined values.

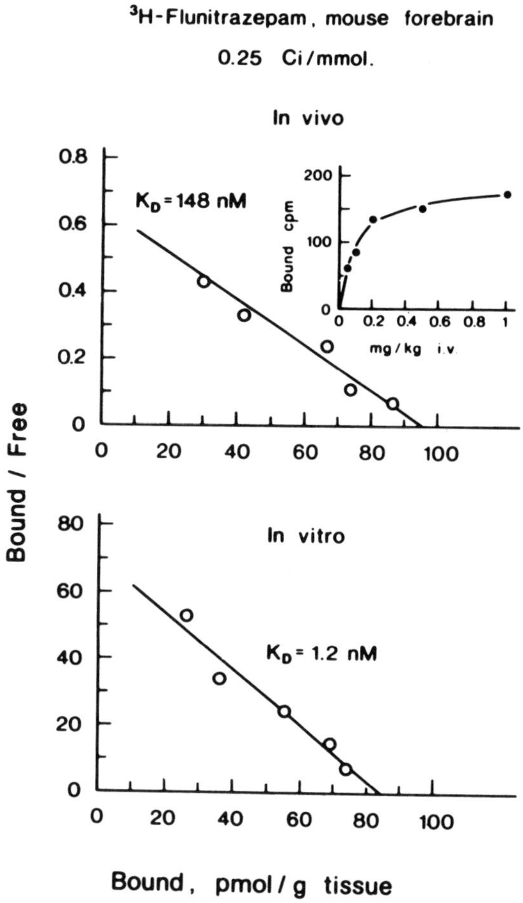


**Fig. 2.** *Groups of three NMRI mice were treated intravenously with increasing doses (0.05–1.0 mg/kg) of $^3$H-flunitrazepam (specific activity, 0.25 Ci/mmol). Parallel groups of three mice received the same doses of $^3$H-flunitrazepam + 10 mg/kg i.p. clonazepam given 50 min before decapitation for determination of non-specific binding. Specific binding as shown on the insert upper graph was determined as described in the legend to Figure 1. The lower graph shows a conventional Scatchard analysis of $^3$H-flunitrazepam binding in crude homogenates of mouse forebrain membranes in vitro.*

## IN VIVO BINDING OF BENZODIAZEPINE RECEPTOR LIGANDS

In the preceding section it was described how the in vivo receptor occupancy can be determined in the brain of living mice, based on methods originally published by Chang and Snyder [1978] and Duka et al. [1979]. The in vivo occupancy of benzodiazepine receptors has been investigated for a number of benzodiazepine receptor ligands. Experimental values are usually expressed as the $ED_{50}$ values which express the dose of a given test substance that inhibits by 50% the amount of specific binding of a benzodiazepine receptor radioligand in vivo. When the $ED_{50}$ value is determined as described above, it is also an estimate of the dose at which the ligand occupies 50% of the brain benzodiazepine receptors. It should be recalled that the $ED_{50}$ value represents the average occupancy throughout the brain area investigated (usually the whole forebrain of mice); in cases where a compound, for example CL 218872, shows selectivity for the $BZ_1$ receptor subclass, then the occupancy of $BZ_1$ receptors will be more than 50% after administration of a dose equal to $ED_{50}$, while the occupancy of $BZ_2$ receptors will be less than 50%. In addition, the dose–response curve will be flattened. Radioligands which selectively label the $BZ_1$ or the $BZ_2$ receptor subclass in vivo are not available.

The benzodiazepine receptor ligands listed in Table I have been divided into three groups, based on the efficacy of each of the ligands. It will be clear from Table I that all three groups include ligands that effectively reach the brain and bind to benzodiazepine receptors after systemic administration. In addition, it is striking that Table I exclusively contains ligands that are believed to interact directly with the benzodiazepine receptor recognition site of benzodiazepine receptors. None of the substances that are known to modulate benzodiazepine receptors indirectly via allosteric mechanisms inhibit benzodiazepine receptor binding in vivo. Substances investigated include bicuculline, picrotoxin, pentylenetetrazol, phenobarbital [Braestrup and Nielsen, 1983]. In vivo enhancements of $^3$H-diazepam or $^3$H-flunitrazepam binding have been reported with the GABA agonist muscimol [Tallman et al., 1979], the purine deriv-

**TABLE I. Inhibitors of Mouse Brain Benzodiazepine Receptor Binding In Vivo**

| Group | Agent | $ED_{50}$[a] mg/kg i.p. |
|---|---|---|
| 1. Agonists, positive efficacy | Triazolam | 0.12 |
| | RO 116896 (B10 (+) ) | 0.14 |
| | Clonazepam | 0.2 |
| | Lorazepam | 0.3 |
| | Flunitrazepam | 0.6 |
| | ZK 93423 | 1.0 |
| | Pinazepam | ca. 1 |
| | Clobazam | 1.1 |
| | Nitrazepam | 1.25 |
| | Diazepam | 1.2 |
| | CGS 9896 | 1.5 |
| | Midazolam | 1.3 |
| | Estazolam | 2.2 |
| | ZK 91296 | 4.5 |
| | Oxazepam | 7 |
| | Zopiclone | 10 |
| | CL 218872 | 18 |
| | Chlordiazepoxide | 21 |
| | PrCC | 300 |
| 2. Receptor antagonists, nil efficacy | CGS 9895 | 1 ( (p.o.), 60 min) |
| | RO-15-1788 | 4 |
| | ZK 93426 | 3 |
| | CGS 8216 | 1.1 |
| 3. Inverse agonists, negative efficacy | FG 7142 | 60 |
| | β-CCE | 90 |
| | β-CCM | 22 |
| | DMCM | 2 |

[a]Administered 35 min before decapitation, see legend to Figure 1. Data from Braestrup and Nielsen [1982b, 1983]; Brown et al., [1984].

ative EMD 28422 [Skolnick et al., 1980], and the anomalous anxiolytic buspirone [Oakley and Jones, 1983]. When enhanced benzodiazepine receptor binding is observed in vivo, care should be exercised to show that the enhancement is not simply due to an impairment of the circulation, excretion, or metabolism of the radioligand. These factors may enhance the amount of tracer present in the brain at the time of decapitation and thus artifactually result in enhanced binding values.

## CORRELATION OF OCCUPANCY TO PHARMACOLOGICAL EFFECT

Benzodiazepines are potent anticonvulsant agents and it has been assumed, since the discovery of benzodiazepine receptors in 1977, that occupancy of benzodiazepine receptors by benzodiazepines elicited an anticonvulsant effect. In consequence there should be a tight correlation between anticonvulsant effect and occupation of benzodiazepine receptors. The first study showing such a correlation with a number of drugs [Duka et al., 1979] has since been extended [Braestrup et al., 1982b]. The results shown in Figure 3 include only such studies where occupancy and effects were determined in one single mouse strain and at comparable dose/time schedules; Duka et al. [1979] used male NMRI mice challenged with pentylenetetrazol, 80 mg/kg s.c., 10 min after the anticonvulsant, while Braestrup et al. [1982a] used

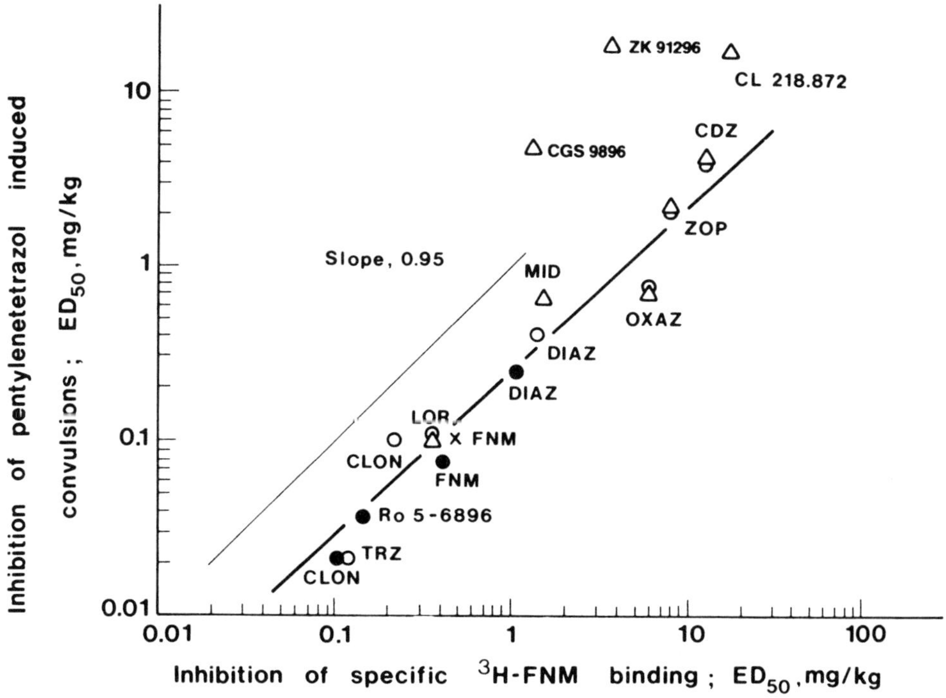


**Fig. 3.** *Correlation between the in vivo occupancy of benzodiazepine receptors and antagonism of pentylenetetrazol induced convulsions in mice. The* $ED_{50}$ *value is the dose of the compound that inhibits by 50% the specific binding of* $^3$*H-flunitrazepam binding in the forebrain of living mice. Test substances were administered by the interperitoneal route 35 min before decapitation and* $^3$*H-flunitrazepam binding was determined as described in the legend to Figure 1. The* $ED_{50}$ *value against pentylenetetrazol is that the dose which abolishes clonic convulsions and death in 50% of mice challenged with 150 mg/kg s.c. of pentylentetrazol 30 min after the protective agent which was administered either i.p., △, or s.c., ○, data from Braestrup et al. [1982b and unpublished]; ●, data from Duka et al. [1979] (CLON, clonazepam; CDZ, chlordiazepoxide; DIAZ, diazepam; FNM, flunitrazepam; LOR, lorazepam; MID, midazolam; OXAZ, oxazepam; TRZ, triazolam; ZOP, zopiclone.)*

male NMRI mice challenged with pentylenetetrazol 150 mg/kg s.c., 30 min after the anticonvulsant. The results in Figure 3 strongly suggest that the anti-pentylenetetrazol activity in mice is elicited by an interaction with benzodiazepine receptors. Application of the mass action law indicates thatapproximately 25% receptor occupancy is sufficient to inhibit by 50% convulsions produced by pentylenetetrazol. We would expect that all benzodiazepine receptor ligands should line up exactly on the correlation line *if* the efficacy of all these ligands was identical and if receptor subtype selectivity was without significance for the anticonvulsant effect. This apparently is not the case. Some benzodiazepine receptor ligands, in particular ZK 91296 and CGS 9896, but also CL 218872, show less efficacy (higher occupancy to obtain a given effect) than most of the benzodiazepine type ligands. Furthermore, ligands in groups 2 and 3 of Table I cannot, at any dose, antagonize pentylenetetrazol convulsions under the present conditions. ZK 91296 and CGS 9896 seem to lack $BZ_1/BZ_2$ selectivity [Petersen et al., 1984; Bennett and Petrack, 1984]. The dislocation observed in Figure 3 is more likely to be due to partial agonist properties for ZK 91296, CGS 9896, and CL 218872 (see further below).

It has been suggested that conventional benzodiazepines do exhibit slight variation in their efficacy at the receptor [Farb et al., 1984], but it is not clear whether such difference can be detected as scatter around the correlation line, Figure 3.

The observation that approximately 25% benzodiazepine receptor occupancy is sufficient for antagonizing pentylenetetrazol convulsions is not to be extrapolated to occupancies for inhibition of convulsions induced by other means. Antagonism of sound-induced seizures in the audiogenic seizure-sensitive mouse strain, DBA/2, occurs after very low doses of benzodiazepines. Parallel determination of receptor occupancy showed that this type of audiogenic seizure was antagonized by receptor occupancies of approximately 2–5% (see Table II). Exact values cannot be given because direct measurement of these low levels of occupancy is not technically feasible and because calculation of the occupancy by extrapolation is confounded with latent error. Note that Ro 15-1788 exerts full anticonvulsant effect against sound-induced seizures in DBA/2 mice, albeit that high receptor occupancies are needed. Several other paradigms reveal that Ro 15-1788 possesses weak agonistic properties [Dantzer and Pério, 1982; Skerritt and MacDonald, 1983; Vellucci and Webster, 1983; Grecksch et al., 1983; Thiebot et al., 1983]. As for pentylenetetrazol convulsions, some partial agonistic ligands must occupy high fractions of receptors before their anticonvulsant effect is obtained in DBA/2 mice. (see Table II).

Comparison of Tables I and II further shows that different mouse strains (NMRI and DBA/2) differ slightly in the level of occupancy obtained after a given dose. This difference is clearly evident for several benzodiazepines that reach the brain and bind to the receptors more easily in DBA/2 mice than in NMRI mice.

Convulsions produced by means other than pentylenetetrazol and sound have not been related in detail to benzodiazepine receptor occupancy. Inhibition of some kinds of convulsions, for example convulsions induced by maximal electroshock, is probably not directly linked to benzodiazepine receptor occupancy.

## BENZODIAZEPINE RECEPTOR OCCUPANCY AND ANTICONFLICT EFFECT

There are a number of animal paradigms which are believed to reflect the anxiolytic effect of benzodiazepines in clinical situations. Anticonflict models are the most popular. In these paradigms animals are punished for carrying out behaviors such as eating, drinking, or exploration. Untreated animals

TABLE II. Effects of Benzodiazepine Receptor Ligands Against Audiogenic Seizures, DBA/2 Mice[1]

| | $ED_{50}$ (mg/kg) | | | |
|---|---|---|---|---|
| | $^3$H-FNM binding | Inhibition of seizures (111 dB) | Facilitation of seizures (83 dB) | Receptor occupancy at $ED_{50}$ (%) |
| Lorazepam | 0.13 | 0.002 | — | <10 |
| Triazolam | 0.15 | 0.003 | — | <10 |
| Clonazepam | 0.08 | 0.005 | — | <10 |
| Diazepam | 0.44 | 0.04 | — | <10 |
| Oxazepam | 4.5 | 0.15 | — | <10 |
| Chlordiazepoxide | 10 | 0.5 | — | <10 |
| ZK 93423 | 0.2 | 0.01 | — | <10 |
| Zopiclone | 2.3 | 0.5 | — | 18 |
| CL 218.872 | 4.6 | 1.5 | — | 25 |
| ZK 91296 | 1.2 | 0.5 | — | 33 |
| CGS 9896 | 0.11 | 0.15 | — | 58 |
| PrCC | 4.3 | 9 | — | 68 |
| Ro 15-1788 | 1.1 | 6 | — | 85 |
| ZK 93426 | 0.3 | 40 | — | >95 |
| FG 7142 | 37 | — | 50 | 57 |
| CGS 8216 | 0.9 | — | 1.0 | 53 |
| β-CCE | 240 | — | 49 | 17 |
| ZK 90886 | 6.6 | — | 0.8 | 11 |
| β-CCM | 3.0 | — | 0.1 | <10 |
| DMCM | 3.1 | — | 0.1 | <10 |

[1]Male DBA/2 mice (8–12 g) were exposed to a 14 kHz tone at either 111 dB (convulsant level) or 83 dB (subconvulsant level) and the number of mice with audiogenic seizures within 30 sec were noted. Dose response curves of the drugs were established and $ED_{50}$-values were calculated. $ED_{50}$-values of in vivo BZ receptor binding to age matched DBA/2 mice were also established and the BZ receptor occupancy of the $ED_{50}$-value for facilitation or inhibition of audiogenic seizures were calculated; from Jensen et al., 1983.

cease to perform these behaviors when punished (by a foot shock or other means), but the conflict situation can be relieved when benzodiazepines are administered before exposure to the conflict situation. The animal then performs as if the punishment were absent. In the so-called four-plate test of Boissier et al. [1968], naive mice are punished for exploring a new cage. The mice receive a foot shock each time they traverse a border between any two of the four metal places that form the cage bottom. Benzodiazepines restore normal exploration. Stephens et al. [1984] determined the minimal effective dose of benzodiazepines and other ligands for normalizing exploration and correlated these doses with receptor occupancy (data from Table I). The correlation is good (Fig. 4); the partial agonists CGS 9896 and ZK 91296 tend to lie at the left; however, no good explanation is offered for the unexpected location of the data point for lorazepam. The correlation line in Figure 4 indicates that roughly 30% benzodiazepine-receptor occupancy is needed for eliciting minimal effective anticonflict effect in this paradigm.

In another study we compared the anticonflict effect in the rat waterlick paradigm with receptor occupancy for three benzodiazepines and found that 60–70% receptor occupancy was needed for lorazepam, diazepam, and triazolam to exert their effect [Braestrup et al., 1982b] (Fig. 5).

The results presented in this and the preced-

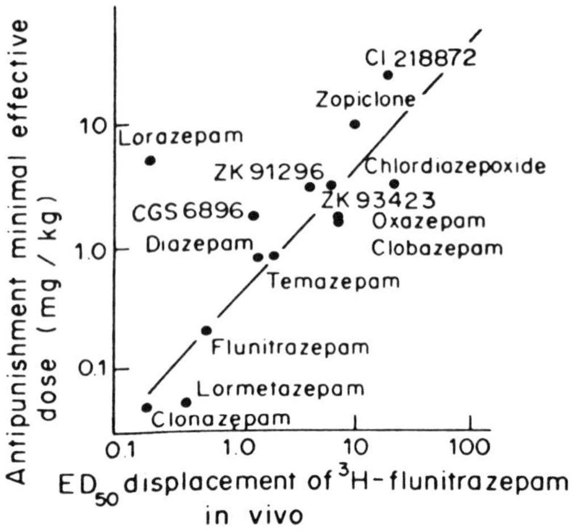


**Fig. 4.** *Correlation between in vivo occupancy of benzodiazepine receptors and anti-punishment activity in the four-plate test in mice. $ED_{50}$ for benzodiazepine binding is the dose that inhibits $^3$H-FNM binding by 50%. Four-plate MED is the smallest dose which significantly enhanced exploratory activity suppressed by foot shock [from Stephens et al., 1984].*

ing section indicate that anticonflict effects need similar or higher receptor occupancies than the anticonvulsant effect. Extrapolation to the clinical situation, however, cannot be made. Anxiolytic effects in humans are obtained at lower doses of benzodiazepines than the antiepileptic effect. The occupancy, therefore, must be less for anxiolytic effects than for anticonvulsant effects. For this reason it is also likely that partial benzodiazepine receptor agonists will be more efficient as anxiolytics than as antiepileptics.

## PARTIAL AGONISTS AT BENZODIAZEPINE RECEPTORS

A partial agonist is a compound which does not elicit the full stimulus of which a receptor is capable, even when the receptor is occupied fully by the partial agonist. The conformational shift in the receptor upon occupation thus appears to be less pronounced for a partial agonist than for a full agonist. It follows that when a partial agonist elicits the same response (stimulus) as a full agonist, then the receptor occupation by the partial agonist is higher than for the full agonist. The partial agonist properties of ZK 91296 [Petersen et al., 1984] and CGS 9896 [Bennett and Petrack, 1984] were first revealed by high occupancies at anti-pentylenetetrazol effective doses. Two other properties of partial agonists will be evident: First, a partial agonist may elicit the full pharmacological response in those animal paradigms where the response is dependent on low receptor stimulations; inhibition of audiogenic seizures in DBA/2 mice is an example of an animal paradigm where effects are elicited at low receptor stimulus. Thus, even though the receptor is only partly stimulated, the overt animal pharmacological response may be fully expressed. Secondly, in other animal paradigms, however, a partial agonist may behave as a pure competitive antagonist—namely, in those tests where a high receptor stimulus is required. In the rotarod test, for example, where a high benzodiazepine receptor stimulus is needed to exert the ataxic effect, ataxia after benzodiazepines is observed only after high doses giving high receptor occupancies. A partial agonist such as ZK 91296 will never, not even after high doses giving more than 96% occupancy, produce the stimulus necessary for producing the ataxic effect [Petersen et al., 1984]. However, ZK 91296 will occupy the receptors and thereby antagonize the effect of diazepam. In this paradigm the partial agonist ZK 91296 behaves as a pure antagonist. The conjecture that CGS 9896 is a partial agonist was recently substantiated by experiments showing that CGS 9896 provided a weak anticonvulsant effect by itself but that it partially antagonized diazepam in the pentylenetetrazol convulsion test [Brown et al., 1984] and in the rotarod ataxia test [Bennett and Petrack, 1984]. A third group of partial agonists, Ro 16-6028/001 and Ro 17-1812, were recently described from the Roche Laboratories [Haefely, 1984].

Seen in this perspective it is not necessary to invoke the existence of multiple receptor subtypes for a compound that is fully active

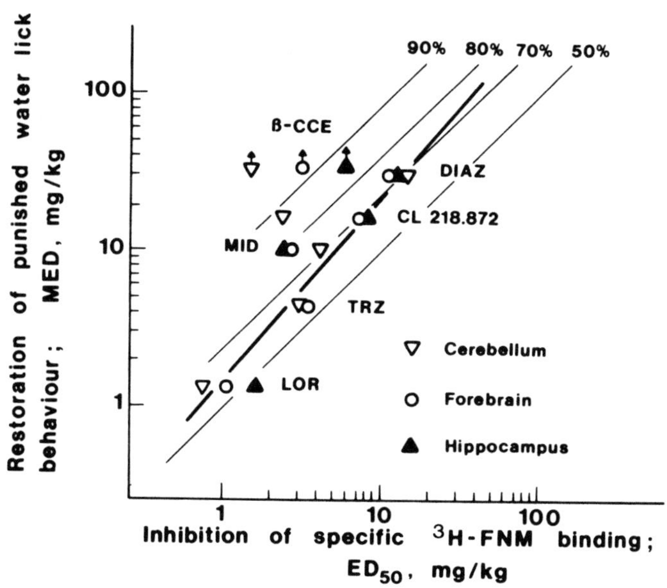


**Fig. 5.** *Correlation between in vivo occupancy of benzodiazepine receptors and anticonflict effects in the water lick paradigm in rats [from Braestrup et al., 1982b].*

as an agonist in some animal paradigms while being a competitive benzodiazepine antagonist in others.

## BIOCHEMICAL AND ELECTROPHYSIOLOGICAL CORRELATIONS TO EFFICACY OF BENZODIAZEPINE RECEPTOR LIGANDS

The benzodiazepine receptor was the first receptor for which it was demonstrated that effects in opposite directions (convulsant/anticonvulsant; anxiolytic/anxiogenic) could be elicited via one single receptor. A similar bidirectional sensitivity has since been described for $Ca^{2+}$ –channels [Schramm et al., 1983; Janis et al., 1984].

Several attempts have been directed at establishing in vitro biochemical or electrophysiological methods that would predict the efficacy of benzodiazepine receptor ligands. The various approaches have been quite successful, even though it has not been possible to select one single type of these various paradigms which correctly classifies all benzodiazepine receptor ligands investigated according to their pharmacological efficacies.

The biochemical paradigms for determining benzodiazepine receptor efficacy all utilize knowledge about the molecular coupling in the GABA/benzodiazepine receptor/chloride channel complex [see Olsen, 1981]. Fig. 6 illustrates that the complex comprises several receptor subunits and at least four neurotransmitter- or drug receptors [see also Olsen, 1981].

The *GABA ratio method* utilizes the fact that stimulation of the GABA receptor by certain GABA agonists (GABA and muscimol) causes a conformational shift in the benzodiazepine receptor in such a way that benzodiazepine receptor agonists bind more tightly while inverse agonists (convulsants such as DMCM) bind less tightly [Ehlert et al., 1981, 1982; Braestrup and Nielsen, 1981; Skerritt et al., 1982; Doble et al., 1982; Fujimoto et al., 1982; Morelli et al., 1982; Braestrup et al., 1982a; Brown and Martin, 1984]. Table III summarizes results obtained

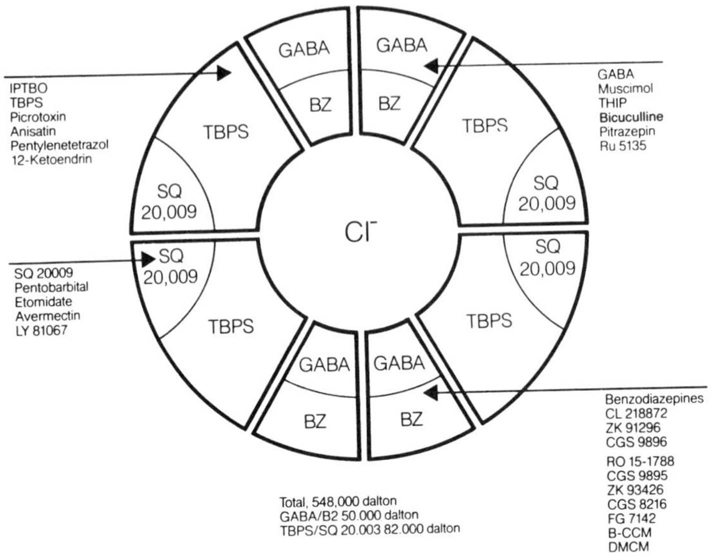


**Fig. 6.** *A highly schematic representation of the interconnections among GABA receptors, benzodiazepine receptors, and the TBPS binding part of the chloride channel [from Nielsen et al., 1985]. Shown are agents believed to interact at specific receptor sites.*

at 0°C. The pharmacological classification of triazolam and of ZK 93426 and CGS 9896 does not correlate well with the GABA ratio. Note that values for the GABA ratio are temperature-dependent. Triazolam, but not ZK 93426 or CGS 9896, attains values which correlate with pharmacological efficacy at 37°C.

The *photo-shift* [Karobath and Supavilai, 1982] utilizes the observation that partial photo-affinity labeling of benzodiazepine receptors with flunitrazepam [Möhler et al., 1980] affects the conformation of remaining unoccupied benzodiazepine receptors. The photo-labeled benzodiazepine receptor conformation binds benzodiazepine receptor agonist 30–100-fold weaker than the unperturbed receptor. The affinity of benzodiazepine receptor antagonists is mostly unchanged [Karobath and Supavilai, 1982; Tallman and Thomas, 1982; Davies et al., 1984; Brown and Martin, 1983; Blanchard et al., 1984; Möhler, 1982; Braestrup et al., 1984]. A detailed investigation of the photo-shift for a great number of ligands, however, reveals many anomalies. The pharmacological efficacies of Zopiclone, Suriclone, ZK 93423, ZK 93426, and others are not correlated to the photo-shift (Table III) [see also Brown and Martin, 1983].

The *barbiturate shift* utilizes the observation [see Leeb-Lundberg and Olsen, 1982] that barbiturates indirectly change the conformation of benzodiazepine receptors in such a way that the affinity for agonists is enhanced while that of inverse agonists is reduced [Honoré et al., 1984; Fischer and Olsen, this volume]. Barbiturates interact with a site related to the chloride channel but this site is distinct from the site where the cage convulsant $^{35}$S-TBPS binds. Probably barbiturates bind at the same site as etazolate, SQ 20, 009 [Ticku and Ramanjaneyulu, 1984; Maksay and Ticku, in press; Supavilai and Karobath, 1984; Nielsen et al., 1985], whereas, for example, pentylenetetrazol [Ramanjaneyulu and Ticku, 1984] and ketoendrin act directly on the TBPS site. As with other efficacy measures, the barbiturate shift mirrors the efficacy of benzodiazepine receptor ligands to some extent. Compounds such as ZK 93423 (a full agonist) and CL 218872, however, exhibit barbiturate shifts which do not reflect the efficacy.

The *TBPS shift*. TBPS (tert-butyl bicyclophosphorothionate) is a potent cage convulsant which produces convulsions by reducing GABA-ergic neurotransmission at the level of the chloride gating mechanism. $^{35}$S-TBPS is a useful radioligand for investigating the properties of the GABA receptor related chloride channel [Squires et al., 1983; 1984]. The TBPS binding site is directly coupled to benzodiazepine receptors (see Fig. 6) and it was observed that benzodiazepine receptor ligands allosterically enhanced or reduced $^{35}$S-TBPS binding in parallel with their efficacy for benzodiazepine receptors [Supavilai and Karobath, 1984a]. We have confirmed and extended these studies (see Table III and Fig. 7). The TBPS-shift method seems to be the most reliable *in vitro* measure of the efficacy of benzodiazepine receptor ligands, even though some ligands are not classified exactly as expected from general pharmacological studies. The increments and decrements in TBPS binding caused by the various benzodiazepine receptor ligands are not spectacular; enhancement of 40–50% is usually observed for agonists, while 40% decreases are observed for inverse agonists. Determination of $^{35}$S-TBPS binding, however, is very reliable and changes of as little as 10% are easily detected.

*Acetylcholine release from preloaded slices*. Recently Supavilai and Karobath [1984b] (see further this volume) observed that benzodi-

**TABLE III. Various Biochemical Parameters Related to Benzodiazepine Receptor Ligand Efficacy**

| Type | Efficacy | Compound | $^3$H-flunitrazepam binding $IC_{50}$ (nM) 0° | GABA ratio[a] 0° | Photo shift[b] 0° | Barbiturate shift[c] 30° | TBPS shift[d] 22° |
|---|---|---|---|---|---|---|---|
| Agonist | ++ | Diazepam | 25 | 2.30 | 0.02 | 3.14 | 1.27 |
| | | Flunitrazepam | 3.9 | 2.45 | 0.03 | 1.96 | 1.46 |
| | | Clonazepam | 4[f] | 2.1 | — | — | 1.38 |
| | | Triazolam | 2[f] | 1.00 | — | 1.83 | 1.56 |
| | | Zopiclone | 80[f] | 1.53 | 0.29 | 2.03 | 1.24 |
| | | ZK 93423 | 1[f] | 2.2 | 1.3 | 1.22 | 1.54 |
| Partial agonist | + | CL 218.872 | 320 | 1.98 | 0.37 | 1.03 | 0.97 |
| | | CGS 9896 | 0.3[g] | — | 1.23[e] | — | 1.03 |
| | | ZK 91296 | 0.6 | 1.4 | 0.6 | 1.13 | 1.16 |
| Receptor antagonist | 0 | Ro 15-1788 | 1.9 | 1.22 | 0.85 | 1.09 | 1.01 |
| | | ZK 93426 | 0.4 | 1.37 | 0.66 | 0.81 | 0.88 |
| | | CGS 8216 | 0.6 | 1.15 | 1.23[e] | — | 0.88 |
| Partial inverse agonist | — | FG 7142 | 300 | 0.87 | 1.0 | 0.51 | 0.79 |
| Inverse agonist | —— | β-CCM | 2.4 | 0.61 | 1.4 | 0.50 | 0.77 |
| | | DMCM | 9.0 | 0.46 | 2.6 | 0.56 | 0.57 |

For methodological details see:
[a]Braestrup et al. [1982a].
[b]Braestrup et al. [1984].
[c]Honoré et al. [1984].
[d]See legend, Fig. 7.
Value(s) from:
[e]Brown and Martin [1983].
[f]Braestrup and Nielsen [1983]; $^3$H Diazepam as radioligand.
[g]Brown and Martin [1984]; $^3$H Ro 15-1788 as radioligand.

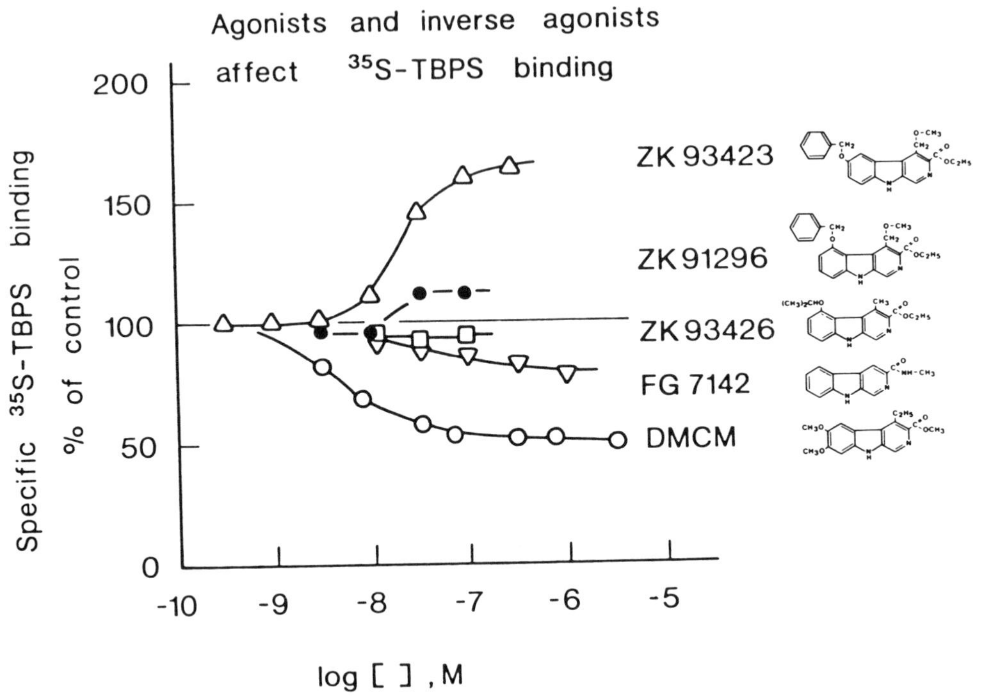


**Fig. 7.** *Illustration of experiments used for calculation of TBPS shifts. Frozen rat cortex membranes were thawed and homogenized in 2 × 5 ml ice-cold 50 mM Tris-citrate, pH 7.1, by an Ultra Turrax homogenizer and centrifuged at 14,000 × g for 10 min at 0°C. The membranes were washed twice by rehomogenization and centrifugation as before. The pellet was frozen (−20°C) overnight, thawed and washed twice by rehomogenization and centrifugation. The freezing and washing procedure was repeated once again. Before assays, the pellet was resuspended in 50 mM Tris-citrate, pH 7.1, in a volume of 100 ml/g original tissue. Aliquots of 0.5 ml were incubated for 60 min at 25°C with 25 μl* $^{35}$*S-TBPS working solution (*$^{35}$*S-butyl bicyclophosphorothionate, tertiary; 15-29 Ci/mmol, NEN Boston; prepared fresh in water) to give a final concentration of 0.9 nM. Test substances were added in volume of 25 μl aqueous solution prior to the radioligand. After incubation, 10 ml ice-cold 50 mM Tris-citrate, pH 7.1, was added and the sample was immediately vacuum filtered through Whatman GF/C glass fiber filters followed by another 10 ml wash. Radioactivity on the filters was measured by conventional scintillation counting at about 95% efficiency. Non-specific binding identified as binding in the presence of picrotoxinin (Sigma, 10 μg/ml), was always subtracted from total binding to give specific binding. The TBPS shift is the specific binding in the presence of the benzodiazepine receptor ligand at a concentration of 30 ×* $IC_{50}$ *for* $^{3}$*H-FNM binding divided by specific TBPS binding in the absence of added substances. Specific binding in the absence of added substances typically amounts to 2000 cpm, non-specific binding approximately 8%.*

azepine receptor agonists would enhance, while benzodiazepine receptor inverse agonists would reduce, GABA regulation of acetylcholine release from preloaded rat striatal slices. This system is the first functional biochemical system where benzodiazepine receptor agonists and inverse agonists can be detected. The sensitivity of the method is not very high, and the method is not yet suitable for classification of partial agonists.

*Miscellaneous other approaches* utilizing the coupling illustrated in Figure 6 for determination of benzodiazepine receptor ligand efficacy have been described. Only restricted series of benzodiazepine receptor ligands have been included in these studies, precluding firm conclusions as to their applicability [Borea et al., 1983; Skerritt et al., 1983].

*Electrophysiological paradigms*. It is well known that benzodiazepines enhance GABA-

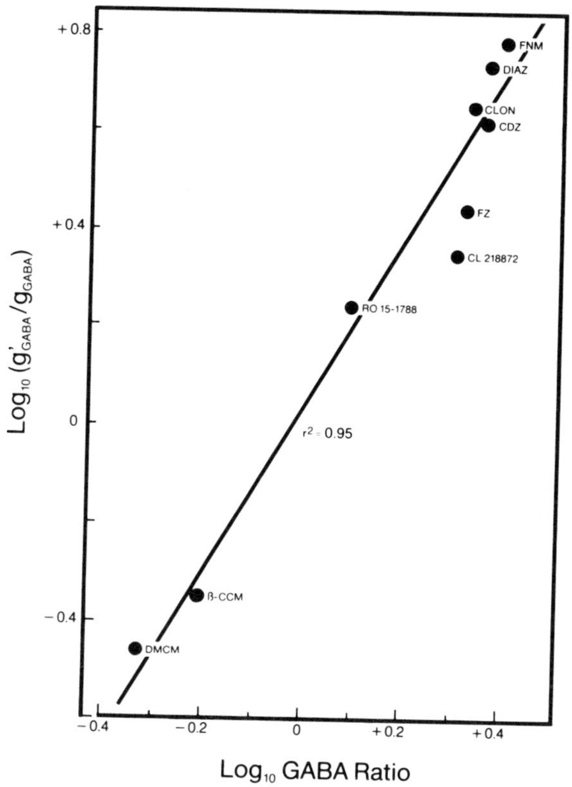


**Fig. 8.** *Correlation of efficacies for modulation of GABA effect on cultured chick spinal neurons with the GABA ratios determined in vitro in benzodiazepine receptor assays.* $g'_{GABA}/g_{GABA}$ *expresses the maximal enhancement or reduction (depending on the benzodiazepine receptor ligand) in the conductance for chloride induced in the spinal neuron by GABA. The GABA ratio expresses the effect of a supramaximal concentration of GABA on the affinity of each particular benzodiazepine receptor ligand for benzodiazepine receptors. Reproduced from Farb et al. [1984].*

ergic neurotransmission (for review see Haefely and Polc, this volume). Conversely it has been demonstrated that inverse agonists reduce GABA-ergic neurotransmission [Skerritt and Macdonald, in press; Farb et al., 1984; Chan et al., 1982]. Isolated neurons grown in culture are well suited for studies relating occupancy to effects because the occupancy can be determined by radioligand binding studies directly in the cultured neurons and the *effect* of the benzodiazepine receptor ligand can be detected by measuring the chloride flux in single neurons with voltage clamp techniques. In a detailed series of experiments, Farb and colleagues [1984] demonstrated first that there is a correlation between the occupancy of benzodiazepine receptors by benzodiazepines in the enhancement in GABA-stimulated chloride flux. Secondly, Farb observed that the maximal enhancement or reductions in GABA-stimulated chloride flux induced by individual benzodiazepine receptor ligands varied from one ligand to another. The degree to which a benzodiazepine receptor ligand enhanced or reduced chloride flux is an independent measure of the agonist efficacy and inverse agonist efficacy, respectively. It turned out that the efficacy as determined by the chloride flux method correlated remarkably well with the independently observed GABA ratio (Fig. 8), thus giving a strong further corroboration of the concept that a continuum of ligands produce effects on the benzodiazepine receptor and, that the coupling to the GABA receptor (GABA ratio) is ultimately linked to the agonist/inverse agonist property.

## REFERENCES

Asano T, Ogasawara N (1981): Chloride-dependent stimulation of GABA and benzodiazepine receptor binding by pentobarbital. Brain Res 225:212–216.

Bennett DA, Petrack B (1984): CGS 9896: A nonbenzodiazepine, nonsedating potential anxiolytic. Drug Development Res 4:75–82.

Blanchard JC, Boireau A, Garret C, Julou L (1979): *In vitro* and *in vivo* inhibition by zopiclone of benzodiazepine binding to rodent brain receptors. Life Sci 24:2417–2420.

Blanchard JC, Zundel JL, Julou L (1984): Differences between cyclopyrrolones (Suriclone and Zopiclone) and benzodiazepines binding to rat hippocampus photolabelled membranes raise a question about the differential interactions of agonists and antagonists with benzodiazepine receptors.

Boissier JR, Simon P, Aron C (1968): A new method for rapid screening of minor tranquilisers in mice. Eur J Pharmacol 4:145–150.

Borea PA, Supavilai P, Karobath M (1983): Differential modulation of etazolate or pentobarbital enhanced $^3$H-muscimol binding by benzodiazepine agonists and inverse agonists. Brain Res 280:383–386.

Braestrup C, Honoré T, Nielsen M, Petersen EN, and Jensen LH (1984): Ligands for benzodiazepine receptors with positive and negative efficacy. Biochem Pharmacol 33:859–862.

Braestrup C, Nielsen M (1981): GABA reduces binding of $^3$H-methyl-$\beta$-carboline-3-carboxylate to brain benzodiazepine receptors. Nature 294:472–474.

Braestrup C, Nielsen M (1983): Benzodiazepine receptors. In Iversen L, Iversen SD, Snyder SH (eds): "Handbook of Psychopharmacology," Vol. 17, Plenum Publishing Corporation, pp 285–384.

Braestrup C, Nielsen M, Olsen CE (1980): Urinary and brain $\beta$-carboline-3-carboxylates as potent inhibitors of brain benzodiazepine receptors. Proc Natl Acad Sci USA 77:2288–2292.

Braestrup C, Schmiechen R, Neef G, Nielsen M, Petersen EN (1982a): Interaction of convulsive ligands with benzodiazepine receptors. Science 216:1241–1243.

Braestrup C, Schmiechen R, Nielsen M, Petersen EN (1982b): Benzodiazepine receptor ligands, receptor occupancy, pharmacological effect and GABA receptor coupling. In Usdin E, Skolnick P, Tallman JF, Greenblatt D, Paul SM (eds): "The Pharmacology of Benzodiazepines." London: MacMillan Press, pp 71–85.

Brown CL, Martin IL (1983): Photoaffinity labelling of the benzodiazepine receptor cannot be used to predict ligand efficacy. Neuroscience 35:37–40.

Brown CL, Martin IL (1984): Modification of pyrazoloquinolinone affinity by GABA predicts efficacy at the benzodiazepine receptor. Eur J Pharmacol (in press).

Brown C, Martin I, Jones B, Oakley N (1984): In vivo determination of efficacy of pyrazoloquinolinones at the benzodiazepine receptor. Eur J Pharmacol 103:139–143.

Chan CY, Gibbs T, Farb DH (1982): Action of Beta-carboline in flunitrazepam-photolinked cultures. In "Abstracts Society for Neuroscience 8," Part 2, 572, No. 158.6.

Chang RSL, Snyder SH (1978): Benzodiazepine receptors: Labeling in intact animals with $^3$H-flunitrazepam. Eur J Pharmacol 48:213–218.

Cowen PJ, Green AR, Nutt DJ, Martin IL (1981): Ethyl $\beta$-carboline carboxylate lowers seizure threshold and antagonizes flurazepam-induced sedation in rats. Nature 290:54–55.

Czernik AJ, Petrack B, Kalinsky HJ, Psychoyos S, Cash WD, Tsai C, Rinnehart RK, Granat FR, Lovell RA, Brundish DE, Wade R (1982): CGS 8216: Receptor binding characteristics of a potent benzodiazepine antagonist. Life Sci 30:363–372.

Dantzer R, Pério A (1982): Behavioural evidence for partial agonist properties of Ro 15-1788, a benzodiazepine receptor antagonist. Eur J Pharmacol 81:655–658.

Davies LP, Chow SC, Johnston GAR (1984): Interaction of purines and related compounds with photoaffinity-labelled benzodiazepine receptors in rat brain membranes. Eur J Pharmacol 97:325–329.

Doble A, Martin IL, Richards DA (1982): GABA modulation predicts biological activity of ligands for the benzodiazepine receptor. British Pharmaceutical Society Meeting (abstract).

Drexler G, Sieghart W (1984): [$^{35}$S]t-butylbicyclophosphorothionate and avermectin bind to different sites associated with the GABA-benzodiazepine receptor complex. Neurosci Lett 50:273–277.

Duka T, Höllt V, Herz A (1979): In vivo receptor occupation by benzodiazepines and correlation with the pharmacological effect. Brain Res. 179:147–156.

Ehlert FJ, Roeske WR, Braestrup C, Yamamura SH, Yamamura HI (1981): $\gamma$-Aminobutyric acid regulation of the benzodiazepine receptor: Biochemical evidence for pharmacologically different effects of benzodiazepines and propyl $\beta$-carboline-3-carboxylate. Eur J Pharmacol 70:593–596.

Ehlert FJ, Ragan P, Chen A, Roeske WR, Yamamura HI (1982): Modulation of benzodiazepine receptor binding: Insight into pharmacological efficacy. Eur J Pharmacol 78:249–253.

Farb DH, Borden LA, Chan CY, Czajkowski CM, Gibbs TT, Schiller GD (1984): Modulation of neuronal function through benzodiazepine receptors:

Biochemical and electrophysiological studies on neurons in culture. "First Colloquium in Biological Sciences," Ann NY Acad Sci (in press).

Fehske KJ, Müller WE (1982): β-Carboline inhibition of benzodiazepine receptor binding in vivo. Brain Res 238:286–291.

File SE, Lister RG, Nutt DJ (1982): Intrinsic actions of benzodiazepine antagonists. Neurosci Lett 32:165–168.

Fujimoto M, Hirai K, Okabayashi T (1982): Comparison of the effects of GABA and chloride ion on the affinities of ligands for the benzodiazepine receptor. Life Sci 30:51–57.

Grecksch G, de Carvalho LP, Venault P, Chapouthier G, Rossier J (1983): Convulsions induced by submaximal dose of pentylenetetrazol in mice are antagonized by the benzodiazepine antagonist Ro 15-1788. Life Sci 32:2579–2584.

Haefely W (1984): Pharmacological profile of two benzodiazepine partial agonists: Ro 16-6028 and Ro 17-1812. "Collegium Internationale Neuro-Psychopharmacologicum, 14th C.I.N.P. Congress. Book of Abstracts."

Hirsch JD, Lydigsen JL (1981): Binding of β-carboline-3-carboxylic acid ethyl ester to mouse brain benzodiazepine receptors in vivo. Eur J Pharmacol 72:357–360.

Honoré T, Nielsen M, Braestrup C (1984): Barbiturate shift as a tool for determination of efficacy of benzodiazepine-receptor ligands. Eur J Pharmacol 100:103–107.

Hunkeler W, Möhler H, Pieri L, Polc P, Bonetti EP, Cumin R, Schaffner R, Haefely W (1981): Selective antagonists of benzodiazepines. Nature 290:514–516.

Janis RA, Rampe D, Sarmiento JG, Triggle DJ (1984): Specific binding of a calcium channel activator, $^3$H-bay k 8644, to membranes from cardiac muscle and brain. Biochem Biophys Res Commun 121:317–323.

Jensen LH, Petersen EN, Braestrup C, Honoré T, Kehr W, Stephens DN, Schneider H, Seidelmann D, Schmiechen R (1984): Evaluation of the β-carboline ZK 93 426 as a benzodiazepine receptor antagonist. Psychopharmacology 83:249–256.

Jensen MS, Lambert JDC (1983): The interaction of the β-carboline derivative DMCM with inhibitory amino acid responses on cultured mouse neurones. Neurosci Lett 40:175–179.

Jones BJ, Oakley NR (1981): The convulsant properties of methyl β-carboline-3-carboxylate in the mouse. Proceedings of the BPS, 884P–885P.

Karobath M, Supavilai P (1982): Distinction of benzodiazepine agonists from antagonists by photoaffinity labelling of benzodiazepine receptors in vitro. Neurosci Lett 31:65–69.

Leeb-Lundberg F, Olsen RW (1982): Interactions of barbiturates of various pharmacological categories with benzodiazepine receptors. Mol Pharmacol 21:320–328.

Maksay G, Ticku MK: The dissociation of [$^{35}$S]-t-butylbicyclophosphorothionate binding differentiates convulsant and depressant drugs that modulate GABAergic transmission. J Neurochem (in press).

Mennini T, Cotecchia S, Caccia S, Garattini S (1982a): Benzodiazepines: Relationship between pharmacological activity in the rat and in vivo receptor binding. Pharmacol Biochem Behav 16:529–532.

Mennini T, Cotecchia S, Caccia S, Garattini S (1982b): Does ethyl-β-carboline-3-carboxylate interact with mouse brain benzodiazepine receptors in vitro? J Pharm Pharmacol 34:394–395.

Mennini T, Garattini S (1982): Minireview. Benzodiazepine receptors: Correlation with pharmacological responses in living animals. Life Sci 31:2025–2035.

Mennini T, Gobbi M, Garattini S (1984): Increased number of brain benzodiazepine receptors after in-vivo administration of estazolam to rats. J Pharm Pharmacol 36:621–622.

Minchin MCW, Nutt DJ (1983): Studies on $^3$H-diazepam and $^3$H-ethyl-β-carboline carboxylate binding to rat brain *in vivo*. I. Regional variations in displacement. J Neurochem 41:1507–1512.

Mitchell R, Martin I (1980): Ethyl β-carboline-3-carboxylate antagonises the effect of diazepam on a functional GABA receptor. Eur J Pharmacol 68:513–514.

Möhler H (1982): Benzodiazepine receptors: Differential interaction of benzodiazepine agonists and antagonists after photoaffinity labeling with flunitrazepam. Eur J Pharmacol 80:435–436.

Möhler H, Battersby MK, Richards JG (1980): Benzodiazepine receptor protein identified and visualized in brain tissue by a photoaffinity label. Proc Natl Acad Sci USA 77:1666–1670.

Möhler H, Okada T (1977): Properties of $^3$H-diazepam binding to benzodiazepine receptors in rat cerebral cortex. Life Sci 20:2101–2110.

Morelli M, Gee KW, Yamamura HI (1982): The effect of GABA on *in vitro* binding of two novel non-benzodiazepines, PK 8165 and CGS 8216 to benzodiazepine receptors in the rat brain. Life Sci 31:77–81.

Nielsen M, Honoré T, Braestrup C (1985): Radiation inactivation of brain [$^{35}$S] t-butylbicyclophosphorothionate binding sites reveals complicated molecular arrangements of the GABA/benzodiazepine receptor chloride channel complex. Biochem Pharmacol 34:3633–3642.

Nutt DJ, Cowen PJ, Little HJ (1982): Unusual interactions of benzodiazepine receptor antagonists. Nature 295:436–437.

Oakley NR, Jones BJ (1983): Buspirone enhances $^3$H-flunitrazepam binding in vivo. Eur J Pharmacol 87:499–500.

Olsen RW (1981): GABA-benzodiazepine-barbiturate

interactions. J Neurochem 37:1–13.

Petersen EN, Jensen LH, Honoré T, Braestrup C, Kehr W, Stephens DN, Wachtel H, Seidelman D, Schmiechen R (1984): ZK 91296, a partial agonist at benzodiazepine receptors. Psychopharmacology 240–248.

Polc P, Ropert N, Wright DM (1981): Ethyl β-carboline-3-carboxylate antagonizes the action of GABA and benzodiazepines in the hippocampus. Brain Res 217:216–220.

Ramanjaneyulu R, Ticku M (1984): Interactions of pentamethylenetetrazole and tetrazole analogues with the picrotoxinin site of the benzodiazepine-GABA receptor-ionophore. Eur J Pharmacol 98:337–345.

Schramm M, Thomas G, Towart R, Franckowiak G (1983): Novel dihydropyridines with positive inotropic action through activation of $Ca^{2+}$ channels. Nature 303:535–537.

Skerritt JH, Davies LP, Chow SC, Johnston GAR (1982): Contrasting regulation by GABA of the displacement of benzodiazepine antagonist binding by benzodiazepine agonists and purines. Neurosci Lett 32:169–174.

Skerritt JH, Johnston GAR, Braestrup C (1983): Modulation of GABA binding to rat brain membranes by alkyl β-carboline-3-carboxylate esters. Eur J Pharmacol 86:299–301.

Skerritt JH, Macdonald RL (1983): Benzodiazepine Ro 15-1788: Electrophysiological evidence for partial agonist activity. Neurosci Lett 43:321–326.

Skerritt JH, Macdonald RL (1984): Benzodiazepine (BZ) receptor ligand actions on GABA responses: II. β-carbolines, purines. Eur J Pharmacol 101:135–41.

Skolnick P, Lock K-L, Paugh B, Marangos P, Windsor R, Paul S (1980): Pharmacologic and behavioural effects of EMD 28422: A novel purine which enhances $^{3}H$-diazepam binding to brain benzodiazepine receptors. Pharmacol Biochem Behav 12:685–689.

Squires RF, Braestrup C (1977): Benzodiazepine receptors in rat brain. Nature 266:732–734.

Squires RF, Casida JE, Richardson M, Saederup E (1983): [$^{35}S$]t-Butylbicyclophosphorothionate binds with high affinity to brain-specific sites coupled to γ-aminobutyric acid-A and ion recognition sites. Molecular Pharmacol 23:326–336.

Squires RF, Saederup E, Crawlev JN, Skolnick P, Paul SM (1984): Convulsant potencies of tetrazoles are highly correlated with actions on GABA/benzodiazepine/picrotoxin receptor complexes in brain. Life Sci 35:1439–1444.

Stephens DN, Kehr W, Schneider HH, Braestrup C (1984): Bidirectional effects on anxiety of β-carbolines acting as benzodiazepine receptor ligands. Neuropharmacology 23:879–880.

Supavilai P, Karobath M (1984a): [$^{35}S$]-t-butylbicyclophosphorothionate binding sites are constituents of the γ-aminobutyric acid benzodiazepine receptor complex. J Neurosci 4:1193–1200.

Supavilai P, Karobath M (1984b): Modulation of acetylcholine release from rat striatal slices by the GABA/benzodiazepine receptor complex. Life Sci (in press).

Tallman JF, Thomas JW (1982): Agonist–antagonist interaction of the benzodiazepine binding site. In Paul SM, Skolnick P, Greenblatt D, Tallman F, Usdin E (eds): "Pharmacology of Benzodiazepine," London: MacMillan Press, pp 133–139.

Tallman JF, Thomas JW, Gallager DW (1979): Identification of diazepam binding in intact animals. Life Sci 24:873–880.

Tenen SS, Hirsch JD (1980): β-carboline-3-carboxylic acid ethyl ester antagonizes diazepam activity. Nature 288:609–610.

Thiébot M-H, Childs M, Soubrié P, Simon P (1983): Diazepam-induced release of behaviour in an extinction procedure: Its reversal by Ro 15-1788. Eur J Pharmacol 88:111–116.

Ticku MK, Ramanjaneyulu R (1984): Differential interactions of GABA agonists, depressant and convulsant drugs with [$^{35}S$]-t-butylbicyclophosphorothionate binding sites in cortex and cerebellum. Pharmacol Biochem Behav 21:151–158.

Vellucci SV, Webster RA (1983): Is Ro 15-1788 a partial agonist at benzodiazepine receptors? Eur J Pharmacol 90:263–268.

Williamson MJ, Paul SM, Skolnick P (1978): Labelling of benzodiazepine receptors *in vivo*. Nature 275:551–553.

Yokoyama N, Ritter B, Neubert AD (1982): 2-Arylpyrazolo (4,3-c)quinolin-3-ones: Novel agonist, partial agonist and antagonist of benzodiazepines. J Med Chem 25:337–339.

Benzodiazepine/GABA Receptors and Chloride Channels:
Structural and Functional Properties, pages 185–193


# 8

# Depressant Drug Interactions With Benzodiazepine/GABA Receptors

**Manfred Karobath and Porntip Supavilai**

*Preclinical Research, Sandoz Ltd., CH-4002 Basel, Switzerland*

There exist several classes of receptors for GABA, one of the most abundant transmitter substances in the mammalian brain [see Roberts, this volume]. According to one proposed classification, $GABA_A$ and $GABA_B$ receptor types can be distinguished [Hill and Bowery, 1981; Bowery et al., 1984]. $GABA_B$ receptor types are not affected by the classic GABA antagonists bicuculline or picrotoxin; they are accepted by (-)baclofen but not by GABA-mimetics such as isoguvacine, THIP (4,5,6,7-tetrahydroisoxazolo-[4,5*C*]pyridin-3-ol), or muscimol. $GABA_B$ receptors do not appear to exert their effects by regulating chloride ion permeability through neuronal membranes [for review see Bowery et al., 1984]. In this chapter $GABA_B$ receptors will not be discussed, but a subclass of $GABA_A$ receptors, the GABA/benzodiazepine receptor complex. $GABA_A$ receptors in mammals, like those in invertebrates, have been characterized initially in electrophysiological investigations [see also Barker and Owen, this volume], where it has been shown that they are recognized by the GABA-mimetics muscimol, THIP, and isoguvacine, but not by (-) baclofen [see also Krogsgaard-Larsen, this volume]. Bicuculline is a competitive-like GABA inhibitor at these receptors [Simmonds, 1982], whereas picrotoxin inhibits noncompetitively [Takeuchi and Takeuchi, 1969; Simmonds, 1982]. The activation of $GABA_A$ receptors leads to an increase of chloride ion permeability through neuronal membranes [see Barker and Owen, this volume, and Haefely and Polc, this volume]. The $GABA_A$/benzodiazepine receptor complex is a subtype of $GABA_A$ receptors which emerged late during phylogeny, appeared first in bony fish [Nielsen et al., 1978], and appears to be the most abundant GABA receptor in a mammalian brain. It should, however, be pointed out that perhaps other "phylogenetically old" $GABA_A$ receptors which do not appear to contain high-affinity benzodiazepine binding sites exist in several areas of the mammalian brain like the cerebellum [Placheta and Karobath, 1979; Olsen et al., 1984] or the thalamus [Palacios et al., 1980]. This notion is based mainly on the finding that the cerebellum as compared to other brain areas, contains more binding sites for GABA, muscimol [Zukin et al., 1974; Greenlee et al., 1978; Placheta and Karobath, 1979; Leeb-Lundberg, et al., 1981], bicuculline methiodide [Olsen and Snowman, 1983], $^{35}$S-t-butylbicyclophosphorothionate($^{35}$S-TBPS)[Squires et al., 1983], or avermectin $B_{1a}$ [Drexler and Sieghart, 1984] than for benzodiazepines. There are also regional differences in localization of these binding sites within the cerebellum [Palacios et al., 1980; Wamsley et al., this volume; but see Möhler and Richards, this volume]. These postulated $GABA_A$ receptors appear to contain a recognition site for GABA as well as drug receptors for CNS depressants and CNS convulsants, but no high-affinity binding sites for benzodiazepine [Supavilai et al., unpublished data]. Although these $GABA_A$ receptors are not the subject of this chapter, their very existence in different brain regions can explain why the properties of $^3H$-muscimol or of $^{35}S$-TBPS binding can exhibit regional differences. For example the allosteric effects of benzodiazepines on $^{35}S$-TBPS binding are stronger in cerebral cortex than in cerebellum [Supavilai and Karobath, unpublished].

The basic electrophysiological findings that demonstrated that drugs like muscimol, bicuculline, or picrotoxin in appropriate concentrations have selective actions on $GABA_A$ receptor function have provided the basis for the development of suitable ligands for the characterization of the $GABA_A$/benzodiazepine receptor complex with binding studies. Current concepts conceive this $GABA_A$ receptor as a supramolecular complex that contains recognition sites for the endogenous ligand GABA and drug receptors like the benzodiazepine receptor and another drug receptor which is commonly referred to as the picrotoxin/barbiturate receptor [Leeb-Lundberg et al., 1980]. Only recently, evidence has been generated that this "picrotoxin/barbiturate receptor" may actually consist of two separate, although functionally closely related, drug receptors [Squires et al., 1983; Supavilai and Karobath, 1984; Ramanjaneyulu and Ticku, 1984; Drexler and Sieghart, 1984a]. In the following the development of this concept is summarized.

### The Development of the Concept of the GABA/Benzodiazepine/Picrotoxin/ Barbiturate Receptor Complex

When GABA-induced stimulation of benzodiazepine receptor binding was discovered by Tallman et al. [1978] and independently by G. Sperk and myself [Karobath and Sperk, 1979], by Martin and Candy [1978], and by Briley and Langer [1978] it became immediately apparent that benzodiazepine receptors are intimately associated with $GABA_A$ receptors. Thus it was to be expected that drugs like picrotoxin, which are thought to inhibit GABA effects at a site close to the chloride ion channel of $GABA_A$ receptors [Takeuchi and Takeuchi, 1969; Simmonds, 1982] should also alter benzodiazepine receptor binding. Several groups who investigated this found that picrotoxin did not markedly alter benzodiazepine receptor binding in the assay conditions they used (0°C) [Squires and Braestrup, 1977; Möhler and Okada, 1977]. However, it was observed that chloride and some other "Eccles" anions could stimulate benzodiazepine receptor binding (by a Kd effect) which was postulated as evidence for an association of benzodiazepine receptors with chloride ion channels [Costa et al., 1979]. This effect of anions was perhaps indirect and related to the presence of endogenous GABA. Martin and Candy [1978], in their initial paper on GABA stimulation of benzodiazepine receptor binding, described that chloride ions and other "Eccles" anions potentiated the allosteric effects of GABA.

At this time it was reported that the purine analogues etazolate (SQ 20009) and cartazolate (SQ 65396), two CNS depressants, which can restore behavior suppressed by punishment, can enhance benzodiazepine receptor binding [Williams and Risley, 1979; Beer et al., 1978]. Since etazolate, which is chemically not related to GABA, was reported not to interfere with muscimol for its binding sites [Williams and Risley, 1979], an allosteric effect, distinct from that of GABA, may have been responsible for the stimulating actions of these drugs on benzodiazepine receptor binding. The observation that the effect of etazolate on benzodiazepine receptor binding was chloride ion dependent and sensitive to inhibition by picrotoxin [Supavilai and Karobath, 1979] provided the first direct link between benzodiazepine receptor binding and a picrotoxin-sensitive site of $GABA_A$ receptors. This notion was strengthened when it was found that etazolate could also stimulate

$^3$H-muscimol binding when chloride ions were added to the incubation medium [Placheta and Karobath, 1980]. Leeb-Lundberg et al. [1981] were able to show that etazolate and cartazolate could inhibit $^3$H-dihydropicrotoxin ($^3$H-DHP) binding in the low micromolar range. They could also demonstrate that several hypnotic barbiturates—which all had previously been shown to inhibit $^3$H-DHP binding [Ticku and Olsen, 1978]—could enhance benzodiazepine receptor binding in a chloride-ion-dependent and picrotoxin-sensitive mode [Leeb-Lundberg et al., 1980]. These observations provided the basis for the concept of a picrotoxin/barbiturate receptor site as a constituent of the $GABA_A$ benzodiazepine receptor [Leeb-Lundberg et al., 1980]. In the following the interactions between the various components of the $GABA_A$/benzodiazepine receptor complex are described. Emphasis will be given to the actions of the CNS depressants.

## EFFECTS OF CNS DEPRESSANTS ON THE PICROTOXININ/TBPS DRUG RECEPTOR OF THE $GABA_A$/BENZODIAZEPINE RECEPTOR COMPLEX

Those CNS depressants that are now thought to act on the chloride ion channel of the $GABA_A$/benzodiazepine receptor complex have been found to inhibit $^3$H-DHP binding [Ticku and Olsen, 1978] or $^{35}$S-TBPS binding [Squires et al., 1983] stereoselectively and with a potency that correlates with their anesthetic efficacy [see also Ticku, Table II, this volume]. In contrast to the convulsants picrotoxin or isopropyl bicyclophosphate (IPTBO), the depressants etazolate or pentobarbital have biphasic effects on $^{35}$S-TBPS binding in 200 mM NaCl, with stimulation followed by inhibition [Supavilai and Karobath, 1984]. The inhibition of $^{35}$S-TBPS binding by the CNS depressants but not by picrotoxin or by IPTBO has a Hill number $>1$, which suggests multiplicity of $^{35}$S-TBPS binding sites in a $GABA_A$/benzodiazepine receptor complex and cooperativity between these sites and/or the existence of different drug receptors for CNS depressants and CNS convulsants [Supavilai and Karobath, 1984]. Ticku and Ramanjaneyulu [1984] have demonstrated that etazolate and pentobarbital decrease primarily the $B_{max}$ of $^{35}$S-TBPS binding, and that these drugs, in contrast to unlabeled TBPS and picrotoxinin, facilitate the dissociation of $^{35}$S-TBPS from its binding site. Also, the inhibition of $^{35}$S-TBPS binding by the depressants, but not by the convulsants, can be noncompetitively reversed by the GABA antagonist 3$\alpha$-hydroxy-16-imino-5$\beta$-17-aza-androstan-11-one (R 5135) (Squires et al., 1983). All these observations as well as the results from the irradiation inactivation studies of $^{35}$S-TBPS binding [Nielsen et al., 1985] are compatible with the concept that CNS depressants and CNS convulsants recognize different drug receptors of the $GABA_A$/benzodiazepine receptor complex.

The affinity of the CNS depressants for their postulated drug receptor, as it is suggested by their indirect inhibitory effects on $^3$H-DHP [Leeb-Lundberg et al., 1980], and $^{35}$S-TBPS binding [Squires et al., 1983] is rather low with cartazolate being the most potent drug with an $IC_{50}$ around 0.5 $\mu$M. Due to this rather low affinity, these drugs are probably not suitable as radiolabeled ligands for the direct investigation of their drug receptors on the $GABA_A$/benzodiazepine receptor complex. In addition, the actions of these drugs are not specific for the $GABA_A$/benzodiazepine receptor complex and they have numerous other pharmacological actions in concentrations where they act on $GABA_A$ receptors [Beer et al., 1972; Hirsch, 1983; Lohse et al., 1984; Teichberg et al., 1984]. A possible exception may be the avermectins.

### The Avermectin $B_{1a}$ Binding Site

Avermectin $B_{1a}$ (AVM) is a derivate of anthelmintic and insecticidal macrocyclic lactones which has been demonstrated to increase in nanomolar concentrations $GABA_A$ receptor-associated chloride conductance in an invertebrate preparation, the lobster stretcher muscle [Fritz et al., 1979]. AVM enhances the muscle-relaxant effects of diazepam in rats in vivo, and it stimulates the binding of benzodiazepine receptor agonists in nanomolar concentrations to rat brain membrane fractions in vitro [Williams and Yarbrough, 1979; Paul et al., 1980; Supavilai and Karobath, 1981]. AVM in higher concentrations ($>1$ $\mu$M) also stimulates $^3$H-GABA binding by increasing the apparent $B_{max}$ of these binding sites [Pong and Wang, 1982]. The specificity of this effect has recently been questioned [Drexler and Sieghart, 1984c]. At lower concentrations AVM decreases high-affinity GABA agonist binding and increases antagonist ($^3$H-bicuculline) binding [Olsen and Snowman, 1985]. AVM stimulates in low nanomolar concentrations and inhibits in higher nanomolar concentrations $^{35}$S-TBPS binding to membranes of cerebral cortex [Drexler and Sieghart, 1984a], which is qualitatively similar to the actions of etazolate or pentobarbital on $^{35}$S-TBPS binding in very similar assay conditions [Supavilai and Karobath, 1984]. These observations are compatible with the notion that AVM interacts with a drug receptor that has many similarities to the proposed drug receptor for the pyrazolopyridines and barbiturates. On the other hand, these latter drugs allosterically *inhibit* GABA antagonist and benzodiazepine inverse against binding (Wong et al., 1984a).

Specific binding of $^3$H-AVM has been investigated in rat and in dog brain and the affinities of various analogues of AVM for this binding site correlated well with the anthelmintic properties [Pong and Wang, 1980]. $^3$H-AVM binding in rat cerebellum was independent of the presence of chloride ion and was partially inhibited by several GABA receptor agonists; this effect was blocked by R 5135 [Drexler and Sieghart, 1984b], a potent $GABA_A$ receptor antagonist [Hunt and Clements-Jewery, 1981]. Etazolate and pentobarbital slightly inhibited $^3$H-AVM binding in the presence of chloride ions, whereas picrotoxin and IPTBO stimulated $^3$H-AVM binding [Drexler and Sieghart, 1984b]. Ligands for the benzodiazepine receptor did not modify $^3$H-AVM binding to membranes of rat cerebellum [Drexler and Sieghart, 1984c]. These observations seem to imply that AVM is recognized by a site of the $GABA_A$ receptor, and—as its allosteric actions on benzodiazepine receptor binding suggest, also by the $GABA_A$/benzodiazepine receptor complex—which is similar but perhaps not identical to the postulated CNS depressant drug receptor. However, AVM has potent effects in invertebrate $GABA_A$ receptors [Fritz et al., 1979] which do not contain mammalian CNS-type benzodiazepine binding sites [Nielsen et al., 1985]. Thus its actions—like those of the other CNS depressants or CNS convulsants—are not selective to the $GABA_A$/benzodiazepine receptor complex but may be related to the chloride channel of several types of $GABA_A$ receptors. This may also explain the high density of $^3$H-AVM binding sites in cerebellum [Pong and Wang, 1980; Drexler and Sieghart, 1984b], an area of the brain that appears to contain a considerable number of $GABA_A$ receptors that do not appear to contain benzodiazepine binding sites [Placheta and Karobath, 1979].

Despite this limited selectivity, perhaps to chloride channels of many types of $GABA_A$ receptors, high-affinity $^3$H-AVM binding could, at least in selected brain areas like the cerebral cortex, be a useful tool to study the properties of its drug receptor on the $GABA_A$/benzodiazepine receptor complex. However, $^3$H-AVM is highly hydrophobic, binds to glass and plastic surfaces and is lost from aqueous solutions during storage and incubations [Pong and Wang, 1980; Drexler and Sieghart, 1984c]. These properties are a hindrance to a widespread use of AVM as a radiolabeled ligand.

## EFFECTS OF CNS DEPRESSANTS ON BENZODIAZEPINE RECEPTOR BINDING

The pyrazolopyridines and the depressant barbiturates decrease the dissociation rate of tritiated benzodiazepine receptor agonists, and by this they increase the affinity of benzodiazepine receptor agonists for their binding sites [Beer et al., 1978; Supavilai and Karobath, 1979 and 1981; Williams and Risley, 1979; Leeb-Lundberg et al., 1980, 1981 and 1982; Skolnick et al., 1980; Ehlert et al., 1982]. This allosteric action is dependent on the presence of chloride and other Eccles anions, and it can be antagonized by picrotoxin, cage convulsants, bicuculline, and R 5135 [Supavilai and Karobath, 1979; Olsen, 1981; Leeb-Lundberg et al., 1980; Skolnick et al., 1980]. Conversely, the CNS depressants decrease the affinity of inverse benzodiazepine receptor agonists like βCCM or DMCM for their binding sites, whereas the affinity of benzodiazepine receptor antagonists is not altered by these drugs [Braestrup et al., 1984; Wong et al., 1984a]. These CNS depressant–induced affinity changes of benzodiazepine receptor ligands are additive with those of GABA [Leeb-Lundberg et al., 1981; Supavilai and Karobath, 1982], and they have been proposed as an in vitro method to determine the intrinsic activity of drugs acting primarily at benzodiazepine receptors [Honoré et al., 1984; see also Braestrup and Nielsen, this volume].

## EFFECTS OF CNS DEPRESSANTS ON $GABA_A$ RECEPTOR BINDING

$^3$H-GABA agonist binding can be enhanced [Placheta and Karobath, 1980; Willow and Johnston, 1980; Leeb-Lundberg et al., 1981] by the pyrazolopyridines and by several, but not all, barbiturates that have been demonstrated to interfere with $^3$H-DHP [Leeb-Lundberg et al., 1980; Olsen, 1981] or $^{35}$S-TBPS binding [Squires et al., 1983; Maksay and Ticku, 1984; Supavilai and Karobath, 1984]. This allosteric action is chloride-dependent [Placheta and Karobath, 1980] and other Eccles anion–dependent, and it can be inhibited by picrotoxin, anisatin, and cage convulsants [Asano and Ogasawara, 1982; Olsen and Snowman, 1982; Supavilai et al., 1982a]. The kinetic analysis of this CNS depressant–induced enhancement of the binding of tritiated GABA agonists indicates an increase in the number of high- and low-affinity binding sites, perhaps due to the conversion of low-affinity GABA binding sites (which cannot be detected in conventional binding experiments) to sites with higher affinity [Olsen and Snowman, 1983]. In support of this concept is the observation that picrotoxin and IPTBO can inhibit $^3$H-muscimol binding, mainly by decreasing the apparent number of low-affinity binding sites [Supavilai et al., 1982]. Recently it has been found that those barbiturates or CNS depressants that enhance GABA binding can induce the opposite allosteric effect on the binding of the GABA receptor antagonist $^3$H-bicuculline methochloride [Wong et al., 1984a]. The effect of the CNS depressants was chemically selective (dimethylbutylbarbituric acid > pentobarbital = secobarbital > phenobarbital), stereospecific ((−)mephobarbital>(+) mephobarbital, and (+)hexobarbital>(−) hexobarbital)) and blocked by picrotoxin [Wong et al., 1984a].

This allosteric modulation of GABA receptor binding by the CNS depressants is perturbed by the occupancy of benzodiazepine receptors with its ligands. For example, etazolate or pentobarbital stimulation of $^3$H-muscimol binding in the presence of a benzodiazepine receptor agonist shows a higher affinity, and in the presence of an inverse benzodiazepine receptor agonist shows a lower affinity. Ro 15-1788 can inhibit this action of benzodiazepine receptor agonists or inverse benzodiazepine receptor agonists [Borea et al., 1983]. This observation can also be used to investigate the intrinsic activity of benzodiazepine receptor ligands in vitro.

## DISCUSSION

The multiple effects of CNS depressants on the binding of ligands to the GABA/benzodiazepine receptor complex are summarized in Table I. These findings demonstrate that the interaction of the CNS depressants with their drug receptor can induce conformational

**TABLE I. Effect of CNS Depressants and CNS Convulsants on the Binding of Ligands to Their Recognition Site of the $GABA_A$/Benzodiazepine Receptor Complex**

| | GABA recognition site | | Benzodiazepine receptor | | | |
|---|---|---|---|---|---|---|
| | Agonist ligands | Antagonist ligands | Agonist ligands | Inverse agonist ligands | Convulsant binding site | AVM binding site |
| CNS depressants | ↑ | ↓ | ↑ | ↓ | ↓ | ↓ |
| CNS convulsants | ↓ | 0 | (↑) | (↑) | ↓ | ↑ |

↑: stimulation; ↓: inhibition; 0: no effect; (↑): weak effect.

changes of all other binding sites of the $GABA_A$/benzodiazepine receptor complex (Fig. 1). As a consequence of this interaction the function of $GABA_A$ receptors appears to be altered. In a rat hippocampal slice model it could be shown that the activation of $GABA_A$ receptors can increase the efflux of preloaded $^{36}Cl$ [Wong et al., 1984b]. Barbiturates stereoselectively enhance the functional response to the stimulation of this $GABA_A$ receptor [Wong et al., 1984b]. In intact cultured embryonic chick neurons, the GABA-mediated increase of $^{36}Cl$ permeability is also enhanced by CNS depressants [Thampy et al., 1984]. In a subcellular particle fraction derived from $P_1$ pellets of rat cerebral cortex $^{36}Cl$-efflux has also been investigated [Schwartz et al., 1984]. In this model, barbiturates have been found to stimulate and picrotoxin to inhibit the efflux of preloaded $^{36}Cl$ [Schwartz et al., 1984]. It is currently not known whether the functional effects on $^{36}Cl$-fluxes in the aforementioned models are triggered by a $GABA_A$/benzodiazepine receptor complex or by a $GABA_A$ receptor. In another functional model it has been shown that a $GABA_A$/benzodiazepine receptor complex can modulate $^3H$-acetylcholine release from rat striatum slices [Supavilai and Karobath, 1985]. In this test system (+)etomidate has effects compatible with an enhancement of GABAergic activity, but pentobarbital leads to a general, local anesthetic–like, depression of spontaneous and of electrically induced $^3H$-acetylcholine release [Supavilai and Karobath, unpublished]. These phenomena are compatible with results from electrophysiological studies [see Barker and Owen, this volume] where it could be demonstrated that CNS depressant barbiturates can prolong the GABA-induced opening time of chloride ion channels. On the other hand, the CNS depressant barbiturates have numerous other pharmacological effects in those concentrations where they effect GABAergic neurotransmission. Nevertheless, it appears likely that some of their pharmacological actions are related to their effects on $GABA_A$ receptors.

## REFERENCES


Asano T, Ogasawara N (1981): Chloride-dependent stimulation of GABA and benzodiazepine receptor binding by pentobarbital. Brain Res 225:212–216.

Beer M, Chasin M, Clody DE, Vogel JR, Horovitz ZP (1972): Cyclic adenosine monophosphate phosphodiesterase in brain: Effect on anxiety. Science 176:428–430.

Beer B, Klepner CA, Lippa AS, Squires RF (1978): Enhancement of $^3H$-diazepam binding by SQ 65-396: a novel anti-anxiety agent. Pharmacol Biochem Behav 9:849–851.

Borea PA, Supavilai P, Karobath M (1983): Differential modulation of etazolate or pentobarbital enhanced [$^3H$]muscimol binding by benzodiazepine agonists and inverse agonists. Brain Res 280:383–386.

Bowery NG, Price GW, Hudson DR, Hill GP, Wilkin GP, Turnbull MJ (1984): GABA receptor multiplicity: Visualization of different receptor types in the mammalian CNS. Neuropharmacology 23:219–231.

Braestrup C, Nielsen M, Honoré T (1983): Binding of [$^3H$]DMCM, a convulsive benzodiazepine ligand, to rat brain membranes. Preliminary studies. J Neurochem 41:454–459.

Braestrup C, Honoré T, Nielsen M, Petersen EN, Jensen LH (1984): Ligands for benzodiazepine receptors with positive and negative efficacy. Biochem

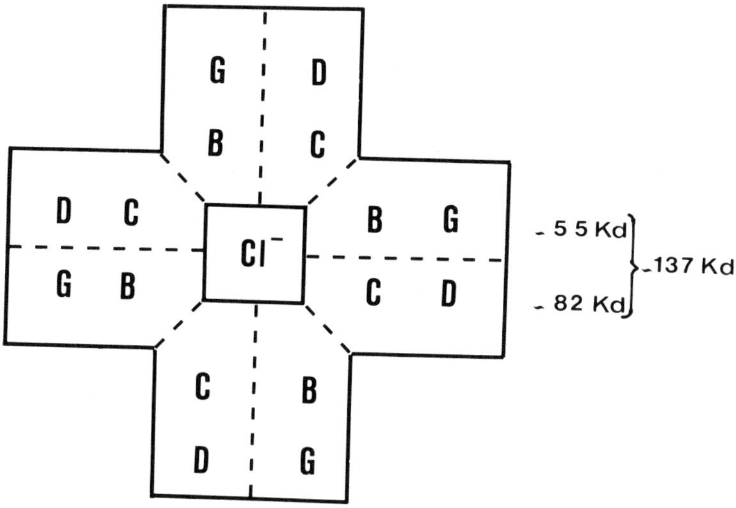


**Fig. 1.** *Diagrammatic representation of a $GABA_A$/benzodiazepine receptor complex adapted from the results of Nielsen et al. [1985]. These authors attempted the determination of the molecular mass of the $GABA_A$/benzodiazepine receptor complex and of its components using a radiation inactivation technique [see also Braestrup and Nielsen, this volume]. The abbreviations used are: G: GABA binding site; B: benzodiazepine receptor; D: depressant drug receptor; C: convulsant drug receptor; Cl: chloride ion channel; Kd: molecular mass in kilodaltons.*

*The tetrameric nature shown here is also supported from results of photoaffinity labeling studies [Möhler et al., 1980]. In this diagrammatic representation it is hypothesized that GABA and benzodiazepine receptors reside on one protein of 51–55 Kd, whereas the convulsant and the depressant drug receptors reside on another protein of 82 Kd. The terms depressant or convulsant drug receptor are used because the drugs that are currently known to primarily interact with these sites are CNS depressants, or CNS convulsants, respectively. (A possible exception may be the convulsant pyrethroids of type 2, which in mammalians produce a syndrome similar to that of picrotoxinin, and which have been found to inhibit $^{35}$S-TBPS binding noncompetitively with a rank order corresponding to their toxicity [Lawrence and Cassida, 1983]. These convulsant drugs could perhaps interact primarily with the CNS "depressant" drug receptor). The chloride ion channel may perhaps be the site where "Eccles" anions act to modulate bindings.*

*The total molecular mass of the $GABA_A$/benzodiazepine receptor complex, according to this concept, is $4x(82+50$ to $55) \approx 530$-$550$ Kd [Nielsen et al., 1985]. Heterogeneity of the $GABA_A$/benzodiazepine receptor complex [see Lawrence et al., this volume] has not been considered. Different values for the molecular mass of the $GABA_A$/benzodiazepine receptor complex have been obtained with detergent solubilized receptors [see Fischer and Olsen, this volume].*

Pharmacol 33:859–862.

Briley MS, Langer SZ (1978): Influence of GABA receptor agonists and antagonists on the binding of $^3$H-diazepam to the benzodiazepine receptor. Eur J Pharmacol 52:129–132.

Costa T, Rodbard D, Pert CB (1979): Is the benzodiazepine receptor coupled to a chloride anion channel? Nature 277: 315–317.

Drexler G, Sieghart W (1984a): [$^{35}$S]tert.-Butylbicyclophosphorothionate and avermectin bind to different sites associated with the $\gamma$-aminobutyric acid-benzodiazepine receptor complex. Neurosci Lett 50: 273–277.

Drexler G, Sieghart W (1984b): Properties of a high affinity binding site for [$^3$H]avermectin $B_{1a}$. Eur J Pharmacol 99:269–277.

Drexler G, Sieghart W (1984c): Evidence for association of a high affinity avermectin binding site with the benzodiazepine receptor. Eur J Pharmacol 101:201–207.

Ehlert FJ, Regan P, Chen A, Roeske WR, Yamamura HI (1982): Modulation of benzodiazepine receptor binding: Insight into pharmacological efficacy. Eur J Pharmacol 78:249–253.

Fritz LC, Wang CC, Gorio A (1979): Avermectin $B_{1a}$ irreversibly blocks postsynaptic potentials at the lobster neuromuscular junction by reducing muscle membrane resistance. Proc Nat Acad Sci USA 76:2062–2066.

Greenlee DV, Olsen RW (1972): Solubilization of gamma-aminobutyric acid receptor protein from mammalian brain. Biochem Biophys Res Commun 88:380–387.

Hill DR, Bowery NG (1981): $^3$H-GABA binds to bicuculline insensitive $GABA_B$ sites in rat brain. Nature 290:149–152.

Hirsch MD (1983): Barbiturate competition for TRH receptors in mouse brain: neuromodulation of anesthesia. Peptide 4:255–260.

Honoré T, Nielsen M, Braestrup C (1984): Barbiturate shift as a tool for determination of efficacy of benzodiazepine receptor ligands. Eur J Pharmacol 100:103–107.

Hunt P, Clements-Jewery S (1981): A steroid derivative R 5135, antagonizes the GABA/benzodiazepine receptor interaction. Neuropharmacology 20:357–361.

Karobath M, Sperk G (1979): Stimulation of benzodiazepine receptor binding by γ-aminobutyric acid. Proc Nat Acad Sci USA 76:1004–1006.

Lawrence LJ, Casida JE (1983): Stereospecific action of pyrethroid insecticides on the γ-aminobutyric acid receptor–ionophore complex. Science 221:1399–1401.

Leeb-Lundberg F, Snowman A, Olsen RW (1980): Barbiturate receptors are coupled to benzodiazepine receptors. Proc Nat Acad Sci USA 77:7468–7472.

Leeb-Lundberg F, Snowman A, Olsen RW (1981): Perturbation of benzodiazepine receptor binding by pyrazolopyridines involves picrotoxinin/barbiturate receptor sites. J Neurosci 1:471–477.

Leeb-Lundberg F, Olsen RW (1982): Interaction of barbiturates of various pharmacological categories with benzodiazepine receptors. Mol Pharmacol 21:320–328.

Lohse MJ, Lenschow V, Schwabe V (1984): Interaction of barbiturates with adenosine receptors in rat brain. Naunyn-Schmiedeberg's Ark Pharmacol 326:69–74.

Maksay G, Ticku MK (1984): The dissociation of [$^{35}$S]t-butylbicyclophosphorothionate binding differentiates convulsant and depressant drugs that modulate GABAergic transmission. J Neurochem 44:480–486.

Martin IL, Candy JM (1978): Facilitation of specific benzodiazepine binding in rat brain membrane fragments by a number of anions. Neuropharmacology 19:175–179.

Möhler H, Okada T (1977): Benzodiazepine receptors: demonstration in the central nervous system. Science 198:849–851.

Möhler H, Battersby MK, Richards JG (1980): Benzodiazepine receptor protein identified and visualized in brain tissue by a photoaffinity label. Proc Nat Acad Sci USA 77:1666–1670.

Nielsen M, Braestrup C, Squires R (1978): Evidence for a late evolutionary appearance of brain specific benzodiazepine receptors: an investigation of 18 vertebrate and 5 invertebrate species. Brain Res 141:342–346.

Nielsen M, Honoré T, Braestrup C (1985): Radiation inactivation of brain $^{35}$S-TBPS binding sites reveals complicated molecular arrangements of the GABA/benzodiazepine receptor chloride channel complex. Biochem Pharmacol 34:3633–3642.

Olsen RW (1981): GABA-benzodiazepine barbiturate receptor interactions. J Neurochem 37:1–13.

Olsen RW, Snowman AM (1982): Chloride-dependent enhancement by barbiturates of GABA receptor binding. J Neurosci 2:1812–1823.

Olsen RW, Snowman AM (1983): [$^3$H]Bicuculline-methochloride binding to low affinity GABA receptor sites. J Neurochem 41:1653–1663.

Olsen RW, Snowman AM (1985): Avermectin $B_{1a}$ modulation of γ-aminobutyric acid/benzodiazepine receptor binding in mammalian brain. J Neurochem 44:1074–1082.

Olsen RW, Wong EHF, Stauber GB, Murakami D, King RG, Fischer JB (1984): Biochemical properties of the GABA/barbiturate/benzodiazepine receptor-chloride ion channel complex. Adv Exp Med Biol 175:205–219.

Palacios JM, Young WS, Kuhar MJ (1980): Autoradiographic localization of γ-aminobutyric acid (GABA) receptors in the rat cerebellum. Proc Natl Acad Sci USA 77:670–674.

Paul SM, Skolnick P, Zatz M (1980): Avermectin $B_{1a}$: an irreversible activator of γ-aminobutyric acid-benzodiazepine-chloride-ionophore receptor complex. Biochem Biophys Res Communication 96:632–638.

Placheta P, Karobath M (1979): Regional distribution of $Na^+$-independent GABA and benzodiazepine binding sites in rat CNS. Brain Res 178:580–583.

Placheta P, Karobath M (1980): In vitro modulation by SQ 20009 and SQ 65396 of GABA receptor binding in rat CNS membranes. Eur J Pharmacol 62:225–228.

Pong SS, Wang CC (1980): The specificity of high affinity binding of avermectin $B_{1a}$ to mammalian brain. Neuropharmacology 19:311–317.

Pong SS, Wang CC (1982): Avermectin $B_{1a}$ modulation of γ-aminobutyric acid receptors in rat brain. J Neurochem 38:375–379.

Ramanjaneyulu R, Ticku MK (1984): Binding characteristics and interactions of depressant drugs with [$^{35}$S]t-butylbicyclophosphorothionate, a ligand that binds to the picrotoxinin site. J Neurochem 42:221–229.

Simmonds MA (1982): Classification of some GABA antagonists with regard to site of action and potency in slices of rat cuneate nucleus. Eur J Pharmacol 80:347–358.

Schwartz RD, Skolnick P, Hollingsworth EB, Paul SM

(1984): Barbiturate and picrotoxin-sensitive chloride efflux in rat cerebral cortical synaptoneurosomes. FEBS Lett 175:193–196.

Skolnick P, Paul SM, Barker JL (1980): Pentobarbital potentiates GABA-enhanced $^{3}$H-diazepam binding to benzodiazepine receptors. Eur J Pharmacol 65:125–127.

Squires R, Braestrup C (1977): Benzodiazepine receptors in rat brain. Nature 266:732–734.

Squires RF, Casida JE, Richardson M, Saederup E (1983): [$^{35}$S]t-Butylbicyclophosphorothionate binds with high affinity to brain specific sites coupled to γ-aminobutyric acid-A and ion recognition sites. Mol Pharmacol 23:326–336.

Stephenson FA, Watkins AE, Olsen RW (1982): Physicochemical characterization of detergent-solubilized γ-aminobutyric acid and benzodiazepine receptor proteins from bovine brain. Eur J Biochem 123:291–298.

Supavilai P, Karobath M (1979): Stimulation of benzodiazepine receptor binding by SQ 20009 is chloride-dependent and picrotoxin sensitive. Eur J Pharmacol 60:111–113.

Supavilai P, Karobath M (1981): In vitro modulation by avermectin $B_{1a}$ of the GABA/benzodiazepine receptor complex of rat cerebellum. J Neurochem 36:798–803.

Supavilai P, Mannonen A, Karobath M (1982a): Modulation of GABA binding sites by CNS depressants and CNS convulsants. Neurochem Int 4:259–268.

Supavilai P, Mannonen A, Collins JF, Karobath M (1982b): Anion-dependent modulation of [$^{3}$H]muscimol binding and of GABA-stimulated [$^{3}$H]flunitrazepam binding by picrotoxin and related CNS convulsants. Eur J Pharmacol 81:687–691.

Supavilai P, Karobath M (1984): [$^{35}$S]t-Butylbicyclophosphorothionate binding sites are constituents of the γ-aminobutyric acid benzodiazepine receptor complex. J Neurosci 4:1193–1200.

Supavilai P, Karobath M (1985): Modulation of acetylcholine release from rat striatal slices by the GABA/benzodiazepine receptor complex. Life Sci 36:417–426.

Takeuchi A, Takeuchi T (1969): A study of the action of pictrotoxin on the inhibitory neuromuscular junction of the crayfish. J Physiol 205:377–391.

Tallman JF, Thomas JW, Gallager DW (1978): GABAergic modulation of benzodiazepine binding site sensitivity. Nature 274:384–385.

Teichberg VI, Tal N, Goldberg O, Luini A (1984): Barbiturates, alcohols and the CNS excitatory neurotransmission: Specific effects on the kainate and quisqualate receptors. Brain Res 291:285–292.

Thampy KG, Barnes EM Jr (1984): gamma-Aminobutyric acid-gated chloride channels in cultured cerebral neurons. J Biol Chem 259:1753–1757.

Ticku MK, Olsen RW (1978): Interaction of barbiturates with dihydropicrotoxinin binding sites related to the GABA receptor-ionophore system. Life Sci 22:1643–1651.

Ticku MK, Ramanjaneyulu R (1984): Differential interactions of GABA agonists, depressant and convulsant drugs with [$^{35}$S]t-butylbicyclophosphorothionate binding sites in cortex and cerebellum. Pharm Biochem Behav 21:151–158.

Williams M, Risley EA (1979): Enhancement of binding of [$^{3}$H]flunitrazepam to rat brain membranes in vitro by SQ 20009, a novel anxiolytic, γ-aminobutyric acid (GABA) and muscimol. Life Sci 24:833–841.

Williams M, Yarbrough G (1979): Enhancement of in vitro binding and some of the pharmacological properties of diazepam by a novel anthelmintic agent, avermectin $B_{1a}$. Eur J Pharmacol 56:273–276.

Willow M, Johnston GAR (1980): Enhancement of GABA binding by pentobarbitone. Neurosci Lett 18:323–327.

Wong EHF, Snowman AM, Leeb-Lundberg F, Olsen RW (1984a): Barbiturates allosterically inhibit GABA antagonist and benzodiazepine inverse agonist binding. Eur J Pharmacol 102:205–212.

Wong EHF, Leeb-Lundberg LMF, Teichberg VI, Olsen RW (1984b): γ-Aminobutyric acid activation of $^{36}$Cl-flux in rat hippocampal slices and its potentiation by barbiturates. Brain Res 303:267–275.

Zukin SR, Young AB, Snyder SH (1974): Gamma-aminobutyric acid binding to receptor sites in rat central nervous system. Proc Natl Acad Sci USA 71:4802–4807.

Benzodiazepine/GABA Receptors and Chloride Channels:
Structural and Functional Properties, pages 195–207


# 9

# Convulsant Binding Sites on the Benzodiazepine/ GABA Receptor

**Maharaj K. Ticku**

*Division of Molecular Pharmacology, Department of Pharmacology, The University of Texas Health Science Center at San Antonio, San Antonio, Texas 78284*

The role of γ-aminobutyric acid (GABA) as a potential inhibitory neurotransmitter was first described at the crustacean synapses and, subsequently, in the mammalian CNS. GABA is known to produce both pre- and postsynaptic inhibition in a variety of preparations, primarily by increasing the conductance to chloride ions [Takeuchi and Takeuchi, 1967; McBurney and Barker, 1978]. One of the first compounds that was shown to block GABA responses at the inhibitory neuromuscular junction of crayfish was picrotoxin [Elliot and Florey, 1956; Takeuchi and Takeuchi, 1969], whose convulsant effect was known since 1875. Later studies confirmed that picrotoxin also antagonized GABA responses on vertebrate neurons [e.g., Simmonds, 1980; Nicoll, 1980].

Picrotoxin is a non-nitrogenous polycyclic lactone found in the plants of *Menispermaceae* family and is made up of picrotoxinin and picrotin in an equimolar ratio. Jarboe et al. [1968] demonstrated that picrotoxinin, which differs from picrotin by hydration (in picrotin) of the isopropenyl group (a substituent at the β-position of γ-butyrolactone), was ~ 50-fold more potent than picrotin. This was an interesting observation and suggested a strict structural requirement for picrotoxin-like convulsants [for review, see Ticku and Maksay, 1983]. Picrotoxin inhibits GABA responses noncompetitively, suggesting the possibility of a different site of action [Takeuchi and Takeuchi, 1969; Simmonds, 1982]. Several other convulsants that inhibit GABA responses noncompetitively include cage-convulsant bicyclophosphate esters, pentamethylenetetrazole, bemegride, anisatin, and a benzodiazepine, RO5-3663 (Table I). Figure 1 shows structural similarity of some of these antagonists (for discussion, see below).

Other convulsants and GABA antagonists that lack structural similarity to that of picrotoxin and related cage convulsants incude bicuculline, R-5135, d-tubocurarine, strychnine, and penicillin. It is most likely that these convulsants act at the level of GABA recognition sites, binding preferentially to the antagonist state of the receptor [Table I; Olsen and Snowman, 1983; Maksay and Ticku, 1984b].

## Convulsant Binding Sites

In order to determine if there existed a distinct site for picrotoxin-like convulsants at the GABA synapse, a radioactive analogue of picrotoxinin, [$^{3}$H]α-dihydropicrotoxinin (DHP) was synthesized [Ticku et al., 1978; Olsen et al., 1978]. DHP, like picrotoxin, produces convulsions, blocks GABA-induced $^{36}Cl^-$-influx in crayfish muscle and is a GABA antagonist [Ticku et al., 1978; Olsen et al., 1978].

## Properties of DHP Binding

Initial studies with [$^{3}$H]DHP revealed that the specific binding of this radioligand to brain

**TABLE I. Antagonists of GABAergic Transmission**

| Convulsant | Structural similarity for picrotoxin site [see Ticku and Maksay 1983] | $IC_{50}$ [$^{35}$S]-TBPT binding ($\mu$M) | Antagonism of GABA responses | References |
|---|---|---|---|---|
| Picrotoxinin | Yes | 0.40 | Noncompetitive | Takeuchi & Takeuchi, 1969; Simmonds, 1982 |
| Isopropyl bicyclophosphate ester | Yes | 0.16 | Noncompetitive | Simmonds, 1982 |
| Pentylenetetrazole | Yes | 700 | Noncompetitive | Simmonds, 1982 |
| Bemegride | Yes | 200 | Like pentylene tetrazole | |
| Anisatin | Yes | 0.10 | Noncompetitive | Kudo et al., 1981 |
| RO5-3663 | Yes | 22 | Like picrotoxin | Harrison & Simmonds, 1983 |
| Bicuculline | No | NE[b] | Competitive | Simmonds, 1982 |
| R-5135 | No | NE | Competitive | Hunt & Clements-Jewery, 1981 |
| Penicillin[a] | No | NE | ND[c] | Olsen et al., 1978 |
| d-Tubocurarine | No | NE | Competitive | Simmonds, 1982 |

[a]Inhibits [$^{3}$H]GABA binding.
[b]NE refers to not effective in inhibiting [$^{35}$S]TBPT binding.
[c]ND refers to not determined.

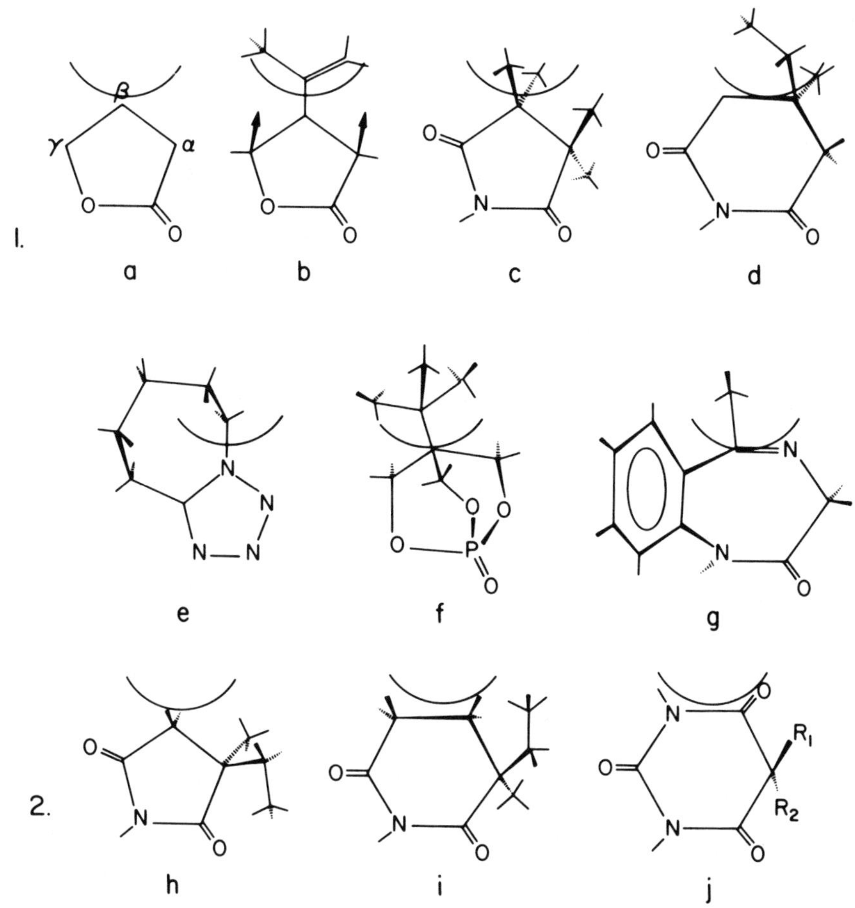


**Fig. 1.** *Schematic representation of the steric structure of drugs which interact with the picrotoxin binding sites and their possible relationship to a hydrophobic allosteric regulatory site (represented by a semicircle). [From Ticku and Maksay, 1983.] 1a: γ-butyrolactone; 1b: the corresponding lactone of α-picrotoxinin; 1c: tetramethylsuccinimide; 1d: bemegride; 1e: pentylenetetrazol; 1f: TBPT; 1g: RO5-3663; 2h: ethosuximide; 2i: glutethimide; 2j: barbiturates.*

membranes was only 10–15% of the total binding [Ticku et al., 1978; Ticku and Olsen, 1978]. In spite of an unacceptable signal:noise ratio, further studies with [$^3$H]DHP were carried out, since at that time there was no better ligand available for the picrotoxin site. [$^3$H]DHP binding to mammalian brain membranes and crayfish membranes was specific and saturable. The binding of [$^3$H]DHP was specific to neurons, since specific binding could be detected in primary neuronal cell lines and to cloned neuronal cell lines, but not in glia or non-neuronal cell lines or peripheral tissues that lacked GABAergic innervation [Ticku et al., 1978; Ticku, 1980b; Napias et al., 1980]. The brain regional distribution of DHP binding (cortex = cerebellum > hippocampus > striatum > hypothalamus > brainstem) is similar, but not identical, to that of [$^3$H]GABA and [$^3$H]diazepam binding [Olsen, 1981].

The apparent $K_D$ of [$^3$H]DHP was ~ 1–2 μM, which is similar to its biologically effective concentrations. Specific [$^3$H]DHP binding to brain and crayfish membranes was inhibited by biologically active (e.g., picrotoxinin, tutin, and α-DHP), but not inactive (e.g., picrotoxinin acetate and alkaline-hydrolyzed picrotoxinin), analogues of picrotoxin, with potencies that correlated with their pharmacologically effective concentrations [Olsen et al., 1978; Ticku et al., 1978].

[$^3$H]DHP binding was also inhibited by bicyclophosphate esters [R-C-$(CH_2O)_3$-P = O], which are convulsants and GABA antagonists [Bowery et al., 1976; Casida et al., 1976; Ticku and Olsen, 1979]. The order of potency of [$^3$H]DHP inhibition (R = *t*-butyl > isopropyl = *n*-propyl >> methyl) was similar to their GABA antagonistic profile. DHP binding to brain membranes was also inhibited by a convulsant benzodiazepine, RO5-3663, and pyrethroid insecticides [Olsen and Leeb-Lundberg, 1981].

Studies with [$^3$H]DHP also revealed that besides convulsants, the binding was also inhibited by sedative/hypnotic/anticonvulsant barbiturates that facilitate GABAergic transmission [Ticku and Olsen, 1978]. This led to the suggestion that the picrotoxin site may represent a receptor site for barbiturates [Ticku and Olsen, 1978; Ticku, 1980a]. Subsequently, nonbenzodiazepine anxiolytic pyrazolopyridine compounds (like etazolate) were also shown to inhibit DHP binding [Leeb-Lundberg et al., 1981]. All these studies suggested that the picrotoxin site at the GABA synapse was a target of drug action for several classes of centrally acting drugs that either inhibit or enhance GABAergic transmission.

### [$^{35}$S]t-Butylbicyclophosphorothionate (TBPT): A New Probe for Picrotoxin Site

Recently, Squires et al. [1983] developed a new radioligand, [$^{35}$S]*t*-butyl bicyclophosphorothionate (TBPS or TBPT) for the picrotoxin site. TBPT is structurally similar to other cage convulsant bicyclophosphates that have been shown to inhibit [$^3$H]DHP binding [Ticku and Olsen, 1979]. Figure 2 shows structural similarities between TBPT and picrotoxinin. The greatest advantage with TBPT is a highly superior signal:noise ratio, as compared to [$^3$H]DHP. TBPT binding to brain membranes is specific, saturable, dependent upon $Cl^-$ ions, and coupled to GABA and benzodiazepine (BZ) receptors [Squires et al., 1983; Ramanjaneyulu and Ticku, 1984a,b; Ticku and Ramanjaneyulu, 1984b; Wong et al., 1984; King and Olsen, 1984; Supavilai and Karobath, 1984]. Subcellular distribution and brain regional distribution of TBPT binding are similar to those of DHP and benzodiazepine binding [Squires et al., 1983; Wamsley et al., 1983].

Table II lists the $IC_{50}$ values for a series of convulsants, depressants, anticonvulsants, and anxiolytics that inhibit TBPT binding. The drug specificity of TBPT and [$^3$H]DHP binding is similar. Thus, TBPT binding is inhibited by picrotoxin and related cage convulsants, anisatin, PTZ, and RO5-3663. A series of tetrazoles, including pentylenetetrazole, inhibit TBPT binding potently [Ramanjaneyulu and Ticku, 1984b]. The $IC_{50}$ values of pentylenetetrazole (700 μM) and bemegride (200 μM) are consistent with their biological concentrations and the potency differences with which they inhibit GABAergic transmission. All the tetrazoles tested were more potent inhibitors of TBPT binding than diazepam binding. Further, like picrotoxin, pentylenetetrazole inhibits TBPT binding competitively. However, tetrazoles have also been reported to inhibit benzodiazepine binding competitively [Rehavi et al., 1982]. Based on structural similarities between pentylenetetrazole and picrotoxin, competitive inhibition of TBPT binding by pentylenetetrazole and identical dissociation profiles of TBPT binding by pentylenetetrazole and picrotoxin, it has been suggested that tetrazoles produce convulsions by binding to the picrotoxin site [Ramanjaneyulu and Ticku, 1984b; Maksay and Ticku, 1984c].

RO5-4864 has recently been shown to produce convulsions and block GABAergic transmission [Weissman et al., 1983, 1984; Ticku and Ramanjaneyulu, 1984a; Rastogi and Ticku, unpublished observations]. RO5-4864 is also proconvulsant with bicuculline

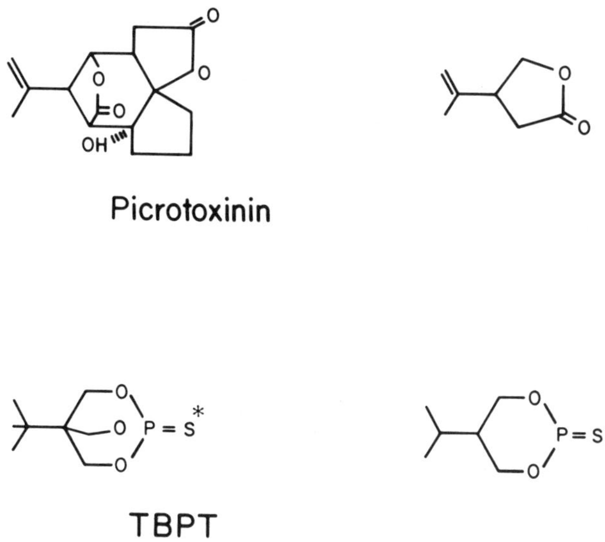


**Fig. 2.** *Structural similarities of picrotoxinin and TBPT.*

and picrotoxin, and its convulsions could be blocked by facilitators of GABAergic transmission like benzodiazepines, pentobarbital, ethanol, and amino-oxyacetic acid. We have recently demonstrated that RO5-4864 inhibits [$^{35}$S]TBPT binding at the concentrations at which it produces convulsions [Ticku and Ramanjaneyulu, 1984a]. Thus, it appears that this benzodiazepine may produce convulsions and block GABAergic transmission by acting at the picrotoxin site.

Initial reports on TBPT binding also confirmed that several classes of depressant/convulsant/anticonvulsant/anxiolytic drugs that had been shown to inhibit DHP binding also inhibited TBPT binding [Squires et al., 1983; Ramanjaneyulu and Ticku, 1984a; Ticku and Ramanjaneyulu, 1984b; Wong et al., 1984]. Recently, Supavilai and Karobath [1984] have shown that benzodiazepine agonists enhance and inverse benzodiazepine agonists inhibit basal [$^{35}$S]TBPT binding. These investigators also reported that muscimol, pentobarbital, and etazolate had biphasic effects on TBPT binding in buffer containing 200 mM NaCl. This effect was represented by a low-dose hook (i.e., stimulation followed by inhibition) which is usually observed for positively cooperative interactions.

### Equilibrium and Dissociation Studies With [$^{35}$S]TBPT Binding

Although both convulsant and depressant drugs inhibit TBPT binding, such inhibition does not indicate whether the inhibition is direct (i.e., competitive) or indirect (i.e., allosteric). Equilibrium Scatchard analysis indicated that convulsants like picrotoxinin and pentylenetetrazole decrease the affinity of TBPT binding, without altering the $B_{max}$ [Ramanjaneyulu and Ticku, 1984a,b; Ticku and Ramanjaneyulu, 1984b]. These results suggest that these convulsants apparently bind to the same sites. This is also consistent with the structural similarities of these convulsants [Figure 1; see below; Ticku and Maksay, 1983]. In contrast, GABA agonists and depressants like pentobarbital and etazolate decrease primarily the $B_{max}$ of TBPT binding, suggesting a noncompetitive type of interaction [Ticku and Ramanjaneyulu, 1984b]. This is also consistent with the lack of structural similarities of these drugs to picrotoxin-like convulsants [see Ticku and Maksay, 1983].

**TABLE II. $IC_{50}$ Values of Various Ligands for [$^{35}$S]TBPT Binding**

| Ligand | $IC_{50}$ ($\mu$M) |
|---|---|
| Convulsants: | |
| Anisatin | 0.10 |
| Isopropylbicyclophosphate | 0.16 |
| Picrotoxinin | 0.40 |
| Ethylbicyclophosphate | 1.20 |
| 6,6′-Dichloropentamethylenetetrazole | 10 |
| RO5-4864 | 10 |
| RO5-3663 | 22 |
| 1-Cyclohexyl-5-methyl-tetrazole | 40 |
| Quinolinic acid | 120 |
| 7-Methyl-10-isopropylpentamethylenetetrazole | 145 |
| 7-Methylpentamethylenetetrazole | 160 |
| 7-Methyl-9-isopropylpentamethylenetetrazole | 200 |
| Bemegride | 200 |
| Heptamethylenetetrazole | 230 |
| 6-0-Chlorophenyltetrazole | 310 |
| 1-Isobutyl-5-methylenetetrazole | 480 |
| 1-Methyl-5-cyclohexyltetrazole | 500 |
| Pentamethylenetetrazol (PTZ) | 700 |
| (+)N-Methyl-5-phenylpropyl barbituric acid (MPPB) | 760 |
| l-Kynurenine sulfate | >1000 |
| Methyl-6,7-dimethyoxy-4-ethyl-β-carboline-3-carboxylate | >1000 |
| Ethyl-β-carboline-3-carboxylate (β-CCE) | >2000 |
| N-Methyl-β-carboline-carboxamide (FG7142) | >2000 |
| Depressants, anticonvulsants, anxiolytics, and other drugs: | |
| GABA | 0.45 |
| Cartazolate | 0.52 |
| Etazolate | 5 |
| (+)Etomidate sulfate | 9 |
| RO11-6896 | 16 |
| (±)Secobarbital | 20 |
| (−)Secobarbital | 40 |
| Cannabidiol | 40 |
| Levonantradol HCl | 40 |
| Dextronantradol HCl | 50 |
| (±)DMBB | 50 |
| (+)Secobarbital | 70 |
| (−)Pentobarbital | 70 |
| (±)Pentobarbital | 90 |
| Methaqualone | 100 |
| (−)Etomidate | 100 |
| (−)N-Methyl-5-phenylpropyl barbituric acid (MPPB) | 116 |
| (+)Pentobarbital | 125 |
| (+)Hexobarbital | 140 |
| Metharbital | 145 |
| Tofizopam | 160 |
| (±)Mephobarbital | 210 |
| (±)Hexobarbital | 220 |
| (−)Hexobarbital | 380 |
| MK-801 | >200 |
| Phenobarbital | 480 |
| Ethanol | 500 |
| (+)N-Methyl-5-phenylpropyl barbituric acid (MPPB) | 760 |

**TABLE II. $IC_{50}$ Values of Various Ligands for [$^{35}$S]TBPT Binding (continued)**

| Ligand | $IC_{50}$ ($\mu$M) |
| --- | --- |
| γ-Butyrolactone | >2000 |
| Buspirone HCl | >2000 |
| (−)Ketamine HCl | >2000 |
| (+)Ketamine HCl | >2000 |

Drugs that inhibited the specific TBPT binding by less than 25% at $10^{-4}$ M include: diphenylhydantoin, Ru-5135, bicuculline methiodide, furosemide, acetylcholine, inosine, hypoxanthine, adenosine, and 1-methyl-isoguanosine.

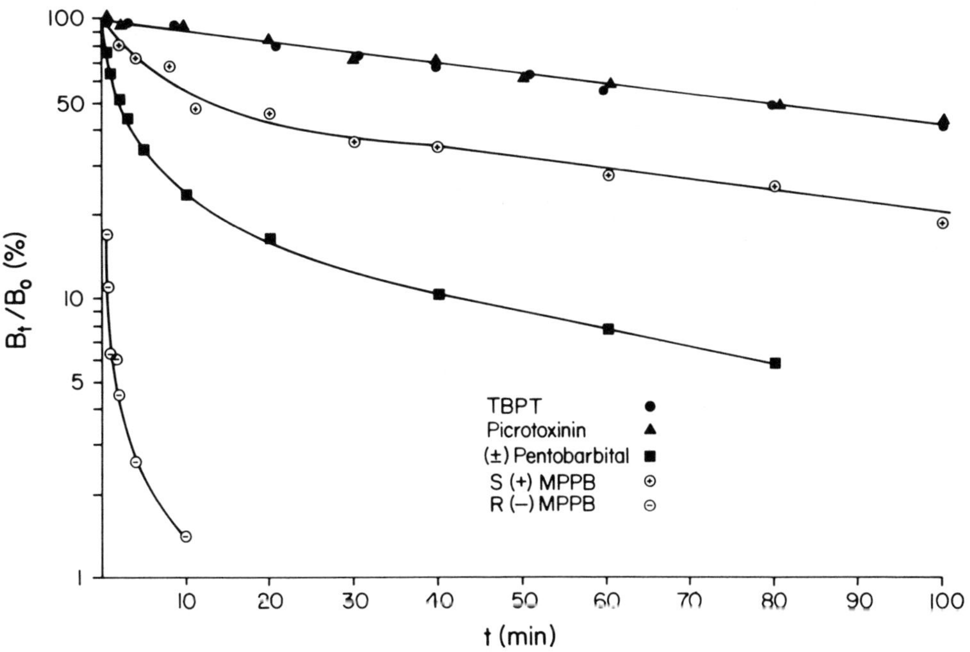


**Fig. 3.** *Representative dissociation plots of [$^{35}$S] TBPT binding in freeze-thawed membranes elicited by various convulsants and depressants. [Modified from Maksay and Ticku, 1984c.]*

Dissociation studies with TBPT binding further substantiated the concept that convulsants and depressants bind to distinct but coupled sites [Ticku and Maksay, 1984; Maksay and Ticku, 1984c]. Figure 3 shows the dissociation of TBPT observed in the presence of convulsants and depressants. Convulsants like TBPT, picrotoxin, and pentylenetetrazole (data not shown) gave identical monophasic dissociation patterns, suggesting that they bind to the same (convulsant) sites. This also indicates that [$^{35}$S]TBPT labels the same sites as picrotoxin. Conversely, depressants like (±)pentobarbital and etazolate greatly facilitated the dissociation rate of [$^{35}$S]TBPT binding, suggesting that these ligands bind to a distinct but coupled site [Ticku and Maksay, 1984; Maksay and Ticku, 1984c].

We also used stereoisomers of N-methyl-5-phenylpropyl barbituric acid (MPPB), which have opposite pharmacological activities, to determine their dissociation patterns. Thus,

R(−)MPPB is a depressant, whereas S(+) MPPB is a convulsant [Grossman et al., 1974]. The racemic MPPB produces pre-anesthetic excitement and uncoordinated motor activity during anesthesia, symptoms that are not observed with R(−)MPPB. We have also recently confirmed these observations behaviorally. R(−)MPPB, at 30 mg/kg, produces a transient loss of righting reflex and ataxia. Higher doses (~ 90 mg/kg) produce immediate (within 2–3 min) loss of righting reflex [Rastogi and Ticku, unpublished observations]. Conversely, S(+)MPPB produces convulsions in a dose-dependent manner. Further, S(+)MPPB was proconvulsant with subconvulsive doses of picrotoxin, and the convulsant effects of S(+)MPPB were blocked by R(−)MPPB and pentobarbital [unpublished observation].

These isomers could also be distinguished by their dissociation profile of [$^{35}$S]TBPT binding [Maksay and Ticku, 1984c]. Figure 3 shows that, like etazolate, depressant R(−)MPPB accelerated the dissociation of [$^{35}$S]TBPT. In contrast, S(+)MPPB gave a dissociation profile (the slower phase) very similar to that observed with other convulsants. Pentobarbital, which has a dual depressant/stimulatory effect [Downes et al., 1970], gave a dissociation profile that fell in between the convulsants and depressants (Fig. 3). Further, the Hill coefficient of pentobarbital inhibition of TBPT binding was ~ 1.5, whereas for other convulsants, etazolate and (+)etomidate, it was ~ 1.0 [Ticku and Ramanjaneyulu, 1984b]. It is feasible that the (+)isomer of pentobarbital, which possesses excitatory activity, may compete with the convulsant (TBPT) site, whereas the depressant (−)isomer interacts allosterically via the depressant site. These results are consistent with our notion that convulsants and depressants bind to distinct but coupled sites at the benzodiazepine-GABA receptor–ionophore complex. The possibility exists that convulsants and depressants may bind to the same sites in a negative cooperativity manner; however, there are no electrophysiological or other studies that favor this notion at this time.

## Topography of the Convulsant Site

Jarboe et al. [1968] were the first to suggest there may exist strict structural requirements for a series of picrotoxin analogues and related compounds to produce convulsions. One of the critical structural requirements for these convulsants was to possess the 5-isopropenyl (or isopropyl) group coplanar to the lactone ring and *trans* to the -OH at $C_6$. Recently, Klunk et al. [1983] suggested similarity between γ-butyrolactones and the essential lactone ring (i.e., the ring bridging $C_3$ and $C_5$ at the cyclohexane ring) of picrotoxin analogues. These authors postulated that, since an alkyl substituent at the β-position of the lactone ring was critical for activity, these convulsants may plug the chloride channel by inserting the β-substituent into it. This concept is similar to the one previously proposed for tetrodotoxin plugging of sodium channels [Hille, 1975]. The plugging theory may hold true for the sodium channel, since the guanidinum group is charged and has a hydrated diameter similar to $Na^+$, and there apparently is a negative charge (ionized carboxyl group) present in the $Na^+$-channel [Hille, 1975]. Further, tetrodotoxin and saxitoxin binding is inhibited by only those cations which bind to the negative charge in the channel [Henderson et al., 1974]. In contrast, picrotoxin is uncharged, doesn't look like $Cl^-$, and one would expect the chloride channel to have a positive charge. Additionally, anions that permeate the chloride channel are needed for [$^{35}$S]TBPT binding and do not inhibit it. This is a situation opposite to that observed for the $Na^+$-channel and tetrodotoxin action. Based on these, it is difficult to reconcile how the β-substituent (alkyl group) can plug the chloride channel. Furthermore, thermodynamic considerations also make this hypothesis unlikely. In addition, there is no evidence that the γ-butyrolactones studied by Klunk et al. [1983] actually inhibit GABA function or picrotoxin/TBPT binding sites.

We have recently modified the above hypothesis, based on steric requirements of the picrotoxin-like convulsants that are GABA antagonists, and the available binding data [Ticku and Maksay, 1983]. Figure 1 shows the orientation and fit of some of the convulsants and depressants that inhibit picrotoxinin/TBPT binding to the convulsant site. It has been previously reported that both the convulsant activity and the displacing potencies of bicyclophosphates on DHP/TBPT binding increase, with an increase in the chain length and branching of the alkyl substituents up to *t*-butyl group [Ticku and Olsen, 1979; Squires et al., 1983]. Conversely, introduction of a -OH group at $C_{12}$ of dihydropicrotoxinin decreases drastically both the convulsant activity and the binding affinity of the resulting compound, picrotin [Ticku et al., 1978]. These observations would be consistent with the notion that the chloride channel per se should be hydrophilic and could not be plugged by hydrophobic alkyl substituents. We have previously postulated that the $\beta$-substituents of the convulsants bind to an allosteric hydrophobic regulatory site which is in close proximity to the channel pore [Ticku and Maksay, 1983]. The compounds that lack the alkyl $\beta$-substituent, like $\gamma$-butyrolactone, do not inhibit TBPT binding and are not convulsants. Thus, it appears that binding to the hydrophobic regulatory site will dictate the closing of the chloride channel. Figure 1 also shows that, besides picrotoxin, other convulsants that are GABA antagonists (e.g., bemegride, pentylenetetrazole, TBPT, and RO5-3663) possess alkyl substituents that correspond to the $\beta$-orientation requirement. Ethosuximide and glutethimide lack this critical group, and they do not inhibit [$^{35}$S]TBPT binding [Squires et al., 1983]. Figure 1 also shows that barbiturates lack an analogous alkyl group which could interact with the regulatory site. Further equilibrium and disassociation studies with [$^{35}$S]TBPT binding have demonstrated an indirect or allosteric interaction of depressant barbiturates, etazolate, and (+)etomidate with the convulsant site (see above).

In summary, based on structural differences between convulsants and depressants, and their differential effects on equilibrium and dissociation binding of [$^{35}$S]TBPT, it is difficult to reconcile how depressants can bind to the same sites as convulsants. However, we have previously suggested that some barbiturates with substituents at C-5 position of either a longer chain than ethyl (since gluethimide possesses an ethyl group and does not inhibit TBPT binding) or a longer flexible saturated ring may occupy the hydrophobic regulatory site (e.g., (+)MPPB; (+)pentobarbital). The effects on the dissociation rate of [$^{35}$S]TBPT by (+)MPPB and racemic pentobarbital also favor this interpretation. It is known that aliphatic chain influences both the potency and qualitative properties of barbiturates [Büch et al., 1973]. Figure 4 shows a hypothetical version of the GABA macromolecular receptor complex with potential sites of drug action.

## Solubilization of the Picrotoxin/TBPT Binding Sites

Solubilization of the picrotoxin binding sites has been reported with detergents like Lubrol-PX [Davis and Ticku, 1981a,b] or CHAPS [King and Olsen, 1984]. The ligand specificity of [$^{3}$H]DHP or [$^{35}$S]TBPT binding in the crude solubilized fraction was similar to membrane homogenate. However, Davis and Ticku [1981b] reported separation of the DHP binding fraction (185,000 daltons) from the diazepam binding fraction (61,000 daltons). Further, pentobarbital, while inhibiting the binding of [$^{3}$H]DHP to the 185,000 dalton fraction, did not enhance the binding of [$^{3}$H]diazepam to the 61,000 dalton fraction. This led to the suggestion that picrotoxin and benzodiazepine bind to two distinct sites, and that pentobarbital sites appear to be tightly associated with the picrotoxin sites. Recent equilibrium and dissociation studies with [$^{35}$S]TBPT binding seem to concur with the notion that the barbiturate sites are closely coupled to the TBPT sites (see above).

King and Olsen [1984] reported solubilization of [$^{35}$S]TBPT binding activity with CHAPS, which retained the ligand specificity as observed in membranes. These investigators reported comigration of TBPT/GABA/benzodiazepine binding complex by gel filtra-

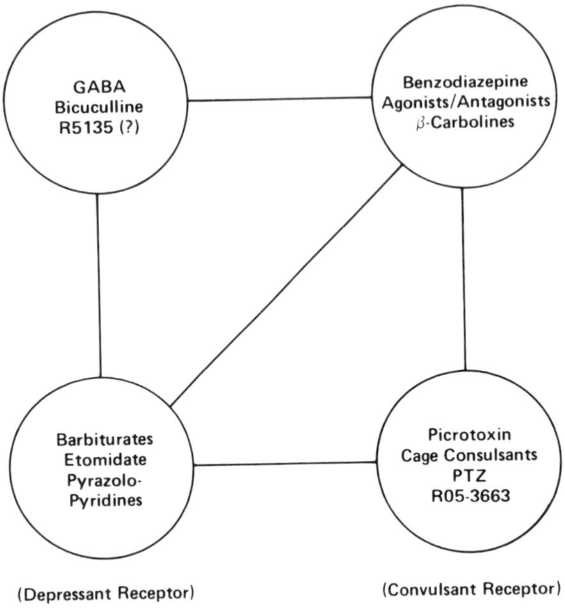


**Fig. 4.** *Hypothetical representation of benzodiazepine/GABA receptor/ionophore complex with potential sites of drug action.*

tion column chromatography, which retains the relevant allosteric interactions [King and Olsen, 1984; Olsen et al., 1984].

### Allosteric Interaction at the Benzodiazepine-GABA Receptor-Ionophore Complex

Drugs that either facilitate or inhibit GABAergic transmission bind to one of the sites at the macromolecular receptor complex, and these modulate the binding of ligands to other sites. Thus, pyrazolopyridines, depressant barbiturates, and (+)etomidate inhibit the binding of [$^{35}$S]TBPT [Squires et al., 1983; Ramanjaneyulu and Ticku, 1984a] but enhance the binding of [$^{3}$H]-GABA and [$^{3}$H]benzodiazepine agonists in vitro [Supavilai and Karobath, 1979; Leeb-Lundberg et al., 1980, 1981; Ashton et al., 1981; Olsen, 1981; Willow 1981; Skolnick et al., 1981; Ticku, 1981; Supavilai et al., 1982; Johnston, 1983; Thyagarajan et al., 1983]. The enhancing effects of these facilitators are blocked by picrotoxin and related cage convulsants at concentrations at which they inhibit DHP/TBPT binding. Barbiturates, which have stereospecific behavioral effects, also exhibit a similar stereoselectivity in enhancing benzodiazepine and GABA binding and in inhibiting DHP/TBPT binding. Pentobarbital increases the benzodiazepine binding by increasing the affinity of benzodiazepine agonists for their binding sites, apparently by a decreased rate of dissociation [Leeb-Lundberg and Olsen, 1982]. Barbiturate enhancement of benzodiazepine binding is anion-dependent and can be blocked by picrotoxin, related cage convulsants, and also by bicuculline [Olsen, 1981; Ticku, 1981; Skolnick et al., 1981; Supavilai et al., 1982].

It is of interest that all barbiturates that are depressants do not give the same maximal enhancement of benzodiazepine binding [Ticku, 1981]. Further, barbiturates like phenobarbital do not enhance benzodiazepine binding while some convulsant barbiturates do [Ticku, 1981; Leeb-Lundberg et al., 1980]. These observations raise some concern about the relationship of benzodiazepine enhancement by various barbiturates to their pharmacological effects. However, it may be pointed out that barbiturates have several pharmacological effects, and it is feasible that one or more molecular mechanisms are responsible for various effects [Rastogi and Ticku, 1985].

Barbiturate enhancement of GABA binding has been demonstrated by several groups, and is reviewed elsewhere [Olsen, 1981, 1983; Johnston, 1983]. The barbiturate enhancement of GABA binding is stereospecific, dependent upon anions like chloride [Asano and Ogasawara, 1982; Olsen and Snowman, 1982; Olsen, 1983] or independent of chloride [see Johnston, 1983], blocked by picrotoxin-like cage convulsants [Thyagarajan et al., 1983; Olsen, 1983], but not by RO15-1788 [Skerritt et al., 1983]. This is an interesting observation and suggests that the picrotoxin site but not the benzodiazepine site is crucial for pentobarbital to enhance GABA binding. These results are consistent with electrophysiological evidence that both bicuculline and picrotoxin can block pentobarbital effects on GABAergic transmission [Nicoll, 1980; Simmonds, 1982]. Pentobarbital and (+)etomidate increase the number of both high- and low-affinity GABA receptors [Olsen and Snowman, 1983; Thyagarajan et al., 1983]. It has been suggested that the increase in the number of receptors observed with pentobarbital may be due to its ability to shift the super-low-affinity GABA receptors to higher-affinity states [Olsen and Snowman, 1982]. This speculation would be consistent with the electrophysiological evidence since, in the presence of pentobarbital, GABA dose–response curves are shifted to the left [Barker and Ransom, 1978]. Other lines of evidence that suggest interconvertibility of GABA receptor populations include 1) ability of picrotoxin and related cage convulsants to decrease the $B_{max}$ of low-affinity [$^3$H]muscimol binding [Supavilai et al., 1982] and 2) ability of pretreatment of membranes at 37°C but not at 0–4°C with GABA or other facilitators of GABAergic transmission, followed by extensive washing, to increase subsequent [$^3$H] GABA binding [Maksay and Ticku, 1984d]. Interestingly, GABA pretreatment increased the $B_{max}$ of both high- and low-affinity GABA receptors, a situation analogous to that seen with pentobarbital in vitro. Further, the GABA pretreatment enhancing effect could be blocked by simultaneous presence of picrotoxin or bicuculline, suggesting that the super-low-affinity GABA sites are functional [Maksay and Ticku, 1984d].

Finally, studies to demonstrate which of the three GABA receptor sites (high, low, and super-low) are involved in pharmacological and physiological actions at the GABA synapse have yet to be defined. Although the ligand specificities of the high- and low-affinity GABA receptors appear to be similar [Olsen et al., 1981], the topography of these two receptors is not similar [Maksay and Ticku, 1984a]. The low- and possibly super-low-affinity GABA recognition sites, but not the high-affinity sites, contain tyrosine and lysine residues [Maksay and Ticku, 1984a,b]. Further, low- (but not high-) affinity receptor sites are coupled to both benzodiazepine and TBPT binding sites [Browner et al., 1981; Burch et al., 1983; Maksay and Ticku, 1984a; Ticku and Ramanjanayulu, 1984b]. These results suggest that the low and super-low GABA receptors have a similar chemical topography and may be functionally relevant. The pharmacological significance of the high-affinity sites is not known, but may very well represent the desensitized state of GABA receptors, a situation analogous to the acetylcholine nicotinic receptor system.

## ACKNOWLEDGMENTS

I gratefully acknowledge my colleagues G. Maksay, S.K. Rastogi, and R. Thyagarajan for helpful discussions, and Marilyn Wilson for excellent secretarial assistance. Supported by NIH grant NS-15339 and NIAAA grant AA-04090.

## REFERENCES

Asano T, Ogasawara N (1981): Chloride-dependent stimulation of GABA and benzodiazepine receptor binding by pentobarbital. Brain Res 225:212–216.

Ashton D, Geerts R, Waterkeyn C, Leysen JE (1981): Etomidate stereospecifically stimulates forebrain but not cerebellar [$^3$H]diazepam binding. Life Sci 29:2631–2636.

Barker JL, Ransom BR (1978): Pentobarbital pharmacology of mammalian central neurons grown in tissue culture. J Physiol 280:355–372.

Bowery NG, Collins JF, Hill RG, Pearlson S (1976): GABA antagonism as a possible basis for convulsant action of series of bicyclophosphorous esters. Br J Pharmacol 58:435p–436p.

Browner M, Ferkany JW, Enna SJ (1981): Biochemical identification of pharmacologically and functionally distinct GABA receptors in rat brain. J Neurosci 1:514–518.

Büch HP, Schneider-Affeld F, Rummel W, Knabe J (1973): Sterochemical dependence of pharmacological avtivity in a series of optically active N-methylated barbiturates. Naunyn-Schmeideberg's Arch Pharmacol 277:191–198.

Burch TP, Thyagarajan R, Ticku, MK (1983): Group-selective reagent modifications of the benzodiazepine-GABA receptor–ionophore complex reveals that low affinity GABA receptors stimulate benzodiazepine binding. Mol Pharmacol 23:52–59.

Casida GE, Eto M, Moscioni AD, Engel JL, Milbrath DS, Verkade JG (1976): Structure-toxicity relationship of 2,6,7-trioxyabicyclo(2.2.2)-octane and related compounds. Toxicol Appl Pharmacology 36:261–279.

Davis WC, Ticku MK (1981a): Solubilization of the picrotoxinin binding receptor from mammalian brain. J Neurochem 36:1572–1579.

Davis WC, Ticku MK (1981b): Picrotoxin and diazepam bind to two distinct proteins: Further evidence that pentobarbital may act at the picrotoxinin site. J Neurosci 1:1036–1042.

Downes H, Perry RS, Ostlund E, Karler R (1970): A study of the excitatory effects of barbiturates. J Pharm Exp Ther 175:692–699.

Elliott KAC, Florey E (1956): Factor I—Inhibitory factor from brain. J Neurochem 1:181–182.

Grossman W, Jurna I, Theres C (1974): The site of action of the optical isomers of 1-methyl-5-phenyl-5-propylbarbituric acid. Naunyn-Schmiedeberg's Arch Pharmacol 282:367–377.

Harrison NL, Simmonds MA (1983): The picrotoxin-like action of a convulsant benzodiazepine, RO5-3663. Eur J Pharmacol 87:155–158.

Henderson R, Ritchie JM, Stricharz GR (1974): Evidence that tetrodotoxin and saxitoxin act at a metal cation binding site in the sodium channel of nerve membranes. Proc Natl Acad Sci USA 71:3936–3940.

Hille B (1975): The receptors for tetrodotoxin and saxitoxin. Biophys J 15:615–619.

Hunt P, Clements-Jewery S (1981): A steroid derivative, R-5135, antagonizes the GABA/benzodiazepine receptors. Neuropharmacology 20:357–361.

Jarboe CH, Porter LA, Buckler RT (1968): Structural aspects of picrotoxinin action. J Med Chem 11:729–731.

Johnston GAR (1983): Regulation of GABA receptors by barbiturates and by related sedative-hypnotics and anticonvulsant drugs. In Enna SJ (ed): "GABA Receptors," NJ: Humana Press, pp 107–128.

King RG, Olsen RW (1984): Solubilization of convulsant/barbiturate binding activity on the γ-aminobutyric acid/benzodiazepine receptor complex. Biochem Biophys Res Commun 119:530–536.

Klunk WE, Kalman BL, Ferrendelli JA, Covey DF (1983): Computer-assisted modeling of the picrotoxinin and γ-butyrolactone receptor site. Mol Pharmacol 23:511–528.

Kudo Y, Oka J, Yamada K (1981): Anisatin, a potent GABA antagonist, isolated from *Illicium anisaltem.* Neurosci Lett 25:83–88.

Leeb-Lundberg F, Snowman A, Olsen R (1980): Barbiturate receptor sites are coupled to benzodiazepine receptors. Proc Natl Acad Sci USA 77:7462–7472.

Leeb-Lundberg F, Snowman A, Olsen RW (1981): Perturbation of benzodiazepine receptor binding by pyrazolopyridines involves picrotoxinin/barbiturate receptor sites. J Neurosci 1:471–477.

Leeb-Lundberg F, Olsen RW (1982): Interaction of barbiturates of various pharmacological categories with benzodiazepine receptors. Mol Pharmacol 21:320–328.

Maksay G, Ticku MK (1984a): Characterization of GABA-benzodiazepine receptor complex by protection against inactivation by group-specific reagents. J Neurochem 42:1715–1727.

Maksay G, Ticku MK (1984b): Diazotization and thiocyanate differentiate agonists from antagonists for the high- and low-affinity receptors. J Neurochem 43:261–268.

Maksay G, Ticku MK (1984c): The dissociation of [$^{35}$S]*t*-butyl-bicyclophosphorothionate binding differentiates convulsant and depressant drugs that modulate GABAergic transmission. J Neurochem 44:480–486.

Maksay G, Ticku MK (1984d): Pretreatment with GABA and modulatory ligands enhances GABA receptor binding. Eur J Pharmacol 98:337–345.

McBurney RN, Barker JL (1978): GABA-induced conductance fluctuations in cultured spinal neurones. Nature 274:596–597.

Napias C, Olsen RW, Schubert D (1980): GABA and picrotoxin binding in cloned nerve cells. Nature 283:298–299.

Nicoll RA (1980): Sedative-hypnotics: Animal pharmacology. In Iversen LL, Iversen SD, Snyder SH (eds): "Handbook of Psychopharmacology: A Generation of Progress, Volume 12, Drugs of Abuse" New York: Plenum Press.

Olsen RW, Ticku MK, VanNess PC, Greenlee D (1978): Effects of drugs on γ-aminobutyric acid receptor uptake, release and synthesis *in vitro*. Brain Res 139:277–294.

Olsen RW (1981): GABA-benzodiazepine barbiturate receptor interactions. J Neurochem 37:1–13.

Olsen RW, Leeb-Lundberg F (1981): Convulsant and anticonvulsant drug binding sites related to GABA-regulated chloride ion channels. In Costa E, Di-Chiara D, and Gessa GL (eds): "GABA and Benzodiazepine Receptors" (Adv. Biochem. Psychopharm., Vol. 26) New York: Raven Press, pp 93–102.

Olsen RW, Bergman MO, VanNess PC, Lummis SC, Watkins AE, Napias C, Greenlee DV (1981): γ-Aminobutyric acid receptor binding in mammalian brain: Heterogeneity of binding sites. Mol Pharmacol 19:217–227.

Olson RW, Snowman AM (1982): Chloride-dependent enhancement by barbiturates of GABA receptor binding. J Neurosci 2:1812–1823.

Olson RW (1983): Biochemical properties of GABA receptors. In Enna SJ (ed): "GABA Receptors." NJ: Humana Press, pp 107–128.

Olsen RW, Snowman AM (1983): [$^3$H]Bicuculline-methochloride binding to low affinity GABA receptor sites. J Neurochem 41:1653–1663.

Olsen RW, Wong EHF, Stauber GB, Murakami D, King RG, Fischer JB (1984): In Kito S, Segawa T, Kuriyama K, Yamamura HI, Olsen RW (eds): "Neurotransmitter Receptors: Mechanisms of Action and Regulation." New York: Plenum Press, pp 205–219.

Ramanjaneyulu R, Ticku MK (1984a): Binding characteristics and interactions of depressant drugs with [$^{35}$S]*t*-butylbicyclophosphorothionate, a ligand that binds to the picrotoxinin site. J Neurochem 42:221–229.

Ramanjaneyulu R, Ticku MK (1984b): Interactions of pentamethylenetetrazole and tetrazole analogues with the picrotoxinin site of the benzodiazepine-GABA receptor-ionphore complex. Eur J Pharmacol 98:337–345.

Rastogi S, Ticku MK (1985): Involvement of a GABAergic mechanism in the anticonvulsant effect of pentobarbital against maximal electroshock-induced seizures in rats. Biochem Pharm Behav 22:141–146.

Rehavi MP, Skolnick P, Paul SM (1982): Effects of tetrazole derivatives on [$^{3}$H]diazepam binding *in vitro*: Correlation with convulsant potency. Eur J Pharmacol 78:353–358.

Simmonds MA (1980): Evidence that bicuculline and picrotoxin act at separate sites to antagonize γ-aminobutyric acid in rat cuneate nucleus. Neuropharmacology 19:39–45.

Simmonds MA (1982): Classification of some GABA antagonists with regard to site of action and potency in slices of rat cuneate nucleus. Eur J Pharmacol 80:347–358.

Skerritt JH, Johnston GAR, Katsikas T, Tabar J, Nicholson GM, Andrews PR (1983): Action of pentobarbitone and derivatives with modified 5-butyl substituents on GABA and diazepam binding to rat brain synaptosomal membranes. Neurochem Res 8:1337–1350.

Skolnick P, Moncada V, Barker JL, Paul SM (1981): Pentobarbital: Dual actions to increase brain benzodiazepine receptor affinity. Science 211:1448–1450.

Squires RF, Casida JE, Richardson M, Saederup E (1983): [$^{35}$S]-*t*-Butylbicyclophosphorothionate binds with high affinity to brain specific sites coupled to γ-aminobutyric acid $_{A}$ and ion recognition sites. Mol Pharmacol 23:326–336.

Supavilai P, Karobath M (1979): Stimulation of benzodiazepine receptor binding by SQ20009 is chloride-dependent and picrotoxin-sensitive. Eur J Pharmacol 60:111–113.

Supavilai P, Mannonen A, Karobath M (1982): Modulation of GABA binding sites by CNS depressants and CNS convulsants. Neurochem Int 4:259–268.

Supavilai P, Karobath M (1984): [$^{35}$S]-*t*-Butylbicyclophosphorothionate binding sites are constituents of the γ-aminobutyric acid–benzodiazepine receptor complex. J Neurosci 4:1193–1200.

Takeuchi A, Takeuchi T (1967): Anion permeability of the inhibitory post-synaptic membrane of the crayfish neuromuscular junction. J Physiol 191:575–590.

Takeuchi A, Takeuchi T (1969): A study of the action of picrotoxin on the inhibitory neuromuscular junction of the crayfish. J Physiol 205:377–391.

Thyagarajan R, Ramanjaneyulu R, Ticku MK (1983): Enhancement of diazepam and γ-aminobutyric acid binding by (+)etomidate and pentobarbital. J Neurochem 41:578–585.

Ticku MK, Ban M, Olsen RW (1978): Binding of [$^{34}$H]α-dihydropicrotoxinin, a γ-aminobutyric acid synaptic antagonist, to rat brain membranes. Mol Pharmacol 14:391–402.

Ticku MK, Olsen RW (1978): Interaction of barbiturates with dihydropicrotoxinin binding sites related to the GABA receptor-ionophore system. Life Sci 22:1643–1652.

Ticku MK, Olsen RW (1979): Cage convulsants inhibit picrotoxinin binding. Neuropharmacology 18:315–318.

Ticku MK (1980a): Is picrotoxin binding site at the GABA synapse a site of action for barbiturates? Brain Res Bull 5 (Suppl. 2):919–923.

Ticku MK (1980b): [$^{3}$H]-α-Dihydropicrotoxinin binds to primary neuronal but not glial cells. Eur J Pharmacol 65:135–136.

Ticku MK (1981): Interaction of depressant, convulsant and anticonvulsant barbiturates with [$^{3}$H]diazepam sites at the benzodiazepine-GABA-receptor-ionophore complex. Biochem Pharmacol 30:2573–2579.

Ticku MK, Maksay G (1983): Convulsant/depressant site of action at the alllosteric benzodiazepine-GABA receptor-ionophore complex. Life Sci 33:2363–2375.

Ticku MK, Maksay G (1984): Dissociation and equilibrium studies with [$^{35}$S]*t*-butyl-bicyclophosphorothionate (TBPT) binding indicates that convulsant and depressant drug's bind to different but coupled sites at the GABA receptor-ionophore complex. Neuroscience Abstracts 10:389 (#117.11).

Ticku MK, Ramanjaneyulu R (1984a): RO5-4864 inhibits the binding of [$^{35}$S]*t*-butylbicyclophosphorothionate to rat brain membranes. Life Sci 34:631–638.

Ticku MK, Ramanjaneyulu R (1984b): Differential interactions of GABA agonists, depressant and convulsant drugs with [$^{35}$S]*t*-butylbicyclophosphorothionate binding sites in cortex and cerebellum. Pharm Biochem Behav 21:151–158.

Wamsley JK, Gee KW, Yamamura HI (1983): Comparison of the distribution of convulsant/barbiturate and benzodiazepine receptors using light microscopy autoradiography. Life Sci 33:2321–2329.

Weissman BA, Cott J, Paul SM, Skolnick P (1983): RO5-4864, a potent benzodiazepine convulsant. Eur J Pharmacol 90:149–150.

Weissman BA, Cott J, Hammer D, Paul SM, Skolnick P (1984): Electophysiological and pharmacological actions of the convulsant benzodiazepine RO5-4864. Eur J Pharmacol 97:257–263.

Willow M (1981): A comparison of the action of pentobarbital and etomidate on [$^{3}$H]GABA binding to crude synaptosomal rat brain membranes. Brain Res 220:427–431.

Wong DT, Threlkeld PG, Bymaster FP, Squires RF (1984): Saturable binding of $^{35}$S-*t*-butylbicyclophosphorothionate to the sites linked to the GABA receptor and the interaction with GABAergic agents. Life Sci 34:853–860.

Benzodiazepine/GABA Receptors and Chloride Channels:
Structural and Functional Properties, pages 209–224


# 10

# Ligand and Ion Site Interactions in GABA and Benzodiazepine Receptor Complexes

**Richard F. Squires**

*Nathan S. Kline Institute for Psychiatric Research, Orangeburg, New York 10962*

It now appears that there is a greater variety of GABA receptors in vertebrate species than had been generally appreciated. This multiplicity is well illustrated by the presence of at least four types of GABA receptors in crayfish leg muscles [Dudel and Hatt, 1976] and at least five types of GABA receptors on *Aplysia* neurons, four of which regulate cation channels [Yarowsky and Carpenter, 1978]. In addition to the well-known GABA-regulated chloride channels in vertebrate neurons and crustacean muscle, there is increasing evidence that the so-called peripheral, or non-neuronal, benzodiazepine binding sites are coupled to calcium channels [Cantor et al., 1984 Mestre et al., 1984; Le Fur et al., 1984], and there is evidence that both GABA receptors and benzodiazepine binding sites evolved from ancestral cation channels, similar to those present in *Paramecia.* In addition to the well-established coupling of divalent cation (calcium) recognition sites to bicuculline-resistant $GABA_B$ receptors, recent electrophysiological evidence indicates that $GABA_B$ receptors also regulate a potassium channel in some neuronal membranes [Newberry and Nicoll, 1984].

Neuronal benzodiazepine binding sites, most studied in mammalian brain, represent a heterogeneous subpopulation of GABA receptor complexes made up of four or five separate benzodiazepine-binding subunits which may be combined in different ways to yield functional polymers [Sieghart and Drexler, 1983; Sieghart et al., 1983; Squires, 1983]. Most, if not all, neuronal benzodiazepine binding sites are probably coupled to both anion and cation recognition sites [Squires and Saederup, 1982a,b]. The preponderance of GABA-regulated cation channels in invertebrate species [e.g., Yarowsky and Carpenter, 1978] and their continuing presence in mammalian species support the hypothesis that the "ancestral" GABA receptors were coupled to cation channels earlier than to anion channels. Benzodiazepine and picrotoxin binding sites have also been described in invertebrate species. Below, we will review some evidence for the coupling of GABA, benzodiazepine, picrotoxin, and barbiturate recognition sites with anion and cation channels. It is probable that all neuronal benzodiazepine and picrotoxin binding sites are coupled to $GABA_A$ receptors. The evidence supporting this conclusion has been reviewed elsewhere [Olsen, 1982; Squires et al., 1983; Squires, 1984].

## COUPLING OF GABA RECEPTORS TO ANION RECOGNITION SITES

### Electrophysiological Studies

Many electrophysiological studies have demonstrated that GABA, especially when applied to perikarya, hyperpolarizes (or partially depolarizes) the nerve and increases chloride conductance [Adams and Brown,

1973; Nishi et al., 1974; Krnjević, 1976; Barker and Ransom, 1978; Gallager et al., 1978; Study and Barker, 1981]. Both diazepam and pentobarbital enhance the increase in chloride ion permeability, elicited by GABA, across the cell membrane of mouse spinal neurons grown in culture [Study and Barker, 1981]. Since diazepam and pentobarbital act at separate sites in GABA receptor complexes [Leeb-Lundberg et al., 1980; Squires et al., 1983], it is interesting that diazepam acts by increasing the frequency of chloride channel openings, while pentobarbital increases the open-time of the chloride channel without greatly affecting opening frequency [Study and Barker, 1981]. Similarly, application of GABA to crustacean muscle also results in increased permeability of the cell membrane to chloride cion [Takeuchi and Takeuchi, 1967; Constanti and Nistri, 1976].

In addition to chloride channels, there is much evidence that at least some types of GABA receptors and benzodiazepine binding sites regulate cation channels in nerve and muscle cell membranes of both invertebrate and vertebrate species. This evidence will be discussed in more detail below.

**Biochemical Studies**

Two early studies [Costa et al., 1979; Martin and Candy, 1980] indicated that $^3$H-diazepam binding to brain-specific sites [coupled to $GABA_A$ receptors] was enhanced by the addition of certain anions. Lo and Snyder [1978] reported that chloride, bromide, and iodide, but no other anions tested, selectively enhanced the binding of $^3$H-flunitrazepam to solubilized type 2 binding sites but had no effect on type 1 binding sites. Although all three studies found the halide ions bromide, iodide, and chloride to be active, the rank-order of potencies for the anions tested did not appear to be the same. All three studies are flawed by the use of high concentrations of buffers—50 mM Tris maleate [Costa et al., 1979], 100 mM Tris citrate [Martin and Candy, 1980], and 25 mM phosphate [Lo and Snyder, 1983]—which themselves potently enhance the binding of $^3$H-flunitrazepam to brain-specific sites [Squires, 1981a,b].

The binding of $^3$H-flunitrazepam [Squires, 1981a,b] and $^3$H-CGS 8216 [Squires and Saederup, 1982a] was reported to be absolutely dependent on the presence of ions, with anions playing the predominant role. The concentrations of buffer required for 50% of maximum binding ranged from 0.7 mM for Tris phosphate to 14 mM for Tris glutamate. The piperidine-derived $GABA_A$ receptor agonist 4,5,-6,7-tetrahydroisoxazolo [4,5c]pyridin-3-ol (THIP) [Krogsgaard-Larsen et al., 1977] dramatically increased the concentrations of buffers required to give 50% of maximum $^3$H-flunitrazepam binding [Squires, 1981a], but the magnitude of the increase varied from 3.4-fold for sodium EDTA (pH 7.4) to 12-fold for Tris acetate [Squires, 1981a]. Two other piperidine-derived $GABA_A$ receptor agonists, isoguvacine [Krogsgaard-Larsen et al., 1977] and piperidine-4-sulfonate [Krogsgaard-Larsen et al., 1980], also greatly decreased the affinity for chloride ion (as Tris-HCl, pH 7.5) in enhancing $^3$H-flunitrazepam binding, while two nonpiperidine $GABA_A$ receptor agonists, imidazole-acetate and 3-amino propane sulfonate, also decreased the affinity of chloride, but to a smaller extent [Squires, 1981b). The effects of THIP and piperidine-4-sulfonate on chloride affinity for the $^3$H-flunitrazepam binding system could be competitively reversed by GABA or muscimol [Squires, 1981a,b]. It is clear that there is no correlation between the ability of the anions to enhance $^3$H-flunitrazepam binding, alone or in the presence of THIP (200 $\mu$M), and their ability to substitute for chloride electrophysiologically [Squires, 1981a]. However, the ability of $GABA_A$ receptor agonists to modulate the affinities of virtually all anions tested, despite large differences in their affinities, suggests that the anions bind to a physiologically relevant site in benzodiazepine/GABA receptor complexes. Originally Eccles [1964] proposed that only anions with hydrated sizes near that of chloride ion could substitute for chloride in generating inhibitory postsynaptic potentials in cat motoneurons; thus common anions like bicarbonate, sulfate, and phosphate would be too large and therefore inactive.

A more thorough study of a large number of anions, tested for ability to reverse inhibitory postsynaptic potentials, when injected into cat cerebral cortical neurons, revealed that all monovalent anions tested (with the one exception of iodate ion) were active in at least 33% of the neurons tested, although bromide, iodide, and chloride were active in virtually all neurons tested. The larger anions tended to be active on a smaller proportion of neurons [Kelly et al., 1969]. With the exception of chromate ion, which could substitute for chloride in 85% of the neurons tested, polyvalent anions tended to have little or no activity [Kelly et al., 1969]. There seemed to be a fairly good correlation between the activity of an anion and its conductivity in water. The time required for an anion to reach electrical equilibrium was also fairly well correlated with its conductivity. It was also concluded that the activity of an anion was partly determined by its ability to interact with the relevant membrane recognition (or binding) sites [Kelly et al., 1969]. Thus many of the anions that promote the binding of $^3$H-flunitrazepam may be poor chloride substitutes electrophysiologically because of their slow movement through the "chloride channel" or their ability to antagonize chloride.

Since the absolute requirement of anions for the binding of benzodiazepine receptor ligands has not been studied extensively, it would be of interest to determine, for example, whether benzodiazepine binding to solubilized sites is also anion dependent.

Alone, NaCl can almost completely protect $^3$H-flunitrazepam binding sites from heat inactivation (30 min at 60°C) [Squires, 1981a; 1982; Squires and Saederup, 1982a] with 50% of maximum protection near 240 mM NaCl. This value can be reduced to about 38 mM by GABA (500 $\mu$M), which alone has no protective effect [Squires and Saederup, 1982b]. Similarly, picrotoxin also reduces the concentration of NaCl required to protect 50% of the $^3$H-flunitrazepam binding sites [Squires and Saederup, 1982b], an effect shared by several picrotoxin-like convulsants, including anisatin, isopropyl bicyclophosphate, isopropyl bicyclophosphorothionate, tetramethylene disulfotetramine, and *t*-butyl bicyclophosphate [Squires and Saederup, 1982a].

The protection of $^3$H-flunitrazepam binding sites by NaCl is characterized by a Hill coefficient between 2 and 3, suggesting that simultaneous occupation of recognition sites for both sodium and chloride ions may be required for protection. Similar protective effects are seen with LiCl, KCl, and CsCl. In contrast, sodium or potassium phosphate seems to protect maximally only about 70% of the $^3$H-flunitrazepam binding sites protected by NaCl. However, phosphate ion has a higher affinity for its binding site than chloride ion, with 50% of maximum protection at 50 mM sodium phosphate, pH 7.5, and a Hill number near one. GABA reduced the half-maximum protective concentration of phosphate from 50 mM to 14 mM and the Hill number to near 0.8 [Squires, 1982], again suggesting that phosphate ion, while not being able to substitute for chloride ion electrophysiologically, does bind to a physiologically relevant anion recognition site in the benzodiazepine/GABA/ion receptor complex.

Certain pyrazolopyridines (e.g., SQ 20,009 [etazolate] and SQ 65,396 [cartazolate]) enhance the binding of $^3$H-flunitrazepam or $^3$H-diazepam to brain-specific sites in the presence of chloride, bromide, iodide, nitrite, and nitrate, but not in the presence of fluoride, acetate, formate, or sulfate [Supavilai and Karobath, 1981]. Similarly, several barbiturates enhance $^3$H-diazepam binding in the presence of bromide, iodide, chloride, nitrate, thiocyanate, or formate, but not in the presence of sulfate, bicarbonate, fluoride, phosphate, and several other anions reported by Eccles [1964] to be unable to substitute for chloride electrophysiologically [Leeb-Lundberg et al., 1980]. The enhancement of $^3$H-flunitrazepam binding by the pyrazolpyridines in the presence of chloride is antagonized by picrotoxinin and isopropyl bicyclophosphate [Supavilai and Karobath, 1981]. Similarly, the enhancement of $^3$H-diazepam binding by pentobarbital in the presence of 200 mM NaCl is competitively reversed by picrotoxinin [Leeb-Lundberg et al., 1980]. The pyrazolopyridines, barbiturates, and picrotoxin-like convulsants all displace $^{35}$S-*t*-butyl bicyclophosphorothionate (TBPS) from brain-specific picrotoxinin binding sites [Squires et al., 1983].

$^{35}$S-TBPS binding, like $^3$H-flunitrazepam binding, is entirely dependent on ions, with anions again playing the predominant role. GABA and 15 $GABA_A$ receptor agonists (but not baclofen, a $GABA_B$ receptor agonist) completely block $^{35}$S-TBPS binding in the presence of 200 mM KBr, showing that all TBPS binding sites are coupled to $GABA_A$ receptors. GABA, at 0.5 $\mu$M, greatly increases the concentrations of bromide and chloride, but not sulfate, phosphate, fluoride, or bicarbonate, required for 50% of maximum binding (which varies, depending on the anion) [Squires et al., 1983]. Also, like $^3$H-flunitrazepam binding sites, $^{35}$S-TBPS binding sites are almost completely protected against heat inactivation (30 min, 60°C) by NaCl alone with half-maximal protection near 190 mM NaCl. In contrast to $^3$H-flunitrazepam binding sites, however, GABA has a biphasic effect on $^{35}$S-TBPS binding sites with potentiation of the protective effects of low (< 100 mM) NaCl concentrations and antagonism of higher (> 200 mM) NaCl concentrations [Squires et al., 1983]. Several ligands for benzodiazepine binding sites have little or no effect on $^{35}$S-TBPS binding to brain-specific sites. However, suriclone, a new cyclopyrrolone, which potently displaces $^3$H-flunitrazepam from brain-specific sites with a $K_i$ value near 1.6 nM [Blanchard and Julou, 1983], partially displaces $^{35}$S-TBPS from brain sites with great potency ($IC_{50}$ near 1 nM in 200 mM KBr and 40% maximum displacement of $^{35}$S-TBPS). This partial displacement by suriclone was seen in KBr and, to a smaller extent, in NaCl (maximum 20% displacement), but was absent in 200 mM NaF and $Na_2SO_4$. Suriclone slightly enhanced the binding of $^{35}$S-TBPS in 200 mM Tris citrate or 200 mM sodium phosphate pH 7.5 [Squires and Saederup, unpublished observations].

GABA has been reported to accelerate the uptake of $^{36}Cl^-$ ion into cultured chick cerebral neurons [Thampy and Barnes, 1984] and efflux from rat hippocampal slices [Wong et al., 1984]. Uptake into cultured chick cerebral neurons was increased 3-fold by GABA during the first 20 sec after adding GABA or muscimol, which produced 50% of their maximal effects at 6 $\mu$M and 8 $\mu$M, respectively. Half-maximal uptake of $^{36}Cl^-$ was obtained at a concentration of 105 mM. Bicuculline and picrotoxin both proved to be noncompetitive inhibitors of this GABA effect. About half of the basal uptake of $^{36}Cl^-$, in the absence of GABA, was blocked by the nonspecific anion transport inhibitors furosemide and 4-acetamido-4′-isothiocyano-2,2′stilbene sulfonate, substances which had little or no effect on GABA-induced $^{36}Cl^-$ uptake [Thampy and Barnes, 1984]. Efflux of $^{36}Cl^-$ from rat hippocampal slices was also increased by GABA, muscimol, and 3-aminopropane sulfonate, and these increases were blocked by bicuculline and picrotoxinin. Pentobarbital, 1,3 dimethylbutyl barbiturate, secobarbital, and (+)hexobarbital also increased $^{36}Cl^-$ efflux, and the effect of these barbiturates was also blocked by bicuculline and picrotoxinin. The stimulating effects of the barbiturates were potentiated by GABA. Phenobarbital was inactive alone, or in combination with GABA [Wong et al., 1984].

## COUPLING OF CATION RECOGNITION SITES TO GABA, BENZODIAZEPINE, AND PICROTOXIN RECOGNITION SITES

### Vertebrate Neurons

Mackerer and Kochman [1978] were the first to report the enhancing effect of several divalent cations on $^3$H-diazepam binding, with $Ni^{2+}$ producing the largest effect. $Ni^{2+}$ increased the affinity of $^3$H-diazepam for its binding sites on rat brain membranes, with no effect on the number of receptors ($B_{max}$). Squires [1981a] reported that $Ni^{2+}$ ion (1 or 2 mM) selectively antagonized the inhibitory effect of beta-carboline-3-carboxylate methyl ester on $^3$H-flunitrazepam binding. This effect of $Ni^{2+}$ ion was highly specific for beta-carboline-3-carboxylate esters, since the affinities of other GABA-neutral or GABA-negative ligands for benzodiazepine receptors, such as Ro 15-1788 [Hunkeler et al., 1983] and CGS 8216 [Czernik et al., 1982] are unaffected by $Ni^{2+}$ [Squires and Saederup, unpublished]. Hirsch and Kochman [1983] reported that $Ca^{2+}$ blocks both the enhancing effects of $Ni^{2+}$ on $^3$H-diazepam binding and the inhibitory effect of $Ni^{2+}$ on the binding of beta-carboline-3-carboxylate propyl ester. Lo and Snyder [1983] found that eight divalent cations ($Cu^{2+}$, $Ni^{2+}$, $Co^{2+}$, $Zn^{2+}$, $Mn^{2+}$, $Ca^{2+}$, $Ba^{2+}$, and $Mg^{2+}$) stimulated the binding of $^3$H-flunitrazepam to both type 1 and type 2 solubilized benzodiazepine sites. Mizuno et al. [1982] found that $Co^{2+}$, $Ni^{2+}$, $Cu^{2+}$, and $Zn^{2+}$ (but not $Li^+$, $Na^+$, $K^+$, $Mg^{2+}$, $Ca^{2+}$, $Cr^{3+}$, $Mn^{2+}$, $Fe^{2+}$, or $Fe^{3+}$) enhanced $^3$H-diazepam binding to sites in rat brain membranes prepared by Polytron homogenization. In the presence of 1 mM $Co^{2+}$, $Ni^{2+}$, $Cu^{2+}$, or $Zn^{2+}$, there were two distinct binding sites with $K_d$ values near 1 nM and 6 nM. The high-affinity site, which constituted about 20% of the $^3$H-diazepam binding sites, was not detected in the absence of the divalent cations.

Silver ion ($Ag^+$) was found to increase the affinity of $^3$H-muscimol for $GABA_A$ receptors coupled to benzodiazepine binding sites [Supavilai et al., 1982]. Three days after an oral dose of methyl mercury, the number of $^3$H-diazepam binding sites increased in rat cerebral cortex, cerebellum, and retina without affecting the number of $^3$H-GABA or $^3$H-spiperone binding sites [Corda et al., 1981]. $^3$H-flunitrazepam binding sites in EDTA/water–dialyzed rat forebrain $P_2$ membranes were protected against heat inactivation (60°C for 30 min) by a combination of $Ca^{2+}$ and GABA, but by neither substance alone. In the presence of saturating concentrations of GABA (500 $\mu$M), $CaCl_2$ produced 50% of its maximal protective effect near 600 $\mu$M. Conversely, in the presence of saturating concentrations of $CaCl_2$ (5 mM), GABA produced 50% of its maximal protective effect near 20 $\mu$M. Ten $GABA_A$ receptor agonists (but not the $GABA_B$ receptor agonist baclofen) could substitute for GABA with different efficacies and 50% protective concentrations ranging from 5 $\mu$M (muscimol) to 150 $\mu$M (imidazole acetate). Four divalent cations could substitute for $Ca^{2+}$ ($Mg^{2+}$, $Ba^{2+}$, $Sr^{2+}$, and $Mn^{2+}$) with half-maximal protective concentrations ranging from 200 $\mu$M ($Mn^{2+}$) to 930 $\mu$M ($Ba^{2+}$). Eight other di- or tri-valent cations were inactive and antagonized the protective effect of $Ca^{2+}$ in mainly noncompetitive ways. Time courses of heat inactivation at 60°C in the presence of saturating $CaCl_2$ (5 mM) and $GABA_A$ receptor agonists were all polyphasic, with slow components ranging from 42–48% (THIP, isoguvacine, and piperidine-4-sulfonate) to 66–69% (muscimol, GABA). Picrotoxin enhanced the protective effects of $CaCl_2$ and NaCl [Squires and Saederup, 1982b].

CGS 8216, a pyrazoloquinoline GABA-neutral, or slightly GABA-negative, ligand for benzodiazepine binding sites, completely displaces $^3$H-flunitrazepam from its binding sites with an $IC_{50}$ value near 0.3 nM [Czernik et al., 1982]. However, a comparison of Scatchard plots of $^3$H-flunitrazepam and $^3$H-CGS 8216 binding revealed that the total number of binding sites for $^3$H-flunitrazepam was greater than the total number of $^3$H-CGS 8216 binding sites, suggesting that the occupation of one binding site by CGS 8216 can block more than one flunitrazepam binding site. Further, Scatchard analysis of $^3$H-flunitrazepam binding in the presence of two CGS 8216 concentrations (0.2 and 0.4 nM) revealed mixed competitive, noncompetitive inhibition (i.e., CGS 8216 decreases the $B_{max}$ values for $^3$H-flunitrazepam and simultaneously increases its $K_d$ values) [Czernik et al., 1982].

Protection of $^3$H-CGS 8216 sites against heat inactivation (60°C, 30 min) revealed that

NaCl protected $^3$H-CGS 8216 and $^3$H-flunitrazepam binding sites in almost identical fashion: 50% maximal protection with 250 mM NaCl and Hill numbers near 2.0. Both GABA and picrotoxin reduced the 50% effective NaCl concentration to the same extent [Squires and Saederup, 1982a]. However, a comparison of the protection provided by GABA and $CaCl_2$ of $^3$H-CGS 8216 and $^3$H-flunitrazepam binding sites, respectively, revealed some interesting differences: the protection of $^3$H-CGS 8216 sites provided by $CaCl_2$ plus GABA was greater than the protection of $^3$H-flunitrazepam sites, expressed as a percentage of each ligand's own unheated control binding. Time courses of heat inactivation at 60°C in the presence of saturating $CaCl_2$ (5 mM) and GABA (500 μM) revealed that the time course for $^3$H-CGS 8216 was almost monophasic in contrast to the polyphasic time course for $^3$H-flunitrazepam reported earlier [Squires and Saederup, 1982b]. The half-life of the $^3$H-CGS 8216 sites (96 min at 60°C) was almost identical to the slow component of the $^3$H-flunitrazepam time course. The "fast" component of the CGS 8216 time course constituted at most 5% of total binding at zero-time, in contrast to 31% for $^3$H-flunitrazepam. It can therefore be concluded that $^3$H-CGS 8216 binds preferentially to a subset of flunitrazepam binding sites which are selectively protected by GABA plus $CaCl_2$ [Squires and Saederup, 1982a].

Diazepam and GABA individually inhibit $Ca^{2+}$-dependent, $K^+$-induced release of $^3$H-serotonin from rat hippocampal synaptosomes, and diazepam enhanced the inhibitory effect of GABA on $K^+$-induced $^3$H-serotonin release [Balfour, 1980]. GABA also inhibited the $Ca^{2+}$-dependent $K^+$-evoked release of $^3$H-glycine from rat substantia nigra slices, an effect which was reversed by picrotoxin [Kerwin and Pycock, 1980]. GABA and isoguvacine *increased* the spontaneous release of $^3$H-norepinephrine from rat hippocampal synaptosomes, effects which were blocked by bicuculline and picrotoxinin, while baclofen and chlordiazepoxide alone had no effect. The $Ca^{2+}$-dependent, $K^+$-evoked release of $^3$H-norepinephrine from the same synaptosomal preparation was inhibited by GABA, baclofen, and chlordiazepoxide, but not by isoguvacine. The inhibitory effects of GABA here were not reversed by bicuculline, picrotoxin, or chlordiazepoxide, which also failed to potentiate the inhibitory effect of GABA [Fung and Fillenz, 1983]. The $Ca^{2+}$-dependent, potassium-induced release of $^3$H-GABA from preloaded slices of rat cerebral cortex was inhibited in a concentration-dependent way by the $GABA_A$ receptor agonists muscimol and THIP, which presumably act on GABA autoreceptors [Brennan, 1982]. The inhibitory effects of muscimol and THIP were prevented by picrotoxin and bicuculline but were not potentiated by flunitrazepam, indicating that GABA autoreceptors, which have several of the properties of $GABA_A$ receptors, are *not* coupled to benzodiazepine receptors [Brennan, 1982]. The $Ca^{2+}$-dependent release of somatostatin from primary cultures of dispersed hypothalamic cells *in vitro* was inhibited by GABA in a bicuculline-reversible way [Gamse et al., 1980]. The $Ca^{2+}$-dependent $K^+$-evoked release of $^3$H-D-aspartate (a non-metabolizable glutamate substitute) from cerebellar granule cells in primary culture *in vitro* was inhibited by the $GABA_A$ receptor agonists GABA, muscimol, and THIP (but not by baclofen) in a bicuculline-sensitive way [Schousboe et al., 1984].

Cutler et al. [1974] reported that pentobarbital and amobarbital (but not hexobarbital or phenobarbital) inhibited the spontaneous, as well as the ouabain-induced and potassium-induced release of $^3$H-GABA from preloaded rat cerebral cortex slices. Similarly, pentobarbital inhibits $Ca^{2+}$-dependent, $K^+$-evoked release of $^3$H-norepinephrine and $^3$H-GABA from mouse forebrain synaptosomes [Haycock et al, 1977]. Four anesthetics, including pentobarbital, inhibited the $Ca^{2+}$-dependent, $K^+$-induced release of GABA and D-aspartate from rat brain slices [Minchin, 1981]. The convulsant barbiturate CHEB was found to inhibit the $Ca^{2+}$-dependent, potassium-induced release of $^{14}$C-GABA and stimulate potassium-induced release of 3H-D-aspartate from preloaded rat brain slices, while phenobarbital inhibited potassium-induced release of $^{3}$H-D-aspartate more than that of $^{14}$C-GABA [Skerritt et al., 1983].

Tozalakian and Csillag [1982] reported that GABA reduced the uptake of $K^+$ by rat cortical synaptosomes, decreased the synaptosomal content of $K^+$ and increased the $Na^+$ content. When rat hippocampal slices are perfused with a low $Ca^{2+}$ medium (0.2 mM), synaptic transmission from Schaffer collateral/commissural fibers is blocked, but electrical stimulation of afferent fibers resulted in a decrease in the external $Ca^{2+}$ concentration that can be blocked by GABA [Konnerth and Heinemann, 1983]. Barbiturates and benzodiazepines have also been reported to inhibit $Ca^{2+}$ uptake into synaptosomes from rabbit and mouse brains [Friedman et al., 1979], as well as rat brain [Blaustein and Ector, 1975; Friedman et al., 1979; Leslie et al., 1980a,b; Elrod and Leslie, 1980; Pincus and Hsiao, 1981]. Taft and DeLorenzo [1984] reported that high micromolar concentrations (100–200 $\mu$M) of diazepam and Ro-4864 inhibit $K^+$ or veratridine-induced uptake of $^{45}Ca^{2+}$ into rat brain synaptosomes. Veratridine-induced, but not $K^+$-induced, $^{45}Ca^{2+}$ uptake could be blocked by tetrodotoxin. $K^+$-induced $^{45}Ca^{2+}$ uptake was also inhibited by pre-photolabeling the "micromolar" benzodiazepine site with flunitrazepam (600 $\mu$M), as well as by $Mn^{2+}$, $Co^{2+}$, verapamil, nimodipine, and nitrendipine.

The binding of $^{3}$H-GABA or $^{3}$H-baclofen to $GABA_B$ receptors is $Ca^{2+}$- or $Mg^{2+}$-dependent [Hill and Bowery, 1981]. Kato et al. [1983], confirming and extending these findings, reported that $Mn^{2+}$ and $Ni^{2+}$ promoted $^{3}$H-baclofen binding to a greater extent than $Ca^{2+}$ and $Mg^{2+}$, which were, in turn, more efficacious than $Sr^{2+}$ or $Ba^{2+}$. Corda and Guidotti [1983] found that $Ca^{2+}$ decreased the binding of $^{3}$H-GABA and $^{3}$H-muscimol to $GABA_A$ receptors while increasing the binding of $^{3}$H-GABA to $GABA_B$ receptors. While GABA and (−)-baclofen inhibit the release of various neurotransmitters from peripheral [Bowery et al., 1981] and central neurons [Potashner, 1979; Bowery et al., 1980; Collins et al., 1982] by a bicuculline-resistant mechanism, it is also known that baclofen, applied iontophoretically, depresses neuronal activity [Davies and Watkins, 1974; Curtis et al., 1974; Olpe et al., 1978]. A recent study [Newberry and Nicoll, 1984] demonstrated that (−)-baclofen, acting on bicuculline-resistant, postsynaptic GABA receptors, located on the soma and dendrites of rat hippocampal $CA^1$ pyramidal cells, causes slow neuronal hyperpolarization and inhibition, probably associated with increases in $K^+$ conductance with no effect on chloride conductance. GABA was reported to decrease $Ca^{2+}$ "spikes" in tetrodotoxin-blocked neurons from chick dorsal root ganglia grown in tissue culture [Dunlap and Fischbach, 1978, 1981]. Using the same neuronal preparation, Dunlap [1981] found that both GABA and baclofen reduced the duration of $Ca^{2+}$-dependent action potentials in a bicuculline-resistant way.

The depolarization of desheathed frog sciatic nerves by GABA is accompanied by an increase in the extracellular $K^+$ concentration. Dorsal root fibers are much more sensitive to GABA than ventral root fibers [Morris et al., 1983].

Concentrations of caffeine ($\leqslant$ 50 $\mu$M) which have little effect of their own potentiate the depolarizing effect of GABA on dorsal root fibers of the frog spinal cord. This potentiating effect of caffeine is blocked by flurazepam and the $Ca^{2+}$ antagonists $Mn^{2+}$ and $Cd^{2+}$, but not by bicuculline. Higher concentrations of caffeine (near 1 mM) antagonized GABA responses in a noncompetitive manner [Nistri and Berti, 1983]. This demonstration of caffeine interactions with GABA receptors and cation recognition sites is consistent with the specific localization of $^{3}$H-cyclohexyladenosine binding sites to granule cells in cerebellum [Goodman et al., 1983; Wojcik and Neff, 1983], a cell type also known to possess high densities of $GABA_A$ receptors [Palacios et al., 1981] and binding sites for $^{35}S$-TBPS [Wamsley et al., 1983] which are also known to be coupled to $GABA_A$ receptors [Squires et al., 1983]. It is also interesting, in this respect, that the inhibitory actions of GABA on $^{35}S$-TBPS binding are reversed by theophylline and caffeine [Squires and Saederup, unpublished].

Application of GABA to cat pelvic vesicle ganglia *in vitro* produces a biphasic response: depolarization associated with a conductance increase, followed by hyperpolarization associated with a decrease in conductance [Mayer et al., 1983]. Ion substitution experiments suggested that the conductance decrease is to both chloride ion and other, as yet unidentified, ions.

Spontaneous action potentials in guinea-pig cerebellar cortex slices were inhibited by GABA and related amino acids [Okamoto and Quastel, 1976; Okamoto et al., 1976]. Inhibition brought about by GABA was associated with increased $Cl^-$ and, perhaps, $K^+$ permeability [Okamoto et al., 1976]. Chlordiazepoxide enhanced the inhibitory action of GABA, perhaps by increasing the membrane permeability to $K^+$, but not to $Cl^-$ [Okamoto and Sakai, 1979]. The inhibitory action of GABA was reversed by picrotoxin, and the effect of picrotoxin was blocked by chlordiazepoxide [Okamoto and Sakai, 1979]. GABA was also found to inhibit $Ca^{2+}$ spikes in Purkinje cell dendrites [Okamoto et al, 1983].

GABA depolarized the dorsal roots of frog spinal cord and reduced dorsal root potentials in a way antagonized by bicuculline and picrotoxin. The depolarizing response to GABA was not affected by replacing chloride with impermeant isethionate or sulfate ions, but was abolished by replacing external sodium ion with $Li^+$, choline, or Tris [Barker and Nicoll, 1973].

When applied to the dendrites, GABA also partially depolarizes and inhibits hippocampal pyramidal cells, in contrast to the hyperpolarizing (also inhibitory) effect of GABA when applied to the perikarya [Alger and Nicoll, 1979; Andersen et al., 1980]. Both hyperpolarizing and depolarizing effects of GABA on pyramidal cells are potentiated by the benzodiazepine midazolam [Jahnsen and Laursen, 1981]. In contrast to the *hyperpolarizing* response to GABA, where fluctuation analysis shows a minimum of current fluctuations when the neuron is voltage-clamped at the reversal potential for the GABA effect (which in this case corresponds to the chloride ion potential) the *depolarizing* response to GABA shows no minimum in current fluctuations at the reversal potential, indicating that more than one ion is involved. One of the ions is chloride; the other, probably a cation, was not identified. It did not seem to be $Ca^{2+}$, since the partial depolarizing response to GABA was not blocked by $Mn^{2+}$ or $Mg^{2+}$, which were assumed to be $Ca^{2+}$ antagonists [Djorup et al., 1981]. Substitution of $Na^+$ with choline also failed to affect the GABA response. However, it could not be ruled out that $Mn^{2+}$, $Mg^{2+}$, or choline could pass through GABA-regulated cation channels in the pyramidal cell dendrite membrane.

Using guinea-pig hippocampal slices *in vitro*, it was found that the GABA-negative benzodiazepine Ro 14-7437, when applied to CA1 cells in nanomolar concentrations, caused depolarization, increased spontaneous firing, and a decrease in conductance which were resistant to injection of $Cl^-$ ions, midazolam, or tetrodotoxin application, but were prevented by removing $Ca^{2+}$ from the medium. It was suggested that Ro 14-7437 acted by blocking a $Ca^{2+}$-mediated, $K^+$ conductance [Carlen et al., 1983a]. The GABA-positive benzodiazepine, midazolam, when applied to CA1 hippocampal neurons in the same system, produced long-lasting hyperpolarization and conductance increases which were not affected by tetrodotoxin nor intracellular $Cl^-$ injection, but were blocked by removing $Ca^{2+}$ from the medium. $Ca^{2+}$ spikes in the presence of tetrodotoxin were enhanced by midazolam. It was concluded that low nanomolar concentrations of midazolam, which did not potentiate GABA effects, inhibited neuronal firing by enhancing $Ca^{2+}$-mediated,

$K^+$-conductance [Carlen et al., 1983b].

In myelinated nerve fibers from frog, flurazepam (250 μM) decreased the inward $Na^+$ current and decreased the time required for the $K^+$ current to reach a sustained outward movement [Schwartz and Spielmann, 1983].

There is good evidence for the presence of GABA-containing neurons in the myenteric plexus of the mammalian gut [Jessen et al., 1979; Taniyama et al., 1982]. GABA, muscimol, and 3-aminopropane sulfonate, when applied *in vitro* to guinea-pig ileum myenteric plexus–longitudinal muscle preparations, produced a transient contraction followed by relaxation. Both effects were blocked by atropine, tetrodotoxin, bicuculline, and picrotoxin. On the other hand, GABA and baclofen, but not 3-amino propane sulfonate or muscimol, could relax longitudinal muscle contracted by field stimulation. These relaxing effects of GABA and baclofen were not antagonized by bicuculline, picrotoxin, nor by antagonists of norepinephrine, acetylcholine, or histamine. It was concluded that GABA acts on two types of receptors located on cholinergic neurons in the myenteric plexus: one excitatory (mediating the release of acetylcholine), the other inhibitory (inhibiting the release of acetylcholine) [Kaplita et al., 1982; Kleinrok and Kilbinger, 1983]. The contraction of longitudinal muscle was clearly shown to be due to acetylcholine release. The excitatory effect of GABA on the cholinergic neurons of the myenteric plexus probably represents an action on a novel "$GABA_A$" receptor [Kleinrok and Kilbinger, 1983]. It seems likely that GABA, acting on this receptor, will open a $Na^+$ channel.

GABA and muscimol were reported to stimulate the release of melanocyte-stimulating hormone from pituitary pars intermedia cells maintained *in vitro*, in a $Ca^{2+}$-dependent way, but block the $Ca^{2+}$-dependent, $K^+$-evoked release [Tomiko et al., 1983]. These effects of GABA and muscimol were completely blocked by bicuculline. Demenix et al. [1984] reported that GABA had a biphasic effect on the release of alpha-melanocyte stimulatory hormone from intact rat neurointermediate lobes of the pituitary, with enhancement of release followed by inhibition. The enhancement, but not the inhibition, was blocked by bicuculline, while (−)-baclofen reduced basal and $Ba^{2+}$-induced release. Isoguvacine and GABA potentiated $Ba^{2+}$-induced release in a bicuculline-sensitive way. It was concluded that both $GABA_A$ and $GABA_B$ receptors are present on the endocrine cells of the intermediate lobe.

## CATION EFFECTS MEDIATED BY GABA, BENZODIAZEPINE OR BARBITURATE BINDING SITES IN NON-NEURONAL SYSTEMS

GABA and muscimol were reported to produce dose-dependent depolarizations in about one third of oligodendrocytes in cultures of mouse spinal cord. Replacement of $Na^+$ in the medium completely blocked this effect of GABA, which was also reduced by bicuculline, picrotoxin, pentobarbital, and chlordiazepoxide [Gilbert et al., 1984]. It is of interest that some oligodendrocytes respond to GABA alone, others to glutamate alone, while a third population responded to both GABA and glutamate.

Micromolar concentrations of benzodiazepines affect the passage of both $K^+$ and $Ca^{2+}$ ions across smooth and striate muscle membranes. Ishii et al. [1982] found that diazepam and flurazepam inhibited the tonic component more than the phasic component of the potassium ion-induced contracture of guinea-pig taenia coli, and also inhibited the $Ca^{2+}$-induced contracture of longitudinal muscle excised from the ileum. It was concluded that the two benzodiazepines inhibited the transmembrane influx of $Ca^{2+}$. Hullihan et al. [1983] confirmed and extended these results using longitudinal muscle–myenteric plexus strip preparations of guinea-pig ileum. Diazepam antagonized the contractions of this preparation induced by $K^+$, $Ca^{2+}$, histamine, and carbachol. The inhibition of $Ca^{2+}$-induced contractions was reversed by increasing the concentration of $Ca^{2+}$ in the medium, indicating a reversible blockade of calcium channels in the cell membranes by diazepam.

Diazepam (300 μM) was also reported to inhibit the influx of $^{45}Ca^{2+}$ in rat skeletal muscle *in vitro* [Bihler and Sawh, 1978], to potentiate electrically induced twitches in isolated frog muscle [Degroof et al, 1980; Khan and Edman 1983] and to retard the relaxation of the muscle after electrically induced contraction [Khan and Edman, 1983], an effect which was not antagonized by Ro 15-1788, a benzodiazepine which antagonizes the behavioral and neuronal effects of diazepam [Khan, 1983; Hunkeler et al., 1981]. Diazepam (50 μM) increased the rate of $^{45}Ca^{2+}$ efflux from frog muscle but had no effect on the overall $Ca^{2+}$ content of the muscle tissue [Degroof et al., 1980]. The effects of benzodiazepines on non-neuronal elements are probably mediated by "peripheral-" or "kidney-type" binding sites [Braestrup and Squires, 1977]. Recently, Cantor et al. [1984] have found the binding of $^3$H-Ro 5-4864 (a ligand with high affinity for "peripheral-type" binding sites), to heart, kidney, and brain membranes, was competitively inhibited by the dihydropyridine "calcium channel" blockers nitrendipine and nifedipine, but not by diltiazem and verapamil, which are known to bind to another, allosteric, site in the "calcium channel" complex [Murphy et al., 1983]. Nitrendipine and nifedipine did not block the binding of $^3$H-flurazepam or $^3$H-clonazepam, ligands with high affinity for neuronal binding sites [Cantor et al., 1984].

Ro 5-4864 shared with the calcium channel blockers the ability to decrease the duration of the action potential in isolated guinea-pig papillary muscle and to decrease contractility. This effect of Ro 5-4864 is not affected by GABA and is antagonized by PK 11195 (another ligand with high affinity for peripheral benzodiazepine receptors), as well as by addition of $CaCl_2$. PK 11195 also antagonized the *increase* in the duration of the papillary muscle action potential caused by the calcium channel activator BAY K-8644 [Mestre et al., 1984; LeFur et al., 1984]. Thus, "peripheral" benzodiazepine receptors, which are not coupled to GABA receptors, seem to be coupled to $Ca^{2+}$ "channels," a property they share with at least some GABA receptors. It seems possible that both peripheral benzodiazepine receptors and GABA receptors could have evolved from a common "ancestral" cation channel.

## EFFECTS OF GABA, BARBITURATES, AND BENZODIAZEPINES ON CATION FLUXES IN INVERTEBRATE NERVE AND MUSCLE

GABA is a neurotransmitter in many invertebrate species [Gershenfeld, 1973]. Of five GABA receptors on *Aplysia* neurons [Yarowsky and Carpenter, 1978], three are excitatory and two inhibitory. The excitatory responses to GABA are associated with depolarization. One excitatory response is associated with an increase in inward $Na^+$ conductance. A second is associated with increases in both $Na^+$ and $Cl^-$ conductances, while a third excitatory response was possibly due to a decrease in $K^+$ conductance. Of the two inhibitory responses to GABA, one is associated with fast hyperpolarization and an increase in chloride ion conductance, while the other is associated with slow hyperpolarization and an increase in $K^+$ conductance. The GABA receptors are pharmacologically distinct. Curare selectively blocks fast $Na^+$ and $Cl^-$ channels, while picrotoxin and bicuculline block only fast $Cl^-$ channels. Three of the five GABA receptors were not blocked by any of the drugs tested. This study demonstrates that GABA can regulate five separate ion channels in the neuronal membranes of *Aplysia* [Yarowsky and Carpenter, 1978].

Flurazepam and chlordiazepoxide in the 20 μM–200 μM range, as well as GABA, suppress the excitatory postsynaptic potentials produced by a cholinergic synapse in the abdominal ganglia of *Aplysia* [Tremblay and Grenon, 1980]. The mechanism appeared to be presynaptic. Although GABA and the two benzodiazepines produced similar effects, the benzodiazepines did not potentiate the action of GABA. The ionic mechanism involved is not known, but since presynaptic inhibition of acetylcholine release presumably occurs, inhibition of $Ca^{2+}$ influx may be involved.

The increase in chloride ion conductance induced in lobster muscle by GABA can be selectively inhibited by $Zn^{2+}$ acting on a site separate from the GABA site as well as the picrotoxinin site. Other divalent cations tested ($Mn^{2+}$, $Fe^{2+}$, $Co^{2+}$, $Ni^{2+}$, $Cu^{2+}$) were essentially inactive, except for $Cu^{2+}$ which was weakly active [Smart and Constanti, 1982].

A benzodiazepine Ro 11-3128 which has behavioral effects in mammals similar to diazepam [Hunkeler et al., 1981], produces spastic paralysis in schistosomes by a $Ca^{2+}$-dependent mechanism [Pax et al., 1978]. Ro 11-3128 decreases the uptake of $K^+$ but stimulates the uptake of $Ca^{2+}$ into schistosomes.

Flurazepam (50 $\mu$M–1 mM), applied externally to voltage-clamped squid giant axons, partially inhibited both $Na^+$ and $K^+$ conductances, with $K^+$ being more potently affected than $Na^+$ [Swenson, 1982]. Flurazepam appeared to block $K^+$-channels only when they were open, since repetitive stimulation of the nerve resulted in a cumulative inhibition. Flurazepam was more active when applied inside the axon.

It is of interest that pentobarbital [Barker, 1975] and flurazepam [Nistri and Constanti, 1978] both block $Na^+$-dependent, postsynaptic excitatory responses to glutamate in crustacean muscle without affecting GABA-induced increases in chloride conductance.

## CONCLUSIONS

Several types of GABA receptors, in both invertebrate and vertebrate species, regulate cation channels, and some may regulate both anion (chloride) and cation channels. There is some evidence that both GABA and glutamate receptors may have evolved from the cation channels of unicellular organisms, similar to those described in *Paramecia* [Kung and Saimi, 1982; Eckert and Brehm, 1979; Eckert et al., 1981]. In *Paramecia* certain stimuli, such as touch, result in depolarizing receptor potentials, leading to an action potential which, in turn, opens a $Ca^{2+}$ channel and increases an inward $Ca^{2+}$ flux in the cilia, causing a reversal of ciliary beating and locomotion away from the stimulus [Kung and Saimi, 1982]. Of the six or eight ion channels characterized in *Paramecium* membranes, all are cation channels (for $Ca^{2+}$, $K^+$, and $Na^+$). Receptors for some mammalian neurotransmitters have been found in protozoa: epinephrine and serotonin stimulate adenylate cyclase in *Tetrahymena pyriformis*, and the response to norepinephrine can be selectively blocked by propranolol [Rosensweig and Kindler, 1972]; opiates and opioid peptides alter the feeding behavior of amebas, effects which can be blocked by naloxone [Josefsson and Johansson, 1979]. The existence of GABA recognition sites in protozoa is suggested by our recent finding that emetine (an amebacide), amicarbalide (a babesiacide), pentamidine, propamidine, quinacrine (general antiprotozoals), isometamidium, suramine, and trypan red (antitrypanosomals), reverse the inhibitory effect of GABA on $^{35}$S-TBPS binding [Squires and Saederup, unpublished]. The postulated ancestral GABA recognition site could be a highly conserved amino acid sequence required for the opening and closing of one or more cation channels in the protozoan membrane, which modulate membrane excitability. The $GABA_B$ receptors, located on peripheral nerve terminals where there are no GABA synapses, would thus be similar in this respect to postulated GABA receptors in protozoa. That the ancestral GABA recognition site was associated with a cation, rather than a chloride channel, is suggested by the findings that many bacteria and protozoans [Holz et al., 1962] grow and function normally in essentially chloride-free media, and that the excitability of *Paramecia* membranes is modulated entirely by cation channels [Eckert and Brehm, 1979; Kung and Saimi, 1982].

The ability of GABA and other $GABA_A$ receptor agonists to reduce the $Ca^{2+}$-dependent $K^+$-evoked release of several neurotransmitters and hormones suggests that GABA, acting on "$GABA_A$" receptors, can block the influx of $Ca^{2+}$ into nerve terminals, and pituitary cells containing peptide hormones. This conclusion is supported by the findings that GABA can block "calcium spikes," and that barbiturates and benzodiazepines can block calcium fluxes.

In *Aplysia,* three out of five neuronal responses to GABA are excitatory, while in vertebrate systems the majority of GABA responses are inhibitory. However, a few excitatory responses to GABA have been reported in mammalian systems: GABA stimulates the release of melanocyte-stimulating hormone and prolactin from pituitary, as well as the release of acetylcholine from myenteric plexus neurons. These excitatory responses are presumably due to GABA-induced changes in cation fluxes. The slow hyperpolarization of hippocampal pyramidal cells produced by (−)-baclofen and associated with an increase in $K^+$ conductance has properties remarkably similar to the synaptically evoked dendritic slow inhibitory postsynaptic potentials, which is bicuculline-resistant and considered to be due to a calcium-activated potassium conductance [Newberry and Nicoll, 1984]. One of the inhibitory responses to GABA in *Aplysia* neurons is associated with slow hyperpolarization and an increase in potassium conductance [Yarowsky and Carpenter, 1978]. *Paramecia* also display a slow $Ca^{2+}$-induced, potassium currents [Kung and Saimi, 1982]. The slow activation of a potassium conductance by (−)-baclofen responses may be due to enhancement of neurotransmitter stimulation of adenylate cyclase [Enna and Karbon, 1984; Hill and Dolphin, 1984].

## REFERENCES

Adams PR, Brown DA (1973): Action of γ-aminobutyric acid (GABA) on rat sympathetic ganglion cells. Br J. Pharmacol 47:639–640P.

Andersen P, Dingledine R, Gjerstad L, Langmoen IA, Laursen AM (1980): Two different responses of hippocampal pyramidal cells to application of γ-aminobutyric acid. J. Physiol 305:279–296.

Anderson RA, Mitchell R (1984): Benzodiazepines potentiate the effect of muscimol on prolactin secretion in vitro. Br J Pharmacol 82:343P.

Alger BE, Nicoll RA (1979): GABA-mediated biphasic inhibitory responses in hippocampus. Nature 281:315–317.

Ault B, Evans RH (1981): The depressant action of baclofen on the isolated spinal cord of the neonatal rat. Eur J Pharmacol 71:357–364.

Balfour DJK (1980): Effects of GABA and diazepam on $^3$H-serotonin release from hippocampal synaptosomes. Eur J Pharmacol 68:11–16.

Barker JL (1975): CNS depressants: Effects on postsynaptic pharmacology. Brian Res 92:35–55.

Barker JL, Nicoll RA (1973): The pharmacology and ionic dependency of amino acid responses in the frog spinal cord. J. Physiol 228:259–277.

Barker JL, Ransom BR (1978): Amino acid pharmacology of mammalian central neurones grown in tissue culture. J. Physiol 280:331–354.

Bihler I, Sawh PC (1978): Effects of benzodiazepines on the transport of sugars and ions in rat skeletal muscle in vitro. Mol Pharmacol 14:879–883.

Blanchard J-C, Julou L (1983): Suriclone: A new cyclopyrrolone derivative recognizing receptors labeled by benzodiazepines in rat hippocampus and cerebellum. J. Neurochem 40:601–607.

Blaustein MP, Ector AC (1975): Barbiturate inhibition of calcium uptake by depolarized nerve terminals in vitro. Mol Pharmacol 11:369–378.

Bowery NG, Doble A, Hill DR, Hudson AL, Shaw JS, Turnbull MJ, Warrington R (1981): Bicuculline-insensitive GABA receptors on peripheral autonomic nerve terminals. Eur J Pharmacol 71:53–70.

Bowery NG, Hill DR, Hudson AL, Doble A, Middlemiss DN, Shaw J, Turnbull M (1980): (−)Baclofen decreases neurotransmitter release in the mammalian CNS by an action at a novel GABA receptor. Nature 283:92–94.

Braestrup C, Squires RF (1977): Specific benzodiazepine receptors in rat brain characterized by high affinity $^3$H-diazepam binding. Proc Natl Acad Sci USA 74:3805–3809.

Brennan MJW (1982): GABA autoreceptors are not coupled to benzodiazepine receptors in rat cerebral cortex. J Neurochem 38:264–266.

Cantor EH, Kenessey A, Semenuk G, Spector S (1984): Interaction of calcium channel blockers with non-neuronal benzodiazepine binding sites. Proc Natl Acad Sci USA 81:1549–1552.

Carlen PL, Gurevich N, Polc P (1983a): The excitatory effects of the specific benzodiazepine antagonist Ro14-7437, measured intracellularly in hippocampal CA1 cells. Brain Res 271:115–119.

Carlen PL, Gurevich N, Polc P (1983b): Low-dose benzodiazepine neuronal inhibition: enhanced $Ca^{2+}$-mediated $K^+$-conductance. Brain Res 271:358–364.

Collins GGS, Anson J, Kelly EP (1982): Baclofen: Effects on evoked field potentials and amino acid neurotransmitter release in the rat olfactory cortex slice. Brain Res 238:371–383.
Constanti A, Nistri A (1976): A comparative study of the action of $\gamma$-aminobutyric acid and piperazine on the lobster muscle fibre and the frog spinal cord. Br J Pharmacol 57:347–358.
Constanti A, Quilliam JP (1974): A comparison of the effects of GABA and imidazoleacetic acid on the membrane conductance of lobster muscle fibres. Brain Res 79:306–310.
Corda MG, Concas A, Rossetti Z, Guarneri P, Corongiu FP, Biggio G (1981): Methyl mercury enhances [$^3$H]diazepam binding in different areas of the rat brain. Brain Res 229:264–269.
Corda MG, Guidotti A (1983): Modulation of GABA receptor binding by $Ca^{2+}$. J Neurochem 41:277–280.
Costa T, Rodbard D, Pert CB (1979): Is the benzodiazepine receptor coupled to a chloride anion channel? Nature 277:315–317.
Curtis DR, Game CJA, Johnston GAR, McCulloch RM (1974): Central effects of $\beta$-(p-chlorophenyl)-$\gamma$-aminobutyric acid. Brain Res 70:493–499.
Cutler RWP, Markowitz D, Dudzinski DS (1974): The effect of barbiturates on [$^3$H]GABA transport in rat cerebral cortex slices. Brain Res 81:189–197.
Czernik AJ, Petrack B, Kalinsky HJ, Psychoyos S, Cash WD, Tsai C, Rinehart RK, Grant FR, Lovell RA, Brundish DE, Wade R (1982): CGS 8216: receptor binding characteristics of a potent benzodiazepine antagonist. Life Sci 30:363–372.
Davies J, Watkins JC (1974): The action of $\beta$-phenyl-GABA derivatives on neurones of the cat cerebral cortex. Brain Res 70:501–505.
Degroof RC, Bianchi CP, Narayan S (1980): The effect of diazepam on tension and electrolyte distribution in frog muscle. Eur J Pharmacol 66:193–199.
Demeneix BA, Desaulles E, Feltz P, and Loeffler JP (1984): Dual population of $GABA_a$ and $GABA_b$ receptors in rat pars intermedia demonstrated by release of MSH caused by barium ions. Br J Pharmacol 82:183–190.
Djørup A, Jahnsen H, Laursen AM (1981): The dendritic response to GABA in CA1 of the hippocampal slice. Brain Res 219:196–201.
Dudel J, Hatt H (1976): Four types of GABA receptors in crayfish leg muscles characterized by desensitization and specific antagonist. Pflugers Arch 364:217–222.
Dunlap K (1981): Two types of $\gamma$-aminobutyric acid receptor on embryonic sensory neurones. Br J Pharmacol 74:579–585.
Dunlap K, Fischbach GD (1978): Neurotransmitters decrease the calcium component of sensory neurone action potentials. Nature 276:837–839.
Dunlap K, Fischbach GD (1981): Neurotransmitters decrease the calcium conductance activated by depolarization of embryonic chick sensory neurones. J Physiol 317:519–535.
Eccles JC (1964): "The Physiology of Synapses" Academic Press, New York.
Eckert R, Brehm P (1979): Ionic mechanisms of excitation in *paramecium*. Ann Rev Biophys Bioeng 8:353–383.
Eckert R, Tilllotson DL, Brehm P (1981): Calcium-mediated control of Ca and K currents. Fed Proc 40:2226–2232.
Elrod SV, Leslie SW (1980): Acute and chronic effects of barbiturates on depolarization-induced calcium influx into synaptosomes from rat brain regions. J Pharmacol Exp Ther 212:131–136.
Enna SJ, Karbon EW (1984): $GABA_B$ receptors and transmitter-stimulated cAMP accumulation in rat brain. Neuropharmacology 23:821–822.
Friedman MB, Coleman R, Leslie SW (1979): Barbiturate depression of calcium-mediated stimulus-secretion coupling in synaptosomes: A species and strain comparison. Life Sci 25:735–738.
Fung S-C, Fillenz M (1983): The role of pre-synaptic GABA and benzodiazepine receptors in the control of noradrenaline release in rat hippocampus. Neuroscience Letters 42:61–66.
Gallagher JP, Higashi H, Nishi S (1978): Characterization and ionic basis of GABA-induced depolarizations recorded *in vitro* from cat primary afferent neurones. J. Physiol 275:263–282.
Gamse R, Vaccaro DE, Gamse G, DiPace M, Fox TO, Leeman SE (1980): Release of immunoreactive somatostatin from hypothalamic cells in culture; Inhibition by $\gamma$-aminobutyric acid. Proc Natl Acad Sci USA 77:5552–5556.
Gerschenfeld HM (1973): Chemical transmission in invertebrate central nervous systems and neuromuscular junctions. Physiological Reviews 53:1–119.
Gilbert P, Kettenmann H, Schachner M (1984): $\gamma$-aminobutyric acid directly depolarizes cultured oligodendrocytes. J Neurosci 4:561–569.
Goodman RR, Kuhar MJ, Hester L, Snyder SH (1983): Adenosine receptors: autoradiographic evidence for their location on axon terminals of excitatory neurons. Science 220:967–969.
Haycock JW, Levy WB, Cotman CW (1977): Pentobarbital depression of stimulus-secretion coupling in brain: selective inhibition of depolarization-induced calcium-dependent release. Biochem Pharmacol 26:159–161.
Hill DR, Bowery NG (1981): $^3$H-baclofen and $^3$H-GABA bind to bicuculline-insensitive $GABA_B$ sites in rat brain. Nature 290:149–152.
Hill DR, Dolphin AC (1984): Modulation of adenylate cyclase activity by $GABA_B$ receptors. Neuropharmacology 23:829–830.

Hirsch JD, Kochman RL (1983): Calcium alters divalent cation modulation of brain benzodiazepine receptors. Arch Int Pharmacodyn 265:211–218.

Holz Jr GG, Erwin J, Rosenbaum N, Aaronson S (1962): Triparanol inhibition of tetrahymena, and its prevention by lipids. Arch Biochem and Biophy 98:312–322.

Hullihan JP, Spector S, Taniguchi T, Wang JKT (1983): The binding of [$^3$H-]-diazepam to guinea-pig ileal longitudinal muscle and the *in vitro* inhibition of contraction by benzodiazepines. Br J Pharmacol 78:321–327.

Hunkeler W, Möhler H, Pieri L, Polc P, Bonetti EP, Cumin R, Schaffner R, Haefely W (1981): Selective antagonists of benzodiazepines. Nature 290:514–516.

Ishii K, Kano T, Akutagawa M, Makino M, Tanaka T, Ando J (1982): Effects of flurazepam and diazepam in isolated guinea-pig taenia coli and longitudinal muscle. Eur J Pharmacol 83:329–333.

Jahnsen H, Laursen AM (1981): The effects of a benzodiazepine on the hyperpolarizing and the depolarizing responses of hippocampal cells to GABA. Brain Res 207:214–217.

Jessen KR, Mirsky R, Dennison ME, Burnstock G (1979): GABA may be a neurotransmitter in the vertebrate peripheral nervous system. Nature 281:71–74.

Josefsson J-O, Johansson P (1979): Naloxone-reversible effect of opioids on pinocytosis in *Amoeba proteus*. Nature 282:78–80.

Kapilita PV, Waters DH, and Triggle DJ (1982): $\gamma$-Aminobutyric acid action in guinea-pig ileal myenteric plexus. Eur J Pharmacol 79:43–51.

Karbon EW, Duman R, Enna SJ (1983): Biochemical identification of multiple $GABA_B$ binding sites: association with noradrenergic terminals in rat forebrain. Brain Res 274:393–396.

Kato K, Goto M, Fukuda H (1983): Regulation by divalent cations of $^3$H-baclofen binding to $GABA_B$ sites in rat cerebellar membranes. Life Sci 32:879–887.

Kelly JS, Krnjević K, Morris ME, Yim GKW (1969): Anionic Permeability of Cortical Neurones. Exp Brain Res 7:11–31.

Kerwin R, Pycock C (1980): GABA inhibition of $^3$H-glycine release from slices of rat substantia nigra in vitro. Eur J Pharmacol 64:169–172.

Khan AR (1983): Diazepam receptor in frog muscle fibres. Acta Physiol Scand 118:95–96.

Khan AR, Edman KAP (1983): Diazepam, a highly effective twitch potentiator in isolated muscle fibres of the frog. Acta Physiol Scand 117:533–539.

Kleinrok A, Kilbinger H (1983): $\gamma$-Aminobutyric acid and cholinergic transmission in the guinea-pig-ileum. Arch Pharmacol 322:216–220.

Konnerth A, Heinemann U (1983): Effects of GABA on presumed presynaptic $Ca^{2+}$ entry in hippocampal slices. Brain Res 270:185–189.

Krnjević K (1976): Inhibitory action of GABA and GABA-mimetics on vertebrate neurons. In Roberts E, Chase TN, Tower DB (eds): *GABA in Nervous System Function* New York: Raven Press.

Krogsgaard-Larsen P, Johnston GAR, Lodge D, Curtis DR (1977): A new class of GABA agonist. Nature 268:53–55.

Krogsgaard-Larsen P, Falch E, Schousboe A, Curtis DR, Lodge D (1980): Piperidine-4-sulphonic acid, a new specific GABA agonist. J Neurochem 34:756–759.

Kung C, Saimi Y (1982): The physiological basis of taxes in *Paramecium*. Ann Rev Physiol 44:519–534.

Leeb-Lundberg F, Snowman A, Olsen RW (1980): Barbiturate receptor sites are coupled to benzodiazepine receptors. Proc Natl Acad Sci USA 77:7468–7472.

LeFur G, Mestre M, Carriot T, Uzan A (1984): Are peripheral type benzodiazepine receptors coupled with calcium channels? Abstr Soc Neurosci 10:972.

Leslie SW, Friedman MB, Coleman RR (1980a): Effects of chlordiazepoxide on depolarization-induced calcium influx into synaptosomes. Biochem Pharmacol 29:2439–2443.

Leslie SW, Friedman MB, Wilcox RE, Elrod SV (1980b): Acute and chronic effects of barbiturates on depolarization-induced calcium influx into rat synaptosomes. Brain Res 185:409–417.

Lo MMS, Snyder SH (1983): Two distinct solubilized benzodiazepine receptors: differential modulation by ions. J. Neurosci 3:2270–2279.

Mackerer CR, Kochman RL (1978): Effects of cations and anions on the binding of $^3$H-diazepam to rat brain. Proc Soc Exp Biol Med 158:393–397.

Martin IL, Candy JM (1980): Facilitation of specific benzodiazepine binding in rat brain membrane fragments by a number of anions. Neuropharmacology 19:175–179.

Mayer ML, Higashi H, Gallagher JP, Shinnick-Gallagher P (1983): On the mechanism of action of GABA in pelvic vesical ganglia: biphasic responses evoked by two opposing actions on membrane conductance. Brain Res 260:233–248.

Mestre M, Carriot T, Belin C, Uzan A, Renault C, Dubroeucq MC, Gueremy C, LeFur G (1984): Electrophysiological and pharmacological characterization of peripheral benodiazepine receptors in a guinea pig heart preparation. Life Sci 35:953–962.

Minchin MCW (1981): The effect of anaesthetics on the uptake and release of $\gamma$-aminobutyrate and D-aspartate in rat brain slices. Br J Pharmacol 73:681–689.

Mizuno S, Ogawa N, Mori A (1982): Super-high-affinity binding site for [$^3$H]diazepam in the presence of $Co^{2+}$, $Ni^{2+}$, $Cu^{2+}$, or $Zn^{2+}$. Neurochem Res 7:1487–1493.

Morris ME, DiCostanzo GA, Barolet A, Sheridan PJ (1983): Role of $K^+$ in GABA ($\alpha$-aminobutyric acid)-evoked depolarization of peripheral nerve. Brain Res 278:127–135.

Murphy KMM, Gould RJ, Largent BL, Snyder SH (1983): A unitary mechanism of calcium antagonist drug action. Proc Natl Acad Sci USA 80:860–864.

Newberry NR, Nicoll RA (1984): Direct hyperpolarizing action of baclofen on hippocampal pyramidal cells. Nature 308:450–452.

Nishi S, Minota S, Karczmar AG (1974): Primary afferent neurones: the ionic mechanism of GABA-mediated depolarization. Neuropharmacology 13:215–219.

Nistri A, Berti C (1983): Caffeine-induced potentiation of GABA effects on frog spinal cord: an electrophysiological study. Brain Res 258:263–270.

Nistri A, Constanti A (1978): Effects of flurazepam on amino acid-evoked responses recorded from the lobster muscle and frog spinal cord. Neuropharmacology 17:127–135.

Okamoto K, Kimura H, Sakai Y (1983): Effects of taurine and GABA on Ca spikes and Na spikes in cerebellar Purkinje cells in vitro: intrasomatic study. Brain Res 260:249–259.

Okamoto K, Quastel JH (1976): Effects of amino acids and convulsants on spontaneous action potentials in cerebellar cortex slices. Br J Pharmacol 57:3–15.

Okamoto K, Quastel DMJ, Quastel JH (1976): Action of amino acids and convulsants on cerebellar spontaneous action potentials in vitro: effects of depı vation of $Cl^-$, $K^+$, or $Na^+$. Brain Res 113:147 158.

Okamoto K, Sakai Y (1979): Augumentation by chlordiazepoxide of the inhibitory effects of taurine, $\beta$-alanine and $\gamma$-aminobutyric acid on spike discharges in guinea-pig cerebellar slices. Br J Pharmacol 65:277–285.

Olpe HR, Demiéville H, Baltzer V, Bencze WL, Koella WP, Wolf P, Haas HL (1978): The biological activity of d- and l-baclofen (Lioresal). Eur J Pharmacol 52:133–136.

Olsen RW (1982): Drug interactions at the GABA receptor–ionophore complex. Ann Rev Pharmacol Toxicol 22:245–277.

Palacios JM, Young WE III, Kuhar MJ (1980): Autoradiographic localization of $\gamma$-aminobutyric acid (GABA) receptors in the rat cerebellum. Proc Natl Acad Sci USA 77:670–674.

Pax R, Bennett JL, Fetterer R (1978): A benzodiazepine derivative and praziquantel: Effects on musculature of *Schistosoma mansoni* and *Schistosoma japonicum.* Arch Pharmacol 304:309–315.

Pincus JH, Hsiao K (1981): Calcium uptake mechanisms affected by some convulsant and anticonvulsant drugs. Brain Res 217:119–127.

Postashner SJ (1979): Baclofen: effects on amino acid release and metabolism in slices of guinea pig cerebral cortex. J Neurochem 32:103–109.

Rosensweig Z, Kindler SH (1972): Epinephreine and serotonin activation of adenyl cyclase from *Tetrahymena pyriformis.* FEBS letters 25:221–223.

Schousboe A, Drejer J, Meier E (1984): Physiology and pharmacology of low affinity GABA-receptors on glutamatergic neurons. Neuropharmacology 23:873–874.

Schwartz JK, Spielmann RP (1983): Flurazepam: Effects on sodium and potassium currents in myelinated nerve fibres. Eur J Pharmacol 90:359–366.

Sieghart W, Drexler G (1983): Irreversible binding of [$^3$H]flunitrazepam binding to different proteins in various brain regions. J Neurochem 41:47–55.

Sieghart W, Mayer A, Drexler G (1983): Properties of [$^3$]flunitrazepam binding to different benzodiazepine binding proteins. Eur J Pharmcol 88:291–299.

Skerritt JH, Willow M, Johnston GAR (1983): Contransting effects of a convulsant (CHEB) and an anticonvulsant barbiturate (Phenobarbitone) on amino acid release from rat brain slices. Brain Res 258:271–276.

Smart TG, Constanti A (1982): A novel effect of zinc on the lobster muscle GABA receptor. Proc R Soc Lond B 215:327–341.

Study RE, Barker JL (1981): Diazepam and (−)−pentobarbital: Fluctuation analysis reveals different mechanisms for potentiation of $\gamma$-aminobutyric acid responses in cultured central neurons. Proc Natl Acad Sci USA 78:7180–7184.

Squires RF (1981a): GABA receptors regulate the affinities of anions required for brain specific benzodiazepine binding. Adv Biochem Psychopharmacol 26:129–138.

Squires RF (1981b): Certain GABA mimetics reduce the affinity of anions required for benzodiazepine binding in a GABA-reversible way. Drug Development Res 1:211–221.

Squires RF (1982): Additional evidence for multiple benzodiazepine/anion/GABA receptor complexes in rat cerebellum and forebrain. In Collu R et al. (eds):"Brain Peptides and Hormones" New York: Raven Press, pp 93–106.

Squires RF (1983): Benzodiazepine receptor multiplicity. Neuropharmacology 22:1443–1450.

Squires RF, Casida JE, Richardson M, Saederup E (1983): [$^{35}$S]*t*-Butylbicyclophosphorothionate binds with high affinity to brain-specific sites coupled to $\gamma$-aminobutyric acid-A and ion recognition sites. Mol Pharmacol 23:326–336.

Squires RF, Saederup E (1982a): Evidence for independent cation and anion recognition sites in benzodiazepine-picrotoxin/GABA receptor complexes. In Usdin E et al. (eds): "Pharmacology of Benzodiazepines"London: Macmillan Press, Ltd, pp 567–582.

Squires RF, Saederup E (1982b): $\gamma$-Aminobutyric acid receptors modulate cation binding sites coupled to independent benzodiazepine, picrotoxin, and anion binding sites. Mol Pharmacol 22:327–334.

Squires RF (1984): Benzodiazepine receptors. In Lajtha

A (ed): "Handbook of Neurochemistry." Second edition, Vol 6. New York: Plenum Press, pp 261–307.

Supavilai P, Karobath M (1981): Action of pyrazolopyridines as modulators of [$^3$H]flunitrazepam binding to the GABA/benzodiazepine receptor complex of the cerebellum. J Pharmacol 70:183–193.

Supavilai P, Mannonen A, Karobath M (1982): Modulation of GABA binding sites by CNS depressants and CNS convulsants. Neurochem Int 4:259–268.

Swenson Jr RP (1982): Flurazepam interaction with sodium and potassium channels in squid giant axon. Brain Res 241:317–322.

Taft WC, DeLorenzo RJ (1984): Micromolar-affinity benzodiazepine receptors regulate voltage-sensitive calcium channels in nerve terminal preparations. Proc Nat Acad Sci USA 81:3118–3122.

Takeuchi A, Takeuchi N (1967): Anion permeability of the inhibitory postsynaptic membrane of the crayfish neuromuscular junction. J Physiol 191:575–590.

Taniyama K, Kusunoki M, Saito N, Tanaka C (1982): Release of α-aminobutyric acid from cat colon. Science 217:1038–1040.

Thampy KG, Barnes Jr EM (1984): γ-Aminobutyric acid-gated chloride channels in cultured cerebral neurons. J Bio Chem 259:1753–1757.

Tomiko S, Taraskevich PS, Douglas WW (1983): GABA acts directly on cells of pituitary pars intermedia to alter hormone output. Nature 301:706–707.

Tozalakian PV, Csillag A (1982): The effect of γ-aminobutyric acid on the potassium movements of rat cortical synaptosomes. Exp Brain Res 46:24–28.

Tremblay JP, Grenon G (1980): Benzodiazepines modify synaptic depression, frequency facilitation and PTP of an identified cholinergic synapse of *Aplysia.* Life Sci 27:491–496.

Wamsley JK, Gee KW, Yamamura HI (1983): Comparison of the distribution of convulsant/barbiturate and benzodiazepine receptors using light microscopic autoradiography. Life Sci 33:2321–2329.

Wong EHF, Leeb-Lundberg F, Teichberg VI, Olsen RW (1984): γ-Aminobutyric acid activation of $^{36}$Cl-flux in rat hippocampal slices and its potentiation by barbiturates. Brain Res 303:267–275.

Wojcik WJ, Neff NH (1983): Adenosine Al receptors are associated with cerebellar granule cells. J Neurochem 41:759–763.

Yarowsky PJ, Carpenter DO (1978): Receptors for γ-aminobutyric acid (GABA) on Aplysia neurons. Brain Res 144:75–94.

Benzodiazepine/GABA Receptors and Chloride Channels:
Structural and Functional Properties, pages 225–240


11

# GABA Involvement in Benzodiazepine Receptor Modulation of the Chloride Ionophore

Lowell J. Lawrence, Kelvin W. Gee, and Henry I. Yamamura

*Department of Pharmacology, Arizona Health Sciences Center, Tucson, Arizona 85724*

The pharmacological actions of the 1,4-benzodiazepines (Fig. 1) were recognized shortly after their synthesis in the early 1960s [Sternbach and Reeder, 1961; Sternbach et al., 1964]. In general, the benzodiazepines produce antianxiety (viz., anxiolytic effects) and sedation. They have also been useful as anticonvulsants in the treatment of certain seizure disorders and as muscle relaxants.

In vitro studies have facilitated the identification of specific high-affinity benzodiazepine binding sites in the central nervous systems (CNS) of vertebrates [Squires and Braestrup, 1977; Möhler and Okada, 1977] as well as invertebrates [Abalis et al., 1983]. The in vivo and in vitro potencies of benzodiazepines are highly correlated [Squires and Braestrup, 1977; Möhler and Okada, 1977; Braestrup and Squires, 1978], strongly implicating the benzodiazepine receptor interaction in the production of anxiolytic and anticonvulsant effects. In addition, the distribution of benzodiazepine receptors is closely related to the anatomic regions believed to be involved in their pharmacological actions [Young and Kuhar, 1980].

Evidence is accumulating which suggests that benzodiazepine receptors are heterogeneous and that there are at least two [Klepner et al., 1979; Nielsen et al., 1979; Ehlert et al., 1981] and possibly three [Gee et al., 1984a; Leeb-Lundberg and Olsen, 1983] subtypes based primarily on differential sensitivities to GABA. Benzodiazepine receptor heterogeneity and its possible functional significance has recently been reviewed [Gee et al., 1984b].

The extensive search for compounds with benzodiazepine-related activities has led to the discovery of a number of nonbenzodiazepine compounds which interact specifically and potently with benzodiazepine receptors as agonists, partial agonists, inverse agonists, and antagonists (Fig. 1). These include the beta-carbolines (inverse agonists); the quinoline derivatives, PK 8165 and PK 9084 (partial agonists); and the imidazodiazepine Ro 15-1788 and pyrazoloquinoline CGS 8216, which both act as benzodiazepine receptor antagonists. These compounds provide usful probes of benzodiazepine-receptor function and serve as prototypes for more selective anxiolytic agents [Ehlert et al., 1983].

Considerable evidence has accumulated which indicates that benzodiazepine receptors are functionally coupled to GABA recognition sites and that indeed their pharmacological effects result from their modulation of the GABA receptor. In vitro studies have shown that GABA increases the affinity of benzodiazepine-receptor interactions [Tallman et al., 1978; Briley and Langer, 1978] and con-

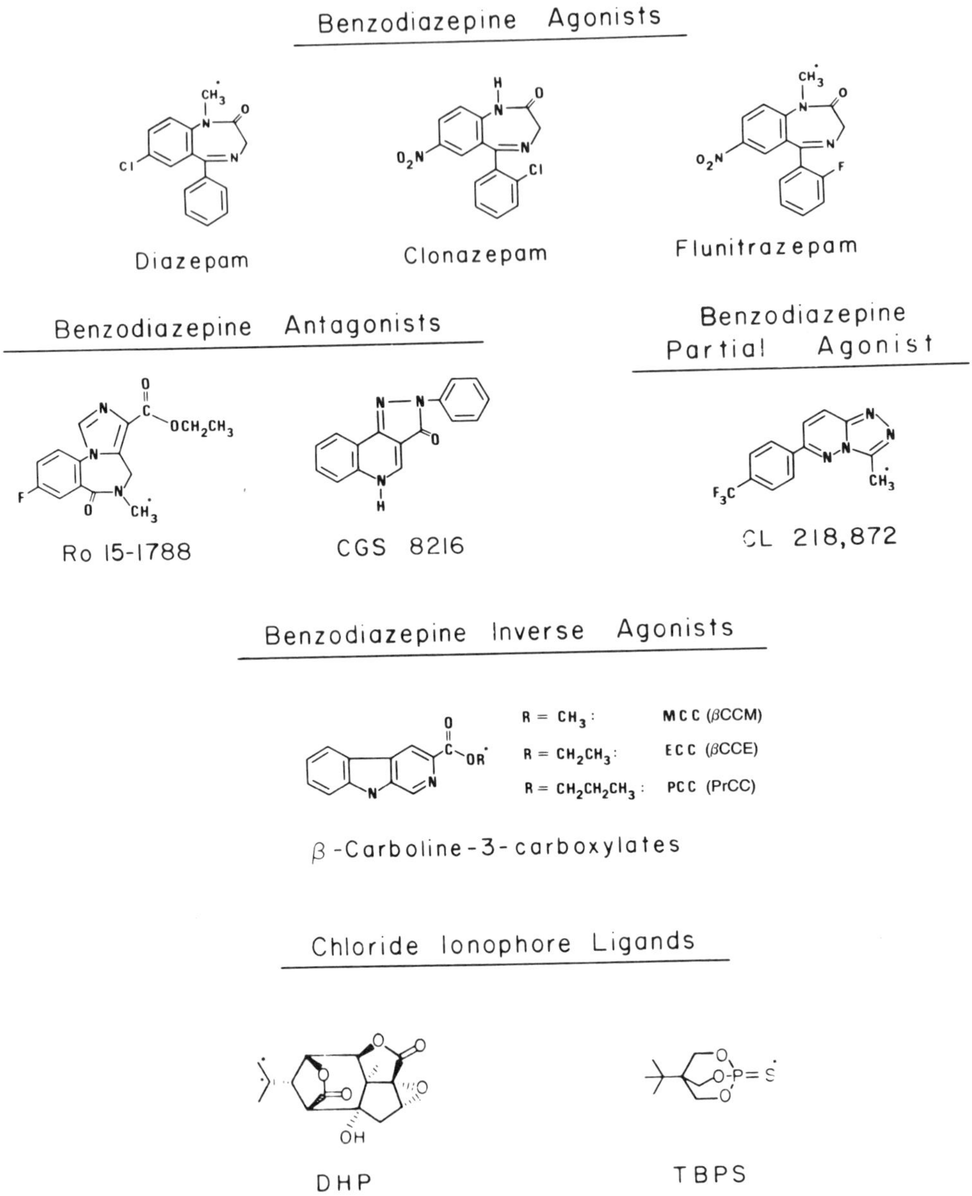


**Fig. 1.** *Chemical structures of various ligands for the benzodiazepine receptor and chloride ionophore. Asterisks indicate positions of radioactive label in the case of radioligands.*

versely that benzodiazepines potentiate the interaction of GABA with its recognition site [Guidotti et al., 1978]. It should be noted however, that not all benzodiazepine receptors respond to GABA. Differential coupling of benzodiazepine receptor subtypes to a GABA site has been observed. The apparent affinity of the benzodiazepine-2 receptors is modified by GABA, however the affinity of benzodiazepine-1 receptors, with the exception of those in cerebellum, is unaltered by the addition of GABA [Gee et al., 1983].

Potentiation of the inhibitory actions of GABA by benzodiazepines has been demonstrated in various neurophysiological studies [Costa et al., 1975; Haefely et al., 1975]. Moreover, electrophysiological studies utilizing voltage-clamped mouse cultured spinal neurons have revealed that diazepam alters the kinetics of ion channel opening rather than the open channel conductance per se [Study and Barker, 1981]. This action of diazepam was shown to quantitatively account for the potentiation by benzodiazepines of the GABA response. Although most experimental findings support an obligatory role of GABA in the anxiolytic and anticonvulsant actions of the benzodiazepines, there are also data which suggest that GABA is of minimal importance in the anxiolytic action of the benzodiazepines. These results are from tolerance and time-dependence studies [Lippa and Regan, 1977]. The precise nature of the relationship between the benzodiazepine receptor and the chloride ionophore and the extent to which the GABA receptor is involved in benzodiazepine–ionophore coupling therefore remain important questions in understanding the mechanism(s) of benzodiazepine action.

Characterization of benzodiazepine-receptor interactions has involved extensive research over the past eight years and has been facilitated largely by the availability of high-affinity radioligands such as [$^3$H]flunitrazepam, [$^3$H]diazepam, [$^3$H]Ro 15-1788, [$^3$H]$\beta$CCM (MCC), [$^3$H]PrCC (PCC), and [$^3$H]CL 218,872 (structures in Fig. 1). This area of research is the subject of a recently published monograph [Biggio and Costa, 1983]. The characterization of specific binding sites associated with the chloride ionophore has proven to be a somewhat more difficult task due to the unavailability of suitable radioligand probes. The recent discovery of [$^{35}$S]*t*-butylbicyclophosphorothionate ([$^{35}$S]TBPS), a high-affinity specific radioligand which interacts with the chloride ionophore [Squires et al., 1983], has, however, facilitated considerable experimentation in this area, which is providing insights into the relationship between the ionophore and its coupled components within the GABA receptor–ionophore complex.

### The Chloride Ionophore

Inhibitory neurotransmission in crustacean muscle involves changes in chloride conductance [Fatt and Katz, 1953; Woodbury, 1980] elicited by GABA [Dudel and Kuffler, 1961]. The interaction of GABA with its receptor sites (which are located pre- and postsynaptically) causes an increase in chloride conductance by modulating (opening) chloride-specific ionophores. This results in either presynaptic inhibition (reduced postsynaptic excitatory potentials), which reduces transmitter release from excitatory neurons [Dudel and Kuffler, 1961] or postsynaptic inhibition (hyperpolarization of motor neurons) [Fatt and Katz, 1953]. Chloride conductance is blocked by several compounds which have no direct action on GABA receptors, e.g., picrotoxin [Takeuchi and Takeuchi, 1969]. The chloride ionophore is therefore a discrete entity within the GABA receptor–ionophore complex regarding its interaction with drugs and toxicants.

Chloride ionophore function has been studied both physiologically and pharmacologically, involving direct measurements of chloride conductance [Takeuchi and Takeuchi, 1969] and membrane chloride flux measurements using $^{36}Cl^-$ [Marchbanks and Campbell, 1976; Ticku and Olsen, 1977; Matsumura and Ghiasuddin, 1983]. Specific chloride ionophore–associated binding sites have been identified using the radiolabeled analogue of picrotoxinin [$^3$H]alpha-dihydropicrotoxinin ([$^3$H]DHP) [Leeb-Lundberg and Olsen, 1980] (Fig. 1), and more recently the potent cage-convulsant [$^{35}$S]TBPS [Squires et al., 1983] (Fig. 1) as radioligands. [$^{35}$S]TBPS is a significant improvement over [$^3$H]DHP due to its much greater specificity and higher signal-to-noise ratio. Both ligands compete for the same binding site [Lawrence and Casida, 1983] and interact to varying degrees with benzodiazepines, GABA, barbiturates, cage convulsants, and a variety of insecticides

[Leeb-Lundberg and Olsen, 1980; Squires et al., 1983; Lawrence and Casida, 1983; Lawrence and Casida, 1984; Casida and Lawrence, 1985]. In general, [$^{3}$H]DHP binding to rat brain membranes is more potently inhibited by benzodiazepines than is [$^{35}$S]TBPS, whereas [$^{35}$S]TBPS is more potently inhibited by GABA-mimetics [Squires et al., 1983]. Evidence that [$^{35}$S]TBPS binding is specific for the GABA-regulated chloride ionophore is several-fold: 1) TBPS and related cage convulants antagonize the GABA-induced chloride conductance in the crayfish claw-opener muscle [Gammon and Casida, 1984] and in mammalian CNS [Bowery et al., 1976] in a manner identical to that of picrotoxin [Takeuchi and Takeuchi, 1969]; 2) [$^{35}$S]TBPS binding is sensitive to Eccles anions (e.g., $Cl^-$, $Br^-$, $SO_4^{=}$) but not to cations such as $Na^+$ or $Ca^{++}$ [Squires et al., 1983]; and 3) the pharmacological profile of [$^{35}$S]TBPS binding is consistent with an interaction with the chloride ionophore [Squires et al., 1983; Casida and Lawrence, 1985]. The use of [$^{35}$S]TBPS as a pharmacological probe of chloride ionophore function and, in particular, its application to the study of benzodiazepine action is the topic of this review.

## The Use of [$^{35}$S]TBPS as a Radioligand

Since the introduction of [$^{35}$S]TBPS (New England Nuclear, Boston, MA) as a radioligand, a considerable number and variety of binding studies have appeared. Although results from these studies are in general agreement regarding the inhibitory potencies of most compounds and the kinetic parameters of [$^{35}$S]TBPS binding, there is considerable variation in the techniques used for the preparation of membranes for assay. Various membrane preparations used in [$^{35}$S]TBPS binding sites are summarized in Table I and results from these studies are compared in Table II. The data summarized in Table II reveal several aspects [$^{35}$S]TBPS binding that are significantly affected by the method of membrane preparation and assay conditions. The $K_d$ of the TBPS-receptor interaction is apparently not a function of the membrane-preparation technique, but it does show temperature dependence, being lower at 25°C than at 37°C. The density of TBPS binding sites, on the other hand, varies considerably as a function of the preparation technique used. Variations in the inhibitory potencies of the competitive inhibitors TBPS and picrotoxin are minimal and appear to be more closely related to temperature than to membrane preparation method. However, the effects of diazepam on [$^{35}$S]TBPS binding (Table II) differ markedly. Diazepam has little or no influence on [$^{35}$S]TBPS specific binding in extensively dialyzed membranes, is inhibitory in unwashed fresh membranes (technique B), and stimulated [$^{35}$S]TBPS-specific binding in an extensively washed 36,000 xg pellet (technique C). The apparent discrepancy between results obtained by another laboratory [Supavilai and Karobath, 1983] and our own findings [Casida and Lawrence, 1985] have prompted a more detailed analysis of the influence of benzodiazepines on [$^{35}$S]TBPS binding [Lawrence et al., 1984].

## Benzodiazepine/[$^{35}$S]TBPS Site Interactions

Using a fresh rat brain synaptic membrane ($P_2$) preparation, we find that benzodiazepines potently modulate TBPS binding sites in a concentration-dependent manner and with potencies that are very closely related to the antimetrazol activities of the benzodiazepines [Lawrence et al., 1984]. The interaction is apparently allosteric, since inhibition curves are shallow and only 25–35% of the specific binding of [$^{35}$S]TBPS is inhibited (Fig. 2). It should also be noted that the weak partial agonist benzodiazepine Ro 15-1788 (Fig. 1) is not an inhibitor of [$^{35}$S]TBPS binding (Fig. 2) (Table III), an observation consistent with its pharmacological actions [Möhler and Richards, 1981; Nutt and Cowen, 1982]. In addition, the beta-carboline inverse agonist βCCM (Fig. 1) enhances [$^{35}$S]TBPS-specific binding (Fig. 2). Table III gives potencies of five full agonists and the partial agonist Ro 15-1788 as inhibitors of "clonazepam-sensitive" [$^{35}$S]TBPS binding and as antagonists

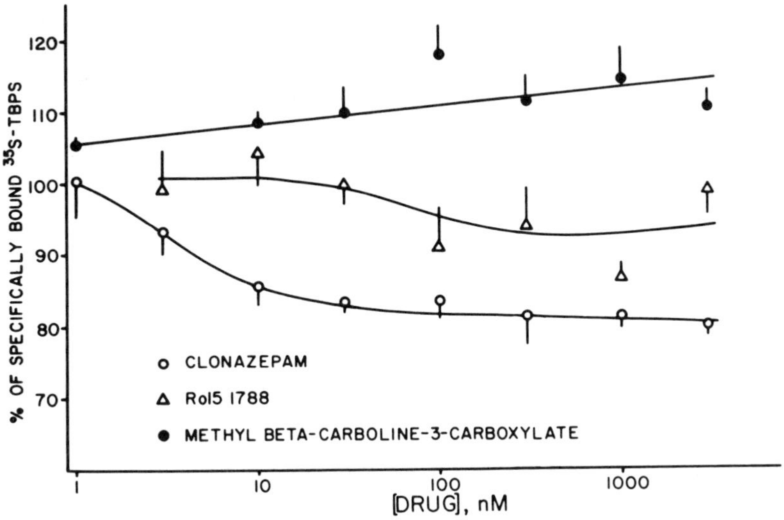


**Fig. 2.** *Modulation of specific [$^{35}$S]TBPS binding in rat brain membranes by clonazepam, Ro 15-1788, and methyl-beta-carboline-3-carboxylate.*

**TABLE I. Various Techniques Used for the Preparation of Rat Brain Membranes for $^{35}$S TBPS Binding Assays**

| Membrane preparation technique | Reference |
|---|---|
| A. 1. Tissue homogenized in 50 volumes in 1 mM EDTA | Squires et al., 1983 |
| 2. $P_2$ fraction prepared and dialyzed 3X against distilled water | |
| 3. Membranes pelleted (25,000 × g, 30 min) and stored frozen | |
| 4. Reconstituted in 5 mM Tris-HCl containing 0.2 M KBr for assay | |
| B. 1. Tissue homogenized in 10 volumes 0.32 M sucrose | Lawrence and Casida, 1983 |
| 2. $P_2$ fraction prepared and resuspended in 50 mM sodium–potassium phosphate buffer for assay | |
| 3. Preparation not stable to freezing | |
| C. 1. Tissue homogenized in 50 volumes 50 mM Tris HCl buffer | Supavilai and Karobath, 1983 |
| 2. Membranes pelleted (36,000 × g, 10 min) and washed 4X in Tris-HCl | |
| 3. Membrane suspension stored frozen and centrifuged just prior to assay | |
| 4. Resuspended in 50 mM Tris-citrate for assay | |
| D. 1. Tissue homogenized in 25 volumes 1 mM EDTA | Ticku and Ramanjaneyulu, 1984 |
| 2. Supernatant from 1000 × g, 10 min centrifugation recentrifuged (140,000 × g, 30 min) | |
| 3. Pellet resuspended in EDTA and dialyzed overnight | |
| 4. Pellet washed and resuspended in Tris-HCl containing 0.2 M KBr for assay | |

of metrazol-induced seizures in rats. Comparison of these data reveal an excellent agreement between the in vivo and in vitro potencies, strongly suggesting that the benzodiazepine/TBPS interaction is relevant to the mechanism of benzodiazepine anticonvulsant action. Since other allosteric modulators of [$^{35}$S]-TBPS specific binding which are known to enhance membrane chloride permeability by activating (opening) chloride

**TABLE II. Comparison of [$^{35}$S]TBPS Binding Characteristics in Four Different Membrane Preparations**

| Parameter | Membrane preparation technique and assay temperature[a] | | | | | |
|---|---|---|---|---|---|---|
| | A, 25°C | A, 37°C[b] | B, 37°C | C, 23°C | D, 24°C | |
| $K_d$ (nM) | 17 | 61–74 | 61–80 | 26[c] | 25 | |
| $B_{max}$ (pmol/mg protein) | 50[d] | 6–7 | 2–3 | 178[c,d] | 1.4 | |
| $IC_{50}$ values (nM) for: | | | | | Cortex | Cerebellum |
| TBPS | 17 | 60 | 80 | — | — | — |
| Picrotoxin | 190 | 340 | 760 | — | 350 | 400 |
| GABA | 340 | 900 | 10,000 | — | 770 | 120 |
| Diazepam | — | — | — | Stimulatory[e] | — | — |

a. Membrane preparation techniques numbered according to Table I.
b. Data are taken from Cole et al., 1984.
c. Values determined in the presence of 0.2 M NaBr. When 0.2 M NaCl was substituted, values for $K_d$ and $B_{max}$ were 60 nM and 193 fmol/mg wet tissue, respectively.
d. Values given as fmol/mg wet tissue.
e. 1 $\mu$M diazepam stimulated [$^{35}$S]TBPS specific binding to 160% of the control value.

**TABLE III. Comparison Between Potencies of Six Benzodiazepines as Inhibitors of [$^{35}$S]TBPS Binding to Fresh Unwashed Rat Brain Cortical Membranes and as Antagonists of Metrazol-Induced Seizures in Rats**

| Benzodiazepine | [$^{35}$S]TBPS inhibitory potency ($IC_{50}$, nM) | Anticonvulsant potency ($ED_{50}$, mg/kg) |
|---|---|---|
| Lorazepam | 14 | 0.1 |
| Clonazepam | 15 | 0.1 |
| Flunitrazepam | 52 | 0.4 |
| Flurazepam | 118 | 4.0 |
| Diazepam | 124 | 1.4 |
| Ro 15–1788 | 10,000 | 50 |

Inhibitory potencies are expressed as the concentration of drug inhibiting 50% of the "clonazepam-sensitive" [$^{35}$S]TBPS specific binding ($IC_{50}$ values) from fresh unwashed rat cortical membranes. The baseline clonazepam-sensitive binding was determined with 10 $\mu$M clonazepam and was equivalent to 76.6% of the total specifically bound [$^{35}$S]TBPS (Fig. 1). The anticonvulsant potencies for lorazepam, clonazepam, flunitrazepam, and diazepam are tabulated as median dosages required to antagonize tonic hindpaw extension following the administration of 80 mg/kg metrazol in rats [personal communication, Hoffman La Roche, 1984] and in the case of Ro 15–1788 the value is the threshold dose in rats [Nutt and Cowen, 1982]. Reproduced with permission from Biochemical and Biophysical Research Communications.

ionophores (e.g., GABA-mimetics) [Squires et al., 1983] also inhibit [$^{35}$S]TBPS binding, this in vitro observation is consistent with the presumed pharmacological actions of the benzodiazepines. In this regard, it is difficult to rationalize the pharmacological relevance of the findings of Supavilai and Karobath [1983 and 1984], i.e., stimulation of [$^{35}$S]TBPS binding by benzodiazepine agonists and inhibition by benzodiazepine inverse agonists. In comparing their results to our own, it became apparent that the major difference in the studies was the technique used to prepare synaptic membranes. Whereas their studies utilized an extensively washed $P_1 + P_2$ fraction, our studies used a fresh, unwashed $P_2$ preparation. Since dialysis or extensive washing of rat brain membranes removes GABA (as well

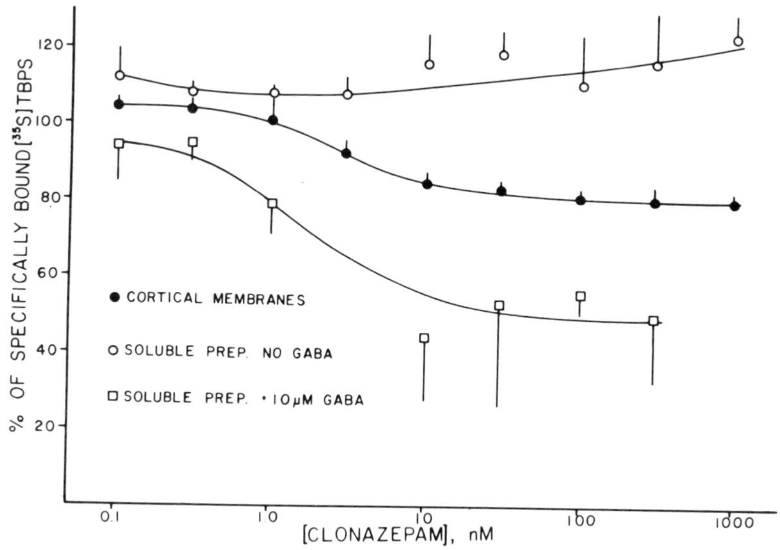


**Fig. 3.** *Effect of clonazepam on [$^{35}$S]TBPS binding to fresh unwashed rat cortical membranes ($P_2$ fraction) and solubilized receptor preparations containing no GABA and 10 μM GABA. Inhibition data are the means ± standard errors from duplicate determinations in eight separate experiments (membranes) or three to four experiments (soluble). [Reproduced with permission from Biochemical and Biophysical Research Communications.]*

as numerous other endogenous substances) we have investigated the role of GABA in benzodiazepine modulation of [$^{35}$S]TBPS binding sites [Lawrence et al., 1984].

Solubilized receptor complex preparations were passed over Sephadex G-50 columns to facilitate the complete removal of GABA from the receptor preparation. This was established by the addition of a trace amount of [$^{3}$H]GABA to the preparation prior to its application on the column, whereby greater than 99% of the applied [$^{3}$H]GABA was retained on the column and the [$^{35}$S]TBPS and [$^{3}$H]flunitrazepam binding activity eluted with the void volume. Figure 3 illustrates the effects of clonazepam on [$^{35}$S]TBPS binding in the absence of GABA, to native membranes and to soluble receptors in the presence of 10 μM GABA. In the absence of GABA, clonazepam enhanced [$^{35}$S]TBPS binding in a concentration-dependent manner, producing a 25% increase in specific binding at 1 μM. This finding is in general agreement with those of Supavilai and Karobath [1983 and 1984]. On the other hand, clonazepam inhibited [$^{35}$S]TBPS binding upon the addition of 10 μM GABA to the assay system, suggesting that the presence of GABA is an absolute requirement for this effect. When 1 μM GABA was added to the soluble "GABA-free" preparation, little or no inhibition of [$^{35}$S]TBPS binding by clonazepam was observed. These results suggest that benzodiazepine modulation of chloride ionophore–associated binding sites (i.e., TBPS-labeled sites) occurs through the low–affinity or micromolar GABA receptor. This premise is in fact supported by the recent autoradiographic demonstration that benzodiazepine and low-affinity GABA binding sites are identically distributed in the CNS [Olsen et al., 1984].

The role of GABA in the modulation of the chloride ionophore by an inverse agonist was also evaluated by observing the effect of βCCM on [$^{35}$S]TBPS binding in the presence or absence of GABA [Gee et al., 1985]. Using a solubilized receptor preparation, free GABA was removed by Sephadex G-50 col-

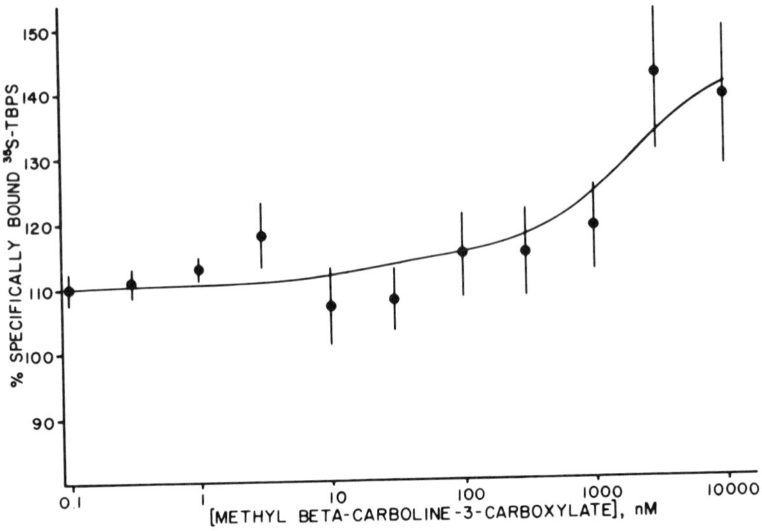


**Fig. 4.** *Effect of methyl-beta-carboline-3-carboxylate on 2 nM [35S]TBPS binding to solubilized receptor preparation from rat cerebral cortex. Soluble preparation was rendered GABA-free by Sephadex G-50 column chromatography. Each point represents the mean ± SEM of four to five independent determinations. [From Gee et al., 1985.]*

umn chromatography. βCCM was incubated with 2 nM [35S]TBPS at 25°C for 90 min in 50 mM Na/K phosphate buffer plus 200 mM NaCl. Similar to that observed with intact membrane preparations, βCCM produced dose-dependent enhancement of [35S]TBPS binding in the absence of GABA (Fig. 4). The addition of 10 μM GABA resulted in enhancement of the potency of βCCM modulation of [35S]TBPS binding (Table IV). These findings suggest that βCCM can modulate [35S]TBPS binding in a similar manner with or without GABA present. It is interesting that the net effect of GABA is to enhance the potency of βCCM, when GABA is known to reduce the apparent affinity of both βCCM and [35S]TBPS for their respective binding sites. These observations are consistent with at least two possible mechanisms underlying βCCM modulation of the chloride ionophore: 1) an allosteric link between the βCCM recognition site and the chloride ionophore which allows βCCM to directly modulate the ionophore independent of the GABA site, or 2) an allosteric link between the βCCM recognition site and the GABA receptor, βCCM modulation of the ionophore occuring through the GABA receptor which is linked to the chloride ionophore. At this time our findings do not rule out either possibility. Yet the experiments demonstrating an effect of βCCM on [35S]TBPS binding in the absence of GABA along with the ability of clonazepam to similarly affect [35S]TBPS binding in a "GABA-free" preparation suggest that a direct link

**TABLE IV. Effect of GABA on Methyl Beta-Carboline-3-Carboxylate (βCCM) Enhancement of [35S]TBPS Binding in the Soluble Fraction From Rat Cerebral Cortex**

| [βCCM], nM | [GABA], μM | % of control [35S]TBPS bound |
|---|---|---|
| 1000 | 0 | 119 ± 7 |
| | 10 | 253 ± 47[a] |
| 3000 | 0 | 142 ± 11 |
| | 10 | 204 ± 14[b] |

Significantly different from the effect in the absence of GABA at [a]p less than 0.05 and [b]p less than 0.02 by Student's t-test. [Data from Gee et al., 1985.]

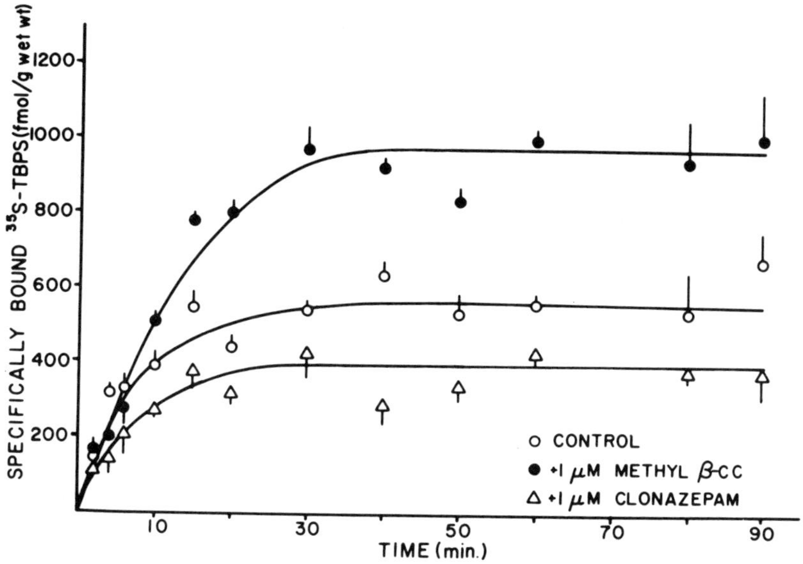


**Fig. 5.** *Association kinetics of 2 nM [$^{35}$S]TBPS binding to rat cortical membranes ($P_2$ fraction) at 25°C in the presence or absence of the agonist, clonazepam (1 μM), or the inverse agonist, methyl-beta-carboline-3-carboxylate (1 μM). All points represent the mean ± SEM of three to four independent determinations [From Gee et al., 1985.]*

may occur between the benzodiazepine site and the chloride ionophore. The pharmacological significance of such an allosteric link remains to be determined. Consistent with previous findings by other investigators, the GABA site associated with the benzodiazepine receptor–chloride ionophore complex appears to be a $GABA_A$ site. This conclusion is based upon the ability of (+)bicuculline (10μM) to block the effect of GABA on both agonist and inverse agonist modulation of [$^{35}$S]TBPS binding. Furthermore, the GABA effect is insensitive to baclofen (10 μM).

In order to determine the possible mechanisms by which benzodiazepine receptor ligands with intrinsic activity modulate [$^{35}$S]TBPS binding, we evaluated the effects of an agonist and an inverse agonist on the kinetics of [$^{35}$S]TBPS binding in rat cerebral cortical $P_2$ fractions [Gee et al., 1985]. Employment of kinetic studies is especially valuable for the determination of changes in the dynamics of ligand–receptor interactions not readily apparent from steady-state binding studies. The association and dissociation kinetics of 2 nM [$^{35}$S]TBPS were examined by incubating clonazepam or βCCM with [$^{35}$S]TBPS at 25°C. Nonspecific binding was defined as that occurring in the presence of 2 μM TBPS. TBPS (10 μM) was used to initiate the dissociation of [$^{35}$S]TBPS. Both association and dissociation of [$^{35}$S]TBPS were carried out in the presence of 10μM GABA. The half lives for association were not significantly different between groups, while the extent of maximal binding was dependent upon the modulator present (Fig. 5). The time to steady-state under all conditions was approximately 30 min at 25°C. βCCM significantly enhanced [$^{35}$S]TBPS binding by approximately 40% above control. In contrast, clonazepam inhibited [$^{35}$S]TBPS binding by approximately 24%. The kinetics of dissociation under control conditions, in the presence of 1 μM clonzepam or 1 μM βCCM, are depicted in Figure 6. The effect of the agonist and inverse agonist on dissociation kinetics was more profound. βCCM and clonazepam were found to increase or decrease the half-life for dissociation, respec-

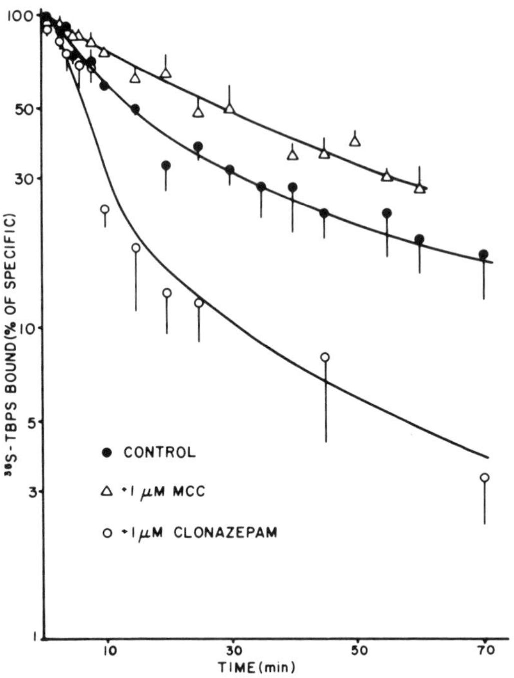


**Fig. 6.** *Dissociation of 2 nM [$^{35}$S]TBPS from rat cortical membranes at 25°C. Dissociation was initiated by the addition of 10 μM TBPS after a 90 min incubation at 25°C. All incubations were carried in the presence or absence of clonazepam (1 μM) or methyl-beta-carboline-3-carboxylate (MCC). All points represent the mean ± SEM of three to four independent determinations. [From Gee et al., 1985.]*

tively (Fig. 6). Interestingly, dissociation of 2 nM [$^{35}$S]TBPS under control conditions appears to reflect dissociation from two conformations or states of the ionophore. Computerized nonlinear regression analysis of the control dissociation curve suggested the presence of two "states" or sites. Dissociation half-lives for the high- and low-affinity sites were approximately 58 min and 4 min, respectively. Twenty-eight percent of the total number of specific sites labeled by [$^{35}$S]TBPS were high-affinity sites. The proportions of high- and low-affinity sites change in the presence of benzodiazepine receptor ligands. This "interconversion" of sites supports the notion that different conformational states of the ionophore exist. In accordance with the observation that βCCM enhances [$^{35}$S]TBPS binding, this inverse agonist reduces the apparent half-life for dissociation. The change in apparent half-life is a reflection of a change in the proportion of the two conformational states present. In the presence of βCCM, no significant difference in the off-rates were observed for each conformation of the [$^{35}$S]TBPS-labeled sites. However, 60% of the sites labled by [$^{35}$S]TBPS in the presence of βCCM were in the high-affinity state, compared to 28% under control conditions. In contrast, less than 5% of the sites were in the high-affinity state when 1 μM clonazepam was present. Thus agonist benzodiazepine receptor ligands appear to induce a conformational change in the ionophore that results in low-affinity for [$^{35}$S]TBPS. The opposite occurs in the presence of an inverse agonist such as βCCM.

Based upon the observed effects of βCCM and clonazepam upon [$^{35}$S]TBPS binding and the pharmacological efficacies of these two compounds, it is tempting to speculate that the low-affinity state of the ionophore is associated with "open" or "activated" chloride channels. GABA-mimetics and benzodiazepine receptor agonists will allosterically "induce" and stabilize "open" state of the ionophore. The high-affinity state of the ionophore may be a reflection of a "closed" or "deactivated" chloride channel with inverse agonist perturbing the ionophore into a conformation with high affinity for [$^{35}$S]TBPS. If such a hypothesis is valid, it may provide a basis for the in vitro study of the molecular events involved in the modulation of chloride channel conductance. Indeed, this phenomenon would provide the first in vitro "response" to occupancy of the benzodiazepine receptor by ligands which possess intrinsic activity.

The sensitivity of [$^{35}$S]TBPS binding to allosteric modulation by sites coupled to a chloride ionophore is well demonstrated by the effect of the phenylquinoline, PK 9084, on [$^{35}$S]TBPS binding in rat cerebral cortex (Fig. 7). Conflicting pharmacological effects of this

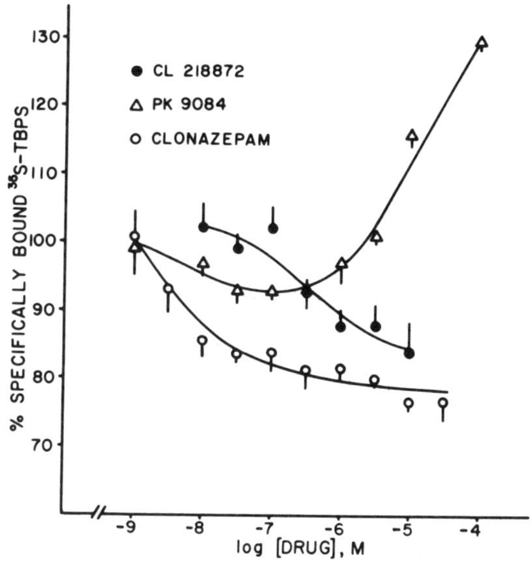


**Fig. 7.** *Modulation of specific [$^{35}$S]TBPS binding in rat cortical membranes ($P_2$ fraction) by CL 218872, PK 9084, and clonazepam. All assays were run at 25°C for 90 min. Each point represents the mean ± SEM of six to nine separate determinations. [From Gee et al., 1985.]*

compound have been reported with different investigators, demonstrating antianxiety effects or the lack of any antianxiety effects [Le Fur et al., 1981; Keane et al., 1984]. It has been proposed that the pharmacological effects, or lack thereof, are related to the low efficacy of PK 9084 at the benzodiazepine receptor relative to the clinically useful agonist benzodiazepines. Perhaps the antianxiety effects devoid of anticonvulsant and sedative potential are related to the weak partial agonist properties of PK 9084 [Gee et al., 1983]. This concept is consistent with the observation that PK 9084 produces a maximal inhibition of [$^{35}$S]TBPS binding of only 5% (Fig. 7). On the other hand, "full agonists" such as clonazepam produce a maximum of 24% inhibition, while "partial agonists" such as CL 218872 produce intermediate effects (Fig. 7). It has also been suggested that PK 9084 acts at a site other than the benzodiazepine receptor [File and Simmonds, 1984]. Yet, the ability of PK 9084 to modulate [$^{35}$S]TBPS binding suggests that its site of interaction is still associated with a chloride ionophore.

At high concentrations (greater than 3 $\mu$M) of PK 9084, dose-dependent enhancement of [$^{35}$S]TBPS binding is actually observed (Fig. 7). Assuming unhindered accessibility of PK 9084 to its target site(s) in the CNS, profound in vivo effects of this compound would be predicted by the apparent enhancement of [$^{35}$S]TBPS binding in vitro. In vivo effects consistent with enhancement of [$^{35}$S]TBPS binding would lend support to the hypothesis that the modulation of [$^{35}$S]TBPS binding is a pharmacologically relevant phenomenon. A recent report by File and Simmonds [1984] supports the predictions based upon the effect of PK 9084 on [$^{35}$S]TBPS binding. These investigators found that PK 9084 in excess of 1 mg/kg was proconvulsant in mice receiving subconvulsant doses of pentylenetetrazol or picrotoxin. Moreover, PK 9084 was found to directly antagonize muscimol induced depolarization in the cuneate nucleus slide preparation, which suggested the possiblity that this phenylquinoline is capable of producing convulsant effects, a prediction consistent with the ability of PK 9084 to enhance [$^{35}$S]TBPS binding in vitro.

Overall, the studies with PK 9084 suggest that modulation of [$^{35}$S]TBPS binding may also be useful for evaluating the pharmacological efficacy and potency of compounds that may modulate the chloride ionophore via sites outside the domain of the GABA and benzodiazepine receptors. Thus compounds that are suspected to alter chloride channel conductance in vivo can be tested for their ability to modulate [$^{35}$S]TBPS binding in vitro.

## Coupling of Benzodiazepine Receptor Subtypes to a Chloride Ionophore

Differential coupling of neurotransmitter/neuromodulator receptors to their effectors is often associated with receptor heterogeneity [Ehlert et al., 1983, Kent et al., 1980]. Consequently, it was of interest to determine if benzodiazepine receptor subtypes discrimi-

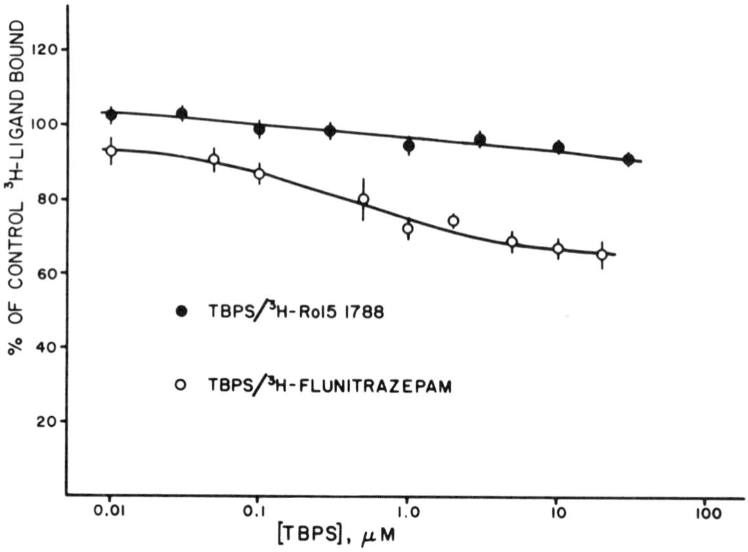


**Fig. 8.** *Effect of TBPS on [$^3$H]flunitrazepam and [$^3$H] Ro15-1788 binding in rat cortical membranes ($P_2$ fraction). Assays were run at 25°C for 90 min. All points represent the mean ± SEM of three to four independent experiments. [From Gee et al., 1985.]*

nated by the beta-carboline, propyl beta-carboline-3-carboxylate (PrCC), were coupled to a chloride ionophore labeled by [$^{35}$S]TBPS. If a chloride ionophore is functionally linked to the benzodiazepine-1 receptor subtype (has high affinity for PrCC), then TBPS would be expected to allostericaly modulate this site. Studies using the benzodiazepine subtype nonselective ligand, [$^3$H]flunitrazepam ([$^3$H]-FLU), have shown that TBPS will allosterically modulate the binding of this radioligand to the benzodiazepine receptor (Fig. 8). The ability of TBPS to modulate the binding of [$^3$H]FLU established the existence of a bidirectional interaction between the chloride ionophore and the benzodiazepine receptor. In addition, the ability of the benzodiazepine-2 receptor ligands to recognize TBPS-induced changes in the conformation of the benzodiazepine-2 receptor is related to ligand efficacy as demonstrated by the minimal effect of TBPS on the binding of the weak partial agonist, [$^3$H]Ro 15-1788. A low concentration of [$^3$H]PrCC (50 pM) was used to selectively label the benzodiazepine-1 receptor subtype in the $P_2$ fraction of rat hippocampus. Previous studies indicated that 50 pM [$^3$H]PrCC labels a single population of high-affinity sites in the hippocampus [Gee et al., 1983]. [$^3$H]PrCC was incubated with various concentrations of TBPS at 25°C for 90 min. TBPS produced a dose-dependent enhancement of [$^3$H]PrCC binding (Fig. 9). The enhancement by TBPS was sensitive to GABA (10 μM), as shown by the shift of the dose–response curve to the right. The binding of [$^3$H]PrCC was insensitive to GABA under the conditions used. Moreover, the GABA effect was (+)bicuculline-sensitive but baclofen-insensitive (data not shown).

The observation that TBPS modulates [$^3$H]PrCC binding to the benzodiazepine-1 receptor subtype in rat hippocampus implies coupling between the benzodiazepine-1 site and a chloride ionophore. In view of our previous findings that the benzodiazepine-1 site is not functionally coupled to a GABA site, it is possible to speculate that the benzodiazepine-1 site may be coupled to an ionophore independent of a GABA site. A picture that emerges is

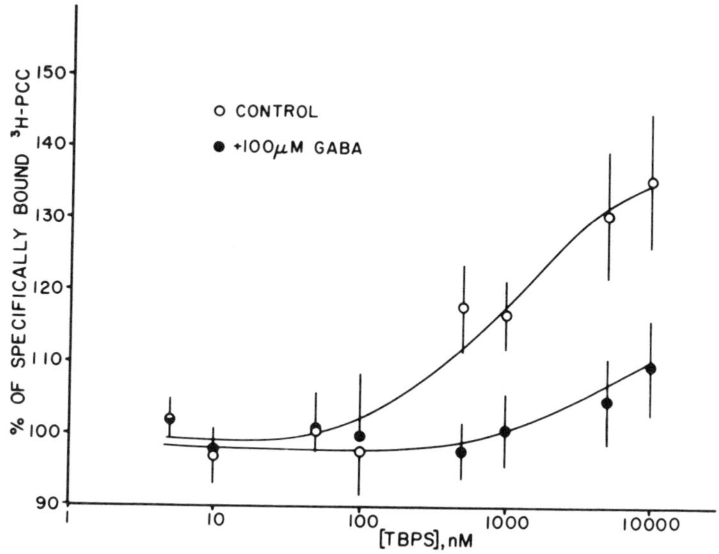


**Fig. 9.** *Dose-dependent enhancement of 50 pM [³H]propyl beta-carboline-3-carboxylate binding in rat hippocampal $P_2$ fractions by TBPS in the presence or absence of 10μM GABA. Each point represents the mean ± SEM of five to six experiments. [From Gee et al., 1985.]*

a chloride ionophore modulated independently by allosteric links to a GABA receptor and a benzodiazepine site. Implicit in this model is the ability of benzodiazepine agonists or inverse agonists to modulate the ionophore independent of GABA. This is a testable hypothesis, since it is possible to isolate the benzodiazepine-1 site along with the GABA receptor–chloride ionophore complex intact. The ability of agonist benzodiazepines to inhibit [$^{35}$S]TBPS binding in a preparation populated exclusively by benzodiazepine-1 sites independent of GABA would be presumptive evidence in support of the hypothesis that agonist benzodiazepines may modulate the chloride ionophore via the benzodiazepine-1 site independent of GABA. Of course, the pharmacological relevance of such interactions can be established only if the rank order of potencies and efficacies of the various benzodiazepine agonists at the benzodiazepine-1 site are correlated with their in vivo pharmacology. If such a benzodiazepine receptor subtype exists, it would be of value to determine the pharmacological effects of benzodiazepine receptor ligands directed specifically at this receptor subtype. Already there is speculation that this receptor subtype may selectively mediate the antianxiety and anticonvulsant effects of agonist benzodiazepines [Klepner et al., 1979]. However, the evidence for this hypothesis at present is equivocal. The answer to this question awaits the development of more selective benzodiazepine receptor subtype agonists.

## CONCLUSIONS

Changes in membrane chloride permeability form the physiological basis of inhibitory neurotransmission in both invertebrates and vertebrates. As modulators of chloride conductance, the benzodiazepines serve as extremely useful agents for the selective treatment of a variety of disorders related to dysfunction of inhibitory neurotransmission. They are therapeutically useful as anxiolytics, anticonvulsants, and muscle relaxants. Research over the past decade has provided considerable insight into the mechanism of benzodiazepine action. Throughout this course of study, a major question in defining the mechanism of action of the benzodiazepines has been the extent to which the GABA receptor is involved in benzodiazepine modulation of chloride conductance. This question is further complicated by the fact that benzodiazepine receptors show heterogeneity on the basis of their differential modulation by GABA.

It is well accepted that benzodiazepine and GABA recognition sites are allosterically coupled, but the degree of direct benzodiazepine–chloride ionophore coupling within the GABA receptor–ionophore complex has proven difficult to establish. This has, in part, been a consequence of the lack of specific and sensitive probes for the chloride ionophore. The recent development of [$^{35}$S]TBPS as a radioligand for a specific chloride ionophore–associated binding site has, however, provided a useful pharmacological probe for the benzodiazepine–ionophore interaction. The specific binding of [$^{35}$S]TBPS is modulated by all classes of compounds known to alter chloride ionophore function. It therefore appears that the conformational changes which accompany ionophore activation/deactivation perturb the TBPS binding domain. In this regard, the findings in "GABA-free" soluble receptor preparations suggest that direct interactions do occur between benzodiazepines and the chloride ionophore, but these effects occur at micromolar concentrations of benzodiazepines. The pharmacologically relevant benzodiazepine effects, on the other hand, appear to be mediated by nanomolar concentrations of benzodiazepines. Benzodiazepine modulation of [$^{35}$S]TBPS binding, and, presumably, chloride ionophore conformation, which occur in the nanomolar range are only observed in the presence of micromolar amounts of GABA and are therefore presumably mediated through low-affinity GABA recognition sites. Inhibition of [$^{35}$S]TBPS binding [Lawrence et al., 1984] rather than enhancement [Supavilai and Karobath, 1983 and 1984] is consistent with this mechanism of action, since GABA and all known GABA-mimetics inhibit [$^{35}$S]TBPS binding. Although the precise molecular nature of the benzodiazepine–GABA–ionophore coupling will only be fully understood through additional pharmacological and biochemical research, [$^{35}$S]TBPS appears to be a valuable probe for studies concerning this very important and interesting receptor-ionophore complex.

## ACKNOWLEDGMENTS

We are grateful to Susan Lawrence and Maria Felix for proofreading and word processing this manuscript. Portions of the research presented herein were supported in part by grants MH 27257 and MH-30636 from the National Institute of Mental Health. Henry I. Yamamura is a recipient of a Research Scientist Development Award (MH-00095) from the National Institute of Mental Health.

## REFERENCES

Abalis IM, Eldefrawi ME, Eldefrawi AT (1983): Biochemical identification of putative GABA/benzodiazepine receptors in house fly thorax muscles. Pestic Biochem Physiol 20:39–48.

Biggio G, Costa E (eds) (1983): Benzodiazepine Recognition Site Ligands: Biochemistry and Pharmacology. New York: Raven Press.

Bowery NG, Collins JF, Hill RG (1976): Bicyclic phosphorus esters that are potent convulsants and GABA antagonists. Nature 261:601–603.

Braestrup C, Squires RF (1978): Pharmacological characterization of benzodiazepine receptors in the brain. Eur J Pharmacol 48:263–270.

Briley MS, Langer SZ (1978): Influence of GABA receptor agonists and antagonists on the binding of $^{3}$H-diazepam to the benzodiazepine receptor. Eur J Pharmacol 52:129–132.

Casida JE, Lawrence LJ (1985): Structure–activity correlations for interactions of bicyclophosphorus esters and some polycholorocycloalkane and pyrethroid insecticides with the brain-specific *t*-butylbicyclophosphorothionate receptor. Environ Health Perspectives (in press).

Cole LM, Lawrence LJ, Casida JE (1984): Similar properties of [$^{35}$S] *t*-butylbicyclophosphorothionate receptor and coupled components of the GABA receptor-ionophore complex in brains of human, cow, rat, chicken and fish. Life Sci 35:1755–1762.

Costa E, Guidotti A, Mao CC, Suria A (1975): New concepts on the mechanism of action of benzodiazepines. Life Sci 17:167–186.

Dudel J, Kuffler SW (1961): Presynaptic inhibition at the crayfish neuromuscular junction. J Physiol 155:543–562.

Ehlert FJ, Roeske WR, Yamamura HI (1981): Multiple benzodiazepine receptors and the regulation by gamma-aminobutyric acid. Life Sci 29:235–248.

Ehlert FJ, Roseke WR, Gee KW, Yamamura HI (1983): An allosteric model for benzodiazepine receptor function. Biochem Pharmacol 32:2375–2383.

Ehlert FJ, Roeske WR, Yamamura HI (1983): The nature of muscarinic receptor binding. In Iversen LL, Iversen SD, and Snyder SH (eds): "Handbook of Psychopharmacology." New York: Plenum Press, pp 241–283.

Fatt P, Katz B (1953): The effect of inhibitory nerve impulses on a crustacean muscle fiber. J Physiol 121:374–389.

File SE, Simmonds MA (1984): Interactions of two phyenylquinolines with picrotoxin and benzodiazepines in vivo and in vitro. Eur J Pharmacol 97:295–300.

Gammon D, Casida JE (1983): Pyrethroids of the most potent class antagonize GABA action at the crayfish neuromuscular junction. Neurosci Lett 40:163–166.

Gee KW, Ehlert FJ, Yamamura HI (1983): Differential effect of gamma-aminobutyric acid on benzodiazepine receptor subtypes labeled by [$^3$H]propyl beta-carboline-3-carboxylate in rat brain. J Pharmacol Exp Ther 225:132–137.

Gee KW, Ehlert FJ, Roeske WR, Yamamura HI (1984a): Heterogeneity of benzodiazepine receptors. In Lajtha A (ed): "Handbook of Neurochemistry, Vol. 6." New York: Plenum Publishing Corp., pp 575–593.

Gee KW, Lawrence LJ, Yamamura HI (1985): Benzodiazepine receptor ligand modulation of the chloride ionophore: Influence of GABA and ligand efficacy. Mol Pharmacol (in press).

Gee KW, Yamamura SH, Roeske WR, Yamamura HI (1984b): Benzodiazepine receptor heterogeneity: possible molecular basis and functional significance. Fed Proc 43:2767–2772.

Guidotti A, Toffano G, Costa E (1978): An endogenous protein modulates the affinity of GABA and benzodiazepine receptors in rat brain. Nature 275:553–555.

Keane PE, Simiand J, Morre M (1984): The quinolines PK 8165 and PK 9084 possess benzodiazepine-like acitivity in vitro but not in vivo. Neurosci Lett 45:89–93.

Kent RS, DeLean A, Lefkowitz RJ (1980): A quantitative analysis of beta-adrenergic receptor interactions: Resolution of high and low affinity states of the receptor. Mol Pharmacol 17:14–23.

Klepner CA, Lippa AS, Benson DI, Sano MC, Beer B (1979): Resolution of two biochemically and pharmacologically distinct benzodiazepine receptors. Pharmacol Biochem Behav 11:457–462.

Lawrence LJ, Casida JE (1983): Stereospecific action of pyrethroid insecticides on the gamma-aminobutyric acid receptor-ionophore complex. Science 221:1399–1401.

Lawrence LJ, Casida JE (1984): Interactions of lindane, toxaphene and cyclodienes with brain-specific *t*-butylbicyclophosphorothionate receptor. Life Sci 35:-171–178.

Lawrence LJ, Gee KW, Yamamura HI (1984): Benzodiazepine anticonvulsant action: gamma-aminobutyric acid-dependent modulation of the chloride ionophore. Biochem Biophys Res Comm 123:1130–1137.

Le Fur G, Mizuole J, Burgevin MC, Ferris O, Heaulme N, Gauthier A, Gueremy C, Vzan A (1981): Multiple benzodiazepine receptors: Evidence of a dissociation between anticonflict and anticonvulsant properties by PK 8165 and PK 9084 (two quinoline derivatives). Life Sci 28:1439–1448.

Leeb-Lundberg F, Olsen RW (1980): Picrotoxinin binding as a probe of the GABA postsynaptic membrane receptor–ionophore complex. In Yamamura HI, Olsen RW, Usdin E (eds): "Psychopharmacology and Biochemistry of Neurotransmitter Receptors." New York: Elsevier, pp 593–606.

Leeb-Lundberg LMF, Olsen RW (1983): Heterogeneity of benzodiazepine receptor interactions with gamma-aminobutyric acid and barbiturate receptor sites. Mol Pharmacol 23:315–325.

Lippa AS, Regan B (1977): Additional studies on the importance of glycine and GABA in mediating the actions of benzodiazepines. Life Sci 21:1779–1784.

Marchbanks RM, Campbell CWB (1976): Sodium and chloride fluxes in syaptosomes *in vitro*. J Neurochem 26:973–980.

Matsumura F, Ghiasuddin SM (1983): Evidence for similarities between cyclodiene type insecticides and picrotoxinin in their action mechanisms. J Environ Sci Health 18:1–14.

Möhler H, Okada T (1977): Benzodiazepine receptor: demonstration in the central nervous system. Science 198:849–851.

Möhler H, Richards JG (1981): Agonist and antagonist benzodiazepine receptor interactions *in vitro*. Nature 294:763–765.

Nielsen M, Schou H, Braestrup C (1981): [$^3$H]Propyl-beta-carboline-3-carboxylate binds specifically to brain benzodiazepine receptors. J Neurochem 36:276–285.

Nutt DJ, Cowen PJ (1982): Unusual interactions of benzodiazepine receptor antagonists. Nature 295:436–438.

Olsen RW, Snowhill EW, Wamsley JK (1984): Autora-

diographic localization of low affinity GABA receptors with [$^3$H]bicuculline methachloride. Eur J Pharmacol 99:247–248.
Squires RF, Braestrup C (1977): Benzodiazepine receptors in rat brain. Nature 266:732–734.
Squires RF, Casida JE, Richardson M, Saederup E (1983): [$^{35}$S]-*t*-Butylbicyclophosphorothionate binds with high affinity to brain-specific sites coupled to gamma-aminobutyric acid-A and ion recognition sites. Mol Pharmacol 23:326–336.
Sternbach LH, Reeder E (1961): Quinazolines and 1,4-benzodiazepines. II. The rearrangement of 6-chloro-2-chloromethyl 4-phenylquinazoline 3-oxide into 2-amino derivates of 7-chloro-5-phenyl-3H-1,4-benzodiazepine 4-oxide. J Organic Chem 26:1111–1118.
Sternbach LH, Randall LO, Gustafson SR (1964): 1,4-Benzodiazepines (chlordiazepoxide and related compounds). In Gordon M (ed): "Psychopharmacological Agents, Vol. 1." New York: Academic Press, pp 137–224.
Study RE, Barker JL (1981): Diazepam and (−)-pentobarbital: fluctuation analysis reveals different mechanisms for potentiation of gamma-aminobutyric acid responses in cultured central neurons. Proc Natl Acad Sci 78:7180–7184.
Supavilai P, Karobath M (1983): Differential modulation of [$^{35}$S]TBPS binding by the occupancy of benzodiazepine receptors with its ligands. Eur J Pharmacol 91:145–146.
Supavilai P, Karobath M (1984): [$^{35}$S]-*t*-Butylbicyclophosphorothionate binding sites are constituents of the gamma-aminobutyric acid benzodiazepine receptor complex. J Neurosci 4:1193–1200.
Takeuchi A, Takeuchi N (1969): A study of the action of picrotoxin on the inhibitory neuromuscular junction of the crayfish. J Physiol 205:377–391.
Tallman JF, Thomas JW, Gallager DW (1978): GABAergic modulation of benzodiazepine binding site sensitivity. Nature 274:383–385.
Ticku MK, Olsen RW (1977): $\gamma$-Aminobutyric acid-stimulated chloride permeability in crayfish muscle. Biochim Biophys Acta 464:516–529.
Ticku MK, Ramanjaneyulu R (1984): Differential interactions of GABA agonists, depressant and convulsant drugs with [$^{35}$S]-*t*-butylbicyclophosphorothionate binding sites in cortex and cerebellum. Pharmacol Biochem Behav 21:151–158.
Woodbury DM (1980): Convulsant drugs: mechanisms of action. In Glaser GH, Penry JK, Woodbury DM (eds): "Antiepileptic Drugs: Mechanisms of Action." New York: Raven Press, pp 249–303.
Young WS, Kuhar MJ (1980): Radiohistochemical localization of benzodiazepine receptors in rat brain. J Pharmacol Exp Ther 212:425–430.

Benzodiazepine/GABA Receptors and Chloride Channels:
Structural and Functional Properties, pages 241–259


# 12

# Biochemical Aspects of GABA/Benzodiazepine Receptor Function

James B. Fischer and Richard W. Olsen

*Department of Pharmacology and Brain Research Institute, University of California, Los Angeles, California 90024*

## INTRODUCTION

The inhibitory neurotransmitter γ-aminobutyric acid (GABA) is widely used throughout the mammalian central nervous system. The most well characterized response of postsynaptic neurons to this transmitter is electrical inhibition due to an increased permeability of the neuronal membrane to chloride ions [Krnjević, 1974; Barker and Owen, this volume]. A great deal of progress has recently been made in understanding the biochemical properties of the receptor complex that mediates this response, and this chapter will summarize some of that progress. A second type of GABA receptor ($GABA_B$) has recently been described [Bowery et al., 1980, 1983] which shows very different pharmacology (stimulated by baclofen but not muscimol, and not blocked by bicuculline) and postsynaptic responses ($Ca^{2+}$ channel and adenylate cyclase regulation) from the "$GABA_A$" receptors described here [Dunlap and Fischbach, 1981; Enna and Karbon, 1984; Hill and Dolphin, 1984].

It has become clear that a receptor for the benzodiazepines is associated with the $GABA_A$ receptor, and that these anxiolytic drugs also affect the chloride ion channel. In addition, many barbiturates affect both the GABA and benzodiazepine receptors as well as the chloride channel in a stereospecific and picrotoxin-sensitive manner, suggesting that these drugs also have an associated receptor in this membrane complex. Although the question of endogenous ligands for these associated binding sites is still an unresolved problem, it is clear that a biochemical description of this receptor must include the entire GABA/benzodiazepine/barbiturate/picrotoxin receptor–chloride channel complex [Olsen, 1982].

## GABA RECEPTORS

The chloride channel–linked $GABA_A$ receptor (hereafter called GABA receptor) can be stimulated by several GABA analogs, such as muscimol, piperidine-4-sulfonic acid, THIP (4,5,6,7-tetrahydroisoxazolo[5,4-*c*] pyri-din-3-o1), and imidazole acetic acid [Barker and Mathers, 1981]. Stimulation by, all these agonists can be blocked by the GABA receptor antagonist bicuculline, and by the "chloride channel antagonist" picrotoxin, one of a family of convulsants found in plants of the family *Menispermaceae* [Takeuchi and Takeuchi, 1969; Barker et al., 1975; Olsen et al., 1978].

Ligand interactions with the GABA receptor can be assayed with various radioligands, including [$^3$H]GABA itself, and the more potent and GABA-specific agonist, muscimol [Snodgrass, 1978; Wang et al., 1979; Beaumont et al., 1978], as well as several other compounds [Olsen, 1983]. These binding sites show the pharmacological specificity expected of the $GABA_A$ muscimol/bicuculline-sensitive receptor sites [Greenlee et al. 1978b; Enna, this volume; Krogsgaard-Larsen, this volume]. Binding of [$^3$H]muscimol to GABA receptor sites in invertebrates has been characterized in crayfish muscle [Meiners et al., 1979] and in insect central ganglia [Lunt et al., 1984].

GABA receptor binding shows multiple affinities when carefully assayed in mammalian brain membranes [Enna and Snyder, 1977; Greenlee et al., 1978a; Winkler et al., 1979]. These data can be interpreted as two or three separate sites or affinity states of the GABA receptor, with $K_d$ values for [$^3$H]GABA of about 0.01, 0.1, and perhaps 1 $\mu$M [Olsen et al., 1981]. Dissociation kinetics for this receptor also show two or three components [Olsen and Snowman, 1982]. Further evidence for the existence of such a "super-low," micromolar affinity state includes the high concentrations (micromolar) of GABA and its analogues necessary to activate the chloride channel [Dichter, 1980; Nowak et al., 1982; Segal and Barker, 1984], enhance benzodiazepine binding [see below; Tallman et al., 1978; Karobath and Sperk, 1979], inhibit cage convulsant binding [Squires et al., 1983], and inhibit bicuculline binding [see below; Olsen and Snowman, 1983]. Because a series of GABA agonists shows the same rank order of potency to inhibit both the high- and lower-affinity binding sites, it seems likely that these sites are different affinity states of the same receptor [Olsen et al., 1984d]. There is also evidence that various membrane treatments (such as exposure to low concentrations of Triton X-100, or freezing) can change the proportion of sites found in the different affinity states, as can the ionic composition of the assay buffer [Toffano et al., 1978; Browner et al., 1981; Supavilai et al., 1982; Olsen and Snowman, 1983]. However, the improvement of GABA agonist binding by these treatments [Enna and Snyder, 1977; Greenlee et al., 1978a] also may be due at least partly to the removal of endogeneous inhibitors of binding, including GABA itself, and possibly a protein [Toffano et al., 1979]. The concentrations of GABA required to activate the membrane chloride channel (about 1–50 $\mu$M) are generally greater than the $K_d$ values for even the lower-affinity receptor populations detected with binding assays ($K_d$ about 0.2 $\mu$M) [Dichter, 1980; Nowak et al., 1982; Segal and Barker, 1984; Thampy and Barnes, 1984]. This suggests that the receptor sites observed with the ligand binding assays may be desensitized or uncoupled states of the receptor, in analogy with what has been learned about the nicotinic acetylchorine receptor [Changeux et al., 1984; Conti-Tronconi and Raftery, 1982].

Recently the GABA antagonist bicuculline has been used as a ligand for receptor binding studies [Möhler and Okada, 1978; Olsen and Snowman, 1983]. Bicuculline is a more potent antagonist of the fast-dissociating (lower-affinity) GABA receptors than the slowly dissociating (high-affinity) ones, with $IC_{50}$ values of 0.8 and 6 $\mu$M, respectively [Olsen and Snowman, 1983]. The $ED_{50}$ for bicuculline to inhibit the physiological response to GABA has been reported to be roughly 1 $\mu$M [Frere et al., 1982]. Also, certain ions (thiocyanate in particular) increase the potency for bicuculline to inhibit GABA binding [Enna and Snyder, 1977], while at the same time abolishing selectively the high-affinity GABA binding, thus leaving only the lower-affinity agonist binding for bicuculline to inhibit [Browner et al., 1981]. In line with this, when used as a radioligand, [$^3$H]bicuculline methochloride binding is significantly enhanced by thiocyanate ion, and it appears to preferentially label the lower-affinity GABA receptors [Olsen and Snowman, 1983; Olsen et al., 1984b]. Bicuculline binding is also inhibited by GABA agonist analogues with an order of potency similar to that seen for GABA binding, but with an absolute potency even lower than for the low-affinity agonist sites ($IC_{50}$

for GABA of 0.6–0.8 $\mu$M). Thus bicuculline may be a specific ligand for the physiologically relevant state of the GABA receptor.

## BENZODIAZEPINE RECEPTORS

In 1977 two groups reported specific, high-affinity receptors for the anxiolytic drugs, the benzodiazepines [Braestrup and Squires, 1977; Möhler and Okada, 1977]. These receptors definitively have been shown to be functionally linked to GABA receptors. GABA and its analogues enhance the binding of benzodiazepines to their receptors [Tallman et al., 1978; Supavilai and Karobath, 1980], and benzodiazepines enhance GABA activation of chloride channels [Gallager, 1978; Study and Barker, 1981; Haefely and Polc, this volume]. Under certain conditions it also appears that benzodiazepines may be able to enhance GABA receptor binding [Guidotti et al., 1978; Skerritt et al., 1982]. Benzodiazepine binding is also enhanced by chloride ion [Costa et al., 1979], and both GABA and benzodiazepine agonists are able to protect the receptors for the other class of compounds against heat inactivation [Squires, this volume; Gavish and Snyder, 1980a].

Several compounds have been found to interact with the benzodiazepine receptors in quite interesting ways. The benzodiazepine Ro 15-1788 appears to antagonize the pharmacological effects of the classical benzodiazepines, such as diazepam, being able to reverse their anticonvulsant, sedative, and ataxic effects [Hunkeler et al., 1981]. The compound CGS 8216 appears to behave in a very similar manner [Czernik et al., 1982]. Another group of compounds, the beta-carboline-3-carboxylate esters, not only antagonize several effects of the benzodiazepines, but can act as convulsants or proconvulsants on their own [Oakley and Jones, 1980; Cowen et al., 1981; Braestrup et al., 1982]. Actions of these "inverse agonists" can also be blocked by the "antagonists" such as Ro 15-1788 [Polc et al., 1982]. It has also been found that GABA, which enhances the binding of benzodiazepine agonists, decreases the binding of the inverse agonists to the benzodiazepine receptor [Braestrup and Nielsen, 1981].

The interactions of some of these compounds with the benzodiazepine receptor have revealed that at least two subtypes of this binding site exist in brain. Specifically, the triazolopyridazine anxiolytics, CL218,872, and some beta-carboline-3-carboxylate esters inhibit benzodiazepine binding in a biphasic manner [Klepner et al., 1979; Nielsen and Braestrup, 1980]. Both types of ligand bind with greater affinity to a receptor subtype that predominates in the cerebellum, termed "type 1" or $BZ_1$. Certain benzodiazepine agonists also bind preferentially to the type 1 receptor subtype, which appears to correspond to a distinct peptide on SDS gels [Sieghart et al., 1983]. The two subpopulations differing in affinity for triazolopyridazines also appear to be two different proteins on the basis of ontogeny, regional localization [Young et al., 1981], detergent solubilization properties, and sensitivity to modulation by ions [Lo et al., 1982]. When used as radioligands, the beta-carbolines directly demonstrate biphasic dissociation curves in those brain regions which show both $BZ_1$ and $BZ_2$ receptors, such as the hippocampus and cerebral cortex [Gee et al., 1983]. GABA was found to enhance the affinity of benzodiazepine agonist binding to the $BZ_1$ receptors of the cerebellum and cortex as well as the $BZ_2$ receptors of the hippocampus and cortex, while the $BZ_1$ receptors in the hippocampus were apparently less affected [Gee et al., 1983]. Several workers [Gee et al., 1983; Braestrup et al., 1982] interpret the apparent heterogeneity of benzodiazepine receptors on the basis of different allosteric states of a single receptor protein, modulated by ligands differing in efficacy, as we have proposed above in discussing the GABA binding sites [Olsen et al., 1984d], and also discussed below.

Insights into the molecular nature of the

benzodiazepine receptor have been provided by the ability of certain benzodiazepines to be used as specific, irreversible photoaffinity labels of the receptor. Benzodiazepines with a nitro group in the 7 position of the conjugated ring, such as flunitrazepam and clonazepam [Möhler et al., 1980; Sieghart and Möhler, 1982], are chemically activated by ultraviolet light and bind covalently to the receptor [Sherman-Gold, 1983]. When maximal amounts of [$^3$H]flunitrazepam are incorporated into membrane-bound benzodiazepine receptors in this way, about 25% of the total number of receptors present are labeled with tritium [Möhler et al., 1980; Thomas and Tallman, 1981; Karobath and Supavilai, 1982]. The remaining receptors cannot be detected upon reassay of the membranes with labeled benzodiazepine agonists (nonradioactive flunitrazepam being used for the photoaffinity labeling). If, however, labeled Ro15-1788, CGS 8216, or a beta-carboline is used to assay the membranes, the receptors can still be detected, and show unaltered affinity and a decrease of only 25% or less in numbers [Hirsch, 1982; Möhler, 1982; Karobath and Supavilai, 1982; Thomas and Tallman, 1983]. Benzodiazepine agonists are able to displace radioactive beta-carboline binding to photoaffinity-labeled membranes, but with greatly reduced affinity compared to their effects in control membranes. Thus it appears that the benzodiazepine agonists interact with these receptors in a way that involves a functional complex of several binding sites, with the photolabeling of one site leading to a decreased affinity for agonists in three more sites. The antagonists or inverse agonists do not seem to invoke such receptor interactions. Indeed, when an antagonist is used for the photolabeling, more than 90% of the sites could be labeled, and sites not labeled had unaltered affinities for both agonists and antagonists [Möhler et al., 1984a]. A further interesting aspect of this apparent subunit–subunit interaction is that the 25% covalent incorporation/75% loss of affinity is not seen in detergent-solubilized extracts, where greater than 50% of the available sites are covalently photolabeled with [$^3$H]flunitrazepam [Thomas and Tallman, 1981].

Another use that has been made for the photoaffinity labeling of benzodiazepine receptors is the determination of the molecular weight(s) of the labeled protein subunit. Using SDS polyacrylamide gel electrophoresis, several groups have reported a major peptide of molecular weight 48–55 kd to be labeled with [$^3$H]flunitrazepam [Möhler et al., 1980; Thomas and Tallman, 1981; Sieghart et al., 1983]. Sieghart et al. [1983] also found several larger peptides labeled (to a lesser degree) in many brain regions, of molecular weights 53 kd, 55 kd, and 59 kd. In the cerebellum only a 51 kd band is labeled, suggesting that this peptide may be associated with the $BZ_1$ receptor subtype which predominates in this brain region, as described above.

## PICROTOXIN/BARBITURATE RECEPTORS

The convulsant picrotoxin has been found by several groups to be a potent inhibitor of the chloride-conductance response to GABA [Takeuchi and Takeuchi, 1969; Ticku and Olsen, 1977; Nowak et al., 1982]. Picrotoxin does not block GABA receptor binding at relevant concentrations, suggesting that it acts through a functionally related but separate site on the receptor complex [Olsen et al., 1978]. When picrotoxin was radioactively labeled and used for receptor binding assays, it was found to label sites that in large part corresponded well to the expected GABA receptor–linked sites [Olsen et al., 1978; Ticku et al., 1978a,b]. These sites show tissue distributions that parallel GABA receptor sites and are inhibited by active, but not inactive, analogues of picrotoxin, as well as by bicyclophosphate "cage convulsants" [Olsen, 1982] which inhibit GABA physiological responses [Bowery et al., 1976].

Another important class of drugs that inhibits picrotoxin binding is the barbiturates. Many of these clinically important anesthetic, hypnotic, and anticonvulsant drugs at micromolar concentrations interact allosterically with GABA and benzodiazepine binding sites as well as the picrotoxin sites [Leeb-Lundberg et al., 1980; Willow and Johnston, 1981; Olsen and Snowman, 1982; Ticku and Maksay, 1983; Olsen et al., 1985a]. Benzodiazepine binding is increased in affinity 2–3 fold by barbiturates [Leeb-Lundberg et al., 1980; Leeb-Lundberg and Olsen, 1982]. Barbiturate enhancement of GABA binding on the other hand appears as an increase in the number of binding sites for both the high- and low-affinity receptors [Olsen and Snowman, 1982]. A possible mechanism by which this change could occur is that the "super-low"-affinity receptor state ($K_d$ of $> 1\ \mu M$) could be shifted into the higher-affinity states by barbiturates [Olsen et al., 1984c,d].

The ability of barbiturates to enhance benzodiazepine and GABA binding can be blocked by picrotoxin and is dependent upon the presence of chloride ions in the incubation medium [Leeb-Lundberg et al., 1980, 1981b; Olsen and Snowman, 1982]. Certain other anions (for example, bromide, iodide, and thiocyanate) can also support these receptor interactions, and they are the same anions that have been shown to be able to move through GABA-activated chloride channels [Eccles et al., 1977]. Barbiturates have also been shown to enhance the GABA activation of chloride channels [Nicoll et al., 1975] by increasing the duration of the response (mean open-channel lifetime) [Study and Barker, 1981]. Interestingly, barbiturates inhibit the binding of the GABA antagonist [$^3$H]bicuculline methochloride and the benzodiazepine "inverse agonist" [$^3$H]beta-carboline-3-carboxylic acid methyl ester, having the opposite effect of what is seen on agonist binding to these sites [Wong et al., 1984b]. These interactions of barbiturates with GABA and benzodiazepine binding sites also can be blocked by picrotoxin.

These effects of the barbiturates show a potency consistent with clinically achieved concentrations of these drugs and also show stereospecificity, indicating specific receptor-type interactions as opposed to nonspecific membrane lipid effects. The chemical specificity is the same for all five of these interactions with the GABA receptor complex, with (−)pentobarbital and (+)hexobarbital showing relatively high potency while phenobarbital is much weaker. This order of potency is the same as that seen for the enhancement of GABA function at the cellular level by these agents and is also well correlated with their in vivo hypnotic actions [Huang and Barker, 1980; Olsen et al., 1985a]. In addition to these barbiturates, several other related depressant drugs have been found to interact with this system in the same manner. These include etomidate (stereospecific, (+)isomer > (−)), several pyrazolopyridines (etazolate, tracazolate and cartazolate), and the diaryltriazine LY81067 [Leeb-Lundberg et al., 1981b; Willow, 1981; Wong et al., 1983, 1984b; Olsen et al., 1985a].

Like picrotoxin itself, several other convulsant drugs inhibit picrotoxin binding. One is a convulsant benzodiazepine, Ro5-3663 [Schlosser and Franco, 1979; Leeb-Lundberg et al., 1981a]. Already mentioned are the bicyclophosphate "cage convulsants" [Ticku and Olsen, 1979]. One derivative of these compounds, [$^{35}$S]*t*-butyl bicyclophosphorothionate (TBPS), has recently been used as a ligand for the picrotoxin binding site [Squires et al., 1983]. Binding of TBPS is enhanced by chloride ion or the other "Eccles anions" mentioned above, and GABA inhibits TBPS binding by decreasing the affinity of these ions to enhance TBPS binding. This ligand has a higher affinity for the barbiturate-sensitive picrotoxin site than picrotoxin itself, gives a much higher signal/noise ratio, and provides much better quantitative data. Many of the same interactions of barbiturates and related depressants as well as convulsants found for picrotoxin binding sites have already been shown for the [$^{35}$S]TBPS binding site [Ramanjaneyulu and Ticku, 1984].

Recent studies with TBPS have suggested that the site of interaction for barbiturates and the related compounds may not be identical to the TBPS binding site. Specifically, the inhibition of TBPS binding by these drugs appears to be noncompetitive, with an apparent decrease in the density of binding sites and an increase in dissociation rate; further, their inhibition can be reduced allosterically by the GABA receptor antagonists bicuculline and RU 5135, while inhibition by other convulsant drugs is not reduced [Squires et al., 1983; Ramanjaneyulu and Ticku, 1984; Trifiletti et al., 1984].

These mutual interactions between drug binding sites have supported a model for the GABA receptor–ion channel complex as shown in Figure 1 [see also Chapters 9 and 10, this volume]. This complex contains at least four binding sites on the same protein complex (not necessarily on four different subunits): the GABA, benzodiazepine, barbiturate, and picrotoxin sites.

## ALLOSTERIC MODEL OF THE GABA/BENZODIAZEPINE RECEPTOR

Figure 2 shows an allosteric model of this receptor complex, which might explain the various site-site interactions reported in the literature.

The GABA receptor–chloride channel is suggested to exist in the three states shown, corresponding to the resting state (channel closed), activated state (channel open), and desensitized state (channel closed). When GABA binds to the receptor in the resting state, it is converted to the open state on a rapid millisecond time scale. If GABA remains on the receptor, it slowly converts to the closed, desensitized state. It is possible that the heterogeneity of GABA affinity seen in binding to membrane homogenates at 0° in nonphysiological buffers is related to these three states of the GABA receptor, which happen to have sufficiently disparate affinities ($K_d$ values of 1 $\mu$M, 0.1 $\mu$M, and 0.01 $\mu$M, respectively), and are *not* interconvertible under the "artificial" in vitro conditions. Note that in such a model, which is similar to one proposed for the nicotinic acetylcholine receptor [Changeux et al., 1984], the highest-

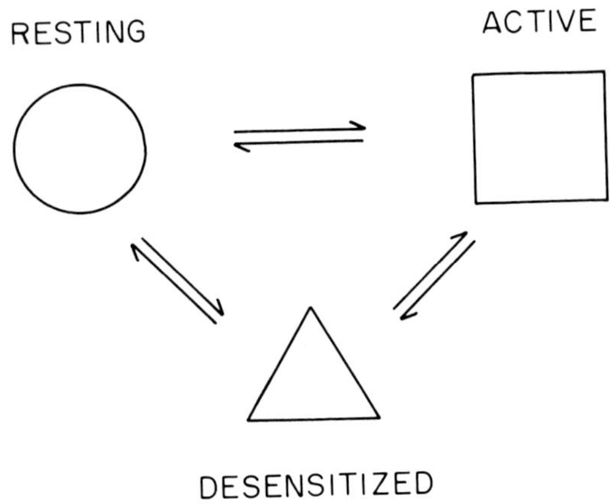


**Fig. 2.** *Three-state model of the GABA receptor. Three functional comformational states are hypothesized: resting, active, and desensitized. These are proposed to correspond to different conformational states detected in vitro, with GABA affinities of desensitized > active > resting (kd values of roughly 10 nM, 100 nM, and 1 $\mu$M). The GABA antagonist bicuculline shows the opposite affinity (resting > active > desensitized). Barbiturates favor a shift to higher GABA affinity (resting → active → desensitized), while picrotoxin opposes this. Benzodiazepines, however, show a different preference: (active > resting > desensitized), with beta-carbolines having the opposite affinity. The desensitized state (high affinity for GABA) is also favored by freezing, Triton-washing, silver ion, and irradiation, while the resting state (low affinity for GABA) is favored by anions and CHAPS detergent.*

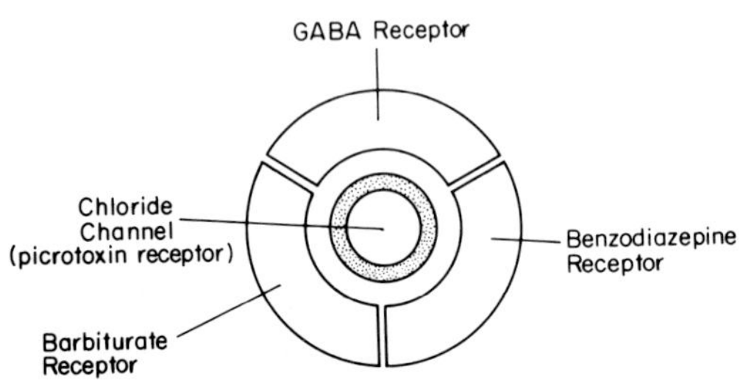


**Fig. 1.** *Donut model of the GABA receptor complex. Four distinct drug binding sites are hypothesized: receptors for GABA, benzodiazepine, barbiturates, and convulsants. [From Olsen et al., 1985a].*

affinity agonist state is not related to channel activation, but is nonfunctional. The different affinity states of GABA binding are all present at equilibrium during in vitro binding assays at 0°C (desensitized > activated > resting for agonist ligands, and the opposite for antagonist ligands). The equilibrium can be altered in vitro by allosteric modulators, such as chloride ion or avermectin (shift to low affinity) or silver ion, barbiturates, or Triton-washing (shift to high affinity). The distribution of GABA receptors between these three states as detected in vitro appears to vary with brain region. This suggests that some physical differences may contribute to the observed heterogeneity, such as variable post-translational modification of the receptor or variable membrane microenvironments.

Experimental support for the three-state model of a single GABA receptor comes from: 1) the need for micromolar concentrations of GABA to activate channels and benzodiazepine binding as described above; 2) the direct detection of micromolar GABA binding sites and their conversion to higher-affinity states by barbiturates [Olsen and Snowman, 1982]; and 3) the reciprocal effect of many treatments on agonist and antagonist binding [Enna and Snyder, 1977; Olsen and Snowman, 1983; Olsen et al., 1984d]. Further, Gallager et al. [1984] have demonstrated a physiological subsensitivity of rat neurons to GABA induced by chronic benzodiazepine administration (tolerance), which is accompanied by a shift in GABA receptor binding to *higher* affinity. In addition, Maksay and Ticku [1984] have shown that incubation of membranes with GABA at 37°C but not at 0°C increases the number of high- and intermediate-affinity binding sites, apparently by converting the low-affinity sites to higher affinity (as seen at 0°C with barbiturates [Olsen and Snowman, 1982]).

The benzodiazepine receptor likewise exhibits apparently heterogeneity [see Chapter 10, this volume] with respect to equilibrium affinity for certain but not all ligands, association–dissociation kinetics [e.g., Chiu et al., 1982], certain physical properties, and allosteric interactions with the GABA, depressant, and convulsant receptor sites in the complex (Fig. 1). The allosteric modulation of benzodiazepine binding suggests the presence of multiple conformational states [Leeb-Lundberg and Olsen, 1983; Braestrup et al., 1982; Gee et al., 1983]; these states may explain at least partially the heterogeneity mentioned. It is difficult to fit in the multiple states of the benzodiazepine receptors with the three-state model proposed here for the GABA receptor (Fig. 2), but we will try.

First of all, while three states are sufficient to explain most observations regarding benzodiazepine binding, these states cannot correspond exactly with those for GABA regarding the low, intermediate, and high affinities for benzodiazepines. For example, avermectin enhances benzodiazepine binding but inhibits GABA binding [Olsen and Snowman, 1985]. However, the three states do not need to have such widely disparate affinities for all ligands as they show for GABA ligands. For example, the affinities of benzodiazepines for the three states may differ by 2-fold or less. Thus, the effects of modulatory drugs and assay conditions on benzodiazepine ligand binding may be more difficult to observe. We propose that the highest-affinity state for GABA agonists (desensitized) has the lowest affinity for depressant (agonist) benzodiazepines; diazepam and analogues would show highest affinity in the order: activated > resting > desensitized, with beta-carbolines like DMCM giving the opposite: desensitized > resting > activated.

This scheme would explain why GABA enhances diazepam binding and inhibits DMCM binding (activated state; GABA cannot convert to desensitized state in vitro at 0°C). The correspondence of a high-affinity beta-carboline binding state with the high-affinity GABA binding state (desensitized) would be consistent with several observations. For example, bicuculline has a low affinity to inhibit barbiturate enhancement of the subpopulation of benzodiazepine receptors showing a high affinity for beta-carbolines, while low-affinity beta-carboline sites have a high affinity for bicuculline [Leeb-Lundberg and Olsen, 1983]. Likewise, Triton-washing, silver ion, and irradiation enhance [$^3$H]DMCM binding, as they enhance GABA binding [Nielsen and Braestrup, 1984].

Since barbiturates favor intermediate- and high-affinity GABA binding, they would be expected to favor desensitization in the long run (a testable hypothesis), but apparently favor activation in the short run. Benzodiazepines, having a low affinity for the high-affinity GABA state, do not dramatically enhance GABA binding. Physiologically, they would be expected to oppose desensitization as their mechanism of action (another testable hypothesis), whereas beta-carbolines like DMCM would promote desensitization.

## CHLORIDE FLUX IN BRAIN SLICES

An assay for the chloride channel response to GABA has recently been developed in this laboratory, based on the method of Teichberg and co-workers [Luini et al., 1981], utilizing $^{36}Cl^-$ as tracer [Wong et al., 1984a; Yang-Ransom et al., 1984]. Brain slices are allowed to equilibrate with $^{36}Cl^-$ in oxygenated Ringer's solution, and then are transferred through a series of vials containing fresh Ringer's at 1-minute intervals. After an initial washout period, the efflux of $^{36}Cl^-$ stabilizes, at which time drugs can be introduced to test their effect on the efflux. GABA, muscimol, and other GABA agonists cause a significant increase in the efflux rate, and these effects are inhibited by bicuculline and picrotoxin. Tetrodotoxin and high-$Mg^{2+}$/low-$Ca^{2+}$ buffer did not reduce the increased efflux, indicating direct receptor activation. Glycine, glutamate, and taurine did not stimulate chloride efflux, nor did nipecotic acid or 2,4-diaminobutyric acid (GABA uptake blockers which do not activate GABA receptors). Barbiturates, especially (−)pentobarbital, are also able to increase the $^{36}Cl^-$ efflux rate in this system, consistent with what is seen with microelectrodes in cultured neurons [Mathers and Barker, 1980]. These responses are also blocked by picrotoxin and bicuculline, indicating their relationship to the GABA receptor complex. In addition, low levels of pentobarbital are able to significantly increase the effect of low levels of GABA on chloride efflux [Yang-Ransom et al., 1984], again showing similar responses to those seen with standard electrophysiological techniques [Nicoll et al., 1975; Study and Barker, 1981]. These findings further support the physiological and neurochemical importance of the picrotoxin and barbiturate sites on the GABA receptor complex [Olsen, 1982]. Most of this work was performed using rat hippocampal slices, but similar responses have been seen in rat cortical and striatal slices, and mouse cortical slices. Earlier, we used an assay of *$^{36}Cl^-$ influx* to assay GABA receptor function in crayfish muscle fibers [Ticku and Olsen, 1977], and this has recently been adapted for cultured vertebrate neurons [Thampy and Barnes, 1984]. GABA-activated $^{36}Cl^-$ flux has not been described in cell-free membrane vesicles, and such an assay would be valuable in the eventual reconstitution of membrane-bound function using purified GABA receptor protein.

## BIOCHEMICAL STUDIES OF THE RECEPTOR COMPLEX: SOLUBILIZATION OF THE RECEPTOR COMPLEX

Many investigators have reported the successful solubilization of specific GABA and/or benzodiazepine binding sites from neuronal membranes. Among those reports showing functional interactions between the solubilized receptors (GABA enhancement of benzodiazepine binding), successful detergents include deoxycholate [Greenlee and Olsen, 1979; Sherman-Gold and Dudai, 1980; Asano et al., 1983; Stephenson et al., 1982], Triton X-100 [Gavish and Snyder, 1980b; Chang and Barnard, 1982], Lubrol-PX [Davis and Ticku, 1981], and Nonidet P-40 [Ito and Kuriyama, 1982]. As summarized in Table I,

**TABLE I. Summary of Literature for Molecular Weight and Subunit Composition of GABA/Benzodiazepine Receptor Protein**

| Source (detergent)[a] | Mol. Wt.[b] | Subunits | References |
|---|---|---|---|
| Bovine brain (DOC) | 900 | — | Greenlee and Olsen, 1979 |
| Rat brain (Lubrol) | 220 | — | Yousufi et al., 1979 |
| Rat cortex (Triton) | 12.8S | — | Lang et al., 1979 |
| Calf cortex (DOC) | 11S | — | Sherman-Gold and Dudai, 1980 |
| Rat, bovine brain (DOC) | 650 (11.3S) | — | Asano et al., 1983 |
| Bovine brain (DOC, Triton) | 6.9 nm; 12.5S[c] | — | Stephenson et al., 1982 |
| Rat brain (NP40) | 270 | — | Ito and Kuriyama, 1982 |
| Bovine brain (CHAPS) | 6.9 nm | — | Stephenson and Olsen, 1982 |
| Bovine, rat cortex (Triton) | 220 (9.4S) | — | Chang and Barnard, 1982 |
| Rat brain (Triton) | 220 | 60 | Martini et al., 1982, 1983 |
| Bovine cortex (DOC, Triton) | 7.3 nm; 11.1s | 53[d], 57[d] | Sigel et al., 1983 |
| Rat brain (NP40) | — | 49[d], 55 | Kuriyama et al., 1984 |
| Bovine cortex (CHAPS) | — | 53[d], 57 | Sigel and Barnard, 1984 |
| Bovine brain (DOC) | — | 50, 55[d], 62 | Möhler et al., 1984b |
| Rat brain (DOC, Triton) | 220 | 49, 55[d], 58 | This chapter |

[a]DOC is deoxycholate; Lubrol is Lubrol-PX; Triton is Triton X-100; NP40 is Nonidet P40; CHAPS is (3-[(3-cholamidopropyl)-dimethylammonio]propanesulfonate).
[b]Molecular weight values are apparent (in kilodaltons) based on gel filtration, or Stokes' radius (in nm) is given, and S values determined from sucrose gradients.
[c]The combined gel filtration and sedimentation data allowed a calculation of molecular weight = 355 kd.
[d]Photoaffinity labeled with $[^3H]$flunitrazepam.

gel filtration chromatography of these solubilized receptor preparations has led to widely differing estimates of the size of the native receptor complex, from about 240 kd [Chang and Barnard, 1982] to 900 kd [Greenlee and Olsen, 1979] when compared to globular standards that do not bind detergents. Stephenson et al. [1982] combined sedimentation data in both $H_2O$ and $D_2O$ with gel-filtration data to correct for detergent binding and molecular shape, and arrived at a molecular weight of 350 kd for the GABA/benzodiazepine receptor–detergent complex solubilized with deoxycholate and chromatographed in Triton X-100 (a procedure which gives both a good solubilization yield and good stability of the activity).

Solubilization of the receptor complex with picrotoxin binding and barbiturate interactions still functionally intact has been reported less often. Davis and Ticku [1981] used Lubrox-PX to solubilize picrotoxin, GABA, and benzodiazepine binding activities, and further found the benzodiazepine site to separate from the other two (smaller-size) upon gel-filtration chromatography. Both barbiturates and GABA enhanced benzodiazepine binding in the crude solubilized extracts, but neither of these substances was active on the benzodiazepine binding fraction after gel filtration. In contrast, deoxycholate/Triton soluble extracts show only poorly reproducible barbiturate enhancement while maintaining the GABA and benzodiazepine binding sites together during gel filtration [Stephenson et al., 1982]. Stephenson and Olsen [1982] have reported that the detergent CHAPS (3-[(3-cholamidopropyl)dimethylammonio]propanesulfonate) was able to solubilize the GABA and benzodiazepine receptors with reproducible barbiturate enhancement of both binding sites, which persisted even after gel filtration chromatography. This enhancement was inhibited by picrotoxin and enhanced by chloride ions, consistent with what is seen in the membrane-bound receptor complex. Gel filtration in CHAPS gave a Stokes' radius for this receptor complex (6.9 nm) identical to that seen in deoxycholate/Triton extracts. King and Olsen [1984] have since shown that [$^{35}$S]TBPS binding to the picrotoxin site is also effectively solubilized with CHAPS, and that this site comigrates upon gel filtration with the GABA/benzodiazepine receptor complex. CHAPS appears to be the best detergent for solubilizing what could be called the complete GABA/benzodiazepine/picrotoxin receptor-complex, or the GABA receptor complex with all subunit interactions intact.

## PURIFICATION OF THE RECEPTOR COMPLEX

The first significant attempts at purification of the GABA/benzodiazepine receptor were described by Gavish and Snyder [1981], who used gel filtration, ion exchange, and concanavalin A and benzodiazepine affinity chromatography, achieving up to a 200-fold purification with the latter technique. The GABA and benzodiazepine binding sites copurified in all of these procedures, and the GABA enhancement of benzodiazepine binding was maintained. Other workers have also used benzodiazepine affinity columns, and greater purifications have been achieved [Martini et al., 1982, 1983; Sigel et al., 1982, 1983; Sigel and Barnard, 1984; Schoch and Möhler, 1983; Olsen et al., 1983a,c]. However, discrepancies exist in these reports as to the specific activities obtained and to the number and size of the subunits of the purified receptor complex as determined by SDS polyacrylamide gel electrophoresis. A single subunit of 60 kd was reported by Martini et al. [1982, 1983], while two subunits were found by Kuriyama et al. [1984] of 48.5 and 54.5 kd, and by Sigel and co-workers [1983, 1984] of 53 and 57 kd (also summarized in Table II). In both of these latter cases the smaller of the two subunits gave considerably more staining density than the larger. Tallman and co-workers [Chapter 14, this volume] have consistently observed three bands in the 50–55 kd range plus a band at approximately 90 kd.

**TABLE II. Properties of Purified GABA/Benzodiazepine/Barbituriate Receptor Protein From Rat Brain**

| |
|---|
| Binds [$^3$H]Muscimol: specific activity 800–12,000 pmol/mg; $K_d$ = 20 nM. Inhibited by muscimol, GABA, bicuculline, 3-aminopropane sulfonate but not DABA or baclofen (10 $\mu$M). Enhanced by pentobarbital (2 mM): 47–56%. |
| Binds [$^{35}$S]TBPS: $K_d$ = 30 nM. Inhibited by ethyl bicyclophosphate, picrotoxinin, pentobarbital, GABA. |
| Binds [$^3$H]Flunitrazepam and clonazepam but under 30% by Ro5-4864 (all at 300 nM). Photoaffinity labeled: 55 kd band. |
| Subunit composition: 49 kd, 55 kd, 58 kd (39 kd, 31 kd, 52–53 kd) plus some minor bands above and below. |
| Native molecular weight: Stokes' radius ~ Catalase. |

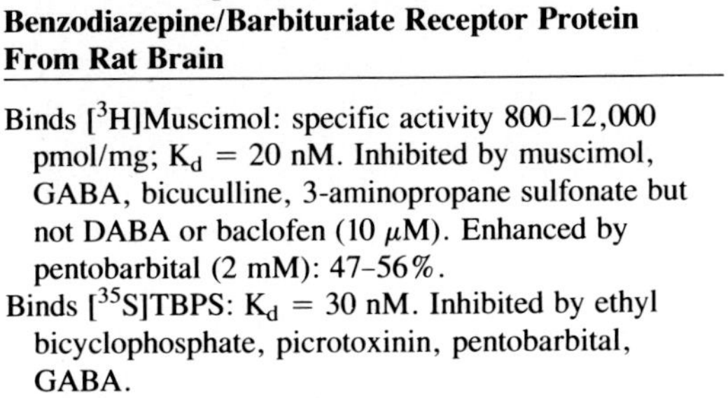


**Fig. 3.** *Structure of a benzodiazepine affinity gel.*

We have also used a benzodiazepine affinity column to isolate the GABA/benzodiazepine receptor complex as well as the TBPS binding site [Olsen et al., 1984a,c]. The benzodiazepine Ro7-1986/1 can be efficiently coupled to a carboxy-terminal agarose gel in a single step reaction using a water-soluble carbodimide at pH 5 (Fig. 3). After thorough washing, these gels consistently bind 90% of the applied benzodiazepine binding activity and 75% of the GABA activity in a soluble preparation. A deoxycholate extract of rat brain membranes passed through a Sepharose 6B gel filtration column in Triton X-100 is usually used in this step. The GABA/benzodiazepine receptor–containing pooled fractions of Sepharose 6B chromatography have the advantages of: 1) the removal of endogenous GABA, allowing more accurate determination of the [$^3$H]muscimol binding activity in the resulting pool that is applied to the affinity gel; 2) a partial purification, especially the separation from the void volume containing unsolubilized material; and 3) the enhanced stability of the receptor activity in Triton X-100 [Stephenson et al., 1982]. However, a crude deoxycholate soluble extract with Triton added may be used directly on the affinity gel with satisfactory results. In addition, several protease inhibitors are added at the time of solubilization [Stephenson and Olsen, 1982] and are maintained throughout the purification: these are phenylmethyl sulfonyl fluoride, benzamidine, bacitracin, benzethonium chloride, and EDTA.

After binding the receptor complex, the affinity column is washed extensively to remove nonspecifically bound proteins, typically with 20–30 column volumes of buffer, including several volumes of high salt (0.2 M KCl) buffer, and several volumes of 1 M urea (all in Triton X-100). Usually only a few percent of the receptor activities are removed by these washes. Specific elution of the receptor is accomplished with 2–4 volumes of 1 mM flurazepam with 1 M urea in buffer over several hours. GABA receptor binding activity (measured with [$^3$H]muscimol) shows up to a 10,000-fold enrichment in the eluant compared to the crude soluble extract. Benzodiazepine binding cannot be measured at this stage due to the presence of high concentrations of flurazepam, but these can be removed by DEAE ion exchange chromatography, giving a preparation with both GABA and benzodiazepine binding in a ratio of about 2:1 [Olsen et al., 1984a,c; Stauber, King, and Olsen, unpublished].

When the purification is carried out in Triton detergent as described above, the TBPS/picrotoxin receptor and the barbiturate enhancement of the GABA and benzodiazepine receptors are not seen reproducibly (since, as described earlier, these activities are very unstable or inhibited in Triton). However, if

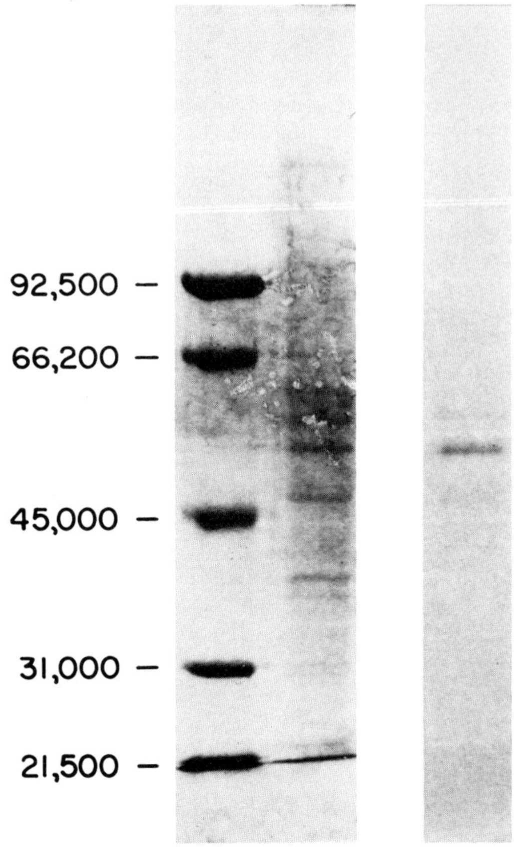


**Fig. 4.** *Fluorograph of SDS gel showing photoaffinity-labeled, affinity-purified GABA/benzodiazepine receptor from rat brain (right lane), compared to Coomassie blue stain of affinity-purified GABA/benzodiazepine receptor (center lane) and standards (left lane).*

CHAPS detergent is used throughout the purification, TBPS binding and the barbiturate-enhancement effects are retained by the purified preparation. The properties are summarized in Table II. This result, also found by Sigel and Barnard [1984], supports the concept that the complete GABA/benzodiazepine/picrotoxin–barbiturate receptor complex exists as tightly linked entity.

After DEAE chromatography, the purified receptor can be irreversibly photoaffinity-labeled with [$^3$H]flunitrazepam. Figure 4 shows tritium-labeled proteins from a purified receptor preparation analyzed by SDS polyacrylamide gels, and shows clearly that a 55 kd protein is the primary site of labeling. Lighter labeling could also be detected in two bands at 58 kd and 49 kd. The latter might represent a proteolytic product of one of the larger bands. The major labeling at 55 kd with minor labeling at 58 kd is consistent with the reports of a microheterogeneity of photoaffinity labeling of membranes by Sieghart et al. [1983].

Figure 4 also shows a Coomassie blue-stained SDS gel containing several molecular weight standards and an aliquot of a purified receptor preparation. In ten different preparations, a major band was seen at 55 kd, corresponding to the photolabeled subunit. A second major band at 49 kd was generally seen, and a third band at 58 kd seemed to correspond to the minor photoaffinity-labeled peptides. Bands at 39 kd, 31 kd, 62 kd, and 52 kd were observed with variable frequency. Many other minor protein components are visible in these gels, especially with silver staining, indicating that further improvements in the purification procedure are needed. However, the high specific activities attained in these preparations suggest that the receptor subunits should constitute the major proteins present. (If the equivalent weight represented one binding site per 50 kd, then the theoretical specific activity of the pure protein would be 20,000 pmol/mg protein.)

We have also analyzed the purified receptor by gel filtration chromatography on Sepharose 6B in Triton X-100 after labeling with [$^3$H]flunitrazepam or $^{125}$Iodine. In both cases, the receptor eluted as a single peak near the standard catalase (Stokes' radius of 5.2 nm; 240 kd), as shown in Figure 5. This is considerably smaller than the apparent size of the freshly solubilized receptor chromatographed in Triton by Stephenson et al. [1982], where a Stokes' radius of 6.8 nm was found. In the presence of Triton, the TBPS binding site is not detectable; it may become dissociated from the GABA/benzodiazepine complex during the purification steps, or a less-aggre-

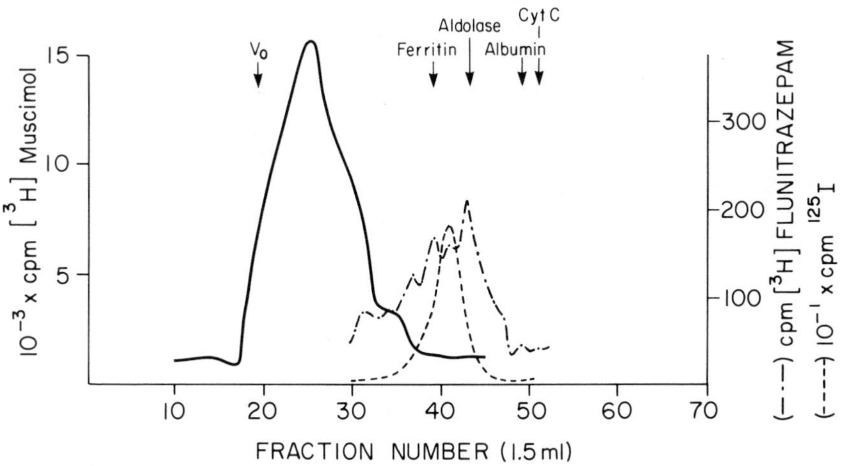


**Fig. 5.** *Gel filtration on Sepharose 6B in Triton X-100 of a crude solubilized extract of the GABA/benzodiazepine receptor, photoaffinity-labeled purified receptor, and $^{125}$I-labeled purified receptor.*

gated state of the receptor containing nevertheless the same peptide subunits may be purified. Alternatively, the Triton-purified samples may contain the convulsant binding sites in a low-affinity conformational state of the same protein seen to possess [$^{35}$S]TBPS binding activity in CHAPS extracts. This possibility is suggested by Sigel and Barnard [1984], who found the same native molecular weight and subunit composition for protein purified in Triton (with undetectable TBPS binding) or in CHAPS (with TBPS binding). Our studies also suggest only slightly different molecular properties for the receptor purified in the two detergents [King and Olsen, unpublished].

We cannot yet specify exactly the subunit structure of the complete GABA/benzodiazepine/picrotoxin receptor–chloride channel complex. Other investigators have reported the purified receptor protein subunits to have molecular weights on SDS gels of 48.5 kd (major) and 54.5 kd (minor) [Kuriyama et al., 1984], 53 kd (major) and 57 kd (minor) [Sigel and Barnard, 1984], and 60 kd [Martini et al., 1983], or 55 kd and 50 kd (major) and 62 kd (minor) [Möhler et al., 1984b]. The photoaffinity-labeled benzodiazepine-binding subunit(s) appeared as the lower (49 kd) [Kuriyama et al., 1984], or both bands (with the lower more heavily labeled [Möhler et al., 1984b; Sigel and Barnard, 1984]. Interestingly, Asano et al. [1983b] have reported that the GABA binding subunit can be photoaffinity-labeled with [$^3$H]muscimol, showing a molecular weight equivalent to the benzodiazepine-binding subunit of 50 kd. Native molecular weights from gel filtration chromatography of the purified receptor have been estimated at about 240 kd [Sigel et al., 1983; Martini et al., 1982], consistent with our own findings. Chang and Barnard [1982] also showed by irradiation inactivation that the target sizes in membranes for benzodiazepine receptor binding and GABA receptor binding were equal and roughly equal to 220 kd.

A possible structure for the receptor complex suggested by these data is one of four or five subunits of about 50 kd each, with one to five (probably four) benzodiazepine and GABA binding sites per oligomer. Taking into account all of the available SDS gel and affinity-labeling data, a multimeric protein based on a *single subunit* is the most likely structure at the moment. However, it is difficult to imagine how all of the binding activities and the chloride channel–forming ability could be present in one subunit. An $\alpha_4$ tetrameric model would be consistent with the benzodiazepine photoaffinity labeling experiments which suggest that photolabeling one site causes conformational changes in three additional binding sites [Karobath and Supavilai, 1982; Möhler et al., 1982; Thomas and Tallman, 1983]. An $\alpha_2\beta_2$ structure, presumably with two binding sites for each ligand, is suggested by Sigel and Barnard [1984]. However, little is known about the picrotoxin/TBPS and barbiturate binding sites. It is possible that some drug binding sites may be formed at the intersection of two or more peptide subunits [Changeux et al., 1984]. Nielsen and Braestrup [1983] have indicated that while the benzodiazepine and GABA binding sites have an irradiation target size of 54 kd, the TBPS binding site has one of 134 kd. What role a subunit or molecular complex of this size might play in the complete GABA receptor complex is currently unknown. Further, these workers have shown that the allosteric interactions between benzodiazepines and other drugs with the TBPS binding site reflex a target size on the order of 350 kd [Nielsen and Braestrup, 1984]. Thus the larger aggregate states seen in some detergent extracts such as CHAPS may more closely resemble the membrane-bound condition.

Clearly there are many unanswered questions about the biochemical structure and functioning of the GABA/benzodiazepine receptor complex. Some that seem very close to resolution are those dealing with the size of the full complex and its component subunits. An understanding of the allosteric interactions of these subunits that give rise to the functional interrelationships seen with the various ligand-binding and physiological measurements is probably somewhat further off. The question of which structures are necessary for a functional chloride channel is a problem that rarely has been addressed. The molecular basis of the heterogeneity seen in both GABA and benzodiazepine receptors should become clear as purification techniques are refined and applied to discrete brain regions. And certainly one of the most exciting areas for future research is that of possible GABA receptor complex dysfunction in neurological and psychiatric diseases, such as epilepsy [Olsen et al., 1985b], Huntington disease [Van Ness et al., 1982], affective disorders [Emrich et al., 1980], anxiety [Tallman et al., 1980], and possibly many more. Finally, it appears likely that the gene for the GABA receptor protein will be isolated in the near future and open up new horizons. Brain RNA injected into *Xenopus* oocytes has been shown to express functional GABA receptors in the frog egg membrane [Smart et al., 1983], and monoclonal antibodies to the benzodiazepine/GABA receptor have recently been produced [Möhler et al., 1984b], laying the groundwork for the molecular genetics.

## ACKNOWLEDGMENTS

Supported by NIH Grants NS20704 and AM07310. The authors are grateful to Greg Stauber, Ronnie King, and Erik Wong for contributions to the research described.

## REFERENCES

Asano T, Yamada Y, Ogasawara N (1983a): Characterization of the solubilized GABA and benzodiazepine receptors from various regions of bovine brain. J Neurochem 40:209–214.

Asano T, Sakakibara J, Ogasawara N (1983b): Molecular sizes of photolabeled GABA and benzodiazepine receptor proteins are identical. FEBS Lett 151:277–180.

Barker JL, Nicoll RA, Padjen A (1975): Studies on convulsants in the isolated frog spinal cord. J Physiol 245:521–548.

Barker JL, Mathers DA (1981): GABA analogues activate channels of different duration on cultured mouse spinal neurons. Science 212:358–360.

Beaumont K, Chilton W, Yamamura HI, Enna SJ (1978): Muscimol binding in rat brain: association with synaptic GABA receptors. Brain Res 148:153–162.

Bowery NG, Collins JF, Hill RG (1976): Bicyclic phosphorus esters that are potent convulsants and GABA antagonists. Nature 261:601–603.

Bowery NG, Hill DR, Hudson AL, Doble A, Middlemiss DN, Shaw J, Turnbull M (1980): (−)Baclofen decreases neurotransmitter release in the mammalian CNS by an action at a novel GABA receptor.

Nature 283:92–94.

Bowery NG, Hill DR, Hudson AL (1983): Characteristics of $GABA_B$ receptor binding sites on rat whole brain synaptic membranes. Br J Pharmacol 78:191–206.

Braestrup C, Squires RF (1977): Specific benzodiazepine receptors in rat brain characterized by high affinity [$^3$H]diazepam binding. Proc Natl Acad Sci USA 74:3805–3809.

Braestrup C, Nielsen M (1981): GABA reduces binding of [$^3$H]methylcarboline-3-carboxylate to brain receptors. Nature 294:472–474.

Braestrup C, Schmiechen R, Neff G, Nielsen M, Petersen EN (1982): Interaction of convulsive ligands with benzodiazepine receptors. Science 216:1241–1243.

Browner M, Ferkany JW, Enna SJ (1981): Biochemical identification of pharmacologically and functionally distinct GABA receptors in rat brain. J Neurosci 1:514–518.

Chang L-R, Barnard EA (1982): The benzodiazepine/GABA receptor complex: molecular size in brain synaptic membranes and in solution. J Neurochem 39:1507–1518.

Changeux J-P, Devillers-Thiery A, Chemouilli P (1984): Acetylcholine receptor: an allosteric protein. Science 225:1335–1345.

Chiu TH, Dryden DM, Rosenberg HC (1982): Kinetics of [$^3$H]flunitrazepam binding to membrane-bound benzodiazepine receptors. Mol Pharmacol 21:57–65.

Conti-Tronconi B, Raftery MA (1982): The nicotinic cholinergic receptor: correlation of molecular structure with functional properties. Ann Rev Biochem 51:491–530.

Costa T, Rodbard D, Pert CB (1979): Is the benzodiazepine receptor coupled to a chloride anion channel? Nature 277:315–317.

Cowen P, Green AT, Nutt DJ, Martin IL (1981): Ethyl $\beta$-carboline-3-carboxylate lowers seizure threshold and antagonizes flurazepam-induced sedation in rats. Nature 290:54–55.

Czernik AM, Petrack B, Kalinsky HJ, Psychoyos S, Cash WD, Tsai C, Rinehart RK, Granat FR, Lovell RA, Brundish DE, Wade R (1982): Receptor binding characteristics of a potent benzodiazepine antagonist. Life Sci 30:363–372.

Davis WC, Ticku MK (1982): Picrotoxinin and diazepam bind to two distinct proteins: further evidence that pentobarbital may act at the picrotoxinin site. J Neurosci 1:1036–1042.

Dichter MA (1980): Physiological identification of GABA as the inhibitory transmitter for mammalian cortical neurons in cell culture. Brain Res 190:111–121.

Dunlap K, Fischbach GD (1981): Neurotransmitters decrease the calcium conductance activated by depolarization of embryonic chick sensory neurones. J Physiol 317:519–535.

Eccles J, Nicoll RA, Oshima T, Rubia FJ (1977): The anionic permeability of the inhibitory postsynaptic membrane of hippocampal pyramidal cells. Proc R Soc Lond (Biol) 198:345–361.

Emrich HM, Zerssen DV, Kissling W, Moller HJ, Windorfer A (1980): Effect of sodium valproate on mania. The GABA-hypothesis of affective disorders. Arch Psychiatr Nervenkr 229:1–16.

Enna SJ, Snyder SH (1977): Influences of ions, enzymes, and detergents on $\gamma$-aminobutyric acid-receptor binding in synaptic membranes of rat brain. Mol Pharmacol 13:442–453.

Enna SJ, Karbon EW (1984): $GABA_B$ receptors and transmitter-stimulated cAMP accumulation in rat brain. Neuropharmacol 23:821–822.

Frere RC, Macdonald RL, Young AB (1982): GABA binding and bicuculline in spinal cord and cortical membranes from adult rat and from mouse neurons in cell culture. Brain Res 244:145–153.

Gallager DW (1978): Benzodiazepines: potentiation of a GABA inhibitory response in the dorsal raphe nucleus. Eur J Pharmacol 49:133–143.

Gallager DW, Rauch SL, Malcolm AB (1984): Alterations in a low affinity GABA recognition site following chronic benzodiazepine treatment. Eur J Pharmacol 98:159–160.

Gavish M, Snyder SH (1980a): Benzodiazepine recognition sites on GABA receptors. Nautre 287:651–652.

Gavish M, Snyder SH (1980b): Soluble benzodiazepine receptors: GABAergic regulation. Life Sci 26:579–582.

Gavish M, Snyder SH (1981): $\gamma$-Aminobutyric acid and benzodiazepine receptors: copurification and characterization. Proc Natl Acad Sci USA 78:1939–1942.

Gee KW, Ehlert FJ, Yamamura HI (1983): Differential effects of $\gamma$-aminobutyric acid on benzodiazepine receptor subtypes labeled by [$^3$H]propyl $\beta$-carboline-3-carboxylate in rat brain. J Pharmacol Exp Ther 225:132–137.

Greenlee DV, Olsen RW (1979): Solubilization of gamma-aminobutyric acid receptor protein from mammalian brain. Biochem Biophys Res Comm 88:380–387.

Greenlee DV, Van Ness PC, Olsen RW (1978a): Endogenous inhibitor of GABA binding in mammalian brain. Life Sci 22:1653–1662.

Greenlee DV, Van Ness PC, Olsen RW (1978b): Gamma-aminobutyric acid binding in mammalian brain: receptor-like specificity of sodium-independent sites. J Neurochem 31:933–938.

Guidotti A, Toffano G, Costa E (1978): An endogenous protein modulates the affinity of GABA and benzodiazepine receptors in rat brain. Nature 275:553–555.

Hill DR, Dolphin AC (1984): Modulation of adenylate cyclase activity by $GABA_B$ receptors. Neuropharmacol 23:829–830.

Hirsch S (1982): Phtolabeling of benzodiazepine receptors spares [$^3$H]propyl β-carboline binding. Pharmacol Biochem Behav 16:245–248.

Huang L-Y M, Barker JL (1980): Pentobarbital: stereospecific actions of the (+) and (−) isomers revealed on cultured mammalian neurons. Science 207:195–197.

Hunkeler W, Möhler H, Pieri L, Polc P, Bonetti EP, Cumin R, Schaffner R, Haefely W (1981): Selective antagonists of benzodiazepines. Nature 290:514–516.

Ito Y, Kuriyama K (1982): Some properties of solubilized GABA receptor. Brain Res 236:351–363.

Karobath M, Sperk G (1979): Stimulation of benzodiazepine receptor binding by gamma-aminobutyric acid. Proc Natl Acad Sci USA 76:1004–1006.

Karobath M, Supavilai P (1982): Distinction of benzodiazepine agonists from antagonists by photoaffinity labelling of benzodiazepine receptors in vitro. Neurosci Lett 31:65–69.

King RG, Olsen RW (1984): Solubilization of convulsant/barbiturate binding activity on the GABA/benzodiazepine receptor complex. Biochem Biophys Res Comm 119:530–536.

Klepner CA, Lippa AS, Benson DI, Sano MC, Beer B (1979): Resolution of two biochemically and pharmacologically distinct benzodiazepine receptors. Pharmacol Biochem Behav 11:457–462.

Kuriyama K, Yoneda Y, Taguchi J, Takahashi M, Ohkuma S (1984): Properties of purified γ-aminobutyric acid (GABA) receptor and modulation of GABA receptor binding by membrane phospholipids. Neuropharmacology 23:839–840.

Krnjević K (1974): Chemical nature of synaptic transmission in vertebrates. Physiol Rev 54:418–540.

Lang B, Barnard EA, Chang LR, Dolly JO (1979): Putative benzodiazepine receptor: a protein solubilized from brain. FEBS Lett 104:149–153.

Leeb-Lundberg F, Snowman AM, Olsen RW (1980): Barbiturate receptors are coupled to benzodiazepine receptors. Proc Natl Acad Sci USA 77:7468–7472.

Leeb-Lundberg F, Napias C, Olsen RW (1981a): Dihydropicrotoxinin binding sites in mammalian brain: interactions with convulsant and depressant benzodiazepines. Brain Res 216:399–408.

Leeb-Lundberg F, Snowman AM, Olsen RW (1981b): Perturbation of benzodiazepine receptor binding by pyrazolopyridines involves picrotoxinin/barbiturate receptor sites. J Neurosci 1:471–477.

Leeb-Lundberg F, Olsen RW (1982): Interactions of barbiturates of various pharmacological categories with benzodiazepine receptors. Mol Pharmacol 21:320–328.

Leeb-Lundberg LMF, Olsen RW (1983): Heterogeneity of benzodiazepine receptor interactions with GABA and barbiturate receptors. Mol Pharmacol 23:315–325.

Lo MMS, Strittmatter SM, Snyder SH (1982): Physical separation and characterization of two types of benzodiazepine receptors. Proc Natl Acad Sci USA 79:680–684.

Luini A, Goldberg O, Teichberg VI (1981): Distinct pharmacological properties of excitatory amino acid receptors in the rat striatum: study by $Na^+$ efflux assay. Proc Natl Acad Sci USA 78:3250–3254.

Lunt GC, Robinson TN, Miller TA, Knowles WP, Olsen RW (1984): [$^3$H]Muscimol binding to GABA receptors in insect CNS. Soc Neurosci Abstr 10:688 #202.2.

Maksay G, Ticku MK (1984): Pretreatment with GABA and modulatory ligands enhances GABA receptor binding. Eur J Pharmacol 104:185–188.

Martini C, Lucacchini A, Ronca G, Hrelia S, Rossi CA (1982): Isolation of putative benzodiazepine receptors from rat brain membranes by affinity chromatography. J Neurochem 38:15–19.

Martini C, Rigacci T, Lucacchini A (1983): [$^3$H]Muscimol binding site on purified benzodiazepine receptor. J Neurochem 41:1183–1185.

Mathers DA, Barker JL (1980): (−)Pentobarbital opens ion channels of long duration in cultured mouse spinal neurons. Science 209:507–509.

Meiners BM, Kehoe P, Shaner DM, Olsen RW (1979): γ-Aminobutyric acid receptor binding and uptake in membrane fractions of crayfish muscle. J Neurochem 32:979–990.

Möhler H, Okada T (1977): Benzodiazepine receptors: demonstration in the central nervous system. Science 198:849–851.

Möhler H, Okada T (1978): Properties of γ-aminobutyric acid receptor binding with (+)-[$^3$H]bicuculline methiodide in rat cerebellum. Mol Pharmacol 14:256–265.

Möhler H, Battersby MK, Richard JG (1980): Benzodiazepine receptor protein identified and visualized in brain tissue by a photoaffinity label. Proc Natl Acad Sci USA 77:1666–1670.

Möhler H (1982): Benzodiazepine receptors: Differential interaction of benzodiazepine agonists and antagonists after photoaffinity labeling with flunitrazepam. Eur J Pharmacol 80:435–436.

Möhler H, Sieghart W, Richards JG, Hunkeler W (1984a): Photoaffinity labeling of benzodiazepine receptors with a partial inverse agonist. Eur J Pharmacol 102:191–192.

Möhler H, Schoch P, Haring P, Takacs B, Stahli C (1984b): A purified GABA/benzodiazepine receptor complex from bovine brain: biochemical and immunological characterization. Abstr IUPHAR 9th Int Cong Pharmacol #615.

Nicoll RA, Eccles JC, Oshima TC, Rubia F (1975): Prolongation of hippocampal inhibitory post-synaptic potentials by barbiturates. Nature 258:625–627.

Nielsen M, Braestrup C (1980): Ethyl-β-carboline-3-carboxylate shows differential benzodiazepine receptor interaction. Nature 286:606–607.

Nielsen M, Braestrup C (1983): The molecular target size of brain TBPS binding sites. Eur J Pharmacol 96:321–322.

Nielsen M, Braestrup C (1984): Target size of brain benzodiazepine receptor, GABA receptor and $^{35}S$-TBPS binding sites. In Vici ES, Magyar K (eds): "Regulation of Transmitter Function, Basic and Clinical Aspects." Budapest: Akademiai Kiado, pp 429–432.

Nowak LM, Young AB, Macdonald RL (1982): GABA and bicuculline actions on mouse spinal cord and cortical neurons in cell culture. Brain Res 244:155–164.

Oakley N, Jones B (1980): The proconvulsant and diazepam-reversing effects of ethyl β-carboline-3-carboxylate. Eur J Pharmacol 68:381–382.

Olsen RW, Ticku MK, Miller T (1978): Dihydropicrotoxinin binding to crayfish muscle sites possibly related to γ-aminobutyric acid receptor ionophores. Mol Pharmacol 14:381–390.

Olsen RW, Bergman MO, Van Ness PC, Lummis SC, Watkins AE, Napias C, Greenlee DV (1981): γ-Aminobutyric acid receptor binding in mammalian brain: heterogeneity of binding sites. Mol Pharmacol 19:217–227.

Olsen RW (1982): Drug interactions at the GABA receptor ionophore complex. Ann Rev Pharmacol Toxicol 22:245–277.

Olsen RW (1983): Biochemical properties of GABA receptors. In Enna SJ (ed): "The GABA Receptors." Clifton, NJ: Humana Press, pp 63–91.

Olsen RW, Snowman AM (1982): Chloride-dependent enhancement by barbiturates of GABA receptor binding. J Neurosci 2:1812–1823.

Olsen RW, Snowman AM (1983): [$^3$H]Bicuculline-methochloride binding to low affinity GABA receptor sites. J Neurochem 41:1653–1663.

Olsen RW, Snowman AM (1985): Avermectin $B_{1a}$ modulation of GABA/benzodiazepine receptor binding in rat brain. J Neurochem 44:1074–1082.

Olsen RW, Fischer JB, King RG, Ransom JY, Stauber GB (1984a): Purification of the GABA/benzodiazepine/barbiturate receptor complex. Neuropharmacology 23:853–854.

Olsen RW, Snowhill EW, Wamsley JK (1984b): Autoradiographic localization of low-affinity GABA receptors with [$^3$H]bicuculline methochloride. Eur J Pharmacol 99:247–248.

Olsen RW, Wong EHF, Stauber GB, Murakami D, King RG, Fischer JB (1984c): Biochemical properties of the GABA/barbiturate/benzodiazepine receptor-chloride ion channel complex. In Kito S, Segawa T, Kuriyama K, Yamamura HI, Olsen RW (eds): "Neurotransmitter Receptors: Mechanisms of Action and Regulation." New York: Plenum, pp 205–219.

Olsen RW, Wong EHF, Stauber GB, King RG (1984d): Biochemical pharmacology of the GABA/benzodiazepine receptor/ionophore protein. Fed Proc 43:2773–2778.

Olsen RW, Fischer JB, Dunwiddie TV (1985a): Barbiturate enhancement of GABA receptor binding and function as a mechanism of anesthesia. In Roth S, Miller K (eds): "Molecular and Cellular Mechanisms of Anaesthetics." New York: Plenum, pp 165-177.

Olsen RW, Wamsley JK, Lee R, Lomax P (1985b): The benzodiazepine/barbiturate/GABA receptor–chloride ionophore complex in a genetic model for generalized epilepsy. In Delgado-Escueta AV, Ward AA, Woodbury DM (eds): "Basic Mechanisms of the Epilepsies." New York: Raven Press (in press).

Polc P, Bonetti EP, Schaffner R, Haefely W (1982): A three-state model of the benzodiazepine receptor explains the interactions between the benzodiazepine antagonist Ro 15-1788, benzodiazepine tranquilizers, β-carbolines, and phenobarbitone. Naunyn-Schmeideberg's Arch Pharmacol 321:260–264.

Ramanjaneyulu R, Ticku MK (1984): Binding characterstics and interactions of depressant drugs with *[$^{35}$S]t*-butylbicyclophosphorothionate, a ligand that binds to the pictrotoxinin site. J Neurochem 42:221–229.

Schlosser W, Franco S (1979): Reduction of gamma-aminobutyric acid (GABA)-mediated transmission by a convulsant benzodiazepine. J Pharmacol Exp Ther 211:290–295.

Schoch P, Möhler H (1983): Purified benzodiazepine receptor retains modulation by GABA. Eur J Pharmacol 95:323–324.

Segal M, Barker JL (1984): Rat hippocampal neurons in culture: properties of GABA-activated $Cl^-$ ion conductance. J Neurophysiol 51:500–515.

Sherman-Gold R, Dudai Y (1980): Solubilization and properties of a benzodiazepine receptor from calf cortex. Brain Res 198:485–490.

Sherman-Gold R (1983): Photolabeling of benzodiazepine-receptors: possible mechanisms of reaction. Neurochem Int 5:171–174.

Sherman-Gold R, Dudai Y (1983): Heterogeneity in the physicochemical properties of deoxycholate-solubilized benzodiazepine receptors from calf cerebral cortex. Neurochem Res 8:853–864.

Sieghart W, Möhler H (1982): [$^3$H]Clonazepam, like [$^3$H]flunitrazepam, is a photoaffinity label for the central type of benzodiazepine receptors. Eur J Pharmacol 81:171–173.

Sieghart W, Mayer A, Drexler G (1983): Properties of [$^3$H]flunitrazepam binding to different benzodiazepine binding proteins. Eur J Pharmacol 88:291–299.

Sigel E, Mamalaki C, Barnard EA (1982): Isolation of a GABA receptor from bovine brain using a benzodiazepine affinity column. FEBS Lett 147:45–48.

Sigel E, Stephenson FA, Mamalaki C, Barnard EA (1983): A $\gamma$-aminobutyric acid/benzodiazepine receptor complex of bovine cerebral cortex. Purification and partial characterization. J Biol Chem 258: 6965–6971.

Sigel E, Barnard EA (1984): A $\gamma$-aminobutyric acid/benzodiazepine receptor complex from bovine cerebral cortex. Improved purification with preservation of regulatory sites and their interactions. J Biol Chem 259:7219–7223.

Skerritt JH, Willow M, Johnston GAR (1982): Diazepam enhancement of low affinity GABA binding to rat brain membranes. Neurosci Lett 29:63–66.

Smart TG, Consanti A, Bilbe G, Brown DA, Barnard EA (1983): Synthesis of functional chick brain GABA-benzodiazepine-barbiturate/receptor complexes in mRNA-injected *Xenopus* oocytes. Neurosci Lett 40:55–59.

Snodgrass SR (1978): Use of $^3$H-muscimol for GABA receptor studies. Nature 273:392–394.

Squires RF, Casida JE, Richardson M, Saederup E (1983): [$^{35}$S]t-Butylbicyclophosphorothionate binds with high affinity to brain-specific sites coupled to $\gamma$-aminobutyric acid-A and ion recognition sites. Mol Pharmacol 23:326–336.

Stephenson FA, Olsen RW (1982): Solubilization by CHAPS detergent of barbiturate-enhanced benzodiazepine-GABA receptor complex. J Neurochem 39:1579–1586.

Stephenson FA, Watkins AE, Olsen RW (1982): Physicochemical characterization of detergent-solubilized $\gamma$-aminobutyric acid and benzodiazepine receptor proteins from bovine brain. Eur J Biochem 123:291–298.

Study RE, Barker JL (1981): Diazepam and (−)-pentobarbital: fluctuation analysis reveals different mechanisms for potentiation of $\gamma$-aminobutyric acid responses in cultured central neurons. Proc Natl Acad Sci USA 78:7180–7184.

Supavilai P, Karobath M (1980): The effect of temperature and chloride ions on the stimulation of [$^3$H] flunitrazepam binding by the muscimol analogues THIP and piperidine-4-sulfonic acid. Neurosci Lett 19:337–341.

Supavilai P, Mannonen A, Karobath M (1982): Modulation of GABA binding sites by CNS depressants and CNS convulsants. Neurochem Int 4:259–268.

Takeuchi A, Takeuchi N (1969): A study of the action of picrotoxin on the inhibitory neuromuscular junction of the crayfish. J Physiol 205:377–391.

Tallman JF, Thomas JW, Gallager DW (1978): GABAergic modulation of benzodiazepine binding site sensitivity. Nature 274:383–385.

Tallman JF, Paul SM, Skolnick P, Gallager DW (1980): Receptors for the age of anxiety: pharmacology of the benzodiazepines. Science 207:274–281.

Thampy KG, Barnes EM, Jr (1984): $\gamma$-Aminobutyric acid-gated chloride channels in cultured cerebral neurons. J Biol Chem 259:1753–1757.

Thomas JW, Tallman JF (1981): Characterization of photoaffinity labeling of benzodiazepine binding sites. J Biol Chem 256:9838–9842.

Thomas JW, Tallman JF (1983): Photoaffinity labeling of benzodiazepine receptors causes altered agonist-antagonist interactions. J Neurosci 3:433–440.

Ticku MK, Olsen RW (1977): $\gamma$-Aminobutyric acid stimulated chloride permeability in crayfish muscle. Biochem Biophys Acta 464–519–529.

Ticku MK, Ban M, Olsen RW (1978a): Binding of [$^3$H]$\alpha$-dihydropicrotoxinin, a $\gamma$-aminobutyric acid synaptic antagonist, to rat brain membranes. Mol Pharmacol 14:391–402.

Ticku MK, Van Ness PC, Haycock JW, Levy WB, Olsen RW (1978b): Dihydropicrotoxinin binding sites in rat brain: comparison to GABA receptors. Brain Res 150:642–647.

Ticku MK, Olsen RW (1979): Cage convulsants inhibit picrotoxinin binding. Neuropharmacology 18:315–318.

Ticku MK, Maksay G (1983): Convulsant/depressant site of action at the allosteric benzodiazepine-GABA receptor-ionophore complex. Life Sci 33:2363–2375.

Toffano G, Guidotti A, Costa E (1978): Purification of an endogenous protein inhibitor of the high affinity binding of $\gamma$-aminobutyric acid to synaptic membranes of rat brain. Proc Natl Acad Sci USA 75:4024–4028.

Trifiletti RR, Snowman AM, Snyder SH (1984): Solubilization and anionic regulation of cerebral sedative/convulsant receptors labeled with [$^{35}$S]*tert*-butylbicyclophosphorothionate (TBPS). Biochem Biophys Res Comm 120:692–699.

Van Ness PC, Watkins AE, Bergman MO, Tourtellotte WW, Olsen RW (1982): $\gamma$-Aminobutyric acid receptors in normal and human brain and Huntington's disease. Neurology 32:63–68.

Wang Y-J, Salvaterra P, Roberts E (1979): Characterization of $^3$H-muscimol binding to mouse brain membranes. Biochem Pharmacol 28:1123–1128.

Willow M (1981): A comparison of the action of pentobarbital and etomidate on [$^3$H]GABA binding to crude synaptosomal rat brain membranes. Brain Res 220:427–431.

Willow M, Johnston GAR (1981): Enhancement by anesthetic and convulsant barbiturates of GABA binding to rat brain synaptosomal membranes. J Neurosci 1:364–367.

Winkler MH, Nicklas WJ, Berl S (1979): The complex

binding of γ-aminobutyric acid to a rat brain synaptic membrane preparation. J Neurochem 32:79–84.
Wong DT, Rathbun RC, Bymaster FP, Lacefield WB (1983): Enhanced binding of radioligands to receptors of γ-aminobutyric acid and benzodiazepine by a new anticonvulsive agent, LY81067. Life Sci 33:917–923.
Wong EHF, Leeb-Lundberg LMF, Teichberg VI, Olsen RW (1984a): γ-Aminobutyric acid activation of $^{36}Cl^-$ flux in rat hippocampal slices and its potentiation by barbiturates. Brain Res 303:267–275.
Wong EHF, Snowman AM, Leeb-Lundberg LMF, Olsen RW (1984b): Barbiturates inhibit GABA antagonist and benzodiazepine inverse agonist binding. Eur J Pharmacol 102:205–212.
Yang-Ransom J, Fischer JB, Olsen RW (1984): γ-Aminobutyric acid receptors in mouse brain: comparison of binding and chloride channel activation. Soc Neurosci Abstr 10:418.
Young WS III, Neihoff D, Kuhar MJ, Beer B, Lippa S (1981): Multiple benzodiazepine receptor localization by light microscopic radiohistochemistry. J Pharmacol Exp Ther 216:425–430.
Yousufi MAK, Thomas JW, Tallman JF (1979): Solubilization of benzodiazepine binding site from rat cortex. Life Sci 25:463–470.

Benzodiazepine/GABA Receptors and Chloride Channels:
Structural and Functional Properties, pages 261–274


# 13

# Purification and Characterization of the Brain GABA/Benzodiazepine Receptor

F. Anne Stephenson and Eric A. Barnard

*MRC Molecular Neurobiology Unit, University of Cambridge, England CB2 2QH*

Studies of the $GABA_A$ receptor are of importance for a number of reasons. Prominent among these, firstly, is the view now generally accepted that in this receptor the binding of the transmitter molecule directly gates an intrinsic ion channel (the receptor chloride channel), placing it in the class of receptors whose function is most directly amenable to analysis. Since for another receptor of that class, the nicotinic acetycholine receptor, the most comprehensive set of structural and functional data for a receptor is now available, it will be highly illuminating to analyse in depth another receptor of that class. Secondly, it has become established that clinically important drugs, the benzodiazepines and related anticonvulsant drugs, act at the level of this receptor, so that it is, in fact, the GABA/benzodiazepine receptor. Thirdly, this receptor differs from the nicotinic acetylcholine receptor by exhibiting a set of natural allosteric regulations of agonist binding and channel gating, due to a complex network of different types of binding sites in one overall receptor complex. The analysis of this receptor regulation is of particular theoretical and practical interest.

In this chapter, we shall describe the studies currently pursued in several laboratories on the purification of the GABA/benzodiazepine receptor, as a necessary step in the study of GABA receptor function and its regulation.

## SOLUBILIZATION OF THE GABA/BENZODIAZEPINE RECEPTOR

The necessary first stage in the purification of the GABA/benzodiazepine receptor is the solubilization of the membrane protein with its ligand-binding properties intact. The $GABA_A$ agonist binding site cannot be defined merely by the binding of GABA, due to the binding of that compound also to the $GABA_B$ receptor [Hill and Bowery, 1981] and (in large amounts in the brain) to the GABA uptake system. The $GABA_A$ binding activity, as defined here, is the binding of muscimol, or the bicuculline-sensitive binding of GABA in the absence of $Na^+$ and $Ca^{2+}$ [Olsen, 1982]. A variety of nondenaturing detergents have been used to solubilize both GABA binding and benzodiazepine binding activities from membrane preparations of bovine, rat and mouse brain; a representative list is given in Table I. This list includes ionic, nonionic and zwitterionic detergents, and although all solubilize the $GABA_A$ and benzodiazepine binding sites, both the maximum number of soluble binding sites recovered following extraction and the allosteric interactions of GABA, benzodiazepine, and $Cl^-$ channel related binding sites, vary with the nature of the detergent. In general, it is found that the maximum extraction of GABA and benzodiazepine receptors is achieved with the ionic detergent, sodium de-

**TABLE I. A Summary of the Detergents Used for Solubilization of GABA and BZ Receptors**

| Solubilization conditions | GABA sites[a] | | Benzodiazepine sites[b] | | Reference |
|---|---|---|---|---|---|
| | $K_D$ (nM) | $B_{max}$ (pmol/mg protein) | $K_D$ (nM) | $B_{max}$ (pmol/mg protein) | |
| 0.25% lysolecithin | (i) 7 | 1 | 1.5 | 0.18 | Gavish et al. [1979] |
| | (ii) 24 | 0.77 | | | |
| 0.4% lysolecithin | (i) 3.5 | 2.8 | ND[c] | ND | Chude [1979] |
| | (ii) 52 | 14 | | | |
| 2% $Na^+$ deoxycholate | 41 | 0.55 | ND | ND | Greenlee and Olsen [1979] |
| 0.5% Lubrol PX | ND | ND | 11 | — | Yousufi et al. [1979] |
| 0.7–2% Triton X-100 | ND | ND | 12 | 1.8 | Lang et al. [1979] |
| 0.5% $Na^+$ deoxycholate | ND | ND | 1.5 | — | Sherman-Gold and Dudai [1980] |
| 0.2% $Na^+$ | 9.3 | 7.58 | 3.1 | 3.44 | Asano and Ogasawara [1981] |
| 20 mM CHAPS | ND | ND | 8 | 1.4 | Stephenson and Olsen [1982] |
| 30 mM CHAPS | ND | ND | 1.4 | 0.55 | Mernoff et al. [1983] |
| 2% Triton X-100 | 5 | 0.3 | 2 | 0.1 | Lo et al. [1982] |

[a]All the values given for GABA receptors were determinied using [$^3$H] muscimol. [b]All the values for benzodiazepine receptors were determined using [$^3$H] flunitrazepam, except for Yousufi et al. [1979] where [$^3$H] diazepam was used.
[c]ND = not determined.

oxycholate: with this, it has been found that 90% of the total recovered benzodiazepine sites and 58% of the total recovered GABA sites are solubilized [Stephenson et al., 1982]. By the total recovered sites is meant the sum of the solubilized and insoluble sites; these percentages are the highest of all those found with a range of detergents. When bovine cortex synaptic plasma membranes are extracted in low concentrations of Triton X-100, the high concentration of GABA uptake carrier sites which is present is removed at below 0.1% detergent, but none of the $GABA_A$ sites [Lester et al., 1981; Chang and Barnard, 1982]. Significant numbers of the $GABA_A$ sites are extracted in active form in 1% Triton X-100 [Chang and Barnard, 1982], but the rest remain active in the membrane, most being solubilized extensively by deoxycholate [Greenlee and Olsen, 1979; Lester et al., 1981; Stephenson et al., 1982].

The inclusion of protease inhibitors in the extraction medium has been shown to increase the number of binding sites recovered after solubilization [Lang et al., 1979; Stephenson et al., 1982], but a significant proportion of both the GABA and the BZ sites is still inactivated upon extraction, presumably due to denaturation by the detergent [Stephenson et al., 1982]. In addition, the binding sites are unstable in the presence of sodium deoxycholate, a loss of 50% of specific binding activity being observed in 24 hours storage at 4°C [Greenlee and Olsen, 1979; Stephenson et al., 1982]. Sodium deoxycholate can readily be exchanged for another detergent in which both the GABA and the benzodiazepine binding sites are stable, e.g., Triton X-100, or Lubrol PX, using either gel filtration [Stephenson et al., 1982] or ammonium sulfate precipitation, resuspension, and dialysis [Asano and Ogasawara, 1981a].

Deoxycholate extracts of brain membranes contain specific high-affinity saturable sites for both GABA and benzodiazepine ligands (Table I). The pharmacological specificity of the respective bindings sites is that expected for $GABA_A$ and "central" benzodiazepine receptors (see also the later discussion of the specificity of the purified GABA/benzodiazepine receptor). The dissociation constants, $K_D$, are up to an order of magnitude higher (lower affinity) in the crude extracts than the values that have been determined for membrane-bound receptor (Table I) [Squires and Braestrup 1977; Olsen, 1981]. The ratio of the number of GABA sites to benzodiazepine sites in the detergent extracts has received much attention. Determinations of this ratio must be carried out only after the removal of endogenous GABA from the extracts; values of either 2 [Asano and Ogasawara, 1981a, Stephenson et al., 1982] or 3 [Lo et al., 1982, Asano and Ogasawara, 1983; Sigel et al., 1983] have been reported. One of the allosteric properties of the receptor, the stimulation of benzodiazepine agonist binding activity, has been demonstrated also in deoxycholate extracts, but it is reduced in magnitude to that found in the membrane [Asano and Ogasawara 1981a; Stephenson et al., 1982]. However, if the zwitterionic detergent, CHAPS, is used for the solubilization of the GABA/benzodiazepine receptor, this GABA allosteric effect is much larger, and a maximum stimulation of 70–80% [Stephenson and Olsen, 1983; Mernoff et al., 1983] over controls in the absence of GABA, with an $EC_{50}$ of 0.1 $\mu$M, has been found for enhancement of benzodiazepine binding activity by muscimol [Stephenson and Olsen, 1984].

It is of interest to note that GABA and benzodiazepine binding sites are not always cosolubilized. For example, a significant proportion (~ 40%) of muscimol binding sites is not extracted by sodium deoxycholate treatment of the brain membranes [Stephenson et al., 1982], in contrast to the benzodiazepine sites, of which less than 10% are nonextractable under the same conditions [Stephenson et al., 1982]. Further differences between the extractability by Triton X-100 of benzodiazepine binding sites and total GABA sites of the membrane, referred to above, have been reported to varying degrees by different investigators [Chiu and Rosenberg, 1979; Massotti et al., 1981; Chang and Barnard, 1982; Stephenson et al., 1982].

Such data on percentage extractions are complicated by the likelihood that a large number of the $Na^+$-independent GABA sites in the mixed cell population of the brain are not those in the GABA/benzodiazepine receptor, that some brain benzodiazepine sites may be present in addition to those in that receptor, that some GABA binding sites can be unmasked by certain detergents [Lester et al., 1981], and that even for those in the $GABA_A$/benzodiazepine complex, it is possible for the two binding activities to suffer differential denaturation or differential modification by bound detergent. The latter case is exemplified by the effect of CHAPS on the pure receptor, where the binding of muscimol is much weakened in affinity (relative to Triton medium), whereas the benzodiazepine binding is unchanged [Sigel et al., 1984a].

At first sight, sodium deoxycholate would appear to be the best detergent for the solubilization of the GABA/benzodiazepine receptor. It can extract to a maximum extent the binding sites from the membrane in an active form, the functional interaction of GABA and benzodiazepine sites is retained, and although the GABA and benzodiazepine binding sites are unstable in the presence of sodium deoxycholate, the detergent can be readily exchanged for another. However, we have not considered a third component of the GABA/benzodiazepine receptor, the binding site for picrotoxinin and related convulsant and depressant drugs. This binding site can be studied in vitro by ligand binding assays using either [$^3$H]dihydropicrotoxinin (DHPTX) [Ticku et al., 1978] or the recently developed convulsant cage compound [$^{35}$S]-*t*-butyl bicyclophosphorothionate (TBPS) [Squires et al., 1983], *or* indirectly by barbiturate stimulation of benzodiazepine binding activity [Leeb-Lundberg et al., 1980] or GABA binding activity [Olsen and Snowman, 1982]. Only one group has reported soluble DHPTX binding activity [Davis and Ticku, 1981] (Table II); this ligand is particularly difficult to use

**TABLE II. Solubilization of the Picrotoxinin and TBPS Binding Sites**

| Solubilization conditions | $[^3H]$DH PTX $K_D$nM | $[^3H]$DH PTX $B_{max}$ pmol/mg protein | $[^{35}S]$ TBPS $K_D$nM | $[^{35}S]$ TBPS $B_{max}$ pmol/mg protein | Reference |
|---|---|---|---|---|---|
| 1% Lubrol | (i) 38 | 1.06 | ND[a] | ND | Davis and Ticku [1981] |
| | (ii) 1850 | 14.6 | ND | ND | |
| 20 mM CHAPS | ND | ND | 26 | 0.4 | King and Olsen [1984] |
| 1.5% CHAPS | ND | ND | 59 | 1.1 | Stephenson et al. [1984] |

[a]ND = not determined.

due to its low affinity. Barbiturate regulation of benzodiazepine and GABA binding activity was absent in deoxycholate extracts [Stephenson et al., 1982], and indeed it was shown that very low concentrations of Triton or Lubrol can inhibit barbiturate stimulation of benzodiazepine binding in the membrane [Asano and Ogasawara, 1981b; Olsen and Snowman, 1982; Skolnick et al., 1982]. In contrast, solubilization of the GABA/benzodiazepine receptor with the zwitterionic detergent CHAPS does preserve the allosteric interactions between the barbiturate binding site and GABA and benzodiazepine receptor sites [Stephenson and Olsen, 1982; Stephenson and Olsen, 1983] (Table II). The barbiturate stimulation is similar to that found in the membrane, i.e., is dose-dependent, stereospecific, and $Cl^-$-dependent. Further, with the development of the ligand $[^{35}S]$TBPS in 1983, the binding site can be demonstrated directly and a high-affinity, saturable-specific site for TBPS and related drugs has been demonstrated in crude CHAPS extracts [King and Olsen, 1984; Stephenson et al., 1984]. The ratio of benzodiazepine to TBPS binding sites in CHAPS crude extracts was 1 [Stephenson et al., 1984]. Thus, in summary, CHAPS solubilizes the fully functional GABA/benzodiazepine/$Cl^-$ ionophore. All the three binding activities are relatively stable in the crude extracts of CHAPS. The maximum recovery, however, of receptor solubilized with the zwitterionic detergent is 74% of the activity recovered by sodium deoxycholate.

It must be emphasized that loss of one or more binding activities of this receptor during solubilization from brain membranes can arise from the differential effects not only of a particular detergent, but also of slight proteolysis. The brain contains a variety of potent proteases, some being released upon homogenization in aqueous medium and much more activity being released on detergent extraction. The optimum results obtained, therefore, depend upon the inclusion of an adequate set of protease inhibitors at every stage, as well as upon the washing away of some proteases in initial homogenization treatments, and also upon minimizing the period spent in every stage.

Another variable is the presence or absence of phospholipid. In the initial detergent extract of the brain membranes, the phospholipid concentration is relatively high, if the usual ratio (0.4 mg/mg) of detergent to protein is used. However, almost all of this phospholipid is washed away in the detergent washes when the receptor is bound on the affinity column. When using CHAPS, we always include phospholipid (0.15% asolectin), to protect at these later stages. However, a comparison of each property in the CHAPS-purified receptor with or without that phospholipid addition has not been fully made, and the requirement for this protection in CHAPS probably varies with the property to be preserved. Such addition of phospholipid has no effect upon those properties retained by the receptor purified in deoxycholate/Triton medium.

## PURIFICATION OF THE GABA/ BENZODIAZEPINE RECEPTOR

It has been shown from studies with crude detergent extracts that the GABA/benzodiazepine receptor is a glycoprotein [Gavish and Snyder, 1981; Stephenson and Olsen, 1983]. It is also an acidic protein with an isoelectric point (pI) of 5–5.5 [Tallman et al., 1981; Chang and Barnard, 1982; Korneyev, 1983] and has a molecular weight of at least 200,000 [Chang and Barnard, 1982; Stephenson et al., 1982]. Thus, the physical properties of the protein are very similar to those of a whole spectrum of other membrane-located receptor proteins in the brain, and therefore, any attempt at purification of the receptor by conventional isolation techniques is of little value. Indeed, the maximum purification that has been achieved by lectin affinity chromatography is a 30-fold purification on a lentil lectin affinity column [Stephenson and Olsen, 1983], whereas to achieve purification of the GABA/benzodiazepine receptor to homogeneity, a purification of at least 5,000-fold is required. This has, to date, been achieved only with the use of affinity chromatography using an immobilized benzodiazepine. The structures of the benzodiazepine affinity columns that have been used for purification are shown in Figure 1.

In our laboratory we have used the benzodiazepine RO7-1986/1 coupled to adipic dihydrazide agarose to purify the GABA/benzodiazepine receptor [Sigel et al., 1982; Sigel et al., 1983]. The purification procedure is outlined in Figure 2. Briefly, the GABA/benzodiazepine receptor is solubilized from cerebral cortex with sodium deoxycholate in the presence of protease inhibitors. The receptor extract is passed through the benzodiazepine affinity column: both the GABA and the benzodiazepine binding sites are retained (to at least 75%) on the resin. The affinity column is washed extensively in a high-salt medium to remove any proteins that may be nonspecifically bound by either hydrophobic or electrostatic interactions. Also, at this stage, the deoxycholate is exchanged for Triton X-100 because, as was mentioned earlier, both GABA and benzodiazepine sites, rapidly become inactivated in the presence of deoxycholate. The GABA/benzodiazepine receptor is specifically eluted from the affinity column with a water-soluble benzodiazepine, chlorazepate or flurazepam. Finally, the latter are removed by ion-exchange chromatography. The GABA/benzodiazepine receptor is bound to a DEAE Sephacel column, which is then washed until all of the chlorazepate is removed and purified GABA/benzodiazepine receptor is recovered by high salt elution. Sucrose is added in the latter stages as this is found to stabilize the binding activities.

The same sequence of purification steps has been followed by the other authors listed in Figure 1. There are, however, differences between the purification procedures employed. Firstly, Martini et al. [1982] and Gavish and Snyder [1981] extract the receptor with Triton-X 100, whereas the other investigators use the ionic detergent deoxycholate.

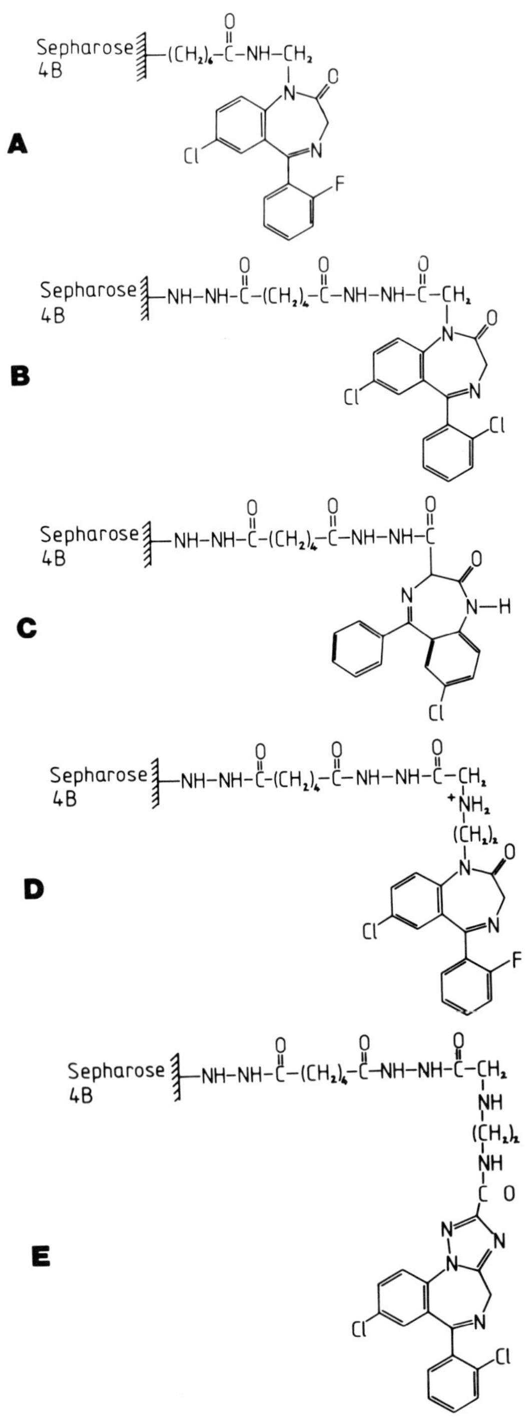


**Fig. 1.** *The structures of BZ affinity gels that have been used for the purification of the GABA/BZ receptor protein. The structures are taken from (A) Gavish and Snyder [1981]; (B) Martini et al. [1982]; (C) Martini et al. [1981]; (D) Sigel et al. [1983] and (E) Taguchi and Kuriyama [1984].*

We have consistently found that the allosteric properties of the receptor are almost all lost in the Triton extract. Secondly, Olsen et al., [1984] and Tallman [1984] use the water-soluble benzodiazepine flurazepam for biospecific elution of the receptor; at submillimolar concentrations of flurazepam, high yields of receptor can be recovered. This difference is not significant however, since we have found that flurazepam or chlorazepate elutions give identical results. Taguchi and Kuriyama [1984] employ a water-insoluble benzodiazepine, 1012S, for elution. Thirdly, the eluting ligand is removed by either ion-exchange [Sigel et al., 1983; Martini et al., 1981], gel filtration [Olsen et al., 1984; Schoch et al., 1984], or diafiltration [Tallman, 1984]. It is important that all traces of the eluting benzodiazepine be removed by the method employed, since the benzodiazepine binding of the product will otherwise be underestimated. This is the reason for employing an eluting benzodiazepine of lower affinity and high water solubility, such as chlorazepate. In all cases it has been shown that a high-affinity GABA binding site is co-purified with the benzodiazepine binding site.

In our purification procedure typically, from 100 g of bovine cerebral cortex, 330 pmol of muscimol and 110 pmol of benzodiazepine binding sites are purified. This corresponds to approximately 150 $\mu$g protein, which is a yield of 0.007% of the total protein present in homogenates of bovine cortex. The benzodiazepine binding site is purified 1,800-fold with respect to the specific activity in the homogenate (Table III). The overall recovery of purified benzodiazepine binding sites is low, i.e., 2% of the starting activity, with the major loss occurring at the affinity column stage. Less than 6% of the total benzodiazepine sites bound to the affinity resin are lost during the washing procedure which removes the non-specifically bound protein; however, only 14–20% of active sites are recovered with chlorazepate elution. The recovery of activity from the ion exchange column is high, in the range of 57–67%, and this procedure minimizes the time involved. No purification of the receptor

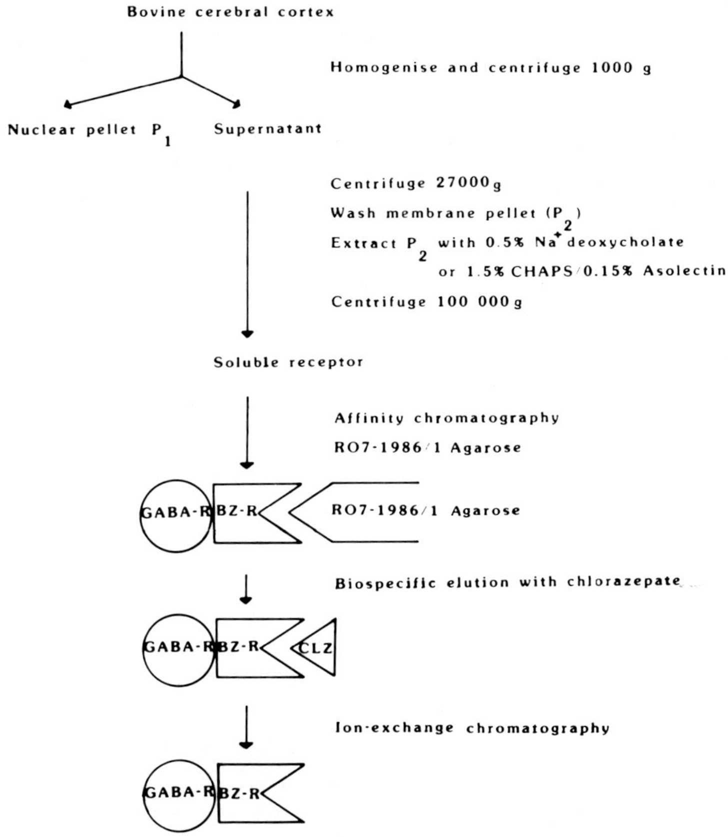


**Fig. 2.** *A schematic representation of the method employed for the isolation of the GABA/BZ receptor protein.*

**TABLE III. Purification of the GABA/Benzodiazepine Receptor Complex**

| Sample | Total protein (mg) | Specific binding activity (pmol/mg protein) | | Yield of [$^3$H] flunitrazepam binding |
|---|---|---|---|---|
| | | [$^3$H]Muscimol | [$^3$H]Flunitrazepam | |
| Homogenate | 14,000 | 0.04 | 0.44 | 100 |
| Crude synaptic membranes | 2,170 | 0.28 | 1.26 | 44 |
| Soluble receptor | 1,010 | 4.4 | 1.46 | 24 |
| Affinity column eluate | 0.45 | 1500 | — | — |
| Ion exchange eluate | 0.15 | 2560 | 800 | 2 |

is achieved at this step; however, the chlorazepate is completely removed and the purified protein is concentrated 10-fold.

The maximum specific activity that has been reported for the purified benzodiazepine receptor is 2.1–3.0 $\mu$mol [$^3$H] flunitrazepam binding sites per g protein by Martini et al. [1982]; this is a purification of 5,200-fold with respect to the specific activity in crude detergent extracts (Table IV). However, even

**TABLE IV. The Binding Properties of the Purified GABA/Benzodiazepine Receptor**

| Species | Muscimol: Specific activity ($\mu$mol/g protein) | Muscimol: $K_D$ (nM) | Flunitrazepam: Specific activity ($\mu$mol/g protein) | Flunitrazepam: $K_D$ (nM) | Muscimol / Flunitrazepam | Reference |
|---|---|---|---|---|---|---|
| Rat | 2 | 23. | 2.1–3 | 8 | 1 | Martini et al. [1982] |
| Bovine | 2.56 | 12 | 0.8 | 10 | 3.5–3.8 | Sigel et al. [1983] |
| Bovine | 1.02 | 13.8 | 0.7 | 2.5 | 1.5 | Schoch et al. [1984] |
| Rat | 1.6 | 20 | 1 | 4 | ND | Olsen et al. [1984] |
| Rat | 0.99 | — | 0.002 | 27.9 | | Taguchi and Kuriyama [1984] |

ND, not determined.

this value is low for the theoretical specific activity of the purified protein with one benzodiazepine binding site per $M_r$ 200,000. Nevertheless, Barnard et al. [1984] showed purification of a single homogeneous protein by isoelectric focussing for purified $^{125}$I-labeled GABA/benzodiazepine receptor, pI 5.5, despite the 25% apparent purity by specific activity. The discrepancy is explained in terms of receptor inactivation during the purification process and by an overestimation of the protein content at the very low protein amounts involved.

## THE BINDING PROPERTIES OF THE PURIFIED GABA/BENZODIAZEPINE RECEPTOR

By this procedure, several groups (Table IV) have purified a benzodiazepine-binding protein which also has a binding site for GABA ligands. Is this protein the GABA/benzodiazepine receptor? The characteristics of the binding properties of the purified receptor are:

1. It has a single high-affinity binding site for benzodiazepine agonists, antagonists [Barnard et al., 1984], and the anxiogenic ligands, the "inverse agonists" or contragonists, e.g., methyl $\beta$ carboline-3-carboxylate [Schoch and Möhler, 1983]. The number of benzodiazepine agonist binding sites equals the number of benzodiazepine antagonist binding sites. The benzodiazepine binding sites have "central" benzodiazepine receptor specificity; for a series of centrally acting benzodiazepines, the rank order of potency at the purified benzodiazepine binding site correlates well with the potency at the membrane-bound receptor site and with in vivo evaluations of both the anxiolytic and anticonvulsant potencies of benzodiazepines (see Fig. 3).

2. By direct binding studies, it has a single high-affinity binding site for the GABA agonist, [$^3$H] muscimol. This binding site has $GABA_A$ receptor specificity, i.e., in the binding assays the activity is sensitive to the

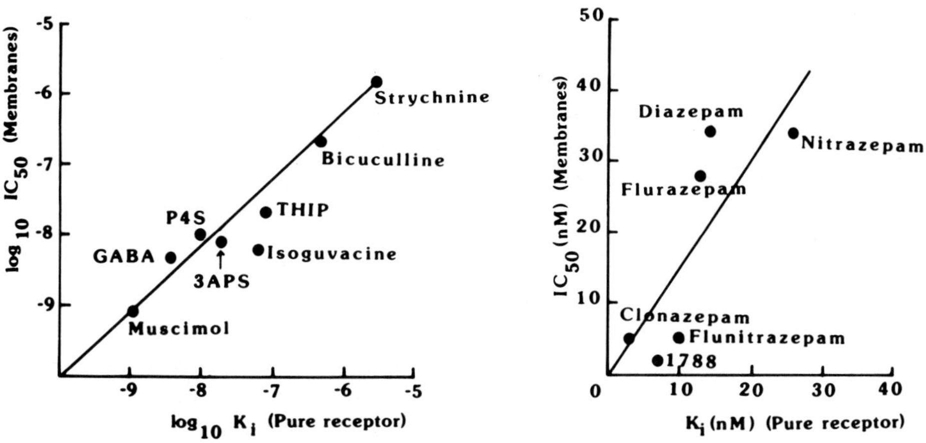


**Fig. 3.** *The pharmacological specificity of the binding sites of the purified GABA/BZ receptor. The left-hand graph is a plot of the* $IC_{50}$ *value, determined by the displacement of [*$^{3}H$*] GABA, at a constant concentration, binding for membrane-bound receptor, versus the* $K_i$ *value, determined by displacement of [*$^{3}H$*] muscimol binding for purified receptor, for a series of agents active at the* $GABA_A$ *receptor. The values are taken from Krogsgaard-Larsen et al. [1981] for membrane-bound receptor and from Sigel et al. [1983] and unpublished results [Stephenson and Barnard] for purified receptor. Linear regression analysis of the plot gave a correlation coefficient,* $r = 0.96$*. The right-hand graph is a plot of the* $IC_{50}$ *value, determined by the displacement of binding at a constant [*$^{3}H$*] diazepam concentration for membrane-bound receptor, versus the* $K_i$ *value, determined by displacement of [*$^{3}H$*] flunitrazepam, for a series of benzodiazepines. The values are taken from Squires and Braestrup [1977] and Möhler and Richards [1981] for membrane-bound receptor and Sigel et al. [1983] and unpublished results [Stephenson and Barnard] for purified receptor. Linear regression analysis of the plot gave a correlation coefficient,* $r = 0.8$*.*

GABA antagonist bicuculline but not to ligands which are specific for the $GABA_B$ binding site or for the $Na^+$ dependent GABA uptake system [Sigel et al., 1983]. Again, for a series of GABA agonists, the rank order of potency at the GABA binding site correlates with the potency at the membrane-bound GABA binding site [Greenlee et al., 1978; Olsen et al., 1981] (i.e., high- and low-affinity sites) and with the neurophysiological activity on the vertebrate nervous system [Krogsgaard-Larsen et al., 1975; Krogsgaard-Larsen and Johnston, 1978, Fig. 3]. The concentrations that are required for receptor binding are up to three orders of magnitude lower than the concentrations of GABA, i.e., $\mu M$, which are needed to evoke GABA-induced conductance changes in neurons. However, the same discrepancies exist for the binding of GABA to the membrane-bound receptor [Olsen, 1982], and the correlation that is found for the series of GABA agonists in both systems is very good. The ratio of GABA to benzodiazepine binding sites is in our hands 3.5–3.8, in deoxycholate extracted receptor purified subsequently in the presence of Triton X-100. This is the same value which we find in crude detergent extracts of bovine cortex. Lower values for this ratio have also been reported (Table IV), and using a different detergent in the isolation we can find a change in the ratio of GABA to benzodiazepine sites [Sigel and Barnard, 1984] (see 4., below).

3. The GABA shift of benzodiazepine affinity is found in the purified protein but the maximum stimulation that is found varies with the detergent used for initial extraction. We found in deoxycholate/Triton receptor a maximum stimulation of 20% [$^{3}$H] flunitrazepam binding activity in the presence of GABA with respect to control values in the absence

of GABA, whereas Martini et al. [1983] report a maximum stimulation of 50% for Triton extracted receptor. The latter finding has not been reproduced by other investigators. The concentrations of GABA that are found for the stimulation of benzodiazepine binding activity to the purified protein are an order of magnitude higher than those required for receptor binding. Thus, this has been called the "low-affinity" GABA site, and reported $EC_{50}$ values are 1$\mu$M [Stephenson et al., 1984] and 0.1$\mu$M [Schoch et al., 1984]. We do not know if the high- and low-affinity GABA sites are topographically related.

4. In all the purified preparations that have been described so far in this article, with the exception of that of Olsen et al. [1984], it has not been shown that the purified protein has a binding site for picrotoxinin, TBPS, or the barbiturates. However, as has already been mentioned in the discussion of the solubilization of the GABA/benzodiazepine receptor, extraction with the zwitterionic detergent CHAPS preserves the barbiturate regulation of both benzodiazepine and GABA binding activities, and in addition, soluble binding activity for [$^{35}$S] TBPS, the cage convulsant can be measured. Sigel and Barnard [1984] have purified the GABA/benzodiazepine receptor by extraction in CHAPS/asolectin with subsequent purification in the presence of these two reagents and they found copurification of a high-affinity, saturable [$^{35}$S]TBPS binding site with the GABA and benzodiazepine sites. Furthermore, pentobarbital regulation of benzodiazepine binding was found with a maximum stimulation of 20–25% (significantly less than found for crude extracts) with an $EC_{50}$ of 0.3 mM. In the presence of CHAPS, the binding characteristics of the benzodiazepine ligands are unchanged. However, initial experiments have shown that the muscimol binding site is shifted to lower affinity and the $B_{max}$ value is lower such that the ratio of GABA to BZ sites approaches 1. This is also the case for crude CHAPS extracts [Stephenson et al., 1984].

## Molecular Properties of Purified GABA/Benzodiazepine Receptor

The molecular weight of the isolated native GABA/benzodiazepine receptor can only be determined accurately by hydrodynamic methods when the appropriate correction is made for the contribution of the bound detergent to the receptor. Thus although the Stokes' radius and the sedimentation coefficient can be determined for the pure protein, the molecular weight cannot be extrapolated directly from these parameters because of the anomalous behavior of membrane proteins in solution with respect to globular proteins. The contribution of the detergent bound to the mass of the receptor–detergent complex can be estimated by the determination of the sedimentation coefficient of the purified receptor in sucrose density gradients of $H_2O$ and $^2H_2O$ and of the Stokes' radius [Meunier et al., 1972]. These types of measurements have been made in crude detergent extracts, where a value of $M_r$ 355,000 was obtained for the GABA/benzodiazepine receptor which was solubilized with sodium deoxycholate and centrifuged in the presence of Triton X-100 [Stephenson et al., 1982]. However, it was found that with the purified receptor, the protein was highly aggregated and eluted in the void volume (i.e., molecular weight $4 \times 10^6$ daltons) in gel filtration columns. On sucrose density gradients the binding activities of both GABA and benzodiazepine ligands were found over several gradient fractions, again an indication that the receptor was in an aggregated form [Sigel et al., 1983; Olsen et al., 1984]. This did not seem to be the case for purified GABA/benzodiazepine receptor extracted initially with Triton X-100, because a single sharp symmetrical peak with the Stokes' radius of catalase was reported [Martini et al., 1982]. We have found that it is possible to reduce the aggregation of the receptor if the size separation is carried out in the presence of high concentrations of salt,

**TABLE V. Molecular Parameters of the Purified GABA/Benzodiazepine Receptor***

| | |
|---|---|
| Isolectric point | 5.6 |
| Stokes radius (nm) | 7.3 ± 0.5 |
| Sedimentation coefficient | 8.7S |
| Partial specific volume (ml $g^{-1}$) | 0.78 |
| Detergent bound per g protein | 0.28 |
| Molecular weight[a] | 230,000 |

*All the parameters were determined for GABA/benzodiazepine receptor extracted by $Na^+$ deoxycholate and subsequently purified in the presence of Triton X-100.

[a]Is the molecular weight of the protein determined by the sedimentation of the purified GABA/benzodiazepine receptor in $H_2O/D_2O$ sucrose density gradients in the presence of 1M KCl and 0.5% Triton X 100, and gel filtration. The data are taken from Barnard et al, [1984] and Mamalaki et al., [unpublished results].

i.e., 0.5–1 M NaCl. However, only when brain membranes are pretreated with low concentrations of Triton X-100 before sodium deoxycholate solubilization is the purified receptor in a monodisperse form in the presence of 1M NaCl. Then the molecular parameters obtained by the $^2H_2O$ method, are as shown in Table V, with the deduction that the molecular weight of the protein is 230,000.

The sedimentation coefficient of the purified GABA/benzodiazepine receptor is smaller than has been reported for crude solubilized receptor, where values of the order 11 S have been reported [Sherman-Gold and Dudai, 1980; Asano and Ogasawara, 1981a; Stephenson et al., 1982]. We have shown that in the presence of 0.5–1 M NaCl concentration this apparent sedimentation coefficient is decreased and that this can be explained by an increase in detergent binding [Barnard et al., 1984]. Nevertheless, the sedimentation coefficient in crude extracts may also be increased by complexing of the receptor with some other component therein.

Radiation inactivation of the purified GABA/benzodiazepine receptor in the frozen state in the presence of Triton X-100 gives a value of $M_r$ 197,000 for both the GABA and benzodiazepine binding components [Nielsen et al., unpublished observations]. This is close to the value of $M_r$ 220,000 found by Chang and Barnard [1982], for the inactivation of the native receptor in the membrane. Thus we conclude that no components of the GABA/benzodiazepine receptor have been lost during the isolation procedure and that the size of the functional receptor is the same in the membrane and when purified. This conclusion is not changed by the observation that in some other conditions the coupling between subunits of the receptor complex can be weakened so that radiation inactivation detects smaller sizes [Nielsen and Braestrup, 1983].

### Receptor Subunits

It is the purpose of this article to review the purification of the GABA/benzodiazepine receptor, and the structure of the protein when purified is outside its scope. However, it should be mentioned briefly that this receptor is a multi-subunit complex [Sigel et al., 1983; Olsen et al., 1984; Schoch et al., 1984]. Furthermore, it has been possible to use a photoaffinity benzodiazepine ligand, flunitrazepam, to identify the benzodiazepine-binding polypeptide [Möhler et al., 1980]. In all the purification studies reported so far, one major subunit($\alpha$), $M_r$ 53,000, is specifically photoaffinity-labeled with [$^3$H] flunitrazepam [Sigel et al., 1983; Barnard et al., 1984; Schoch et al., 1984; Olsen et al., 1984]. At this time, there are no available $GABA_A$ ligands which can be employed as either covalent or photoaffinity reagents for the GABA binding site; hence, a clear identification of the GABA binding subunit has not yet been possible.

It appears, therefore, that the benzodiazepine binding site is located on a subunit of $M_r$ 53,000. The production by prior photoaffinity labeling of a polypeptide migrating at 51,000–53,000 $M_r$ in SDS-polyacrylamide gel electrophoresis is a distinctive property of the GABA/benzodiazepine receptor present, whether the receptor is directly estimated from prelabeled membranes or is in crude extracts or in the pure state. In fact, more than one labeled band can be produced thus, with membranes prepared from different brain regions giving different ratios of second or third labeled bands [Sieghart and Drexler, 1983], but the $\alpha$-subunit band (51,000–53,000 $M_r$) is generally the major one and in some cases the sole one. Hence the ability to produce this band by the photoaffinity labeling is one of the diagnostic features of the GABA/benzodiazepine complex.

## CONCLUSION

We have described the solubilization and purification of the brain $GABA_A$/benzodiazepine receptor. The properties of the purified protein permit the conclusion that all the regulatory binding sites coexist in one macromolecular complex. Thus we can anticipate that, in the near future, it will be possible to study both the isolated GABA-gated chloride ion channel and the molecular biology of the receptor. Already, the GABA/benzodiazepine receptor has been successfully incorporated into liposomes [Sigel et al., 1985] and mRNA encoding the GABA/benzodiazepine receptor subunits from chick or rat brain has been successfully translated in *Xenopus* oocytes to give functional GABA-gated $Cl^-$ ion channels [Smart et al., 1983; Houamed et al., 1984].

## ACKNOWLEDGMENTS

We thank Mr. J. Cavanagh for the illustrations and the Medical Research Council for financial support. F.A. Stephenson holds a Royal Society University Research Fellowship.

## REFERENCES

Asano T, Ogasawara N (1981a): Soluble gamma-aminobutyric acid and benzodiazepine receptors from rat cerebral cortex. Life Sci 29:193–200.

Asano T, Ogasawara N (1981b): Chloride dependent stimulation of GABA and benzodiazepine receptor binding by pentobarbital. Brain Res 225:212–216.

Asano T, Sakakibara J, Ogasawara N (1983): Molecular sizes of photolabelled GABA and benzodiazepine receptor proteins are identical. FEBS Lett 151:277–280.

Barnard EA, Stephenson FA, Sigel E, Mamalaki C, Bilbe G, Constanti A, Smart TE, Brown DA (1984): Structure and properties of the brain GABA/benzodiazepine receptor complex. In Kito S, Segawa T, Kuriyama K, Yamamura HI, Olsen RW (eds): "Neurotransmitter Receptors: Mechanisms of Action and Regulation." New York: Plenum Press, pp 235–254.

Chang L-R, Barnard EA (1982): The benzodiazepine/GABA receptor complex: Molecular size in brain synaptic membranes and in solution. J Neurochem 39:1507–1518.

Chiu TH, Rosenberg HC (1979): Differential effects of Triton X-100 on benzodiazepine and GABA binding in a frozen-thawed synaptosomal fraction of rat brain. Eur J Pharmacol 58:335–338.

Chude O (1979): Solubilization and partial purification of the GABA receptor from mouse brain and a binding assay for the solubilized receptor. J Neurochem 33:621–629.

Davis WC, Ticku MK (1981): Solubilization of the picrotoxinin binding receptor from mammalian brain. J Neurochem 36:1572–1579.

Gavish M, Chang SLR, Snyder SH (1979): Solubilization of histamine H-1 GABA and benzodiazepine receptors. Life Sci 25:783–790.

Gavish M, Snyder SH (1981): $\gamma$-Aminobutyric acid and benzodiazepine receptors: Copurification and characterization. Proc Natl Acad Sci USA 78:1939–1942.

Greenlee DV, Van Ness PC, Olsen RW (1978): Gamma-aminobutyric acid binding in mammalian brain. Receptor-like specificity of sodium-independent sites. J Neurochem 31:933–938.

Greenlee DV, Olsen RW (1979): Solubilization of gamma-aminobutyric acid receptor protein from mammalian brain. Biochem Biophys Res Comm 88:380–387.

Hill DR and Bowery NG (1981): $^3$H-Baclofen and $^3$H-GABA bind to bicuculline-insensitive $GABA_B$ sites in rat brain. Nature 290:149–152.

Hjelmeland LM, Chrambach A (1984): Solubilization of functional membrane-bound receptors. In Venter JC, Harrison LC (eds): "Receptor Biochemistry and Methodology, Vol. 1." New York: Alan R. Liss, Inc., pp 35–46.

Houamed KM, Bilbe G, Smart TG, Constanti A, Brown DA, Barnard EA, Richards BM [1984]: Expression of functional GABA, glycine and glutamate receptors in *Xenopus* oocytes injected with rat brain mRNA. Nature 310:318–321.

King RG, Olsen RW (1984): Solubilization of convulsant/barbiturate binding activity on the $\gamma$-aminobu-

tyric acid/benzodiazepine receptor complex. Biochem Biophys Res Comm 119:530–536.
Korneyev AY (1982): Characterization of solubilized benzodiazepine and muscimol binding sites from rat brain. Neuropharmacology 21:1355–1358.
Krogsgaard-Larsen P, Johnston GAR, Curtis DR, Game CJA, McCullock RM (1975): Structure and biological activity of a series of conformationally restricted analogues of GABA. J Neurochem 25:803–809.
Krogsgaard-Larsen P, Johnston GAR (1978): Structure activity studies on the inhibition of GABA binding to rat brain membranes by muscimol and related compounds. J Neurochem 30:1377–1382.
Krogsgaard-Larsen P, Snowman A, Lummis SC, Olsen RW (1981): Characterization of the binding of the GABA agonist [$^3$H] piperidine-4-sulphonic acid to bovine brain synaptic membranes. J Neurochem 37:401–409.
Lang B, Barnard EA, Chang L-R, Dolly JO (1979): Putative benzodiazepine receptor: A protein solubilized from brain. FEBS Lett 104:149–153.
Lester BR, Miller AL, Peck EJ (1981): Differential solubilization of γ-aminobutyric acid receptive sites from membranes of mammalian brain. J Neurochem 36:154–164.
Leeb-Lundberg F, Snowman AM, Olsen RW (1980): Barbiturate receptors are coupled to benzodiazepine receptors. Proc Natl Acad Sci USA 77:7468–7472.
Lo MMS, Strittmatter SM, Snyder SH (1982): Physical separation and characterization of two types of benzodiazepine receptors. Proc Natl Acad Sci USA 79:680–684.
Martini C, Lucacchini A, Ronca G (1981): Specific absorbents for affinity chromatography of benzodiazepine binding proteins. Preparative Biochem 11:487–499.
Martini C, Lucacchini A, Ronca G, Hrelia S, Rossi CA (1982). Isolation of putative benzodiazepine receptors from rat brain membranes by affinity chromatography. J Neurochem 38:15–19.
Martini C, Rigacci T, Lucacchini A (1983): [$^3$H] Muscimol binding site on purified benzodiazepine receptor. J Neurochem 41:1183–1185.
Massotti M, Guidotti A, Costa E (1981): Characterization of benzodiazepine and γ-aminobutyric recognition sites and their endogenous modulators. J Neurosci 1:409–418.
Mernoff ST, Cherwinski HM, Becker JW, de Blas AL (1983): Solubilization of brain benzodiazepine receptors with a zwitterionic detergent: Optimal preservation of their functional interaction with the GABA receptors. J Neurochem 41:752–758.
Meunier J-C, Olsen RW, Changeux J-P (1972): Studies on the cholinergic receptor protein from *Electrophorus electricus*. FEBS Lett 24:63–68.
Möhler H, Battersby MK, Richards JG (1980): Benzodiazepine receptor protein identified and visualized in brain tissue by a photoaffinity label. Proc Natl Acad Sci USA 77:1666–1670.
Möhler H, Richards JG (1981): Agonist and antagonist benzodiazepine receptor interaction in vitro. Nature 294:763–765.
Nielsen M, Braestrup C (1983): The molecular target size of brain TBPS binding sites. Eur J Pharmacol 96:321–322.
Olsen RW (1981): The GABA postsynaptic membrane receptor-ionophore complex. Molecular and Cellular Biochem 39:261–279.
Olsen RW (1982): Drug interactions at the GABA receptor–ionophore complex. Ann Rev Pharmacol Toxicol 22:245–77.
Olsen RW, Bergman MO, Van Ness PC, Lummis SC, Watkins AE, Napias C, Greenlee DV (1981): γ-Aminobutyric acid receptor binding in mammalian brain. Mol Pharmacol 19:217–227.
Olsen RW, Snowman AM (1982): Chloride-dependent enhancement by barbiturates of GABA receptor binding. J Neurosci 2:1812–1823.
Olsen RW, Wong EHF, Stauber GB, Murakami D, King RG, Fischer JB (1984): Biochemical properties of the GABA/barbiturate/benzodiazepine receptor–chloride ion channel complex. In Kito S, Segawa T, Kuriyama K, Yamamura HI, Olsen RW (eds): "Neurotransmitter Receptors: Mechanisms of Action and Regulation." New York: Plenum Press, pp 205–219.
Schoch P, Möhler H (1983): Purified benzodiazepine receptor retains modulation by GABA. Eur J Pharmacol 95:323–324.
Schoch P, Häring P, Takacs B, Stähli C, Möhler H (1984): A GABA/benzodiazepine receptor complex from bovine brain: purification, reconstitution and immunological characterisation. J Receptor Res 4:189–200.
Sieghart W, Drexler G (1983): Irreversible binding of [$^3$H] flunitrazepam to different proteins in various brain regions. J Neurochem 41:47–55.
Sherman-Gold R, Dudai Y (1980): Solubilization and properties of a benzodiazepine receptor from calf cortex. Brain Res 198:485–490.
Sigel E, Mamalaki C, Barnard EA (1982): Isolation of a GABA receptor from bovine brain using a benzodiazepine affinity column. FEBS Lett 147:45–48.
Sigel E, Stephenson FA, Mamalaki C, Barnard EA (1983): A γ-aminobutyric acid/benzodiazepine receptor complex of bovine cerebral cortex. J Biol Chem 258:6965–6971.
Sigel E, Barnard EA (1984): A γ-aminobutyric acid/benzodiazepine receptor complex from bovine cerebral cortex. Improved purification with preservation of regulatory sites and their interactions. J Biol Chem 259:7219–7223.

Sigel E, Stephenson FA, Mamalaki C, Barnard EA (1984a): The purified GABA/benzodiazepine barbiturate receptor complex: Four types of ligand binding sites, and the interactions between them are preserved in a single isolated protein complex. J Receptor Res 4:175–188.

Sigel E, Mamalaki C, Barnard EA (1985): Reconstitution of the isolated GABA/benzodiazepine receptor complex into phospholipid vesicles. Neurosci Lett 61:165–170.

Skolnick P, Rice KC, Barker JL, Paul SM (1982): Interaction of barbiturates with benzodiazepine receptors in the central nervous system. Brain Res 233:143–156.

Smart TG, Constanti A, Bilbe G, Brown DA, Barnard EA (1983): Synthesis of functional chick brain GABA-benzodiazepine-barbiturate/receptor complexes in mRNA injected *Xenopus* oocytes. Neurosci Lett 40:55–59.

Squires RF, Braestrup C (1977): Benzodiazepine receptors in rat brain. Nature 266:732–734.

Squires RF, Casida JE, Richardson M, Saederup E, (1983): [$^{35}$S]*t*-Butylbicyclophosphorothionate binds with high affinity to brain specific sites coupled to γ-aminobutyric acid-A and ion recognition sites. Mol Pharmacol 23:326–336.

Stephenson FA, Olsen RW (1982): Solubilization by CHAPS detergent of barbiturate-enhanced benzodiazepine-GABA receptor complex. J Neurochem 39:1579–1586.

Stephenson FA, Watkins AE, Olsen RW (1982): Physicochemical characterization of detergent-solubilized γ-aminobutyric acid and benzodiazepine receptor proteins from bovine brain. Eur J Biochem 123:291–298.

Stephenson FA, Olsen RW (1983): Biochemical pharmacology of the GABA receptor–ionophore protein complex. In Mandel P, Defeudis FV (eds): "Advances in Biochemical Psychopharmacology, Vol. 37." pp 71–80.

Stephenson FA, Olsen RW (1984): Soluble GABA/benzodiazepine/barbiturate receptor interactions in mammalian brain. In Reid E, Cook GMW, Morre DJ (eds): "Investigation of Membrane-Located Receptors." New York: Plenum Press, pp 505–507.

Stephenson FA, Sigel E, Mamalaki C, Bilbe G, Barnard EA (1984): The benzodiazepine receptor complex: Purification and characterization. In Vizi ES, Magyar K (eds): "Regulation of Transmitter Function: Basic and Clinical Aspects." Budapest: Akademiai Kiado, pp 437–442.

Taguchi J, Kuriyama K (1984): Purification of γ-aminobutyric acid(GABA) receptor from rat brain by affinity column chromatography using a new benzodiazepine 1012-S, as an immobilized ligand. Brain Res 323:219–226.

Tallman JF (1984): Properties of the purified GABA/benzodiazepine complex. Clin Neuropharmacol 7:558–559.

Tallman JF, Mallorga P, Thomas JW, Gallager DW (1981): Benzodiazepine binding sites: Properties and modulation. In Costa E, Di Chiara G, Gessa Gl (eds): "GABA and Benzodiazepine Receptors." New York: Raven Press, pp 9–18.

Ticku MK, Ban M, Olsen RW (1978): Binding of [$^3$H] α-dihydropicrotoxinin, a gamma-aminobutyric acid synaptic antagonist, to rat brain membranes. Mol Pharmacol 14:391–402.

Yousufi MAK, Thomas JW, Tallman JF (1979): Solubilization of benzodiazepine binding site from rat cortex. Life Sci 25:463–470.

Benzodiazepine/GABA Receptors and Chloride Channels:
Structural and Functional Properties, pages 275–284


# 14

# Characterization of Photolabeled Benzodiazepine Receptors

**John F. Tallman**

*Yale University School of Medicine, Ribicoff Research Facilities,
New Haven, Connecticut 06508*

As this and the other chapters in this book indicate, the molecular mechanism of benzodiazepine action has become, in the last ten years, one of the most intensely investigated areas in neurobiology. This is due to several factors, including the wide use of benzodiazepines as therapeutic agents, their involvement in the action of the brain's major inhibitory transmitter, GABA, and the long-term prospect that studies of the proteins involved in these molecular mechanisms may lead to a greater understanding of the etiology of anxiety and seizures in humans. Since this chapter is a part of the book covering the actions of the benzodiazepines, its scope has been limited to a review of the literature in the area of affinity labeling of the benzodiazepine receptor and related subjects. A more integrated short review of the area has been recently prepared [Tallman and Gallager, 1985].

## PHOTOAFFINITY LABELING FOR THE STUDY OF RECEPTORS

Because receptors frequently have a very high affinity for the drugs with which they interact, these drugs can be used to specifically label the binding sites on the receptor. In the usual case, this binding is reversible and effective study of the receptor requires the ability to rapidly separate free drug from that which is bound. If the drug contains a reactive or potentially reactive group, then it may be possible to form a covalent bond to the receptor protein while it is attached to the binding site. Since reversibly bound ligands of even the highest affinity will dissociate during chromatography or electrophoresis under denaturing conditions, these ligands are not suitable for identification of individual protein species as part of a complex of proteins. Covalently linked ligands allow the use of time-requiring or denaturing conditions and facilitate the study of receptors as proteins. A photosensitive group offers the advantage that until exposure to light the drug behaves in the usual reversible fashion, allowing full characterization of the specificity of the reaction of the drug at the binding site. Recent reviews of photoaffinity labels as pharmacological tools which outline the advantages of different reactive groups are [Fedan et al., 1984; Ruoho et al., this series].

## BENZODIAZEPINE RECEPTORS IN BRAIN

Although the interactions between GABA and benzodiazepine action had been predicted for a number of years, description of the mechanistic aspects of these interactions depended on the discovery of high-affinity, saturable, and stereospecific receptors for both

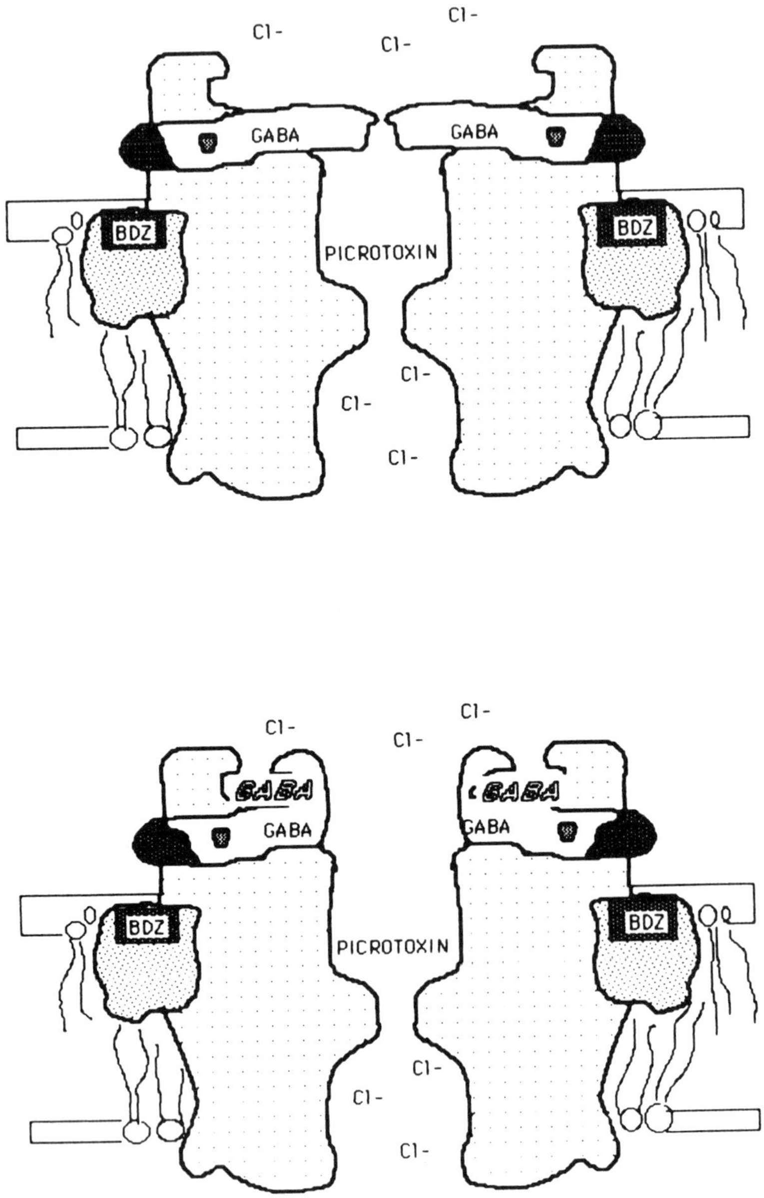


**Fig. 1.** *Domains of the GABA/benzodiazepine/anionophore complex. The complex is shown in a closed and open configuration, depending on whether GABA is present at its binding site. Two moles of GABA per mole of receptor seem to be required. Picrotoxin is shown in its role as a channel blocker. The benzodiazepine binding site is arranged such that it cannot open the anion channel itself but enhance the ability of GABA to carry out this event; benzodiazepine inverse agonists would make the transition from closed- to open-channel conformations unfavorable. No implications about the number of subunits in the receptor complex are to be drawn from this figure.*

GABA and the benzodiazepines [see Möhler et al., this volume]. There was a high degree of correlation between the biological activity of a series of benzodiazepines and their ability to interact with the receptor present on neuronal elements in brain [Möhler and Okada, 1977a, b; Squires and Braestrup, 1977]. Direct biochemical experiments soon indicated that a number of GABAergic agonists were capable of enhancing the affinity of the benzodiazepine binding site in vitro [Tallman et al., 1978] and a large number of analogues possess either full enhancing activity or are only partly active (partial agonists). A number of GABA binding sites with differing affinities exist in brain [Enna, 1983], and it is one of the lower-affinity forms that influences the binding of benzodiazepines [Tallman et al., 1978]. Direct evidence for the involvement of the low-affinity site has been provided through the inactivation of high-affinity binding of GABA by thiocyanate; this inactivation did not alter the modulation of benzodiazepine binding by GABA [Browner et al., 1981].

In addition to the multiple-affinity forms of GABA receptor associated with the benzodiazepine sites, the GABA receptors control channels for chloride ions. The chloride channel is a site of action for a number of different compounds including some of the barbiturates [see Fischer and Olsen, this volume] and a number of convulsant agents. These agents apparently work either by: 1) blocking the anion channel (convulsants) and thus inhibiting the actions of GABA, or 2) prolonging the opening of the channel, thereby enhancing the efficacy of GABA (anxiolytics and sedatives).

Thus the current model for this complex is a tripartite one that encompasses a GABA receptor allosterically linked to a benzodiazepine receptor; both of these control the opening of an anionophore. The stoichiometry of these sites is not known but appears to be more complicated than one-to-one (Fig. 1).

## ALKYLATING BENZODIAZEPINES

The first compound prepared that was shown to form a covalent bond to the benzodiazepine receptor was a derivative of flurazepam containing an isothiocyanyl group as the reactive group [Rice et al., 1979]. This compound was called irazepine (1-(2-isothiocyanatoethyl)-7-chloro-1,3-dihydro-5-(2-fluorophenyl)-2H-1,4-benzodiazepine-2-one). It was a noncompetitive inhibitor of $^3$H-diazepam binding and appeared to be irreversible. Interaction of this compound with model bionucleophiles in vitro indicated substantial reaction with primary and secondary amines as well as mercaptans. The nature of these reactions indicated that the compound probably interacted with a lysine or cysteine near the receptor's active site; the actual fragment of the receptor was not isolated. A related bromoacetyl derivative, called kenazepine, had similar effects [Williams et al., 1980]. These compounds were capable of distinguishing apparent benzodiazepine receptor subtypes and led to a primitive description of apparent receptor domains [Skolnick et al., 1982]. The major limitation of these compounds was a lack of radioactive analogue which would have permitted radiolabeling and identification of the benzodiazepine receptor protein; these derivatives were not developed because, soon after the synthesis of the alkylating benzodiazepines, the photolabeling properties of nitro-containing benzodiazepines were discovered.

## NITRO-CONTAINING BENZODIAZEPINES

Among the large number of compounds developed by Hofmann-LaRoche as potentially useful drugs, a highly active series of derivatives containing a 7-nitro group was synthesized and marketed as hypnotics (flunitrazepam, Rohypnol) or anticonvulsants (clonazepam, Clonopin) [Sternbach, 1983]. These compounds contain a nitro group rather than a halogen at position 7, and the nitro groups are capable of forming a number of resonance structures with the heterocyclic rings. These compounds possess strong absorbances in the near-ultraviolet portion of the spectrum, and in the presence of UV light they can form highly reactive free radicals. When these compounds are localized in a water-free hydrophobic part of the receptor, the free radicals are capable of forming cross-links to the benzodiazepine receptor.

The first to appreciate this phenomenon were chemists working at Hoffman-LaRoche. In their studies of the binding of [$^3$H]-flunitrazepam to brain tissue, they found that the usual reversible binding could be transformed into a covalent bond by exposure to ultraviolet light [Möhler et al., 1980]. Since the formation of this bond depended on occupancy of the benzodiazepine receptor, the photolabeling of the receptor was reversed by occupancy of the receptor by nonradiolabeled benzodiazepines. The potency of these compounds as blockers of photolabeling was directly correlated with their potency as reversible ligands. Photolabeling with flunitrazepam could only be demonstrated with the central-type benzodiazepine receptor; no specific photolabeling was obtained in kidney or $C_6$ glioma cells, both of which possess high-affinity binding sites for the benzodiazepines with high affinity for Ro5-4864 [Thomas and Tallman, 1981]. Ro5-4864 is identical to diazepam, except that it possesses a p-chloro substitution on the phenyl ring, and this single substitution renders it very inactive as an anxiolytic and virtually inactive at central-type benzodiazepine receptors [Thomas and Tallman, 1981]. Another distinction between the central and peripheral sites is in their ability to be solubilized. Although it is quite easy to solubilize the central type of benzodiazepine receptor, the peripheral receptor is extracted from its membrane environment only with great difficulty [Thomas and Tallman, 1981].

Important parameters to be considered in photolabeling are the wavelengths and time of the process. Reaction conditions should be used which maximize the ratio of specific to nonspecific photolabeling while minimizing the destruction of receptor binding sites. The importance of this is emphasized by a recent report which indicates that a more complete incorporation of [$^3$H]-flunitrazepam can be obtained at 254 nm than at 366 nm [Herblin and Mechem, 1984]. However, even in this study, a full incorporation of flunitrazepam into receptor protein was not obtained, and the possible reasons (see below) for lack of 100% incorporation are more important than whether a 25% or 40% incorporation is obtained. The time course of photolabeling should also be examined in any careful study of the process. It is clear in the study of Thomas and Tallman [1981] that continued exposure of tissue to ultraviolet light in the presence of ligand bound at steady state leads to a hyperbolic increase in specific binding and a linear increase in incorporation of nonspecific labeling over time. After 5 minutes, no increase in specific incorporation can be distinguished. Thus, labeling with $^3$H-flunitrazepam for periods greater than about 5 minutes is not justified because the incorporation into nonbenzodiazepine receptor proteins can increase and become a significant part of the total incorporation. This can lead to a large number of radiolabeled bands when the labeled proteins are subsequently separated by SDS gel electrophoresis. This nonspecific incorporation may indeed be blocked by high levels of benzodiazepines because the competitive ligands may act as internal filters and not as competitive inhibitors at the ligand binding sites. It is not simple to determine whether these other sites should be considered "receptors" for benzodiazepines.

## PHOTOAFFINITY LABELING REVEALS LOCALIZATION OF BENZODIAZEPINE RECEPTORS

The covalent incorporation of radiolabeled flunitrazepam into benzodiazepine receptors, using the techniques described above, has allowed the expansion of autoradiographic localization studies from the light- [Young and Kuhar 1980, Young et al., 1981, Niehoff et al., 1982] to electron-microscopic level [Möhler et al., 1981]. These elegant electron-microscopic studies provided evidence for the neuronal localization of benzodiazepine receptors. Immunocytochemical staining for glutamic acid decarboxylase, the committed step in GABA synthesis, frequently colocalized with the radiolabeled flunitrazepam. This finding indicated that at least some of the benzodiazepine receptors were in GABAergic synapses. Such studies do not provide evidence for a pre- or postsynaptic localization for benzodiazepine receptors, because the electron microscopic autoradiography does not provide sufficient resolution. Using lesioning techniques, some differentiation of pre- and postsynaptic receptors has been possible; receptors with a high affinity for triazolopyridazines apparently have a presynaptic localization [Lo et al., 1983].

## APPARENT MULTIPLE BENZODIAZEPINE RECEPTORS

The ability to incorporate radioactively labeled flunitrazepam into the benzodiazepine receptor has allowed the examination of labeled receptor by electrophoresis in the presence of sodium dodecyl sulfate (SDS). All researchers have demonstrated the presence of a major band from rat brain with molecular weight of approximately 50 K daltons [Möhler et al., 1980; Thomas and Tallman, 1981]. An additional band of molecular weight 55 K daltons has been shown to be prominent in hippocampus and striatum [Sieghart and Karobath, 1980]. The 50 K dalton protein has been termed a $BZ_1$ receptor and the 55 K protein the $BZ_2$ receptor. Further minor bands with sizes of 53 K and 59 K have been also demonstrated [Sieghart et al., 1983].

In several of these studies no attempt was made to inhibit proteolysis, and endogenous proteases are apparently present in membrane fractions and could contribute to the number of bands seen on SDS gels [Klotz et al., 1984]. Another hypothesis is that heterogeneity could result from different processing of the precursor forms of the receptor. This may result in slightly different carbohydrate side chains or incomplete removal of leader sequences in pre- versus postsynaptically localized receptors. Altered processing could also account for the changes in molecular weight seen during development; differences in neuronal sprouting could result in differences in proportion of receptors in mature and immature rats. Along with these potential differences, neonatal animals show a much larger effect of GABA on benzodiazepine binding than adult animals [Mallorga et al., 1980].

The most interesting possibility is that the multiple forms of the receptor represent "isoreceptors." Although commonly found for enzymes, genetically distinct forms of receptors have not been described. With more sophisticated probes, it is almost certain that isoreceptors will emerge as important in investigations of the etiology of psychiatric disorders in people.

## PHARMACOLOGICAL DEMONSTRATION OF BENZODIAZEPINE HETEROGENEITY

In pharmacological studies, it has been claimed that the $BZ_1$ receptor has high affinity for a triazolopyridazine (CL 218,872). This compound is of interest because it was the first compound that did not interact with the binding of radioactive benzodiazepines with a Hill coefficient of 1, which would indicate identity in the binding site [Squires et al., 1979; Klepner et al., 1979]. This compound did not distinguish apparent benzodiazepine subtypes in the cerebellum, which contains only $BZ_1$ receptors; cerebral cortex also contains mostly $BZ_1$ receptors. Several unsubstantiated claims were made for the pharmacological properties of the triazolopyridazines; the most important of these is that the triazolopyridazines possess potent anticonflict and anticonvulsant activity with little sedation [Lippa et al., 1979]. More recent appraisal of these compounds indicates that, because they do possess activity at $BZ_2$ receptors, a clear in vivo effect mediated by a benzodiazepine receptor subtype might be hard to distinguish. Under certain circumstances they possess proconflict activity [Melchior et al., 1983]. A smaller GABA shift [Braestrup et al., 1982] and a smaller photoshift [Thomas and Tallman, 1983] than benzodiazepines has also been found for these compounds. The final conclusion is that they may be more appropriately classified as partial benzodiazepine agonists [Gee et al., 1982].

## AGONIST AND ANTAGONIST DOMAINS OF BENZODIAZEPINE RECEPTORS

A series of compounds was isolated from human urine [Nielsen et al., 1979; Braestrup et al., 1980] as inhibitors of benzodiazepine receptor binding. When their structure was determined, it became clear that they had been formed during the isolation procedure. Synthesis of a number of analogues of the basic β-carboline structure allowed a number of investigators to test their activity in behavioral paradigms. It was realized immediately that these compounds could antagonize the activity of diazepam [Tenen and Hirsch, 1980] and possessed their own intrinsic activity opposite to the benzodiazepines. These activities include proconflict, convulsant, anxiogenic, and insomniac properties [see Tallman and Gallager, 1985, for a complete description]. Such activities have led to the naming of these compounds as inverse agonists [Braestrup and Nielsen, 1983]. A biochemical test for these inverse agonist properties emphasizes their interactions with the GABAergic system. In contrast to the benzodiazepines which show increased binding in the presence of GABA, the β-carbolines show a decrease in their affinity for the benzodiazepine site. A third set of compounds show no shift in their affinity in the presence of GABA; these compounds, particularily Ro15-1788, are antagonists of both benzodiazepine and β-carboline binding [Hunkeler et al., 1981; Möhler and Richards, 1981; Möhler et al. 1981]. They also antagonize the behavioral actions of benzodiazepine agonists and inverse agonists but seem to be relatively inactive by themselves. A final set of compounds which include the triazolopyridazines are compounds which have a GABA shift which is to increased affinity but with a magnitude smaller than benzodiazepines. These compounds may be partial agonists and are potentially the most interesting compounds because of their lower potential for the development of tolerance [Tallman and Gallager, 1985].

## PHOTOAFFINITY LABELING OF BENZODIAZEPINE RECEPTORS ALTERS AGONIST DOMAINS

The ability of GABA to differentially alter the binding of benzodiazepine agonists and antagonists means that the loci to which the agonists and antagonists bind cannot be identical. In the presence of GABA, a change in the entropic parameter of benzodiazepine binding appears to be behind the alteration in affinity when the study is carried out at 4°C. At 37°C, an additional enthalpic change also occurs [Thomas and Tallman, 1984].

The separate domains of binding of benzodiazepine agonists and antagonists can also be defined by examining the parameters of ligand binding following photolabeling. During the initial studies of photolabeling, it was discovered that photolabeling with flunitrazepam could cause covalent crosslinks to the receptor, as described above. However, when the total number of binding sites on the membrane were compared to the number of sites being photolabeled, it was discovered that only about 25–40% of the sites were labeled. In addition, the remaining sites were apparently inactivated when reversible binding of benzodiazepines was used to study binding to tissue that had been photolabeled previously. In contrast to the drastic diminution of the agonist binding site, the antagonist binding parameters (studied by Ro15-1788) or inverse agonist binding (studied by β-carboline binding) were relatively unaffected [Möhler 1982; Thomas and Tallman, 1983] by photolabeling. When benzodiazepine agonists are used to displace the binding of antagonists and inverse agonists, the displacement curves are shifted to lower potency and are more shallow. These shallow curves indicate heterogeneity in the sites to which the benzodiazepines bind; the photoshift is quite sensitive to full and partial benzodiazepine agonists; unlike the GABA shift, it is not sensitive to the properties of antagonists or inverse agonists. The triazolopyridazines by this test are partial agonists.

A number of interpretations are possible for the greater inactivation than labeling of sites with flunitrazepam [Thomas and Tallman, 1981]. Among these interpretations, the most interesting is that there are site-to-site interactions between binding sites for benzodiazepines. This is also consistent with the current allosteric models for receptor function and the presence of complex binding curves for GABAergic agonists. While this article focuses on the binding of benzodiazepines to their binding site, it is probably more appropriate to focus on the essentially modulatory role of the benzodiazepines in controlling GABAergic function. It is well known that the electophysiological effects of GABA require the binding of two molecules to initiate the physiological response. The effects of the benzodiazepines are not to cause these interactions by increasing the affinity of GABA for its receptor but to increase the probability that when such an event occurs, opening of a chloride channel will occur [Tallman and Gallager, 1985]. By viewing the system in this way, a greater appreciation of the interactions between the sites in the complex may be obtained. As mentioned above, the stoichiometry of these sites has not been defined but appears to be more complicated than one-to-one and a full description will require the purification of these sites and their reconstitution [see Chapter 13, this volume].

## STRUCTURAL STUDIES OF THE PHOTOLABELED RECEPTOR

It has been shown recently that the membrane-bound photolabeled benzodiazepine receptors are susceptible to degradation by an endogenous trypsin-like activity present in well-washed membrane fractions [Klotz et al., 1984]. The presence of this activity suggested that controlled proteolytic degradation of the membranes might release small fragments of the receptor that would be suitable for further purification. Since [$^3$H]-flunitrazepam can be incorporated into the receptor before proteolysis, it would provide a convenient way to monitor the fragments released from the receptor during purification. A procedure [Klotz, Neale, Guidotti, Conti-Tronconi, and Tallman, in preparation] was developed to release proteolytic fragments of the photolabeled receptor and to purify such small-molecular-weight hydrophilic fragments released from the membrane. This method used sequential filtration to remove substances with molecular weight greater than 10 Kdaltons, gel filtration, and several reverse-phase, high-performance liquid chromatography runs. The method was suitable for the preparation of several micrograms of labeled receptor fragments, and a single major limit peptide with no further proteolytically susceptible sites was obtained by controlled proteolysis by a mixture of trypsin and chymotrypsin.

Amino acid sequence analysis of this peptide was carried out by automated Edman degradation identifying phenylthiohydantoin-derivatized amino acids by HPLC on a IBM cyano column. The sequence of this fraction is Pro-Ala-Pro-Ala-Thr-Thr-Phe(F)-X-Lys(K). X is the amino acid to which flunitrazepam is linked.

## HOMOLOGOUS AMINO ACID SEQUENCES IN OTHER PROTEINS

The sequence PAPA is found in a number of vertebrate proteins that have been recently isolated and sequenced. These include myosin

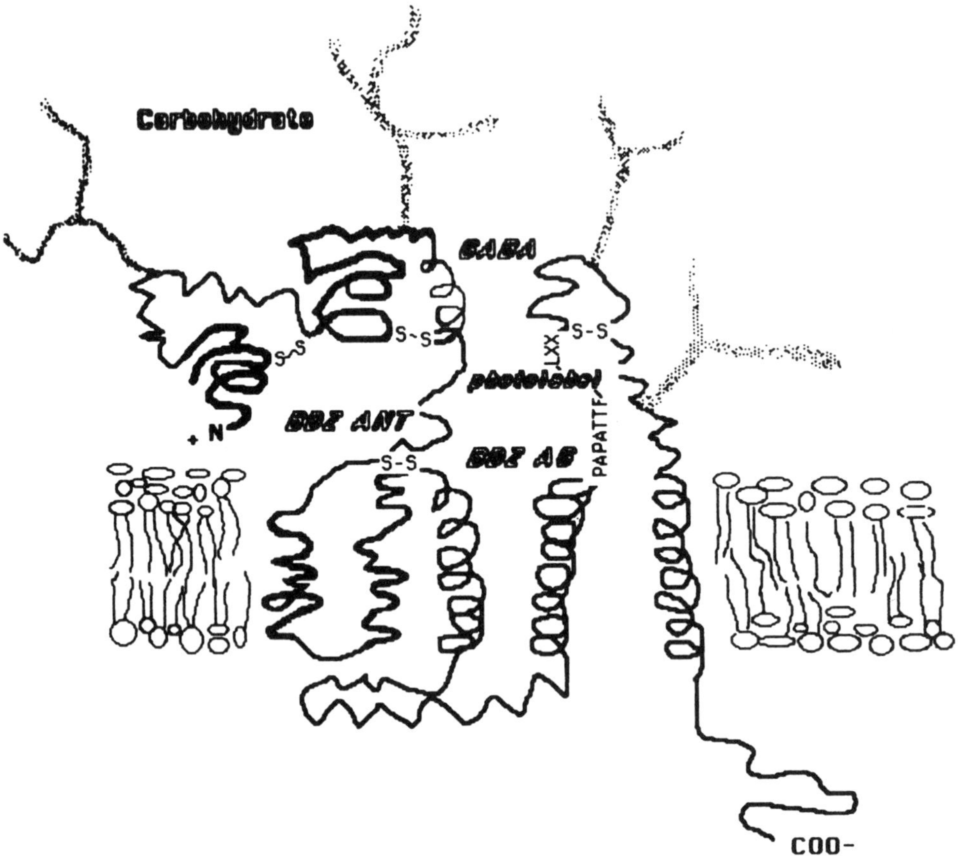


**Fig. 2.** *Theoretical diagram of the structure of the photolabeled benzodiazepine receptor subunit. This structure is based on current models for membrane-spanning proteins. Significant helical and pleated sheet organization is shown, and the fragment of the receptor which has been sequenced has been placed to allow the proline–alanine sequence to assume a role as helical interrupter. Benzodiazepine antagonist (ANT) and agonist (AG) are shown to bind to interactive although not identical sites. The fragment which is photolabeled is a part of the Ag site which would also encompass other amino acids. The membrane-spanning portion of the receptor may form with other subunits part of the associated anion channel (see Fig. 1). Carbohydrate is shown on the outside of the membrane and contributes to the conformational stability of the benzodiazepine binding site.*

light chain, $H_1$ histone, troponin, and immunoglobulin heavy chain. Taking a slightly longer view, the sequence PAPATTFAH has been found in spinach, tobacco, and maize ATP'ases. The sequence PAPAUTF is found in myosin B chain ATP'ase, and a related sequence PSPATTFAH is characteristic of the $\beta$ chain of a protomotive ATP'ase from *E. coli*.

The approach described above for the purification of receptor fragments is generally applicable to any receptor system where a radiolabeled covalent ligand can be linked to the binding site. Using this approach, a partial sequence that has served to identify potentially related proteins was obtained. This sequence is found in a number of ATP'ases and most importantly is in a region of the ATP'ase which is involved in the binding of adenine nucleotides. This may explain why the purines possess activity in displacing benzodiazepine binding and were at one time thought to be potential endogenous ligands for the receptor [see Tallman et al., 1980, for a review of this subject].

Thus we have gone to the limit in determining the sequence of the portion of the receptor to which benzodiazepines crosslink following photolabeling. This site is intimately involved in the active site of the receptor and is near or at the site of action of this widely used group of anxiolytics and hypnotics. Important information about the site of action of these drugs has been obtained and has a sequence to which antibodies may be prepared. Such antibodies have the potential of interaction with the active site of this receptor and may be important pharmacological tools.

We cannot deduce from this structure alone where it is localized in the receptor molecule; however, an interesting picture of the possible form of the receptor may be made [Fig. 2]. The entire photolabeled receptor has an apparent molecular weight of about 50 kD; endoglycosidase H treatment indicates that between 5–8 kD of the receptor is composed of carbohydrate sidechains [Sweetnam and Tallman, in preparation]. Several of the larger fragments released by proteolysis in the paradigm above are glycopeptides (based on their ability to adsorb to concanavalin A agarose). The limit peptide sequenced in this study does not have the proper sequence to contain carbohydrate. The carbohydrate is important in maintaining high-affinity binding, since removal of the carbohydrate by endoglycosidase treatment alters the binding of benzodiazepine ligands.

## CONCLUSIONS

In this article I have attempted to show how knowledge of the molecular properties of a protein that contains a drug binding site can enhance our understanding of drug action. With the purification of the GABA receptor complex to homogeneity and the potential to isolate the genes coding for these proteins, we may soon be in the position to do true molecular pharmacology and study the effects of point alterations in the protein sequence on the function of the GABA/benzodiazepine/anionophore complex.

## REFERENCES

Braestrup C, Nielsen M (1983): Benzodiazepine receptors. In Iversen L, Iversen S, Snyder SH (eds.): "Handbook of Psychopharmacology, Vol. 17." New York: Plenum Publishers, pp 285–384.

Braestrup C, Nielsen M, Olsen CE (1980): Urinary and brain beta-carboline-3-carboxylates as potent inhibitors of brain benzodiazepine receptors. Proc Natl Acad Sci USA 77(4):2288–2292.

Braestrup C, Schmiechen R, Neef G, Nielsen M, Petersen EN (1982): Interaction of convulsive ligands with benzodiazepine receptors. Science 216:1241–1243.

Browner M, Ferkany JW, Enna SJ (1981): Biochemical identification of pharmacologically and functionally distinct GABA receptors in rat brain. J Neuroscience 1:514–518.

Enna SJ (ed.) (1983): "The GABA Receptors" Clifton, NJ: Humana Press.

Fedan J, Hogaboom GK, O'Donnell JP (1984): Photoaffinity labels as pharmacological tools. Biochem Pharmacol. 33:1167–1180.

Gee KW, Morelli M, Yamamura HI (1982): The effect of temperature on CL 218872 and propyl beta-carboline-3-carboxylate inhibition of $^3$H-flunitrazepam binding in rat brain. Biochem Biophys Res Commun 105(4):1532–1537.

Herblin WF, Mechem CC (1984): Short wave ultraviolet irradiation increases photoaffinity labeling of benzodiazepine sites. Life Sci 35:317–324.

Hunkeler W, Möhler H, Pieri L, Polc P, Bonetti EP, Cumin R, Schaffner R, Haefely W (1981): Selective antagonists of benzodiazepines. Nature 290:514–516.

Klepner CA, Lippa AS, Benson DI, Sano MC, Beer B (1979): Resolution of two biochemically and pharmacologically distinct benzodiazepine receptors. Pharmacol Biochem Behav 11:457–462.

Klotz K, Bocchetta A, Neale J, Thomas JW, Tallman JF (1984): Proteolytic degradation of neuronal benzodiazepine binding sites. Life Sci 34:293–299.

Lippa AS, Coupet J, Greenblatt EN, Klepner CA, Beer B (1979): A synthetic non-benzodiazepine ligand for benzodiazepine receptors: a probe for investigating neuronal substrates of anxiety. Pharmacol Biochem Behav 11:99–106.

Lo M, Niehoff D, Kuhar MJ, Snyder SH (1983): Differential localization for type 1 and type 2 benzodiazepine binding sites in substantia nigra. Nature 306:57–60.

Mallorga P, Hamburg M, Tallman JF, Gallager DW (1980): Ontogenetic changes in GABA modulation of brain benzodiazepine binding. Neuropharmacology 19:405–408.

Margules D, Stein L (1968): Increase of "antianxiety" activity and tolerance of behavioral depression during chronic administration of oxazepam. Psychopharmacologia 13:74–80.

Melchior CL, Garrett K, Tabakoff B (1983): Proconvulsant effects of the benzodiazepine agonist CL 218, 872 Soc Neurosci Abstr 9:129.

Möhler H (1982): Benzodiazepine receptors: differential interaction of benzodiazepine agonists and antagonists after photoaffinity labeling with flunitrazepam. Eur J Pharmacol 80:435–436.

Möhler H, Okada T (1977a). Gamma-amminobutyric acid receptor binding with (+) bicuculline methiodide in rat CNS. Nature 267:65–67.

Möhler H, Okada T (1977b): Benzodiazepine receptor: demonstration in the central nervous system. Science 198:845–851.

Möhler H, Richards JG (1981): Agonist and antagonist benzodiazepine receptor interaction in vitro. Nature 294:763–765.

Möhler H, Battersby MK, Richards JG (1980): Benzodiazepine receptor protein identified and visualized in brain tissue by a photoaffinity label. Proc Nat Acad Sci USA 77:1161–1670.

Möhler H, Burkard WP, Keller HH, Richards JG, Haefely W (1981): Benzodiazepine antagonist Ro 15-1788: binding characteristics and interaction with drug-induced changes in dopamine turnover and cerebellar cGMP levels. J Neurochem 37:714–722.

Möhler H, Richards JG, Wu JY (1981): Autoradiographic localization of benzodiazepine receptors in immunocytochemically identified gamma-aminobutyrergic synapses. Proc Natl Acad Sci USA 78:1935–1938.

Niehoff DL, Mashal RD, Horst WD, O'Brien RA, Palacios JM, Kuhar MJ (1982): Binding of a radiolabeled triazolopyridazine to a subtype of benzodiazepine receptor in the rat cerebellum. J Pharmacol Exp Ther 221:670–675.

Nielsen M, Gredal O, Braestrup C (1979): Some properties of $^{3}$H-diazepam displacing activity from human urine. Life Sci 25:679–686.

Rice K, Brossi A, Tallman J, Paul S, Skolnick P (1979): Irazepine, a noncompetitive irreversible inhibitor of diazepam binding to benzodiazepine receptors. Nature 278:854–855.

Skolnick P, Schweri M, Kutter E, Williams E, Paul S (1982): Inhibition of diazepam and $\beta$-carboline binding by irazepine: evidence for multiple domains of the benzodiazepine receptor J Neurochem 39:1142–44.

Sieghart W, Karobath M (1980): Molecular heterogeneity of benzodiazepine receptors. Nature 286:285–287.

Sieghart W, Mayer A, Drexler G (1983): Properties of flunitrazepam binding to different benzodiazepine binding proteins Eur J Pharmacol 88:291–299.

Squires RF, Braestrup C (1977): Benzodiazepine receptors in rat brain. Nature 166:732–734.

Squires RF, Benson DI, Braestrup C, Coupet J, Klepner CA, Myers V, Beer B (1979): Some properties of brain specific benzodiazepine receptors: new evidence for multiple receptors. Pharmacol Biochem Behav 10:825–830.

Sternbach LH (1983): The discovery of CNS active 1,4-benzodiazepines(chemistry). In Usdin E, Skolnick P, Tallman J, Greenblatt D, and Paul S (eds): "The Pharmacology of Benzodiazepines," London: Macmillan Press, pp 7–14.

Tallman J, Gallager D (1985): The GABA-ergic System: a locus of benzodiazepine action Ann Rev Neurosci (in press).

Tallman JF, Paul SM, Skolnick P, Gallager DW (1980): Receptors for the age of anxiety: Pharmacology of the benzodiazepines. Science 207:274–281..

Tallman JF, Thomas JW, Gallager DW (1978): GABAergic modulation of benzodiazepine binding site sensitivity. Nature 274:383–385.

Tenen SS, Hirsch JD (1980): Beta-Carboline-3-carboxylic acid ethyl ester antagonizes diazepam activity. Nature 288:609–610.

Thomas JW, Tallman JF (1981): Characterization of photoaffinity labeling of benzodiazepine binding sites. J Biol Chem 156:9838–9842.

Thomas JW, Tallman JF (1983): Photoaffinity labeling of benzodiazepine receptors causes altered agonist-antagonist interactions. J Neurosci 3:433–440.

Thomas J, Tallman JF (1984): Solubilization and characterization of brain benzodiazepine binding sites. In Marangos et al. (eds): "Brain Neurotransmitter and Neuro Modulator Receptor Methodology." New York Academic Press, pp 95–112.

Williams EF, Rice K, Paul S, Skolnick P (1980): Heterogeneity of benzodiazepine receptors in the CNS demonstrated with kenazepine, an alkylating benzodiazepine. J Neurochem 35:591–597.

Young WS III, Kuhar MJ (1980): Radiohistochemical localization of benzodiazepine receptors in rat brain. J Pharmacol Exp Ther 212:337–346.

Young WS III, Niehoff D, Kuhar MJ, Beer B, Lippa AS (1981): Multiple benzodiazepine receptor localization by light microscopic radiohistochemistry. J Pharmacol Exp Ther 216:425–430.

Benzodiazepine/GABA Receptors and Chloride Channels:
Structural and Functional Properties, pages 285–297


# 15

# Monoclonal Antibodies: Probes for Structure and Location of the GABA Receptor/Benzodiazepine Receptor/Chloride Channel Complex

H. Möhler, P. Schoch, J.G. Richards, P. Häring,
B. Takacs, and C. Stähli

*Pharmaceutical Research Department, F. Hoffmann-La Roche & Co., Ltd., CH-4002 Basel, Switzerland*

## USE OF MONOCLONAL ANTIBODIES IN RECEPTOR RESEARCH

Receptors for bioactive agents such as hormones, neurotransmitters, or drugs are complex protein structures for binding these agents and transducing this event into a biological response. The conventional method of receptor identification in biochemical and morphological studies is the reversible binding of radiolabeled ligands to intact cells, cell membranes, or purified receptors. However, even if a radioligand with sufficiently high affinity and selectivity is available, the analysis is limited. Only one topographic site, which generally occurs only on the native conformation of the receptor, can be monitored by a radioligand. Other stages of the life cycle of the receptor proteins, e.g., their synthesis, conformational maturation, subunit assembly, or receptor degradation, cannot be studied by radioligand binding. The visualization of receptors by autoradiography is usually restricted to the light microscopic level.

Covalently bound radioligands, e.g., photoaffinity ligands, extend the analysis. The subunit polypeptides containing the respective binding sites can be identified biochemically. Morphologically, receptors can be visualized also on the electron-microscopic level, although with an unsatisfactory resolution. Due to radiation scatter, the receptor cannot be attributed to a defined subcellular structure; only the probability of its occurrence in a certain subcellular area can be determined.

Receptor research has gone beyond the frontier of radioligand binding experiments with the availability of monoclonal antibodies, i.e., monospecific antibodies producible in unlimited amounts. They are extremely useful probes for the structure and location of receptors.

### Probes for Receptor Structure and Location

With monoclonal antibodies an array of surface domains of the native receptor can be monitored. For example, the heterogeneity of receptors in different organs or species can be assessed to a degree incomparable to radioligand binding: even single amino acid substitutions and associated changes in the local topography of the protein surface can be detected by antibodies. Also, the topography of receptor subunits with regard to the outer or inner surface of the cell membrane can be assessed. Global conformational states of a receptor, such as might be mediated by allosteric effectors, can be distinguished. Furthermore, antibodies themselves can act as allosteric effectors to influence the functional state of a receptor, e.g., by binding to the ligand binding site, by crosslinking receptors, or by changing the allosteric conformation.

Since monoclonal antibodies frequently interact with the corresponding epitope even when the receptor protein is not in its native state, precursors of the receptor proteins can be identified by monoclonal antibodies to study the synthesis and processing of receptor polypeptides, the conformational maturation and assembly of the subunits, as well as receptor degradation. In addition, receptor subunits separated under denaturing conditions may be identified. In recombinant DNA experiments, fusion proteins, containing only part of a receptor polypeptide, can be detected in expression screening with monoclonal antibodies. Finally, monoclonal antibodies have gained importance in immunoaffinity purification of receptors and their precursors [for reviews see Fraser and Lindstrom, 1984; Goding, 1984; Venter et al., 1984].

The limited resolution of receptor radiohistochemistry, due to radiation scatter and possible diffusion of the ligand, is overcome by immunohistochemistry. The immune reaction is visualized by nondiffusible indicators (stains) such as fluorescein- or peroxidase-coupled secondary antibodies. Thus, on the light-microscopic level, discrete neuronal cell bodies and their processes containing receptors can be identified immunohistochemically but not radiohistochemically. At the electron-microscopic level, receptors can be visualized, by peroxidase- or colloidal gold–coupled secondary antibodies, on discrete subcellular structures, e.g., the pre- or postsynaptic membrane or transport organelles [for reviews see Nghiem et al., 1983; Richards et. al., 1985; Sealock et al., 1984; Sternberger, 1979].

## MONOCLONAL ANTIBODIES TO THE GABA RECEPTOR/BENZODIAZEPINE RECEPTOR/CHLORIDE CHANNEL COMPLEX

Our present biochemical and morphological knowledge of the structure and localization of the GABA receptor/benzodiazepine receptor/chloride channel complex in the central nervous system is based exclusively on radioligand binding experiments. In particular, binding studies with labeled benzodiazepines [for reviews see Möhler and Richards, 1983; Braestrup and Nielsen, 1983; Möhler, 1984; Haefely et al., 1985] 1) led to the discovery of benzodiazepine receptors [Möhler and Okada, 1977; Squires and Braestrup, 1977]; 2) allowed their autoradiographical visualization in brain tissue at the light- and electron-microscopic level [Young and Kuhar, 1980; Möhler et al., 1981; Richards and Möhler, 1984]; 3) demonstrated a receptor complex by allosteric interactions between benzodiazepine binding sites, low-affinity GABA sites, and binding sites for TBPS, a presumptive "chloride channel" ligand [for reviews see Haefely et al., 1985; Karobath and Supavilai, this volume; Squires, this volume]; and 4) allowed the identification of benzodiazepine receptor subunits by photoaffinity labeling [Möhler et al., 1980; Sieghart and Karobath, 1980]. Furthermore, benzodiazepine affinity resins permitted the purification of the GABA receptor complex [Schoch and Möhler, 1983; Sigel et al., 1983; Sigel and Barnard, 1984; Kuriyama et al., 1984; Schoch et al., 1984; Fischer and Olsen, this volume].

To improve the biochemical and morphological analysis of the GABA receptor/benzodiazepine receptor/chloride channel complex, continuous lines of cultured cells secreting monoclonal antibodies directed against the receptor were produced [Häring et al., 1985], following the strategy of Köhler and Milstein [1975].

### Production of Hybridomas

Mice were immunized with receptor preparations purified by affinity chromatography from bovine cerebral cortex [Schoch and Möhler, 1983; Schoch et al., 1984]. These receptor preparations were characterized by the presence of binding sites for benzodiazepines as well as high- and low-affinity sites for GABA. Since the presence of Triton

X-100 precluded binding of TBPS, the presence or absence of its binding site in this preparation could not be assessed. Eight-week-old BALB/cJ mice received 100 $\mu g$ GABA receptor/benzodiazepine receptor preparation in complete Freund's adjuvant by subcutaneous and intraperitoneal (i.p.) injection followed three weeks later by an i.p. injection of the same amount of antigen in incomplete Freund's adjuvant. The mouse with the highest specific antibody titer in serum was boosted 9 weeks later on 2 successive days with 120 $\mu g$ antigen without adjuvant. Two days later hybridomas were produced by fusing the murine spleen cells with PAI myeloma cells in the presence of polyethyleneglycol [Häring et al., 1985]. In the culture medium containing hypoxanthine, aminopterin, and thymidine (HAT), only hybrid cells, but not their parent cells, were able to grow. Aminopterin blocks the de novo synthesis of nucleotides for DNA synthesis, leaving only the salvage pathway with hypoxanthine and thymidine as substrates for nucleotide synthesis. PAI cells, which lack an enzyme of the salvage pathway, cannot grow under these conditions. B-lymphocytes contain the enzymes of the salvage pathway but are short-lived in culture. Thus, hybrid cells are selected, some of which were expected to secrete antibodies against GABA receptor/benzodiazepine receptor/chloride channel complex.

## Primary Screening of Monoclonal Antibodies

The supernatants of various hybridoma cultures were screened for the presence of monoclonal antibodies directed against the GABA receptor/benzodiazepine receptor/chloride channel complex with a rapid and simple solid-phase antibody binding assay [Häring et al., 1985]. Polyvinyl chloride microtiter plates were coated with receptor by incubating them overnight with a purified GABA receptor/benzodiazepine receptor preparation, followed by bovine serum albumin to cover nonspecific protein binding sites on the plastic. The coated wells were then incubated with hybridoma culture supernatant overnight followed by $^{125}$I-labeled sheep antimouse IgG antibodies for 5 hr. After rinsing, the bound radioactivity was counted. With this assay, 28 hybridoma cultures were identified which produced monoclonal antibodies (bd-1-bd-28) that showed binding to GABA receptor/benzodiazepine receptor preparations (7–40-fold background). Of these, 16 monoclonal antibodies were selected for further analysis, based on the stability of the hybridomas and the antibody titers of the supernatants. These monoclonal antibodies were cloned by limited dilution. Twelve were of the $\gamma_1$, two of the $\gamma_{2b}$, and two of $\mu$ heavy chain isotype, and all had $\kappa$ light chains as tested with anti-isotype antibodies (Table I). Since the GABA receptor/benzodiazepine receptor preparations used for immunization and primary screening contained some impurities, the specificity of the monoclonal antibodies for the GABA receptor/benzodiazepine receptor/chloride channel complex had to be verified by other assays.

## Specificity of the Monoclonal Antibodies for the GABA Receptor/Benzodiazepine Receptor/Cl⁻ Channel Complex

The specificity of the sixteen monoclonal antibodies was ascertained by two types of experiments [Häring et al., 1985]: 1) Immunoprecipitation: it had to be shown that the monoclonal antibodies immunoprecipitated the solubilized GABA receptor/benzodiazepine receptor/Cl⁻ channel complex. 2) Immunoblotting: the monoclonal antibodies were expected to interact in immunoblotting experiments with the known subunit polypeptides of the GABA receptor/benzodiazepine receptor/Cl⁻ channel complex.

**Immunoprecipitation.** The immunoprecipitation of the GABA receptor/benzodiazepine receptor/Cl⁻ channel complex was monitored by radioligand binding. Benzodiazepine binding sites were assayed by $^3$H-flunitrazepam binding, high-affinity GABA sites were measured by $^3$H-muscimol binding, and low-affinity GABA sites were tested by the allosteric enhancement of $^3$H-flunitrazepam binding in the presence of 100 $\mu M$ GABA. Receptor

**TABLE I. Summary of Classification of monoclonal antibodies and their Species- and Subunit-Specificities**

| | monoclonal antibodies | | Subunit recognized in different species[c] | |
|---|---|---|---|---|
| Group[a] | No. | Type[b] | Human or bovine receptor | Rat receptor |
| I | bd-3 | IgG | β | β |
| | bd-4 | IgG | | |
| | bd-6 | IgG | | |
| | bd-11 | IgM | | |
| | bd-17 | IgG | | |
| | bd-21 | IgM | | |
| II | bd-9 | IgG | α | n.i.[d] |
| | bd-14 | IgG | | |
| | bd-23 | IgG | | |
| | bd-24 | IgG | | |
| | bd-26 | IgG | | |
| III | bd-8 | IgG | n.d.[e] | n.d.[e] |
| | bd-10 | IgG | | |
| | bd-15 | IgG | | |
| | bd-25 | IgG | | |
| IV | bd-28 | IgG | α | α |

[a]Each group comprises monoclonal antibodies with common epitope specificity.
[b]The heavy chain isotype in IgG was $\gamma_1$ except for bd-6 and bd-15 which contained $\gamma$2b heavy chains; all monoclonal antibodies had $\kappa$ light chains [Häring et al., 1985].
[c]These results are based on immunoblotting experiments with receptor preparations purified from rat cerebral cortex, hippocampus and cerebellum; bovine cerebral cortex and cerbellum; human cerebral cortex [Häring et al., 1985].
[d]Group II monoclonal antibodies did not interact with rat GABA receptor/benzodiazepine receptor in either solid-phase antibody binding assay, immunoprecipitation or immunoblotting (n.i. = no immunoresponse) [Häring et al., 1985].
[e]Although group III monoclonal antibodies were reactive in immunoprecipitation and solid-phase antibody binding assay, their subunit specificity could not be determined (n.d.) since the epitope was sensitive to the denaturing conditions in immunoblot experiments [Häring et al., 1985].

samples purified from bovine or rat brain were incubated with hybridoma culture supernatant for 2 hr, followed by a secondary antibody (goat antimouse IgG coupled to agarose beads) and centrifuged after another 2½ hr. Supernatants and pellets were separately assayed for radioligand binding sites. When the 16 monoclonal antibody culture supernatants were incubated with receptor samples, the binding sites for benzodiazepines and the low- and high-affinity GABA sites were immunoprecipitated completely or at least to a large extent, depending on the monoclonal antibody concentration (Fig. 1) [Häring et al., 1985; Schoch et al., 1985]. Thus, all 16 monoclonal antibodies were specific for the GABA receptor/benzodiazepine receptor complex. Furthermore, to identify the possible presence of TBPS binding sites in the immunoprecipitated protein complex, a crude receptor solubilizate was prepared in CHAPS detergent [Stephenson and Olsen, 1982]. Indeed, $^{35}$S-TBPS binding sites [Squires et al., 1983] were co-precipitated, as tested with bd-17 and bd-24 [Schoch, unpublished results]. Thus, the immunoprecipitation of the GABA receptor/benzodiazepine receptor complex shown for all monoclonal antibodies is likely to represent their interaction with the entire GABA receptor/benzodiazepine receptor/$Cl^-$ channel complex.

**Immunoblotting.** Two subunits of $M_r$ 50,000 (50 k ; α-subunit) and $M_r$ 55 k (β-subunit) are known constituent proteins of the GABA receptor/benzodiazepine receptor/$C1^-$ channel complex (see below). Since these subunits can only be separated under denaturing conditions, receptor samples purified from bovine cerebral cortex [Schoch and Möhler, 1983; Schoch et al., 1984] were subjected to sodium dodecylsulfate polyacrylamide gel electrophoresis, electroblotted to nitrocellulose, and allowed to react with individual monoclonal antibodies. Some monoclonal antibodies (group I) reacted with a protein of $M_r$ 55 k; others (groups II and IV) reacted with a protein of $M_r$ 50 k; the remaining monoclonal antibodies (group III) did not react, presumably because the corresponding epitope was sensitive to the denaturing conditions of immunoblotting (Table I). Thus, at least three of the four groups of monoclonal antibodies (see below) interacted with proteins which correspond to the known subunits of the GABA receptor/benzodiazepine receptor/$C1^-$ channel complex [Häring et al., 1985].

### Distinction of Four Epitopes

The 16 monoclonal antibodies can be divided into four groups (I–IV) specifying different antigenic structures (Table I) [Häring et al., 1985]. The epitope of the group I monoclonal antibodies is localized on the $\beta$-subunit in all three species tested (bovine, human, and rat) as shown by immunoblotting. The epitope of the group II monoclonal antibodies is localized on the $\alpha$-subunit of the human and bovine receptor. This epitope does not occur on the rat receptor, as shown by the lack of immunoreaction in either solid-phase antibody binding assay, immunoprecipitation, immunoblotting, or immunohistochemistry of rat brain. The epitope of the group IV monoclonal antibodies is localized on the $\alpha$-subunit not only of bovine and human brain but also of rat brain. Thus, three groups of subunit specific monoclonal antibodies exist (Table I). They will be valuable tools by which to study the synthesis, conformational maturation, assembly, and degradation of each receptor subunit.

Group III monoclonal antibodies showed no reaction in immunoblots. This epitope appears to be sensitive to the denaturing conditions used in immunoblots, or it might be formed by two subunits. These monoclonal antibodies were active in solid-phase antibody binding assay, immunoprecipitation, and immunohistochemistry in all three species tested. Antibody competition experiments [Häring et al., 1985] were in line with the distinction of four epitopes by the monoclonal antibodies.

None of the monoclonal antibodies interacted with the binding sites for benzodiazepines or the low- and high-affinity GABA sites of the purified receptor [Häring et al., 1985].

## STRUCTURE AND LOCATION OF THE GABA RECEPTOR/BENZODIAZEPINE RECEPTOR/Cl⁻CHANNEL COMPLEX

### Lack of Receptor Heterogeneity

Protein constituents of the GABA receptor/benzodiazepine receptor/$Cl^-$ channel complex were previously identified by photoaffinity labeling of benzodiazepine binding sites in rat brain membranes. With this technique a protein component of $M_r$ 50 k was found in all brain regions [Möhler et al., 1980; Sieghart and Karobath, 1980; Thomas and Tallman, 1981; Sieghart and Drexler, 1983; Möhler et al., 1984]. However, mainly in hippocampus, an additional protein of $M_r$ 55 k was also labeled [Sieghart and Karobath, 1980; Sieghart and Drexler, 1983; Möhler et al., 1984; Eichinger and Sieghart, 1985]. Thus, a molecular heterogeneity of benzodiazepine receptors appeared to exist in the brain which seemed to provide a framework to explain differences in the pharmacological spectrum of benzodiazepine receptor ligands [Gee and Yamamura, 1982; Johnson et al., 1983]. The validity of the hypothesis of a heterogeneous subunit composition of the GABA receptor/benzodiazepine receptor/$Cl^-$ channel complex was now tested with the subunit specific monoclonal antibodies. In immunoblot experiments, not only the $\alpha$-subunit but also the $\beta$-subunit was found to be present in all brain areas and species tested (rat cerebral cortex, hippocampus, cerebellum; bovine cerebral cortex and cerebellum; human cerebral cortex) [Häring et al., 1985]. This result is especially striking in rat cerebellum, where no trace of a $\beta$-subunit (55 k) could be detected by photoaffinity labeling while the $\beta$-subunit could be clearly identified immunologically (Fig. 2). Furthermore, a homogeneous subunit distribution was also found immunohistochemically. In tissue sections using monoclonal antibodies specific for either the $\alpha$- or $\beta$-subunit (bd-24 and bd-17, respectively), an identical pattern of immunostaining was obtained in both bovine and human brain [Schoch et al., 1985]. Thus, immunologically, no heterogeneity in the subunit composition of the GABA receptor/benzodiazepine receptor/$Cl^-$ channel complex in different brain areas could be discovered. Differences in the pharmacological profile of benzodiazepine receptor ligands cannot be attributed to a major structural heterogeneity of the receptor complex.

This finding does not preclude the existence of a conformational heterogeneity. Multiple

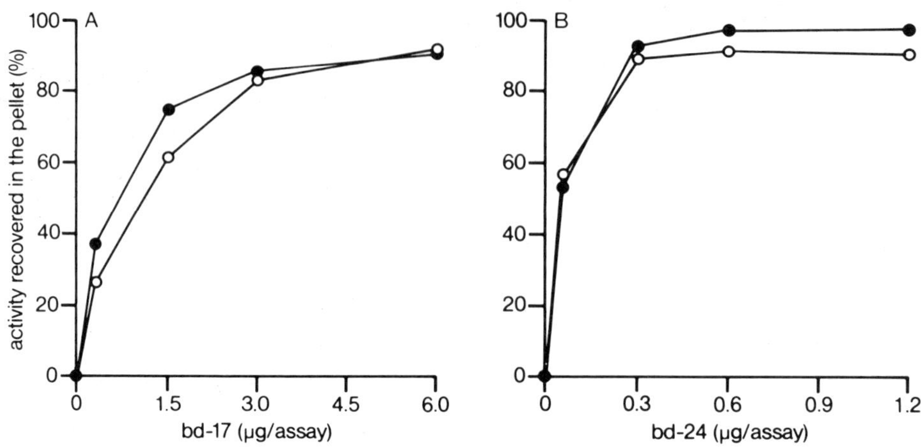


**Fig. 1.** *Immunoprecipitation of binding sites for benzodiazepines and GABA with two subunit-selective monoclonal antibodies (A: bd-17: B: bd-24). Filled and open circles represent specifically bound $^3$H-flunitrazepam and $^3$H-muscimol, respectively. The epitope recognized by bd-17 is localized on the $M_r$ 55 k ($\beta$) subunit, that of bd-24 on the $M_r$ 50 k ($\alpha$) subunit. A GABA receptor/benzodiazepine receptor preparation from bovine cerebral cortex was used. For experimental details see Schoch et al. [1985].*

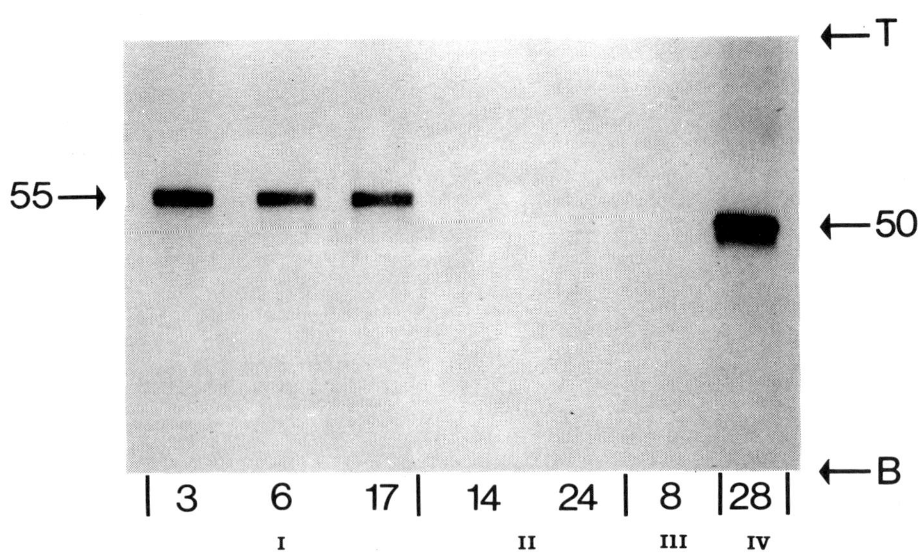


**Fig. 2.** *Autoradiograph of an immunoblot with GABA receptor/benzodiazepine receptor purified from rat cerebellum. Monoclonal antibodies representing all four groups (I–IV) are indicated by the number (bd-x) at the bottom of each strip. Top (T) and bottom (B) of the gel as well as the positions of the $M_r$ 55 k and 50 k bands are indicated by arrows.*

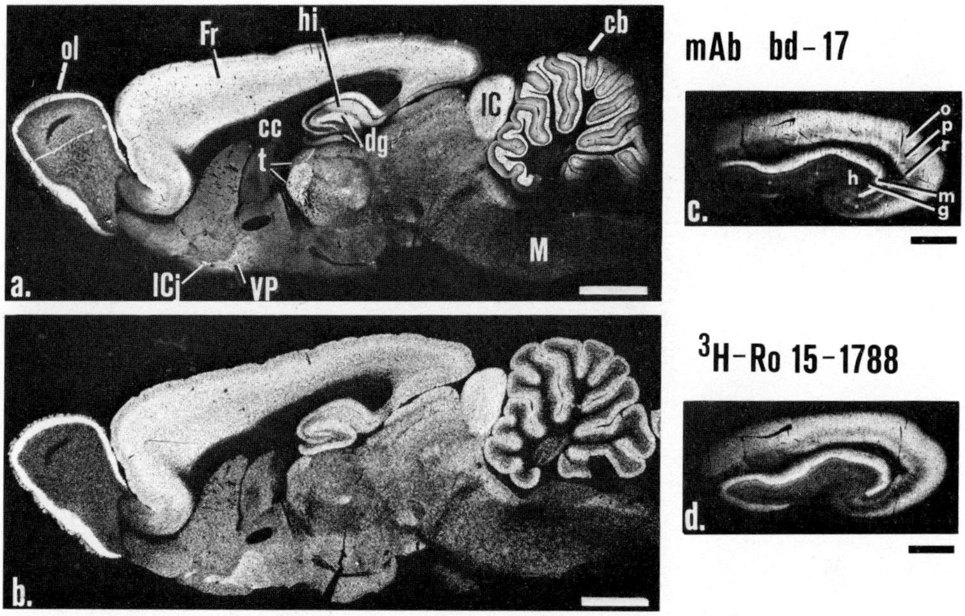


**Fig. 3.** *Immunohistochemical localization of antigenic sites in rat brain (a.: parasagittal section) and in postmortem human hippocampus (c.: coronal section) using monoclonal antibody bd-17 : comparison with the radiohistochemical localization of benzodiazepine receptor sites (b.d. : $^3$H-Ro 15-1788 binding). In these negative images, white areas correspond to high densities of antigenic sites and radioligand binding sites. For the immune staining, the tissue was fixed with 2% formaldehyde + 0.1% glutaraldehyde. Ten-micron cryostat sections were incubated, according to the indirect peroxidase–antiperoxidase (PAP) technique of Sternberger [1979], in sequence with: normal swine serum, hybridoma supernatant bd-17 (5–20 μg monoclonal antibody/ml), swine–antimouse antiserum, PAP complex and finally diaminobenzidine/$H_2O_2$. Benzodiazepine receptor sites were radiolabeled in vitro with 1 nM $^3$H-Ro 15-1788 (b,d). For details see Schoch et al. [1985].*

*Abbreviations: cb, cerebellum; cc, corpus callosum; Fr, frontal cerebral cortex; dg, dentate gyrus; g, stratum granulosum of the dentate gyrus; h, hilus; hi, hippocampus; ICj, islands of Calleja; IC, inferior colliculus; M, medulla; m, stratum moleculare of the dentate gyrus; o, stratum oriens of the hippocampus; ol, olfactory bulb; p, stratum pyramidale of the hippocampus; r, stratum radiatum of the hippocampus; t, thalamus; VP, ventral pallidum. Calibration bars 2 mm.*

conformational states of the receptor may exist, which might explain the apparent heterogeneity detectable in radioligand binding assays in vitro [Haefely et al., 1985; Squires, this volume]. The regional heterogeneity of photoaffinity labeling of a 55 k peptide, possibly an alternate α-subunit or the β-subunit, remains to be clarified. A positive reaction, seen mainly in rat hippocampus, may indicate the presence of a subpopulation containing photoreactive sites.

## Subunit Composition of the GABA Receptor/ Benzodiazepine Receptor/Cl⁻ Channel Complex

Although both α- and β-subunits occur in all brain regions (see above), the arrangement of the subunits in the receptor complex was unknown. Conceivably, a mixture of two receptor populations each containing exclusively either α- or β-subunits could exist. Alternatively, each receptor complex could contain both α- and β-subunits. Immunoprecipitation

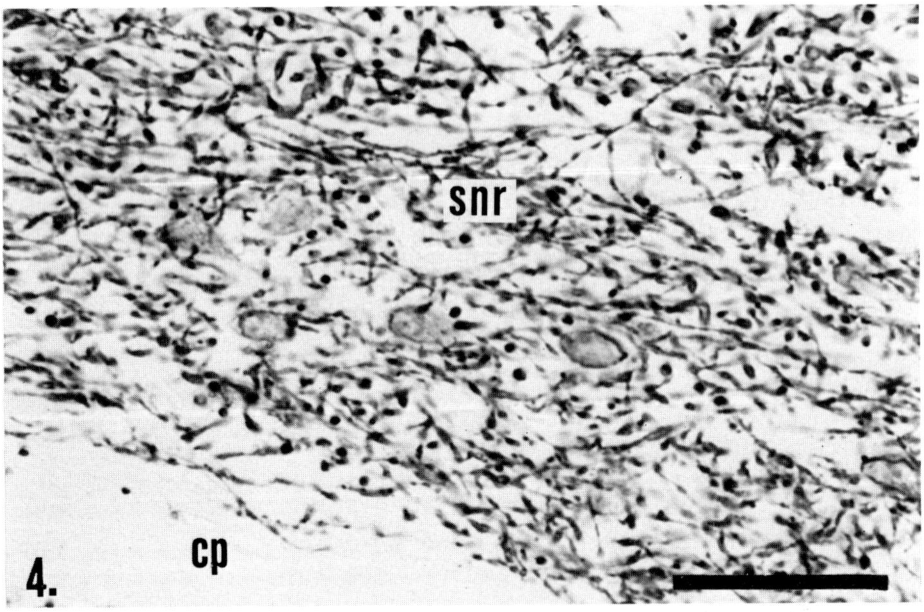


**Fig. 4.** *Light microscopic localization of antigenic sites in rat substantia nigra zona reticulata (snr) (brightfield optics). The peroxidase immune staining surrounds numerous large dendrites and occasional cells. Note the absence of immunoreactivity in the cerebral peduncle (cp). Calibration bar: 50 μm.*

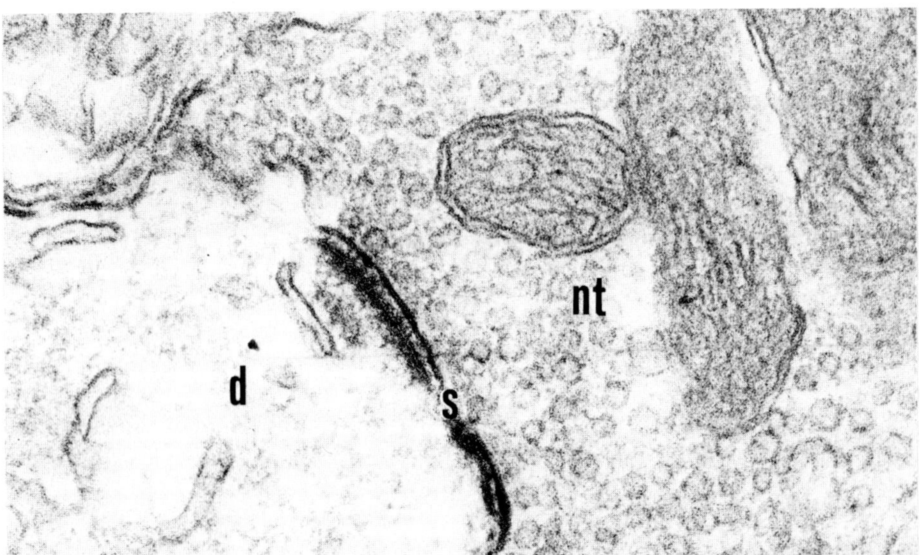


**Fig. 5.** *Electron microscopic localization of antigenic sites in rat substantia nigra zona reticulata. The electron dense peroxidase immune reaction is present exclusively in the pre- and postsynaptic membrane, as shown for an axodendritic contact. This was also observed without counterstaining. nt, nerve terminal; s, synapse; d, dendrite. Calibration bar: 0.5 μm.*

experiments with subunit specific monoclonal antibodies (bd-24, bd-17) clearly point to the latter subunit composition. Both bd-24 and bd-17, specific for either the $\alpha$-subunit or the $\beta$-subunit respectively, were able to precipitate the entire population of binding sites for benzodiazepines and GABA (high- and low-affinity) in purified receptor preparations (Fig. 1). This result clearly indicates a mixed subunit composition of the receptor complex. Thus, a concerted mechanism for gene expression, synthesis, and assembly of $\alpha$- and $\beta$-subunits is postulated.

## Colocalization of GABA and Benzodiazepine Receptors

Although an association of benzodiazepine receptor with the GABA receptor/chloride ionophore in the CNS is evident on electrophysiological and biochemical grounds [for review Haefely and Polc, this volume; Möhler and Richards, 1983], a definitive quantitative assessment of whether all benzodiazepine receptors are part of the GABA receptor complex has not been made. Conceivably, a subpopulation of benzodiazepine receptors might exist which is not associated with the GABA receptor. This possibility would greatly affect our views on the mechanism of action of benzodiazepine receptor ligands.

A previous attempt to quantify the extent of the colocalization of benzodiazepine receptor and the GABA receptor complex was restricted to the autoradiographical assessment of benzodiazepine binding sites and low-affinity GABA sites [Unnerstall et al., 1981]. However, using the monoclonal antibodies, it is now possible to identify the GABA receptor complex as a unit defined by the presence of binding sites for benzodiazepines, GABA (high- and low-affinity) and TBPS. Thus, a comparison of the immunoreaction pattern and the autoradiographical distribution of benzodiazepine binding sites should indicate the extent of the association of benzodiazepine receptors with the GABA receptor complex. In semiadjacent sections of rat brain, the GABA receptor/benzodiazepine receptor/$Cl^-$ channel complex was visualized immunohistochemically with bd-17 and the benzodiazepine binding sites were localized radiohistochemically with $^3$H-Ro 15-1788. A striking similarity was observed between the pattern of the immunostaining and that of the radioligand binding in, e.g., olfactory bulb, ventral pallidum, islands of Calleja, cerebral cortex, dorsal thalamus, hippocampus, dentate gyrus (superior molecular layer), substantia nigra, superior and inferior colliculi, and cerebellar molecular layer (Fig. 3), as well as spinal cord and retina [Schoch et al., 1985]. Thus, most benzodiazepine receptors are part of the GABA receptor complex. This was also observed in postmortem human brain, e.g., hippocampus (Fig. 3), substantia nigra zona reticulata and molecular layer of cerebellum. Areas which lacked immunoreactivity also lacked benzodiazepine binding sites, e.g., the pineal gland and pituitary [Schoch et al., 1985]. These results support the concept that the mechanism of action of benzodiazepine receptor ligands is generally based on the modulation of GABAergic synaptic transmission [Haefely and Polc, 1983; Haefely, 1984; Haefely and Polc, this volume].

An appreciable discrepancy between the immunoreaction and the density of benzodiazepine binding sites was observed only in the granular layer of the cerebellum and in the ventrolateral thalamus, e.g., in the rat (Fig. 3) The intensity of the immune reaction corresponded better to the density of high-affinity GABA sites ($^3$H-muscimol binding) [Unnerstall et al., 1981] and of $^{35}$S-TBPS binding sites [Wamsley et al., 1983; Wamsley et al., this volume]. Thus, in these two regions the receptor complex may occur in a conformation in which high-affinity GABA sites and low-affinity benzodiazepine binding sites predominate.

## Subcellular Localization of the Receptor Complex

The difficulty to assign the radiolabeled binding site to a specific subcellular structure is overcome by receptor immunocytochemistry. So far, only the nicotinic receptor has been localized by monoclonal antibodies at

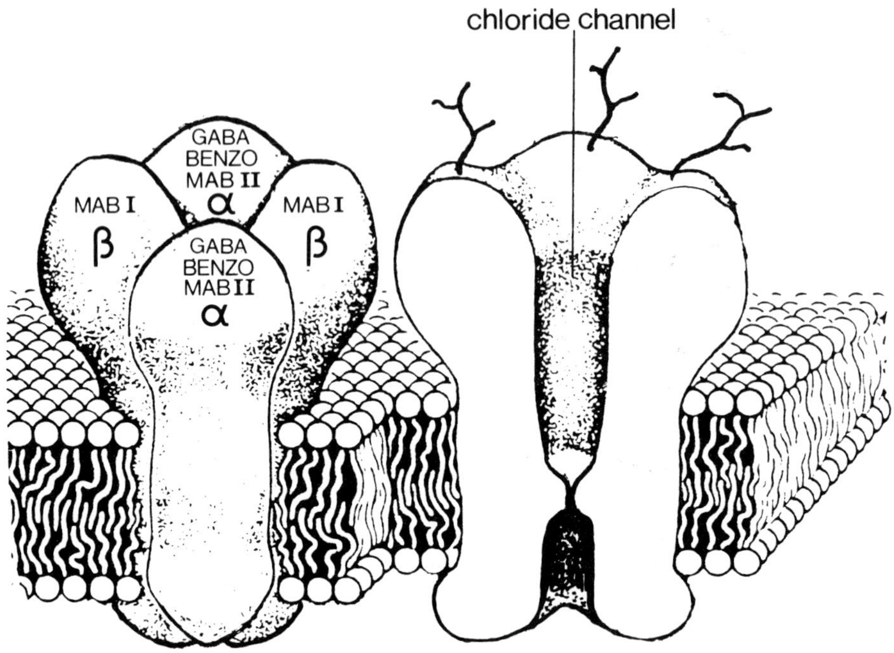


**Fig. 6.** *Preliminary working hypothesis for a model of the GABA receptor/benzodiazepine receptor/Cl⁻ channel complex. For explanation see text. α and β indicate subunits of $M_r$ 50 k and 55 k, respectively. Benzo = benzodiazepine binding sites identified by photolabeling. Benzodiazepine binding sites on the β-subunit (generally not photolabeled) cannot be excluded. GABA = GABA binding sites identified by photolabeling. Nonphotolabeled GABA sites on the β-subunit cannot be excluded. MABI, II indicate subunit-specific epitopes recognized by group I and group II monoclonal antibodies, respectively, on the outer receptor domain. Branched lines = carbohydrate residues.*

the electron-microscopic level [Nghiem et al., 1983; Sealock et al., 1984]. Using our monoclonal antibodies directed against the GABA receptor/benzodiazepine receptor/Cl⁻ channel complex, it was now possible to identify its subcellular distribution using the peroxidase method. In the rat substantia nigra, for example, the immune reaction, visualized light-microscopically on what appear to be large dendrites (Fig. 4), could be resolved electron-microscopically. An exclusive pre- and postsynaptic localization of the immune stain was demonstrated (Fig. 5). The immunoreactivity of postsynaptic membranes is in line with the presence of a postsynaptic GABA receptor/benzodiazepine receptor/Cl⁻ channel complex. It remains to be clarified whether the staining of the presynaptic membrane indicates the presence of presynaptic receptions or is due to diffusion of the stain.

## MODEL OF THE GABA RECEPTOR/ BENZODIAZEPINE RECEPTOR/Cl⁻ CHANNEL COMPLEX

Although many questions about the molecular structure of the GABA receptor complex are still unanswered, the known features can be incorporated into a preliminary working hypothesis (Fig. 6). The receptor has a mixed α/β subunit composition [Häring et al., 1985]. Both subunits show an identical distribution in rat, bovine, and human brain [Schoch et al., 1985]. A tetrameric $\alpha_2/\beta_2$ subunit structure is suggested by the comparison of the molecular weight of the subunits and the native receptor. While the α and β subunits have $M_r$ values of 50 k and 55 k, respectively [Möhler et al., 1980; Sieghart and Karobath, 1980], the target size for benzodiazepine and GABA binding sites in membranes was 220 k determined by radiation inactivation [Chang and Barnard, 1982]. Furthermore, a tetrameric model was also indicated by photoaffinity-labeling experiments in which photoreaction to one benzodiazepine binding site in the membrane caused a conformational affinity change in three additional sites [Möhler et al., 1980; Thomas and Tallman, 1981; Karobath and Supavilai, 1982]. An $\alpha_2/\beta_2$ structure might also accommodate the TBPS binding site, as shown in a receptor preparation purified in CHAPS detergent [Sigel and Barnard, 1984]. It is thus conceivable that the $\alpha_2/\beta_2$ tetramer not only contains all the known binding sites of the receptor complex but also forms the chloride channel (Fig. 6).

Further distinctive features of the receptor complex are the subunit specific epitopes recognized by the monoclonal antibodies (see above). At least the group I and group II epitopes are most likely located on the cell surface [unpublished results]. Thus, both $\alpha$- and $\beta$-subunits are part of the outer receptor structure. A protrusion of the protein is suggested by the proteolytic digestion of the photolabeled benzodiazepine binding site [Klotz et al., 1984]. Binding of the receptor to lectins indicates the presence of carbohydrate residues [Gavish and Snyder, 1981; unpublished results], presumably on the outer surface.

A final allocation of the binding sites for GABA and various drugs to the specific receptor domains is not possible at present. Only the presence of benzodiazepine binding sites on the $\alpha$-subunit is beyond doubt (see above) (Fig. 6). Conceivably, benzodiazepine binding sites might also occur on the $\beta$-subunit, although in a state which is generally not photolabeled. Electrophysiologically, two to four GABA sites were postulated per GABA receptor [Werman, 1979; Hamill et al., 1983; Sakmann et al., 1983]. By $^3$H-muscimol photolabeling, GABA sites were attributed to the $\alpha$-subunit [Asano et al., 1983; Cavalla and Neff, 1985] although the use of short-wave UV light in these studies is viewed with reservation. The possible presence of nonphotolabeled GABA sites on the $\beta$-subunit cannot be excluded. The location of the barbiturate and TBPS binding sites is yet unknown. The existence of binding sites in subunit interfaces might be possible. Some experimental results suggest that the present working hypothesis (Fig. 6) might have to be modified in the future. In our purified receptor preparations an additional protein of $M_r$ 62 k occurs [Schoch et al., 1984]. Furthermore, the target size of the radiation-inactivated TBPS binding site was 134 k [Nielsen and Braestrup, 1983]. Both results might be explained by the presence of an additional subunit with $M_r$ 62 k to about 80 k ($\gamma$-subunit) in the receptor complex. Receptor aggregates may also occur as suggested by the target size of 350 k determined for allosteric interactions with the TBPS binding site [Nielsen and Braestrup, 1984].

## SUMMARY AND OUTLOOK

A collection of 16 monoclonal antibodies to the GABA receptor/benzodiazepine receptor/$Cl^-$ channel complex has been produced, which permits a subunit-selective analysis. The binding sites for benzodiazepines, GABA (high- and low-affinity) and TBPS reside on the same receptor complex. It contains a mixture of $\alpha$- and $\beta$-subunits ($M_r$ 50 k and $M_r$ 55 k, respectively). There was no biochemical or histological evidence for a structural heterogeneity of the receptor; both subunits show an identical distribution in the brain. Furthermore, immunohistochemically, a colocalization of most benzodiazepine receptors with the GABA receptor complex was evident. The receptors were visualized exclusively on synaptic membranes.

In the future, the monoclonal antibodies are expected to 1) contribute further to a detailed biochemical analysis of the receptor complex, including cloning of the receptor gene(s); 2) permit the subcellular receptor localization in defined GABAergic synapses; and 3) identify possible pathological alterations of the GABA receptor/benzodiazepine receptor/$Cl^-$channel complex in human CNS diseases, e.g., in Huntington disease or epilepsy.

## ACKNOWLEDGMENT

We thank Dr. W. Haefely for critical reading of the manuscript.

## REFERENCES


Asano T, Yamada Y, Ogasawara, N (1983): Characterization of the solubilized GABA and benzodiazepine receptor from various regions of bovine brain. J Neurochem 40:209–214.

Braestrup C, Nielsen M (1983): In Iversen LL, Iversen SD, Snyder SH (eds): "Handbook of Psychopharmacology," Vol. 17. New York: Plenum Press, pp 285–384 .

Cavalla D, Neff NH (1985): Photoaffinity labeling of the $GABA_A$ receptor with $^3$H-muscimol. J Neurochem 44:916–921.

Chang L-R, Barnard EA (1982): The benzodiazepine/GABA receptor complex: molecular size in brain synaptic membranes and in solution. J Neurochem 39:1507–1518.

Eichinger A, Sieghart W (1985): Photoaffinity labeling of different benzodiazepine receptors at physiological temperature. J Neurochem 43:1745–1748.

Fraser CM, Lindstrom J (1984): The use of monoclonal antibodies in receptor characterization and purification. In Venter JC and Harrison LC (eds): "Molecular and Chemical Characterization of Membrane Receptors." New York: Alan R. Liss, Inc, pp 1–30.

Gavish M, Snyder SH (1981): γ-Aminobutyric acid and benzodiazepine receptors: copurification and characterization. Proc Natl Acad Sci USA 78:1939–1942.

Gee KW, Yamamura HI (1982): In Usdin E, Skolnick P, Tallman JF, Greenblatt D, Paul SM (eds): "Pharmacology of Benzodiazepines." London: MacMillan Press, pp 93–108.

Goding JW (1984): "Monoclonal Antibodies: Principles and Practice," New York: Academic Press.

Haefely W, Polc P (1983): Electrophysiological studies on the interaction of anxiolytic drugs with GABAergic mechanisms. In Malick JB, Enna SJ, Yamamura HI (eds): "Anxiolytics: Neurochemical, Behavioural and Clinical Perspectives." New York: Raven Press, pp 113–146.

Haefely W (1984): Actions and interactions of benzodiazepine agonists and antagonists at GABAergic synapses. In Bowery NG (ed): "Actions and Interactions of GABA and Benzodiazepines." New York: Raven Press, pp 263–285.

Haefely W, Kyburz E, Gerecke M, Möhler H (1985): Recent advances in the molecular pharmacology of benzodiazepine receptors and in the structure activity relationships of their agonists and antagonists. In "Advances in Drug Research," Vol 14, pp 165–322.

Hamill OP, Bormann J, Sakmann B (1983): Activation of multiple conductance state chloride channels in spinal neurons by glycine and GABA. Nature 305:-805–808.

Häring P, Stähli C, Schoch P, Takacs B, Staehelin T, Möhler H (1985): Monoclonal antibodies reveal structural homogeneity of $GABA_A$/benzodiazepine receptors in different brain areas. Proc Natl Acad Sci USA 82:4837–4841.

Johnson RW, Tallman JF, Squires R, Yamamura HI (1983): In Malick JB, Enna SJ, Yamamura HI (eds): "Anxiolytics: Neurochemical, Behavioural and Clinical Perspectives." New York: Raven Press, pp 93–112.

Karobath M, Supavilai P (1982): Distinction of benzodiazepine agonists from antagonists by photoaffinity labeling of benzodiazepine receptors in vitro. Neurosci Lett 31:65–69.

Klotz KL, Bocchetta A, Neale JH, Thomas JW, Tallman JF (1984): Proteolytic degradation of neuronal benzodiazepine binding sites. Life Sci 34:293–299.

Köhler G, Milstein C (1975): Continuous cultures of fused cells secreting antibody of predefined specificity. Nature 256:495–497.

Kuriyama K, Youeda Y, Taguchi J, Takahashi M, Ohkuma S (1984): Properties of purified γ-aminobutyric (GABA) receptor and modulation of GABA receptor binding by membrane phospholipids. Neuropharmacology 23:839–840

Lindstrom J, Tzartos S, Gullick W, Hochschwender S, Swanson L, Jacob M, Sargent P, Montal M (1983): Use of monoclonal antibodies to study acetylcholine receptors from electric organs, muscle and brain and the autoimmune response to receptor in myasthenia gravis. In "48th Cold Spring Harbor Symposium." New York: Molecular Neurobiology, CSH Laboratory, Cold Spring Harbor, pp 89–100.

Möhler H, Okada T (1977): Demonstration of benzodiazepine receptors in the central nervous system. Science 198:849–851.

Möhler H, Battersby MK, Richards JG (1980): Benzodiazepine receptor protein identified and visualized in brain tissue by a photoaffinity label. Proc Natl Acad Sci USA 77:1666–1670.

Möhler H, Richards JG, Wu J-Y (1981): Autoradiographic localization of benzodiazepine receptors in immunocytochemically identified γ-aminobutyrergic synapses. Proc Natl Acad Sci USA 78:1935–1938.

Möhler H, Richards JG (1983): Receptors for anxiolytic drugs. In Malick JB, Enna SJ, Yamamura HI (eds): "Anxiolytics: Neurochemical, Behavioural and Clinical Perspectives." New York: Raven Press, pp 15–40.

Möhler H (1984): Benzodiazepine receptors and their ligands. In Bowery NG (ed): "Actions and Interactions of GABA and Benzodiazepines." New York: Raven Press, pp 155–166.

Möhler H, Sieghart W, Richards JG, Hunkeler W (1984): Photoaffinity labeling of benzodiazepine receptors with a partial inverse agonist. Eur J Pharmacol 102:191–192.

Nghiem H-O, Cartand J, Dubrenil C, Kordeli C, Buttin G, Changeux J-P (1983): Production and characterization of a monoclonal antibody directed against the 43,000-dalton $\nu_1$ polypeptide from Torpedo marmorata electric organ. Proc Natl Acad Sci USA 80:6403–6407.

Nielsen M, Braestrup C (1983): The molecular target size of brain TBPS binding sites. Eur J Pharmacol 91:321–322.

Nielsen M, Braestrup C (1984): Target size of brain benzodiazepine receptor, GABA receptor and $^{35}S$-TBPS binding sites. In Vici ES, Magyar K (eds): "Regulation of Transmitter Function." Budapest: Akademia Kiado, pp 429–432.

Richards JG, Möhler H (1984): Benzodiazepine receptors. Neuropharmacology 23:233–242.

Richards JG, Schoch P, Möhler H, Haefely W (1985): Benzodiazepine receptors resolved. Experientia (in press).

Sakmann B, Hamill OP, Bormann J (1983): Patch-clamp measurements of elementary chloride currents activated by the putative inhibitory transmitters GABA and glycine in mammalian spinal neurons. J Neural Trans (Suppl)18:83–95.

Schoch P, Möhler H (1983): Purified benzodiazepine receptor retains modulation by GABA. Eur J Pharmacol 95:323–324

Schoch P, Häring P, Takacs B, Stähli C, Möhler H (1984): A GABA/benzodiazepine receptor from bovine brain: purification, reconstitution and immunological characterization. J Receptor Res 4:189–200.

Schoch P, Richards JG, Häring P, Takacs B, Stähl C, Staehelin T, Haefely W, Möhler H (1985): Colocalization of $GABA_A$ and benzodiazepine receptors in the brain. Nature 314:168–170.

Sealock R, Wray BE, Froehner SC (1984): Ultrastructural localization of the $M_r$ 43,000 protein and the acetylcholine receptor in Torpedo postsynaptic membranes using monoclonal antibodies. J Cell Biol 98:2239–2244.

Sieghart W, Karobath M (1980): Molecular heterogeneity of benzodiazepine receptors. Nature 286:285–287.

Sieghart W, Drexler G (1983): Irreversible binding of [$^3$H]flunitrazepam to different proteins in various brain regions. J Neurochem 41:47–55.

Sigel E, Stephenson A, Mamalaki C, Barnard EA (1983): A $\gamma$-aminobutyric acid/benzodiazepine receptor complex of bovine cerebral cortex. J Biol Chem 258:6955–6971.

Sigel E, Barnard EA (1984): A $\gamma$-aminobutyric acid/benzodiazepine receptor complex from bovine cerebral cortex. J Biol Chem 259:7219–7223.

Squires RF, Braestrup C (1977): Benzodiazepine receptors in rat brain. Nature 266:732–734.

Squires RF, Casida JE, Richardson M, Saedrup E (1983): [$^{35}$S]t-Butylbicyclophosphorothionate binds with high affinity to brain-specific sites coupled to $\gamma$-aminobutyric acid-A and ion recognition sites. Mol Pharmacol 23:326–336.

Stephenson FA, Olsen RW (1982): Solubilization by CHAPS detergent of barbiturate-enhanced benzodiazepine-GABA receptor complex. J Neurochem 39:1579–1586.

Sternberger LA (1979): "Immunocytochemistry, 2nd Edition." New York: John Wiley & Sons.

Thomas JW, Tallman JF (1981): Characterization of photoaffinity labeling of benzodiazepine binding sites. J Biol Chem 256:9838–9842.

Unnerstall JR, Kuhar MJ, Niehoff DL, Palacios JM (1981): Benzodiazepine receptors are coupled to a subpopulation of $\gamma$-aminobutyric acid (GABA) receptors: evidence from a quantitative autoradiographic study. J Pharmacol Exp Ther 218:797–804.

Venter JC, Fraser CM, Lindstrom J (eds) (1984): "Monoclonal and Anti-Idiotypic Antibodies: Probes for Receptor Structure and Function." New York: Alan R. Liss, Inc. Vol. 4 this series.

Wamsley JK, Gee KW, Yamamura HI (1983): Comparison of the distribution of convulsant/barbiturate and benzodiazepine receptors using light microscopic autoradiography. Life Sci 33:2321–2329.

Werman R (1979): Stoichiometry of GABA-receptor interactions: GABA modulates the glycine-receptor interaction allosterically in a vertebrate neuron. In Mandel P, De Feudis FV (eds): "GABA—Biochemistry and CNS Function." New York: Plenum Press, pp 287–301.

Young WS III, Kuhar MJ (1980): Radiohistochemical localization of benzodiazepine receptors in rat brain. J Pharmacol Exp Ther 217:337–346.

Benzodiazepine/GABA Receptors and Chloride Channels:
Structural and Functional Properties, pages 299–313


# 16

# The Benzodiazepine/Barbiturate-Sensitive Convulsant/GABA Receptor/Chloride Ionophore Complex: Autoradiographic Localization of Individual Components

James K. Wamsley, Donald R. Gehlert, and Richard W. Olsen

*Departments of Psychiatry and Pharmacology, University of Utah Medical Center, Salt Lake City, Utah 84132 (J.K.W., D.R.G.) Department of Pharmacology and Brain Research Institute, UCLA School of Medicine, Los Angeles, California 90024 (R.W.O.)*

Autoradiographic techniques have been adapted to provide a means of microscopically localizing neurotransmitter and drug receptors in the central nervous system (CNS) [for review see Kuhar, 1981, 1985; Kuhar et al., 1983; Wamsley and Palacios, 1982, 1983]. The use of these techniques allows the localization of receptors with a higher resolution than that possible with any other method currently available. Autoradiographic localization of receptors can be used to determine, visibly and quantifiably, the density of receptors in discrete nuclei of the brain, or in individual layers of laminated structures such as the cerebral cortex. By labeling serial sections with specific ligands, receptor autoradiography can be used to compare directly the density and distribution of different receptors and/or their subtypes in individual brain regions. Thus, receptor autoradiographic techniques, after in vitro labeling of slide-mounted tissue sections, are ideally suited for analysis of the individual components of the benzodiazepine/barbiturate-sensitive convulsant/GABA receptor complex [Olsen, 1981; 1982] associated with a chloride channel [Wamsley and Palacios, 1984].

The benzodiazepines are a class of drugs with a wide use and a variety of applications in psychiatric medicine stemming from the fortuitous discovery of chlordiazepoxide (Librium) by Leo Sternbach and his associates at Roche Laboratories. The depressant action of benzodiazepines on neurons is thought to be mediated via a facilitation of the inhibitory action of the most predominant inhibitory neurotransmitter, gamma-aminobutyric acid (GABA), in the brain [see Haefely, 1983]. The discovery of specific receptors which bind the benzodiazepines allowed the demonstration of the benzodiazepine effect at the GABAergic synapse as opposed to any direct effect of these drugs on the GABAergic receptors [Tallman et al., 1980; Richards and Möhler, 1984]. Benzodiazepine receptors were originally localized under the light microscope by Young and Kuhar [1979; 1980]. Other investigators, using a photoaffinity label technique, have even localized this binding to immunocytochemically identified GABAergic synapses [Möhler et al., 1981]. The presence of a GABA agonist (including GABA itself) will not compete for the receptors occupied by the benzodiazepines; rather, the GABA agonist will enhance the benzodiazepine binding and therefore increase the effects of benzodiazepines tissue [Olsen, 1981; 1982]. Microscopic localization of the sites where benzodiazepine binding can be enhanced by the presence of GABA has been performed using autoradiographic techniques [Unnerstall et al., 1981]. Results obtained in the study led the authors to conclude that the GABA receptors involved in influencing benzodiazepine binding represent only a subset of the entire $GABA_A$ receptor population. Thus, these results provide additional support for the hypothesis that benzodiazepine receptors are coupled to only a subpopulation of GABA receptors [Browner et al., 1981; Olsen and Snowman, 1983].

The inhibitory action of GABA is mediated through its effects on chloride conductance. The facilitation of GABAergic transmission by the benzodiazepines involves augmentation of this chloride conductance [Haefely, 1983]. Allosteric interactions between the various binding sites on the GABA/benzodiazepine receptor complex in vitro are chloride dependent. These observations and others led to the hypothesis that the benzodiazepine receptors and GABA receptors are allosterically coupled to a chloride ionophore [Olsen 1981; 1982]. The presence of GABA receptor agonists causes the opening of the chloride channel, resulting in inhibition of the affected neuron, and this inhibition is prolonged and/or magnified in the presence of benzodiazepines.

Many of the anxiolytic, sedative-hypnotic, anticonvulsant and muscle-relaxant actions of the benzodiazepines are shared by their predecessors, the barbiturates. The CNS depressant action of the benzodiazepines, however, seems to occur in selective neuronal populations, in contrast to the more generalized depressant action of the barbiturates. This could be due to a differential distribution of benzodiazepine and barbiturate receptors–that is, there could be regions of the brain where the barbiturate receptors are present without the benzodiazepine receptor. Barbiturate binding sites have not been localized directly due to the absence of a high-affinity, specific radioactive ligand, but barbiturate-sensitive convulsant binding sites labeled with $[^{35}S]$t-butylbicyclophosphorothionate ($[^{35}S]$TBPS) show a brain regional localization distinct from benzodiazepine receptors as demonstrated autoradiographically [Wamsley et al., 1983]. The action of barbiturates appears again to be associated with a GABA receptor/chloride ionophore complex directly related to that affected by the benzodiazepines [Olsen, 1982]. Thus barbiturates enhance the function of GABA in vivo and modulate the binding of GABA and benzodiazepine receptor ligands in vitro. Barbiturates were shown to inhibit the binding of tritiated picrotoxin to GABA chloride channel sites [Ticku and Olsen, 1978], but this ligand had an unsuitably low affinity for autoradiography. This problem was remedied by the development of the slowly dissociating ligand for the barbiturate-sensitive convulsant sites, $[^{35}S]$TBPS [Squires et al., 1983].

GABA, on the other hand, appears to have many actions independent of those shared or influenced by benzodiazepines and barbiturates. These actions are most likely attributable to the subtypes of GABA receptors which exist in the CNS [Bowery et al., 1984]. Two such GABA receptor subtypes have been discovered: the bicuculline sensitive–baclofen insensitive $GABA_A$ receptor and the baclofen sensitive–bicuculline insensitive $GABA_B$ receptor. The functional significance of the $GABA_B$ receptors is unknown, and the action of GABA at these sites does not appear to involve a chloride ionophore. Therefore, the $GABA_B$ receptor is not considered a component of the macromolecular receptor complex associated with benzodiazepine and barbiturate drug action. The $GABA_A$ receptors can be further subdivided into populations which differ in their affinity for GABA agonist and antagonist compounds [Olsen et al., 1981; Browner et al., 1981; Olsen and Snowman, 1983]. The $GABA_A$ agonist muscimol, for instance, recognizes both high- and low-affinity $GABA_A$ receptors. Likewise, the convulsant compound bicuculline can be used to label multiple $GABA_A$ sites. Results from investigations performed using CNS membrane preparations for binding have generated the hypothesis that the low-affinity $GABA_A$ receptor subpopulation is part of the macromolecular benzodiazepine/barbiturate receptor complex [see Olsen, 1982] and all of the high-affinity $GABA_A$ sites may not be. Initial autoradiographic studies indicated that the high-affinity $GABA_A$ sites, labeled with nanomolar concentrations of [$^3$H]muscimol, existed in many brain regions where the benzodiazepine receptors could not be localized to any great extent and vice versa [Palacios et al., 1980; 1981]. These results indirectly supported the concept that an action at the low-affinity $GABA_A$ receptor must be associated with the macromolecular receptor complex [Browner et al., 1981; Burch et al., 1983; Olsen and Snowman, 1983]. Direct autoradiographic localization of the low-affinity $GABA_A$ receptors was impractical due to the fast off-rate of ligands capable of recognizing the low-affinity site. [$^3$H] Muscimol binds preferentially to high-affinity sites, especially if low ligand concentrations and long post-binding rinse times are used. Lower-affinity sites can be at least partially labeled in addition to high-affinity sites, if higher ligand concentrations are used along with minimal rinse times [see Penney et al., 1981]. Another problem has been the instability of ligands showing some selectivity for the low-affinity site. Binding and autoradiography of tritium-labeled bicuculline methoiodide ([$^3$H]BMC) had been described earlier by Hösli et al. [1980], but this ligand was too unstable chemically for commercial availability. With the introduction of a tritiated form of bicuculline methochloride, however, Olsen and Snowman [1983] were able to show selective labeling of low-affinity $GABA_A$ receptors in a homogenized brain preparation, by limiting the binding of the ligand to the high-affinity site with the anion thiocyanate. Olsen et al. [1984] subsequently made use of these parameters to study the binding of [$^3$H]bicuculline methochloride to slide-mounted tissue sections and demonstrated the first autoradiographic localization in brain tissue of the low-affinity $GABA_A$ receptors.

Autoradiographic techniques, after in vitro receptor labeling, can thus be used to localize the individual components of the benzodiazepine/barbiturate-sensitive convulsant/GABA receptor complex. Benzodiazepine receptors can be labeled with [$^3$H]flunitrazepam [Young and Kuhar, 1979; 1980], high-affinity $GABA_A$ receptors can be labeled with [$^3$H]muscimol [Palacios et al., 1980; 1981], the barbiturate-sensitive convulsant sites can be localized after binding with [$^{35}$S]t-butylbicyclophosphorothionate [Gee et al., 1983; Wamsley et al., 1983], and [$^3$H]bicuculline methochloride in the presence of thiocyanate can be used to obtain labeling to the low-affinity $GABA_A$ receptor site [Olsen et al., 1984]. By labeling serial sections with one of these ligands, the distribution and density of components of the macromolecular complex can be individually examined in the same regions of the brain. This type of direct comparison is not possible with other techniques.

The macromolecular receptor complex consisting of a benzodiazepine/barbiturate-sensitive convulsant/GABA receptor complex associated with a chloride channel appears to be a very important site of drug action in the CNS. Microscopic localization of individual components of this receptor complex provides means of demonstrating where this complex exists in the brain and could supply key information on the potential sites of action of drugs affecting this system.

The receptors associated with other actions of the benzodiazepines and GABA can also be localized using receptor autoradiographic techniques. $GABA_B$ receptors have been localized with [$^3$H]GABA by inhibiting the binding to the high-affinity $GABA_A$ sites with isoguvacine and by including calcium ions which are necessary for $GABA_B$ binding [Wilkin et al., 1981; Bowery et al., 1984], and more recently by direct binding with the selective ligand $[^3H](-)$baclofen [Gehlert et al., 1985]. Subtypes of benzodiazepine receptors also exist [Squires, 1983], and the so-called "peripheral sites" have been autoradiographically localized in the brain after labeling with R05-4864 [Gehlert et al., 1983; 1984; Anholt et al., 1984]. The latter two receptor subtypes ($GABA_B$ and peripheral benzodiazepine sites) do not appear to involve an action at a chloride channel, and their distributions will therefore not be included in this communication.

## MATERIALS AND METHODS

Autoradiographic localization of the individual components of the benzodiazepine/barbiturate-sensitive convulsant/GABA receptor complex associated with a chloride ionophore was accomplished in serial sections of rat brain labeled using compounds and conditions specific for the individual receptor types.

### Tissue Preparation

Male Sprague-Dawley rats (200–250 g) were terminated, while under deep chloroform anesthesia, by intracardial perfusion with 200 ml of ice-cold 0.9% saline. The brains were rapidly dissected and then frozen onto microtome chucks by slow immersion in isopentane at −70°C. Ten-micron-thick frozen sections were cut on a Harris cryostat microtome (North Billerica, MA) and thaw-mounted onto cold chrome-alum/gelatin coated ("subbed") microscope slides. These slide-mounted tissue sections were then stored at −20°C, in the presence of desiccant, for 1–14 days prior to labeling.

### Labeling Procedures

The conditions used to obtain labeling of the respective components of the receptor complex are indicated in Table I. In each case, these parameters were based on the results of previous studies in which the investigators established that these were the optimal conditions for obtaining specific labeling of the individual receptor type. The properties of the binding sites detected by receptor au-

**TABLE I. Labeling Conditions for Receptor Autoradiography**

| | |
|---|---|
| Benzodiazepine receptors | |
| Preincubation | None |
| Incubation | 60 minutes at 4°C in 0.17 M Tris-HCl buffer (pH 7.4) containing 1 nM $[^3H]$flunitrazepam |
| Rinse | 2 minutes at 4°C |
| Blanks | 1 micromolar clonazepam |
| Reference | Young and Kuhar [1980] |
| Low-affinity $GABA_A$ receptors | |
| Preincubation | 15 minutes at 4°C in 10 mM $K_2HPO_4$-$KH_2PO_4$ buffer containing 0.1 M KSCN (pH 7.4) |
| Incubation | 30 minutes at 4°C in buffer with 25 nM $[^3H]$bicuculline methochloride |
| Rinse | 10 seconds at 4°C |
| Blanks | 0.1 mM GABA |
| Reference | Olsen, Snowhill, Wamsley [1984] |
| Barbiturate-sensitive convulsant sites | |
| Preincubation | 30 minutes at room temperature in 50 mM Tris buffer (pH 7.5) containing 200 mM KCl and 1 mM EDTA |
| Incubation | 90 minutes at room temperature in buffer containing 10 nM $[^{35}S]$t-butylbicyclophosphorothionate ($[^{35}S]$TBPS) |
| Rinse | 10 minutes at 4°C |
| Blanks | 10 micromolar picrotoxin |
| Reference | Gee, Wamsley, Yamamura [1983]; Wamsley, Gee, Yamamura [1983] |
| High-affinity $GABA_A$ receptors | |
| Preincubation | 20 minutes at 4°C in 0.31 M Tris-citrate buffer (pH 7.1) |
| Incubation | 40 minutes at 4°C in buffer + 5 nM $[^3H]$muscimol |
| Rinse | 1 minute at 4°C |
| Blanks | 10 micromolar GABA |
| Reference | Palacios, Young, Kuhar [1980]; Palacios, Wamsley, Kuhar [1981] |

toradiography were consistent with literature reports for homogenate binding. The conditions of [$^3$H]muscimol ligand concentration and rinse time were chosen to maximize binding specifically to high-affinity sites. The incubations were performed by immersing the thawed tissue sections, for the indicated period of time, into a Coplin jar containing the radioactive ligand in a buffered solution. For obtaining labeling of the barbiturate-sensitive convulsant sites and the $GABA_A$ receptor subtypes, this incubation was preceded by a preincubation period in buffer to dissociate any endogenous ligand which could be present and subsequently interfere with the binding [Napias et al., 1980]. After incubation, the unbound radioactivity was removed from the tissue sections by immersing them for a specified period in fresh ice-cold buffer. The rinse was followed by a dip in distilled water (to remove the buffer salts which could adhere to the photographic emulsion in subsequent steps), and the labeled tissue sections were rapidly dried on a metal pan over ice by blowing a stream of cool, dry, filtered air over the tissue surface.

### Generation of Autoradiograms

After drying, the slides were stored overnight in desiccant containing slide boxes maintained in a refrigerator (4°C). The following day, the slides were affixed to photographic mounting boards and autoradiograms were generated by apposing sheets of LKB Ultrofilm (Rockville, MD) to the labeled sections in Wolf X-ray cassettes. After allowing an appropriate exposure period at 4°C, the films were removed and developed in D-19 developer (Kodak; Rochester, NY). The latent autoradiographic images were examined and photographed using a Leitz (West Germany) Orthoplan microscope equipped with an Orthomat camera system. Tissue localization was confirmed by comparing the grain distribution of the autoradiogram to the labeled tissue section stained with cresyl violet.

## RESULTS

Autoradiograms generated from tissue sections labeled with each ligand exhibit a high grain density, indicating the presence of specifically bound ligand, in a variety of brain regions. In general, sections incubated in the additional presence of the appropriate displacer produce autoradiograms with a diffuse uniform distribution of grains. Specific binding can be detected in regions containing gray matter, with little or no specific binding in white matter.

For the most part, the *distribution* of the benzodiazepine, low-affinity $GABA_A$, and barbiturate-sensitive convulsant sites overlap in the brain regions examined, although there is a marked difference in the apparent *density* of the receptors in a few areas. Thus, in sections of forebrain, [$^3$H]flunitrazepam, [$^3$H]bicuculline methochloride ([$^3$H]BMC), and [$^{35}$S] t-butylbicyclophosphorothionate ([$^{35}$S]-TBPS) all intensely label such regions as laminae I and IV of the frontoparietal cortex, olfactory tubercle, globus pallidus, substantia innominata, molecular layer of the dentate gyrus, zona incerta, superficial layer of the superior colliculus, substantia nigra-pars reticulata, inferior colliculus, and the dorsal tegmental nucleus.

There are many instances where the common distribution of the benzodiazepine, low-affinity $GABA_A$, and barbiturate-sensitive convulsant receptors differ from that of the high-affinity $GABA_A$ receptors labeled with [$^3$H]muscimol. For instance, in the cortex, high-affinity $GABA_A$ receptors are found in all of the superficial laminae (I–IV) whereas the labeling of the other receptor types is highest in laminae I and IV (Table II). In the olfactory tubercle, the benzodiazepine, low-affinity $GABA_A$, and barbiturate-sensitive convulsant receptors are localized in the internal part of the tubercle in regions including the islands of Calleja (Fig. 1). High-affinity $GABA_A$ receptors are localized predominantly in the outer layers of the tubercle and do not label the islands of Calleja. The contrast in the labeling of these receptor populations is especially evident in the globus pallidus, where the labeling of high-affinity $GABA_A$ receptors is virtually nonexistent whereas the other receptor types show a dense concentration.

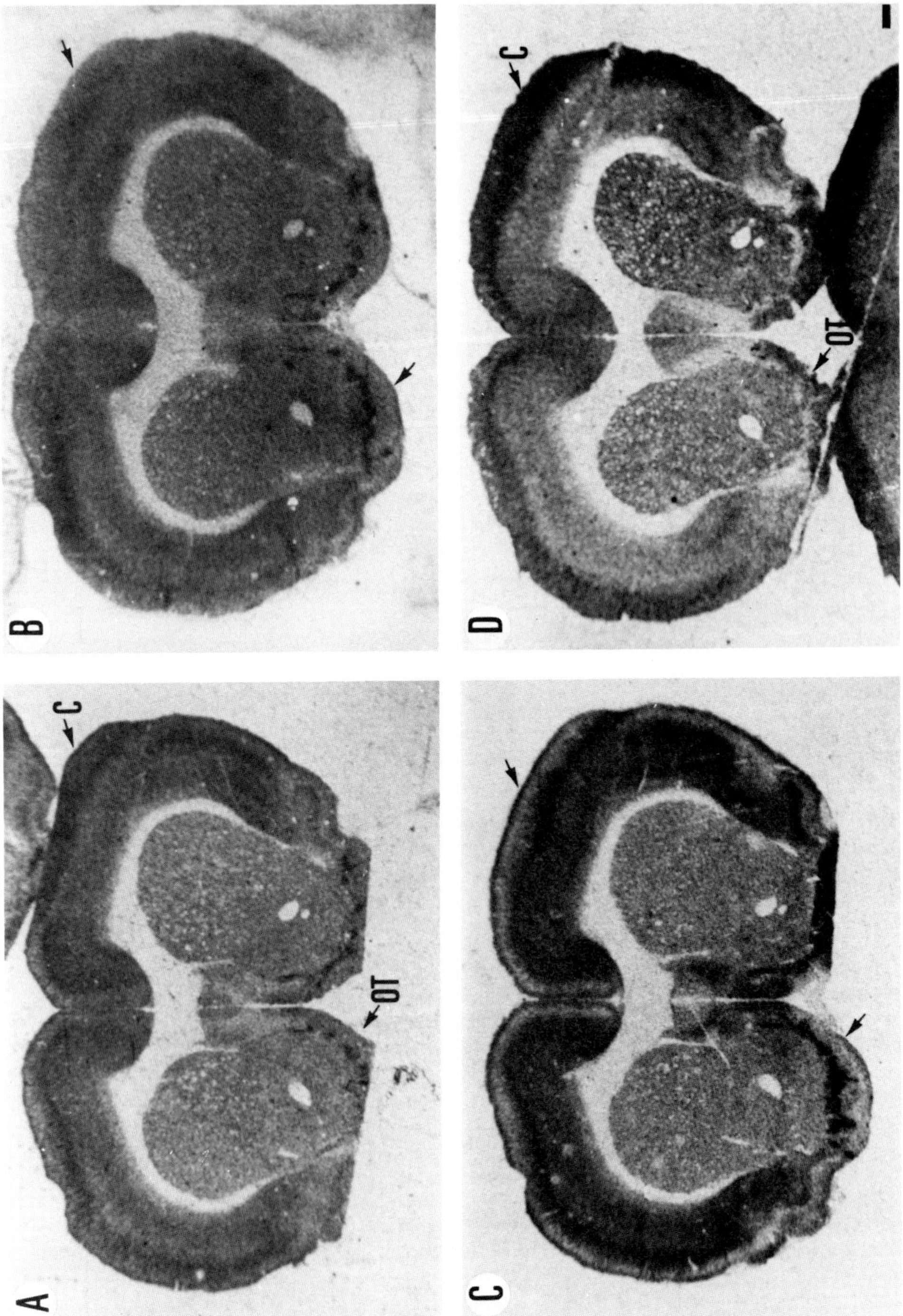
A
B
C
D
C
OT

**TABLE II. Relative Specific Binding***

| Area | [$^3$H]Flu | [$^3$H]Bicuculline | [$^{35}$S]TBPS | [$^3$H]Muscimol |
|---|---|---|---|---|
| Parietal cortex | | | | |
| Lamina I | 73.1 ± 0.6 | 22.5 ± 0.6 | 68.3 ± 1.2 | 68.2 ± 1.0 |
| Lamina II | 66.5 ± 0.8 | 18.0 ± 0.4 | 48.4 ± 0.6 | 67.8 ± 1.0 |
| Lamina III | 67.7 ± 0.5 | 17.9 ± 0.6 | 57.3 ± 0.5 | 67.5 ± 0.8 |
| Lamina IV | 77.1 ± 0.7 | 27.1 ± 0.3 | 71.5 ± 0.3 | 69.7 ± 0.9 |
| Lamina V | 65.9 ± 0.7 | 18.8 ± 0.7 | 63.7 ± 0.5 | 45.7 ± 1.0 |
| Lamina VI | 67.7 ± 0.5 | 21.3 ± 1.1 | 65.3 ± 0.5 | 56.8 ± 0.5 |
| Cerebellum | | | | |
| Granule cell layer | 22.0 ± 0.4 | 17.1 ± 1.8 | 47.0 ± 0.4 | 89.5 ± 0.3 |
| Molecular layer | 59.5 ± 0.5 | 33.6 ± 0.8 | 12.8 ± 0.9 | 40.0 ± 0.9 |

*These measurements (% extinction) were made using an MPV Compact photometer interfaced with an Oki computer attached to a Leitz Orthoplan microscope. The readings represent the mean of four measurements from five different sections, plus or minus the standard error of the mean. Only qualitative data are shown, since different ligand concentrations, different radioactive tags, and different exposure periods are involved. The relative differences between the amount of binding from each ligand can be ascertained in individual laminae. Note that benzodiazepine, low-affinity $GABA_A$ and the barbiturate/convulsant sites are highest in laminae I and IV, whereas the high-affinity $GABA_A$ sites are concentrated in laminae I–IV. In the cerebellum, TBPS and muscimol binding are highest in the granule cell layer while flu and bicuculline binding occurs with the greatest density in the molecular layer.

High densities of benzodiazepine, barbiturate-sensitive convulsant, and low-affinity $GABA_A$ receptors occur in parts of the hippocampal formation, the centromedian nucleus of the thalamus, generalized regions of the hypothalamus (Fig. 2), the superior and inferior colliculi, and the substantia nigra (Fig. 3). Low densities of these receptors are found in the ventrolateral thalamus. The antithesis of these receptor localizations is represented by the distribution of high-affinity $GABA_A$ receptors which have a low density in the hippocampal formation, centromedian nucleus of the thalamus, generalized areas of the hypothalamus, the superior and inferior colliculi, and the substantia nigra, and have a high density in the ventrolateral thalamus.

This gregarious pattern of overlap between the benzodiazepine, low-affinity $GABA_A$, and barbiturate-sensitive convulsant sites in contrast to the high-affinity $GABA_A$ receptors is interrupted in such regions as the medial geniculate body and cerebellum. The density of benzodiazepine and low-affinity $GABA_A$ receptors is low in the medial geniculate body while the density of barbiturate-sensitive convulsant and high-affinity $GABA_A$ sites is high (Fig. 3). Likewise, in the cerebellum the benzodiazepine and low-affinity $GABA_A$ receptors are concentrated in the molecular layer while the barbiturate-sensitive convulsant and high-affinity $GABA_A$ receptors predominate in the granule cell layer (Fig. 4).

**Fig. 1.** *Low-magnification photomicrographs of autoradiograms from rat forebrain sections. Each section is labeled specifically: A, with [3H]flunitrazepam (to demonstrate the distribution and density of benzodiazepine receptors); B, with [3H]bicuculline methochloride (to show low-affinity $GABA_A$ sites); C with [$^{35}$S]TBPS (to label the barbiturate/convulsant sites); and D, with [$^3$H]muscimol (to show the high-affinity $GABA_A$ receptors). The autoradiographic grains appear as dark dots against a light background. Note the similarities in labeling between the layers of cortex (C) and olfactory tubercle (OT) in A, B, and C as opposed to that appearing in D (the labeling in regions of B appears less consistent due to the low affinity of the receptor involved). Bar = 500 microns.*

## DISCUSSION

In vitro receptor autoradiography provides a powerful tool to localize and evaluate both central and peripheral receptor populations.

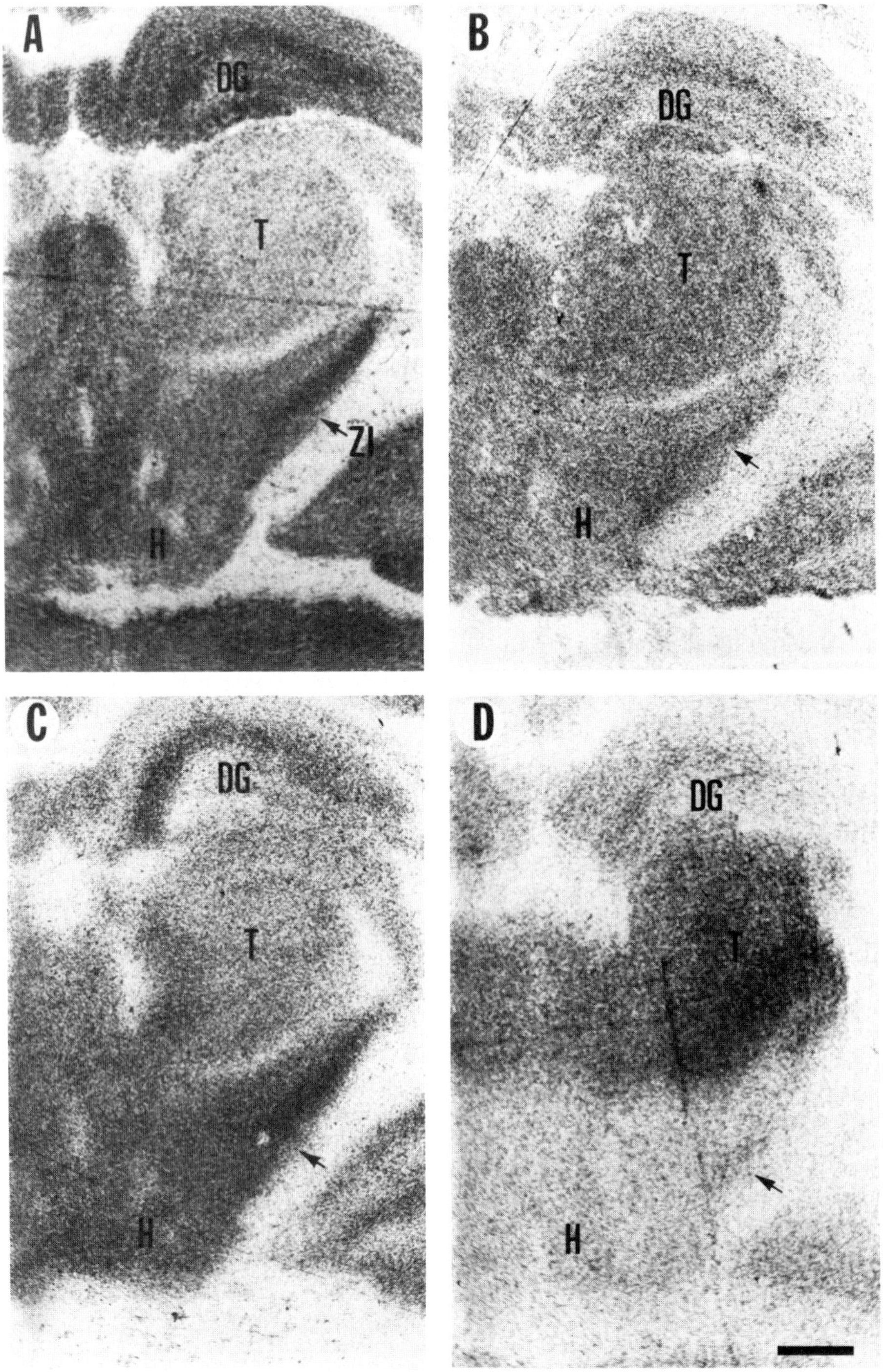
A
DG
T
ZI
H
B
DG
T
H
C
DG
T
H
D
DG
T
H

The advantages of this technique are particulary evident when examining multiple receptor distributions in microscopic brain regions or in laminated structures. By utilizing serial sections of rat brain, we have been able to compare the distribution of $GABA_A$ receptor subtypes with the distribution of benzodiazepine receptors and the barbiturate-sensitive convulsant sites in many brain regions.

The relative grain densities produced on LKB Ultrofilm by each of the radiolabeled ligands, specifically bound to slide-mounted sections of rat brain, were similar in many brain regions. However, the distribution of grain densities for sections labeled with [$^3$H]muscimol, indicating the presence of high-affinity $GABA_A$ receptors, differed greatly from that produced by the other ligands in most of the regions investigated. Particularly striking were the differences seen in the cortex, olfactory tubercle, globus pallidus, hippocampal formation, thalamus, hypothalamus, and inferior colliculus. In most cases, the relative density of benzodiazepine, low-affinity $GABA_A$, and barbiturate-sensitive convulsant receptors were similar, although major differences were noticed between the first two and the last receptor category in the medial geniculate body and cerebellum. The localization of regions where

**Fig. 2.** *Photomicrographs of autoradiograms generated by sections labeled for the demonstration of: A, benzodiazepine receptors ([$^3$H]-flunitrazepam); B, low-affinity $GABA_A$ receptors ([$^3$H]-bicuculline methochloride); C, barbiturate/convulsant sites ([$^{35}$S]-TBPS); and D, high-affinity $GABA_A$ receptors ([$^3$H]-muscimol). Note the similarities between A, B,* and *C* in contrast to the grain distribution and density seen in *D*. Significant densities of benzodiazepine, barbiturate/convulsant and low-affinity $GABA_A$ *receptors occur in the molecular layer of the dentate gyrus (DG), zona incerta (ZI), and in general regions of the hypothalamus (H). Only very low densities of high-affinity $GABA_A$ sites are found in these regions. The latter $GABA_A$ receptor subtype is quite highly concentrated in thalamic regions (T), where only low densities of the other three receptor types are found ([$^3$H]-bicuculline labels this region more intensely than do benzodiazepine or barbiturate/convulsant ligands). Bar = 500 microns.*

the benzodiazepine receptors or barbiturate-sensitive convulsant sites appear to exist independent of each other (and therefore each individual component of the benzodiazepine/barbiturate-sensitive convulsant/GABA receptor complex is not always present) raises the possibility that this differential localization of receptor sites could be responsible for some of the therapeutic differences that exist between these two classes of drugs. It would be interesting, for instance, to determine if the barbiturate-sensitive convulsant receptors are present in the medullary and pontine respiratory control centers without the presence of the benzodiazepine site. This might account for the reasons why the respiratory-depression side-effects of the barbiturates are not shared with the benzodiazepines. Subtypes of the barbiturate-sensitive convulsant receptors might also exist which are not labeled with [$^{35}$S]TBPS [Squires et al., 1983]. These subtypes could bind and respond differently to the presence of convulsant vs. depressant barbiturates. It is also possible that [$^{35}$S]TBPS labels both the benzodiazepine-GABA receptor complex and, in addition, some other sites (possibly even GABA chloride channels) that have no associated benzodiazepine sites, or at least have an undetectably low affinity.

Subtypes of central benzodiazepine receptors (type 1 and type 2) are also thought to exist [Squires, 1983, 1984; Richards and Mohler, 1984] which can be localized differentially by CL218,872 (a triazolopyridazine) displacement of the binding of [$^3$H]flunitrazepam to the $benzodiazepine_1$ receptor [Young et al., 1981; Unnerstall et al., 1982]. This can also be accomplished by quazepam (a 1-N-trifluoroethyl benzodiazepine) displacement of [$^3$H]flunitrazepam binding to the $benzodiazepine_1$ site [Wamsley et al., 1985], and presumably by direct labeling with radioactive forms of the $benzodiazepine_1$ selective agents CL218,872 or the putative endogenous benzodiazepine receptor antagonist, the beta-carboline esters which distinguish two subtypes, the $benzodiazepine_1$ and $benzodiazepine_2$ re-

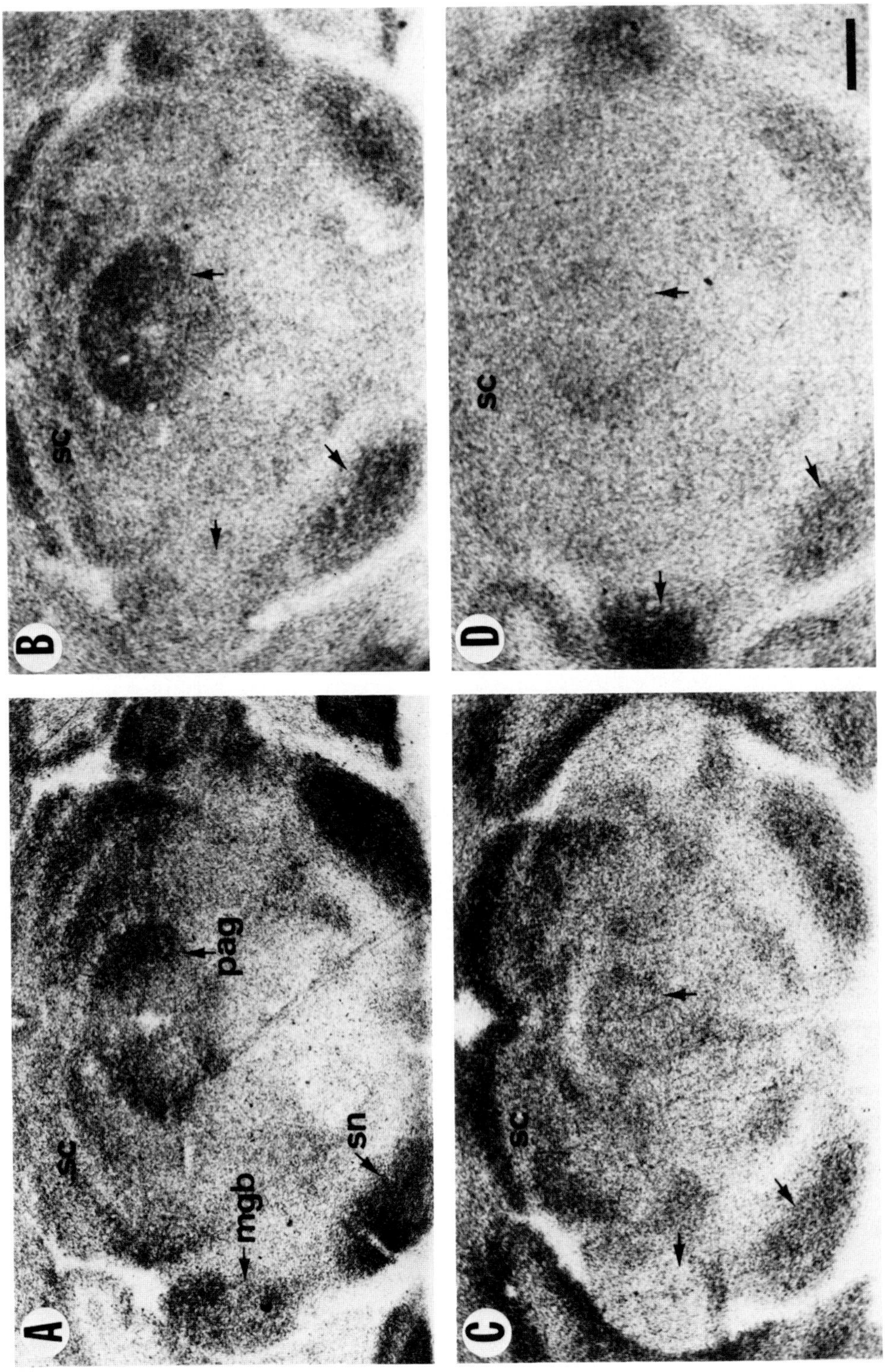
A
sc
pag
mgb
sn
B
sc
C
sc
D
sc

ceptors [see Wamsley and Palacios, 1984]. It may be possible to displace the benzodiazepine$_2$ receptor binding of [$^3$H]flunitrazepam by including fairly substantial amounts of triphenyl methane dyes in the incubation medium [Smith and Squires, 1983]. The functional significance of these benzodiazepine receptor subtypes has yet to be positively identified, however, the highest density of benzodiazepine receptors (which is represented predominantly by the type 1 subtype) overlaps with regions where the highest density of low-affinity $GABA_A$ receptors exists. Thus, one might hypothesize that the type 1 receptor subtype is a very important site of benzodiazepine drug action. On the other hand, Leeb-Lundberg and Olsen [1983] showed that the high-affinity beta-carboline binding sites (benzodiazepine$_1$) had weaker interactions with GABA and bicuculline than did the lower-affinity benzodiazepine$_2$ subpopulation.

The differential localization of high-affinity $GABA_A$ receptors and low-affinity $GABA_A$ receptors may provide some insight into the benzodiazepine/barbiturate-sensitive convulsant/GABA receptor/chloride ionophore complex. As indicated in the autoradiographic study of Unnerstall et al. [1981], the degree of enhancement of benzodiazepine binding by GABA in discrete brain regions had very little relation to the density of muscimol binding sites. In fact, the degree of enhancement was roughly the same in all regions, suggesting that all of the benzodiazepine receptors were associated with a GABA receptor. They concluded that the responsible GABA receptor could be either a subpopulation of high-affinity $GABA_A$ receptors or possibly a receptor that is not labeled by [$^3$H]muscimol. The results shown in the present study would indicate that the low-affinity $GABA_A$ receptors labeled by [$^3$H]BMC are the receptors linked to the benzodiazepine sites due to their similar distribution and correlation in density. The binding of [$^3$H]flunitrazepam and that of [$^3$H]BMC show a virtually identical localization, as would be expected for two components of a single GABA/benzodiazepine receptor complex. In addition, in many of the regions where GABA enhancement of benzodiazepine binding has been detected [Unnerstall et al., 1981], such as the molecular layer of the cerebellum and lamina IV of the cortex, low-affinity $GABA_A$ receptors predominate over high-affinity $GABA_A$ receptors. Also supporting this concept is the observation of the reciprocal effect—that is, the binding of radiolabeled ligands to the low-affinity $GABA_A$ receptor appears to be enhanced by the presence of benzodiazepines in the incubation media [Skerritt et al., 1982].

The functional difference between and underlying purpose for having high- and low-affinity sites (or high- and low-affinity states) of the $GABA_A$ receptors has not been delineated clearly. As demonstrated in the present study, the regions of overlap between the distribution of these two receptor subtypes are few and far between. It is difficult to determine if the overlap between the distributions of high- and low-affinity $GABA_A$ receptors is due to the labeling of [$^3$H]BMC encroaching onto the high-affinity sites (incomplete specificity) or due to some of the [$^3$H]muscimol recalcitrantly usurping itself to a few defiant low-affinity sites in spite of the 1-minute rinse. If either situation turns out to be the case, then the regional separation between the high- and low-affinity receptors may be even more dramatic than indicated here. Certainly, the complete contrast that occurs between the distribution of high- and low-affinity $GABA_A$ receptors in such areas as the globus pallidus

**Fig. 3.** *Receptors in the mesencephalon: A*, benzodiazepine; *B*, low-affinity $GABA_A$; *C*, barbiturate/convulsant; and *D*, high-affinity $GABA_A$ receptors. Fairly high densities of the respective receptors are indicated in the superficial lamina of the superior colliculus (sc), periaqueductal gray matter (PAG), dorsal part of the medial geniculate body (mgb), and in the substantia nigra-pars reticulata (sn). In contrast, the high-affinity $GABA_A$ receptors are absent in the sc and appear quite concentrated in the mgb. All of the receptor types appear to overlap in the PAG and sn. Bar = 500 microns.

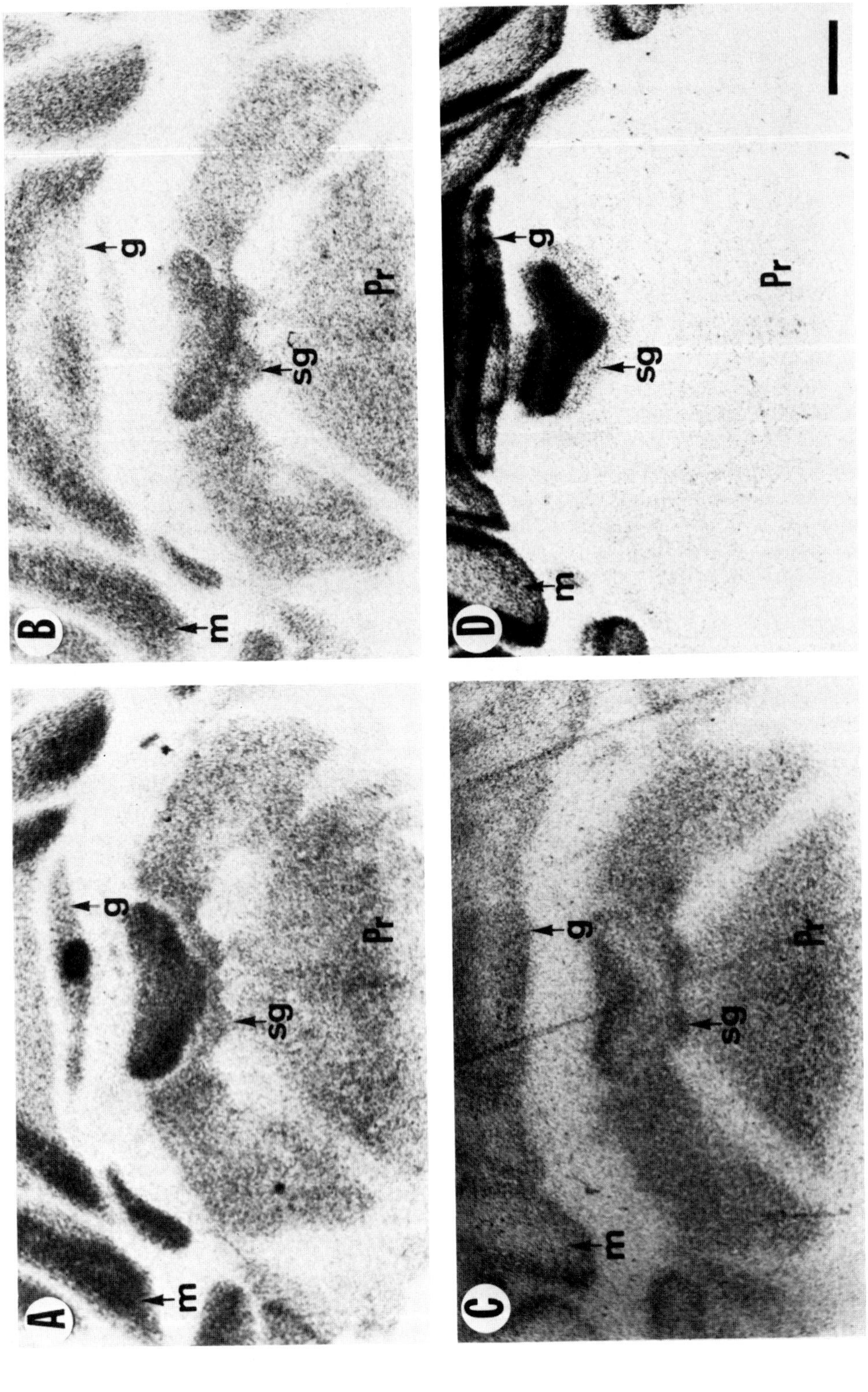
A
m
g
sg
Pr
B
m
sg
g
Pr
C
m
sg
g
Pr
D
m
sg
g
Pr

and cerebellum exemplifies this possibility. It seems likely that biochemical isolation of the receptor proteins (see other chapters in this volume) will resolve the question of this heterogeneity of GABA/benzodiazepine receptors.

In conclusion, the distribution of benzodiazepine, barbiturate-sensitive convulsant, high-affinity $GABA_A$, and low-affinity $GABA_A$ receptors has been localized in serial sections of rat brain. In most brain regions examined, the relative densities of benzodiazepine, barbiturate-sensitive convulsant, and low-affinity $GABA_A$ receptors were similar, while high-affinity $GABA_A$ receptors exhibited a differential distribution. These data provide additional support for the concept of the low-affinity $GABA_A$ receptor as the GABA receptor subtype linked to the benzodiazepine and barbiturate receptors. Autoradiographic localization of these receptors provides a means whereby an investigator can perceive the potential sites of action of compounds affecting the respective receptor subtype. These quantitative localizations also provide a means of detecting precisely where future studies should focus in order to determine the functional differences between receptor subtypes such as the high- and low-affinity $GABA_A$ receptors. The same type of information can be derived for studies aimed at deciphering how benzodiazepines and barbiturates differ in their depressant effects and in which brain areas they are involved in the macromolecular $GABA_A$ receptor/chloride ionophore complex.

**Fig. 4.** *Receptors in the brainstem:* A, benzodiazepine; B, low-affinity $GABA_A$; C, barbiturate/convulsant; and D, high-affinity $GABA_A$ receptors. Both [$^3$H]flunitrazepam and [$^3$H]bicuculline label the molecular layer of the cerebellum (m), while [$^{35}$S]TBPS and [$^3$H]muscimol label the granule cell layer of the cerebellum (g). Benzodiazepine, low-affinity $GABA_A$, and the barbiturate/convulsant sites appear similarly distributed in the pons (sg = suprageniculate nucleus; Pr = pontine reticular nucleus) in contrast to the lack of high-affinity $GABA_A$ sites seen in this region. Bar = 500 microns.

## ACKNOWLEDGMENTS

The authors wish to thank Beth Mitchell for her technical assistance and Margaret Long for secretarial assistance. Some of the work reported here was supported in part by PHS Grant MH36563 and DA00346.

## REFERENCES

Anholt RRH, Murphy KMM, Mack GE, Snyder SH (1984): Peripheral-type benzodiazepine receptors in the central nervous system: Localization to olfactory nerves. J Neurosci 4:593–603.

Bowery NG, Price GW, Hudson AL, Hill DR, Wilkin GP, Turnbull MJ (1984): GABA receptor multiplicity; Visualization of different receptor types in the mammalian CNS. Neuropharmacology 23:219–231.

Browner M, Ferkany JW, Enna SJ (1981): Biochemical identification of pharmacologically and functionally distinct GABA receptors in rat brain. J Neurosci 1:514–518.

Burch TP, Thyagarajan R, Ticku MK (1983): Group-selective reagent modification of the benzodiazepine-gamma-aminobutyric acid receptor-ionophore complex reveals that low affinity gamma-aminobutyric acid receptors stimulate benzodiazepine binding. Mol Pharmacol 23:52–59.

Gee KW, Wamsley JK, Yamamura HI (1983): Light microscopic autoradiographic identification of picrotoxin/barbiturate binding sites in rat brain with [$^{35}$S]-t-butylbicyclophosphorothionate. Eur J Pharmacol 89:323–324.

Gehlert DR, Yamamura HI, Wamsley JK (1983): Autoradiographic localization of "peripheral" benzodiazepine binding sites in the rat brain and kidney using [$^3$H]-R05-4864. Eur J Pharmacol 95:329–330.

Gehlert DR, Yamamura HI, Wamsley JK (1985): Autoradiographic localization of peripheral-type benzodiazepine binding sites in the rat brain, heart and kidney. NS Arch Pharmacol 328:454–460.

Gehlert DR, Yamamura HI, Wamsley JK (1985): $GABA_B$ receptors in the rat brain: Autoradiographic localization using [$^3$H](-)baclofen. Neurosci Lett 56:183–188.

Haefely W (1983): The biological basis of benzodiazepine actions. The benzodiazepines today—two decades of research and clinical experience. J Psychoact Drugs 15:19–39.

Hösli E, Möhler H, Richards JG, Hosli L (1980): Autoradiographic localization of binding sites for [$^3$H]gamma-aminobutyrate, [$^3$H]muscimol, (+) [$^3$H]bicuculline methoiodide and [$^3$H]flunitrazepam in cultures of rat cerebellum and spinal cord. Neurosci 5:1657–1665.

Kuhar MJ (1981): Autoradiographic localization of drug

and neurotransmitter receptors in the brain. Trends Neurosci 4:60–64.

Kuhar MJ (1985): Receptor localization with the microscope. In Yamamura HI, Enna SJ, Kuhar MJ (eds): "Neurotransmitter Receptor Binding, 2nd edition." New York: Raven Press. pp 153–176.

Kuhar MJ, Wamsley JK, Palacios JM, Niehoff DL, Zarbin MA, Unnerstall J (1983): Drug and neurotransmitter receptors: Anatomic localization in the brain. In Yoshida H, Yamamura HI (eds): "Pharmacologic and Biochemical Aspects of Neurotransmitter Receptors." New York: John Wiley & Sons, pp 161–174.

Leeb-Lundberg LMF, Olsen RW (1983): Heterogeneity of benzodiazepine receptor interactions with gamma-aminobutyric acid and barbiturate receptor sites. Mol Pharmacol 23:315–325.

Möhler H, Richards JG, Wu JY (1981): Autoradiographic localization of benzodiazepine receptors in immunocytochemically identified gamma-aminobutyrergic synapses. Proc Nat Acad Sci 78:1935–1938.

Napias C, Bergman MO, Van Ness PC, Greenlee D, Olsen RW (1980): GABA binding in mammalian brain: inhibition by endogenous GABA. Life Sci 27:1001–1011.

Olsen RW (1981): GABA-benzodiazepine-barbiturate receptor interactions. J Neurochem 37:1–13.

Olsen RW (1982); Drug interactions at the GABA receptor ionophore complex. Ann Rev Pharmacol Toxicol 22:245–277.

Olsen RW, Bergman MO, Van Ness PC, Lummis SC, Watkins AE, Napias C, Greenlee DV (1981): Gamma-aminobutyric acid receptor binding in mammalian brain: heterogeneity of binding sites. Mol Pharmacol 19:217–227.

Olsen RW, Snowhill EW, Wamsley JK (1984): Autoradiographic localization of low affinity GABA receptors with [$^3$H]-bicuculline methochloride. Eur J Pharmacol 99:247–248.

Olsen RW, Snowman AM (1983): [$^3$H]-Bicuculline methochloride binding to low affinity GABA receptor sites. J Neurochem 41:1653–1663.

Palacios JM, Young WS III, Kuhar MJ (1980): Autoradiographic localization of gamma-aminobutyric acid (GABA) receptors in the rat cerebellum. Proc Nat Acad Sci 77:670–674.

Palacios JM, Wamsley JK, Kuhar MJ (1981): High affinity GABA receptors—autoradiographic localization. Brain Res 222:285–307.

Penney JB Jr, Pan HS, Young AB, Frey KA, Dauth GW (1981): Quantitative autoradiography of [$^3$H]muscimol binding in rat brain. Science 214:1036–1038.

Richards JG, Möhler H (1984): Benzodiazepine receptors. Neuropharmacology 23:233–242.

Skerritt JH, Willow M, Johnston GAR (1982): Diazepam enhancement of low affinity GABA binding to rat membranes. Neurosci Lett 29:63–66.

Smith TM, Squires RF (1983): Differential inhibition of brain specific [$^3$H]-flunitrazepam binding by several types of dyes. Neurochem Res 8:1177–1183.

Squires RF (1983): Benzodiazepine receptor multiplicity. Neuropharmacology 22:1443–1450.

Squires RF (1984): Benzodiazepine receptors. In Lajtha, A (ed): "Handbook of Neurochemistry, Vol. 6." New York, Plenum Press, pp 261–306.

Squires RF, Casida JE, Richardson M, Saederup E (1983): [$^{35}$S]t-Butylbicyclophosphorothionate binds with high affinity to brain specific sites coupled to gamma-aminobutyric acid-A and ion recognition sites. Mol Pharmacol 23:326–336.

Tallman JF, Paul SM, Skolnick P, Gallager DW (1980): Receptors for the age of anxiety: Pharmacology of the benzodiazepines. Science 207:274–281.

Ticku MK, Olsen RW (1978): Interaction of barbiturates with dihydropicrotoxin binding sites related to the GABA receptor-ionophore system. Life Sci 22:1643–1653.

Unnerstall JR, Kuhar MJ, Niehoff DL, Palacios JM (1981): Benzodiazepine receptors are coupled to a subpopulation of GABA receptors: Evidence from a quantitative autoradiographic study. J Pharmacol Exp Ther 218:797–804.

Unnerstall JR, Niehoff DL, Kuhar MJ, Palacios JM (1982): Quantitative receptor autoradiography using [$^3$H]-Ultrofilm: Application to multiple benzodiazepine receptors, J Neurosci Methods 6:59–73.

Wamsley JK, Gee, KW, Yamamura HI (1983): Comparison of the distribution of convulsant/barbiturate and benzodiazepine receptors using light microscopic autoradiography. Life Sci 33:2321–2329.

Wamsley JK, Golden JS, Yamamura HI, Barnett A (1985): Autoradiographic demonstration of the selectivity of two 1-N-trifluoroethyl benzodiazepines for the BZD-1 receptors in the rat brain, Pharmacol Biochem Behav 23:973–978.

Wamsley JK, Palacios JM (1982): Receptor Mapping by Histochemistry. In Lajtha A (ed): "Handbook of Neurochemistry, Vol 2." New York: Plenum Press, pp 27–51.

Wamsley JK, Palacios JM (1983): Apposition techniques of autoradiography for microscopic receptor localization. In: Barker JL, McKelvy JF (eds): "Current Methods in Cellular Neurobiology." New York: John Wiley & Sons, pp 295–313.

Wamsley JK, Palacios JM (1984): Amino acid and benzodiazepine receptors. In Bjorklund A, Hokfelt T, Kuhar M (eds): "Handbook of Chemical Neuroanatomy, Vol.3." Amsterdam: Elsevier/North Holland, pp 352-385.

Wilkin GP, Hudson AL, Hill DR, Bowery NG (1981): Autoradiographic localization of $GABA_B$ receptors in rat cerebellum. Nature 294:584–587.

Young WS III, Kuhar MJ (1979): Autoradiographic localization of benzodiazepine receptors in the brains of humans and animals. Nature 280:393–394.

Young WS III, Kuhar MJ (1980): Radiohistochemical localization of benzodiazepine receptors in rat brain. J Pharmacol Exp Ther 212:337–346.

Young WS III, Niehoff DL, Kuhar MJ, Beer B, Lippa AS (1981): Multiple benzodiazepine receptor localization by light microscopic radiohistochemistry. J Pharmacol Exp Ther 216:425–430.

Benzodiazepine/GABA Receptors and Chloride Channels:
Structural and Functional Properties, pages 315–335


# 17

# Benzodiazepine/GABA Receptor Binding In Vitro and In Vivo in Analysis of Clinical Disorders

Harry T. Chugani and Richard W. Olsen
*Departments of Neurology, Pediatrics, and Pharmacology, UCLA School of Medicine, Los Angeles, CA 90024*

## INTRODUCTION

Benzodiazepines and GABA-mimetics are widely used clinically and are effective agents in the treatment of various neurological and psychiatric disorders. The mechanisms of action of these drugs were not well understood until the discovery and characterization of a macromolecular receptor complex with distinct, yet interactive, binding sites for benzodiazepines, GABA (and GABA-like agents), and barbiturates [Olsen, 1981]. The receptor complex also binds picrotoxinin and other related convulsants [Ticku, Chapter 9 this volume]. Together, these receptors regulate the influx and efflux of chloride ions through its ionophore. Whereas all benzodiazepine receptors are probably a component of this postsynaptic macromolecular complex, the same apparently is not true for all GABA receptors, which also can occur separate from the complex [Olsen, 1981].

Because of the clinical efficacy of benzodiazepines and GABA-mimetics, it has been postulated that an alteration of one or more components of the benzodiazepine/GABA receptor complex may occur in some of the disease states where these drugs are of therapeutic value. Consequently, an understanding of these receptor changes in the central nervous system (CNS) might provide mechanistic insight into the disease process and lead to more effective therapeutic interventions. Several approaches have been taken to test whether alterations in benzodiazepine or GABA binding are affected in clinical entities. One approach attempts to measure the binding of different ligands to these receptors in animal models of disease. A second approach is to perform similar binding studies in human tissue obtained either postmortem or through surgical resection. In both animal and human studies, receptor binding can be assayed either in homogenates of tissue obtained from different brain regions, or by quantitative autoradiographic techniques, which have the advantage of visualizing the anatomy or localization of tissue binding sites. A third approach, which is currently being developed, is the direct and noninvasive measurement of in vivo receptor binding characteristics in the patient using appropriately labeled ligands and currently available imaging techniques, such a positron emission tomography (PET) [Phelps et al., 1982].

Using one or more of these approaches, altered benzodiazepine and/or GABA receptor properties have been sought in a number of clinical disorders, which include Huntington's disease, various types of epilepsy, Alzheimer's disease, alcoholism, Parkinson's disease, and others. This chapter is a review of these efforts and a discussion of some of the methodological issues encountered in performing these studies. Unless otherwise stated, our discussion will be limited to benzodiazepine and GABA receptors occurring at the postsynaptic level. A review of baclofen-sensitive, bicuculline-insensitive, and benzodiazepine-insensitive GABA receptors is beyond the scope of this chapter but can be found described in the chapters by Enna and Karbon and by Johnston in this volume. Finally, recent progress made in the field of benzodiazepine and GABA receptor imaging with positron emission tomography, both by other investigators and ourselves, will be discussed.

## HUNTINGTON'S DISEASE

Huntington's disease (HD), or Huntington's chorea, is an autosomal dominantly inherited progressive neurological degenerative disorder characterized by dementia, severe choreiform movements, and psychiatric symptoms [Martin, 1984]. Neuropathological features include neuronal loss in the caudate nucleus, putamen, and globus pallidus; more specifically, there is degeneration of striatal interneurons [Bird and Iversen, 1974] and striatofugal projections [Perry et al., 1973; Bird, 1976; Kanazawa et al., 1977]. Biochemical analysis of striatal tissue obtained from terminal patients with HD revealed an 85% decrease in activity of glutamic acid decarboxylasc, thc GABA-synthesizing enzyme [Bird and Iversen, 1974; Stahl and Swanson, 1974], in addition to a deficiency of GABA in striatum, globus pallidus, and substantia nigra [Perry et al., 1973].

Although it is now generally accepted that changes in other neurotransmitter systems besides GABA can be seen in HD [reviewed by Martin, 1984], these early biochemical studies suggested, for the first time, that there might be disruption of a neurotransmitter system within the brains of patients with HD, in a manner analogous to that previously reported for Parkinson's disease [Bernheimer et al., 1973]. As a result, the potential for finding new therapeutic approaches led to a number of pharmacological studies of the GABA system in HD. Using brain homogenates obtained from postmortem tissue of patients dying with HD, Enna et al. [1976a, b] found no differences in sodium-independent [$^3$H]GABA binding in the caudate nucleus, globus pallidus, putamen, and cortex between patients and similarly obtained control brain. However, in the substantia nigra, sodium-independent [$^3$H]GABA binding in choreic brains was almost twice that of control [Enna et al., 1976b]. The authors postulated this increase in binding to be the result of a "denervation supersensitivity" phenomenon occurring as a consequence of degeneration of the striatonigral GABAergic pathway. A possible alternate explanation which had to be considered was that the increased [$^3$H]GABA binding may simply have reflected diminished volume of substantia nigra in an atrophied choreic brain. The question of whether receptor alterations occurred as a result of medications used in these patients was not discussed, and the medications taken prior to death were not listed [Enna et al., 1976a, b). Nevertheless, the "denervation supersensitivity" interpretation was supported in subsequent studies performed by Waddington and Cross [1980], where kainic acid lesioning of the caudate nucleus in the rat produced a rise in the number of high-affinity [$^3$H]GABA binding sites in nigral homogenates with no change in the affinity constant. Similarly, postmortem nigral tissue from patients with HD displayed an elevation of high-affinity [$^3$H]GABA binding sites, with no change in affinity [Waddington and Cross, 1980] (see also discussion of autoradiography studies on tissue slice binding described below).

These findings were in sharp contrast to a study [Lloyd et al., 1977a] in which [$^3$H]GABA binding was 70–80% diminished in the caudate nucleus and putamen of choreic brains as compared to control. In addition, although no differences were observed in the parietal cortex, [$^3$H]GABA binding in the cerebellar cortex of patients was almost twice that of control, largely as a result of higher affinity of the cerebellar binding sites for [$^3$H]GABA in patients with HD [Lloyd and Davidson, 1979]. The discrepancy between this study and previous studies was difficult to explain, and led to the postulation of a subgroup of patients with HD where striatal [$^3$H]GABA binding is decreased [Lloyd et al., 1977a], although another possible explanation is that different procedures for membrane preparation and washing were employed. Indeed, when extensively washed membranes were used, Van Ness et al. [1982] were able to confirm both the decrease of [$^3$H]GABA binding sites in caudate-putamen and an increase in the number of these sites in the substantia nigra of HD brains. However, the increase in binding affinity in the cerebellum was not detectable in thoroughly washed membranes, suggesting that the GABA receptors themselves are not altered. The difference in binding affinity in unwashed cerebellar membranes was ascribed to differences in membrane phospholipid environment between HD and normal cells [Lloyd and Davidson, 1979], or merely to differences in levels of endogenous GABA itself [Olsen et al., 1980]. On the other hand, the decrease in striatal GABA receptor density, using washed or unwashed membranes, has been confirmed repeatedly [Iversen et al., 1979; Reisine et al., 1979a; Olsen et al., 1979].

In order to ascertain whether benzodiazepine receptor binding might also be altered in HD, Reisine et al. [1979b] measured [$^3$H]flunitrazepam ([$^3$H]flu) binding in tissue homogenates obtained from postmortem choreic brains and control brains. In the putamen, there was both a decrease in $B_{max}$ as well as affinity of benzodiazepine receptors for [$^3$H]flu in HD brain when compared to control. In contrast, benzodiazepine receptor density was increased in cerebellum and frontal cortex of HD patients. Furthermore, whereas the $K_D$ for [$^3$H]flu binding was lower in the putamen than in cerebellum and frontal cortex of control brain, such regional differences in affinity were not observed in HD brains. The authors postulated that decreased affinity for [$^3$H]flu in choreic putamen could be explained by GABA depletion [Perry et al., 1973], since GABA is known to increase the affinity of benzodiazepine receptors for [$^3$H]flu [Wastek et al., 1978]. In support of this interpretation is the observation that in brain regions such as cerebellum and frontal cortex of HD patients, where GABA levels are not much lower than in controls [Perry et al., 1973], the affinity of benzodiazepine receptors for [$^3$H]flu was not different from control [Reisine et al., 1979b]. Furthermore, the influence of regional GABA concentration on benzodiazepine receptor affinity could be eliminated by extensive washing of the tissue homogenates prior to incubation with [$^3$H]flu [Reisine et al., 1980]. However, despite extensive washing, the HD brains continued to display increased benzodiazepine receptor numbers in the substantia nigra and decreased numbers in the putamen [Reisine et al., 1980]. These findings are consistent with the loss of benzodiazepine receptors in the striatum of HD patients as a result of cell loss, and the development of "denervation supersensitivity" in substantia nigra, as previously postulated [Enna et al., 1976b]. The finding of increased benzodiazepine receptor density in cerebellum and frontal cortex, however, was more perplexing. Thus it was speculated that decreased levels of an endogenous ligand for the benzodiazepine receptor may have played a role [Reisine et al., 1979b]. An important potential therapeutic strategy that emerged from these studies was the combined use of GABA mimetics and benzodiazepines in patients with HD. This approach was supported by a later study showing that the ability of GABA to potentiate [$^3$H]flu binding in putamen and substantia nigra of control brain is not lost in choreic brains [Reisine et al., 1980].

Recently, Penney and Young [1982] employed quantitative autoradiography to study benzodiazepine and GABA receptors in HD. This technique permits a higher resolution of regional receptor alterations that is not possible with assays of tissue homogenates, where dissection of the tissue sample into macroscopic regions is required. In a single patient, they demonstrated decreased [$^3$H]muscimol and [$^3$H]flu binding in ventrolateral thalamus and striatum compared to a control brain, and increased binding in both the medial and lateral globus pallidus [Penney and Young, 1982]. These findings are consistent with the notion that the loss of striatal cells can lead to a supersensitivity of benzodiazepine and GABA receptors in lateral and medial pallidum, both of which are striatal projection areas. Further studies using similar techniques in two early cases of HD revealed that the increased [$^3$H]muscimol and [$^3$H]flu bind ing in lateral pallidum and decreased binding in putamen can occur even prior to any extensive cell loss and atrophy in these structures [Walker et al., 1984]. Whereas alterations in receptor binding in HD formerly had been thought to be a consequence of striatal neuronal loss, this study raises the possibility of early dysfunction of striatal neurons prior to their death [Walker et al., 1984]. If this were indeed the case, it might be possible to manipulate pharmacologically the diseased neurons even prior to their demise.

Finally, brief mention must be made of the kainic acid animal model of HD [Coyle et al., 1977; 1978a]. This model employs the administration of small doses of kainic acid, an excitotoxic rigid analog of glutamic acid, into rat striatum, resulting in destruction of cell bodies and dendrites near the injection site but sparing axons and terminals arising from cells located in other brain regions [Coyle et al., 1978b; 1981]. A number of biochemical and pharmacological features that exist in HD brains can also be found in the kainic acid model [Sperk and Schlogl, 1979; Pan et al., 1983]. With the use of quantitative autoradiography and this animal model, one can learn much about both topographical and temporal aspects of receptor alterations in HD. For example, in the study by Pan et al. [1983], one week following striatal kainic acid lesion, the number of [$^3$H]muscimol receptors was decreased in the striatum and in the ipsilateral anteroventrolateral and ventromedial thalamic nuclei, but increased in the ipsilateral globus pallidus, entopeduncular nucleus, and substantia nigra pars reticulata. No changes in binding affinities were observed. The authors postulated that, whereas increased GABA receptor numbers in the striatal projection areas were probably due to denervation hypersensitivity, the decreased receptor numbers observed in the thalamus presumably resulted from disinhibition of GABA projections from entopeduncular nucleus and substantia nigra to thalamus. It is clear that the combined use of the kainic acid model and postmortem HD brain tissue in receptor binding studies offers an opportunity to study temporally related receptor alterations in HD, with the potential of providing a correlation between symptoms and altered neuronal circuitry in HD. Such studies might eventually offer a rational approach to pharmacological intervention in both the early and late stages of HD.

## EPILEPSY

The role of GABA neurotransmission in human epilepsy and animal seizure models has received widespread attention in epilepsy research. It has long been known that compounds that elevate brain GABA concentration tend to be anticonvulsants, whereas those that interfere with GABA neurotransmission are often convulsants [Löscher et al., 1983; Löscher, 1984]. The protective role of GABA in seizures is supported by the finding that following a convulsion, there is an increase in potassium-evoked GABA release in rat brain [Liebowitz et al., 1977], although it was not determined whether the release was synaptic or preterminal. Increased seizure threshold to GABA antagonists also follows a seizure [Nutt, 1981]. These and other findings have led several authorities in epileptology to postulate that GABA disinhibition could provide a mechanism underlying the development of epilepsy [Meldrum, 1975; Wood, 1975; Tower, 1976]. Consequently, there has been an effort to document GABA deficiency in human epilepsy and animal seizure models. For example, in animal studies using the alumina gel model of seizures in sensorimotor cortex of monkeys, immunocytochemical techniques have revealed a notable decrease in glutamic acid decarboxylase (GAD), the GABA synthesizing enzyme [Ribak et al., 1979]. The degree of GAD reduction observed in these experiments is thought to be sufficient to affect the inhibitory synaptic control of pyramidal neurons from normal excitatory input, thus resulting in a hypersensitivity within the circuit. Pharmacological studies using the same epilepsy model showed decreased GAD activity and GABA concentration in epileptic cortex compared to the contralateral side; GABA receptor binding was also diminished, being more marked and widespread than other neurochemical markers [Bakay and Harris, 1981]. In the cobalt-induced focal epilepsy model, GABA concentration was decreased and GABA receptor binding increased within the epileptic foci in rat cortex [Ross and Craig, 1981]. Human temporal lobe tissue identified as epileptic foci by stereo-EEG was shown to have lower GAD and GABA receptor binding [Lloyd et al., 1984]. Such studies suggest that there may be a deficit of GABA function presynaptically (GAD) or postsynaptically (receptor) in focal epilepsies.

The association of benzodiazepine receptors with GABA receptors and the chloride ionophore has provided a convenient explanation for the fact that many benzodiazepines are potent anticonvulsants. However, other benzodiazepine receptor ligands (inverse agonists) produce an opposite action on the benzodiazepine binding site, i.e., they are convulsants, or potentiate the convulsant effect of other convulsants [Oakley and Jones, 1980; Cowen et al., 1981; Braestrup et al., 1982; Jensen and Petersen, 1983; Braestrup et al., 1984]. Recent efforts have focused on the development of more specific benzodiazepine and GABA receptor agonists for anticonvulsant use in clinical epilepsy. One such drug, progabide, appears to be promising from animal data [Joy et al., 1984] and is currently undergoing clinical trials [Morselli and Bossi, 1984]. The subject of GABA replenishment as treatment for epilepsy has been reviewed recently [Fariello and Ticku, 1983].

With the advent of pharmacological techniques for measuring receptor binding properties, benzodiazepine and GABA receptor binding studies dealing with mechanistic issues in animal models of epilepsy have been performed. The idea that decreased GABA or benzodiazepine binding, implying either decreased receptor numbers or decreased affinity of receptor for the ligand, may emerge as a common finding in epilepsy, seemed attractive. This would suggest an inadequacy in neuronal inhibition, and susceptibility to seizure, thus explaining the efficacy of GABA or benzodiazepine in the treatment of epilepsy. As will be seen below, the observations that have emerged from GABA and benzodiazepine receptor binding studies have turned out to be much more complicated, and in some cases conflicting.

One of the earliest studies of benzodiazepine receptor binding in epilepsy showed that in rats, following an acute seizure induced by electroshock or pentylenetetrazole, there is an increase in the number of benzodiazepine binding sites in cortical synaptic membranes as measured by [$^3$H]diazepam binding [Paul and Skolnick, 1978]. The increase peaks at 15–30 minutes after the seizure and returns to baseline by 60 minutes. However, these changes were not as transient in another study, where 24 hours after a seizure induced in rats either by amygdala-kindling or electroshock, there was an increase in the number of hippocampal benzodiazepine receptors. The increase correlated with the number of seizures in the kindling sequence or electroshock series [McNamara et al., 1980]. Further studies using the same seizure models localized the increased benzodiazepine numbers to the dentate granule cell and molecular layers within the hippocampus [Valdes et al., 1982]. The measured increase was thought to be associated with the postictal refractoriness to further seizure following a kindled seizure. $[^3H]$Muscimol binding in the dentate gyrus is also increased 24 hours after the third class 5 kindled seizure, but neither benzodiazepine nor $[^3H]$muscimol binding was different from controls 7 days or 24 days after the seizure [Shin et al., 1983]. These investigators therefore believe that the observed benzodiazepine and GABA receptor changes are a direct result of the seizure itself and cannot be used to explain altered seizure thresholds in the kindling process per se [McNamara, 1984]. However, the long-term effect (2 weeks and 2 months) of amygdala-kindling on benzodiazepine receptors, according to authors from another laboratory [Niznik et al., 1983a; Burnham et al., 1983], is to decrease the number of these sites in ipsilateral cortex and hypothalamus, perhaps accounting for the seizure-susceptible state in kindled animals. Still another laboratory found that, 2 weeks after the last amygdala-kindled seizure in rats, there were no changes in $[^3H]$flu binding in cortex and hypothalamus, but increased receptor numbers were detected in amygdala and hippocampus [Tuff et al., 1983a]. These results are consistent with those of McNamara [1984]. The increase in receptor binding was shown to be accompanied by increased inhibition using electrophysiological measurements [Tuff et al., 1983b]. One possible reason for the discrepancy between the study of Tuff et al. [1983a] and those of Niznik et al. [1983a] and Burnham et al. [1983b] is that the former study used well-washed membranes but the latter two studies used unwashed membranes. Indeed, it appears that both patterns of receptor alterations are reproducible depending upon whether washed or unwashed membranes are used [Niznik et al., 1983b]. It is possible that some of the results with unwashed membranes are due to the presence of endogenous GABA. While other potentially meaningful differences may be involved in the studies with unwashed membranes, such studies have the obligation of demonstrating that the simple explanation of altered GABA is not responsible. Clearly, further work is indicated to clarify the mechanistic issues of kindling as related to alterations in benzodiazepine and GABA receptors.

In other models of epilepsy, decreased numbers of benzodiazepine receptors were found in substantia nigra and periaqueductal gray matter of a seizure-susceptible strain of Mongolian gerbil, a genetic model of generalized epilepsy [Olsen et al., 1985a]. Similarly, decreased GABA receptor binding has been seen in several brain regions, especially brainstem, of audiogenic seizure mice (DBA/2J) compared to other strains which have a much lower susceptibility to audiogenic seizures [Ticku, 1979; Horton et al., 1982]. Benzodiazepine binding assessed by autoradiography was low in DBA/2J mice in substantia nigra, mesencephalic and pontine gray, laterodorsal tegmental nucleus, and inferior colliculus central nucleus [Olsen et al., 1985b]. This deficit in benzodiazepine/GABA receptor levels in two genetic animal models of generalized seizures is consistent with the hypothesis of diminished GABA-mediated inhibition in at least some forms of epilepsy. Furthermore, the lower benzodiazepine/GABA receptor density in substantia nigra of these two animal models is of particular interest in light of evidence that the substantia nigra may be the site of action of GABA-augmenting anticonvulsants and is an essential component of the brain circuitry involved in the spread of seizures generated by electroshock, convulsant drugs, or kindling [Gale, 1984; McNamara et al., 1984].

Benzodiazepine receptor numbers were found to be decreased in the dentate gyrus of rats with electrolytic lesions of entorhinal cortex [Kraus et al., 1983], a model of limbic seizures. The decreases were transient and occurred during the period of maximum epileptiform activity. Furthermore, since the receptor alterations could be prevented with phenobarbital treatment, it appears that they are a result of the seizure itself rather than entorhinal cortex lesioning [Kraus et al., 1983]. However, since GABA mechanisms mediate recurrent inhibition in dentate stratum granulosum, the net result of decreased benzodiazepine receptor numbers in this area is to increase excitability of the dentate granule cells, thus promoting seizure activity during the period of maximum epileptiform activity in the entorhinal lesion model [Kraus et al., 1983].

In our own laboratory, we are using quantitative autoradiography to evaluate whether high-affinity postsynaptic GABA receptors (labeled with [$^3$H]muscimol) and benzodiazepine receptors (labeled with [$^3$H]flu) are altered in human epileptic tissue removed by en bloc anterior temporal lobectomy. These patients suffer from intractable hippocampal seizures, and, prior to surgery, in vivo physiological recordings are made during several typical clinical seizures, enabling us to define a unilateral seizure focus [Babb and Brown, 1985]. The resected tissue is immediately deep-frozen in the operating room, and typically consists of epileptic hippocampus (including the dentate gyrus) and temporal neocortex. Control brain tissue is obtained from cadavers having minimal autolysis time prior to deep-freezing. Our preliminary findings demonstrate that this approach provides accurate and reproducible measures of receptor densities and binding affinities compared to published values. However, at the time of this writing, we have not entered sufficient patients into the study to make statistical comparisons between epileptic and control tissue. As more data are accumulated, we will be able to test whether "decreased inhibition" occurs in the seizure focus and adjacent cortex. As mentioned earlier, quantitative autoradiography permits a higher resolution of receptor binding than is possible with the tissue homogenate method. The resolution of the autoradiograms in our preliminary studies is illustrated in Figures 1 and 2.

## OTHER DISEASES

### Alcoholism

Because of the cross-tolerance between benzodiazepines and alcohol, and the clinical effectiveness of benzodiazepines in the treatment of alcohol withdrawal states, benzodiazepine and GABA receptor alterations have been sought in alcoholism. In one study performed on postmortem brain tissue obtained from alcoholics and nonalcoholic controls, no differences in [$^3$H]flu binding were observed between the two groups, but GABA receptors (determined by [$^3$H]muscimol binding) were increased in alcoholic brain [Tran et al., 1981]. Scatchard analysis revealed the [$^3$H]muscimol binding increase to be due to increased receptor numbers rather than a change in affinity. Since chronic ethanol treatment in laboratory animals results in diminished brain GABA levels [Sytinsky et al., 1975], the observed increase in numbers of GABA receptors in alcoholics may be a response to decreased availability of GABA [Tran et al., 1981].

A number of investigations of ethanol effects on GABA and benzodiazepine receptor binding have yielded conflicting results. For example, [$^3$H]diazepam binding was enhanced by ethanol in some in vitro studies [Burch and Ticku, 1980; Davis and Ticku, 1981], but benzodiazepine receptor binding was unchanged in other studies [Marangos and Martino, 1981; Volicer and Biagioni, 1982]. Similarly, ethanol enhanced GABA receptor binding (assayed in vitro) when acutely administered to rats in vivo [Ticku 1980; Ticku and Burch, 1980], an effect not reproducible in vitro [Williams and Risley, 1979; Marangos and Martino, 1981].

The exact nature of ethanol interaction with the benzodiazepine/GABA system remains unclear, and further studies are indicated to determine whether such an interaction may play a role in alcohol addiction. A recent study, however, has suggested that the interaction between ethanol and benzodiazepine/GABA mechanisms may occur not at the level of the benzodiazepine/GABA receptor complex, but at a step subsequent to receptor occupation [Greenberg et al., 1984].

### Alzheimer's Disease

Alzheimer's disease is currently the most common cause of senile dementia, and because of the rising prevalence of this disorder,

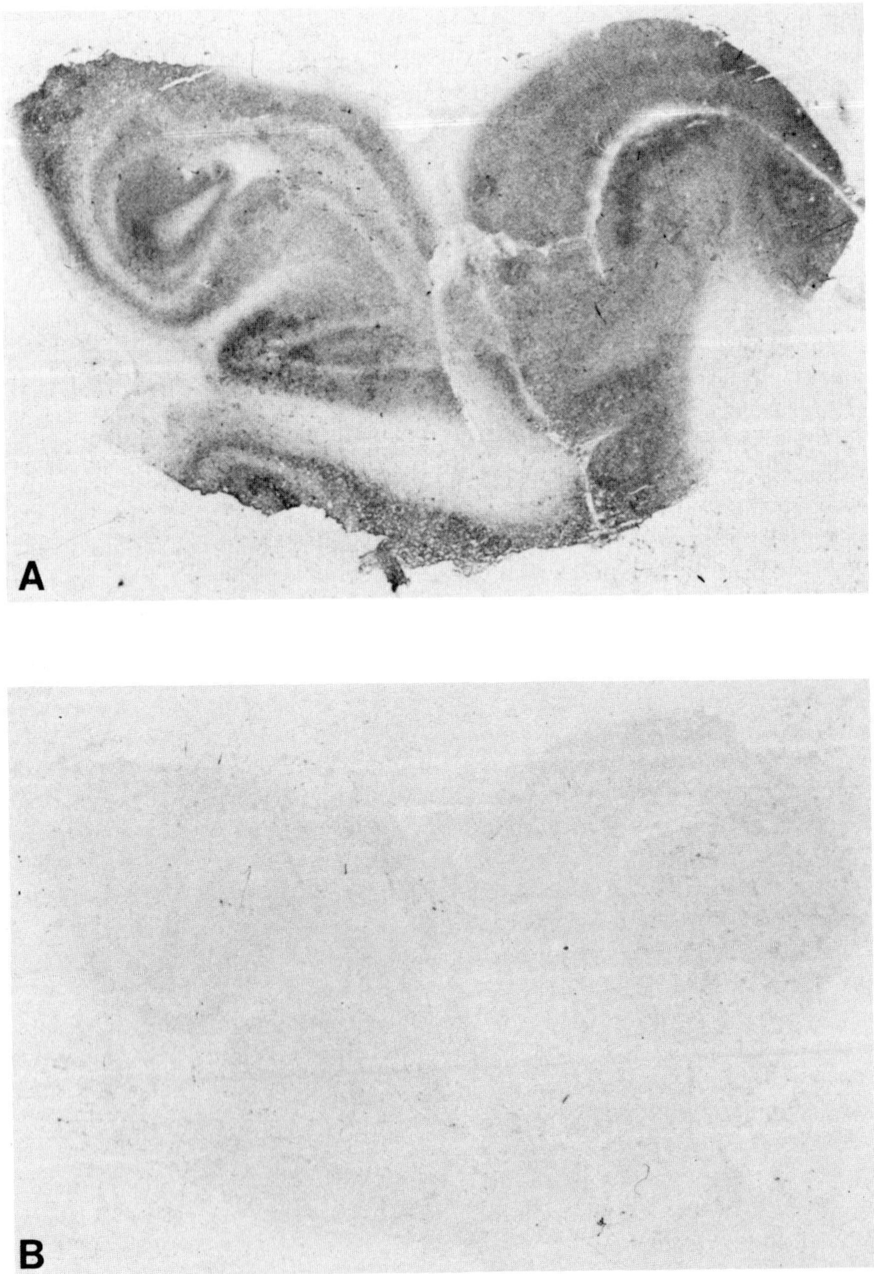

**Fig. 1.** *Autoradiograms generated by incubation of postmortem human brain sections for 40 minutes in 0.17 M Tris-HCl buffer (pH 7.4 at 4°C) containing 2 nM [$^3$H]flunitrazepam, followed by exposure to tritium-sensitive film. (a) Total benzodiazepine receptor binding. Note high uptake in stratum granulosum of dentate gyrus and in the adjacent cerebral cortex. (b) Nonspecific binding, in which specific binding was displaced by coincubation of [$^3$H]flunitrazepam with 1 μM clonazepam. By using various concentrations of [$^3$H]flunitrazepam and quantitative densitometry, a Scatchard analysis for different brain regions can be performed.*

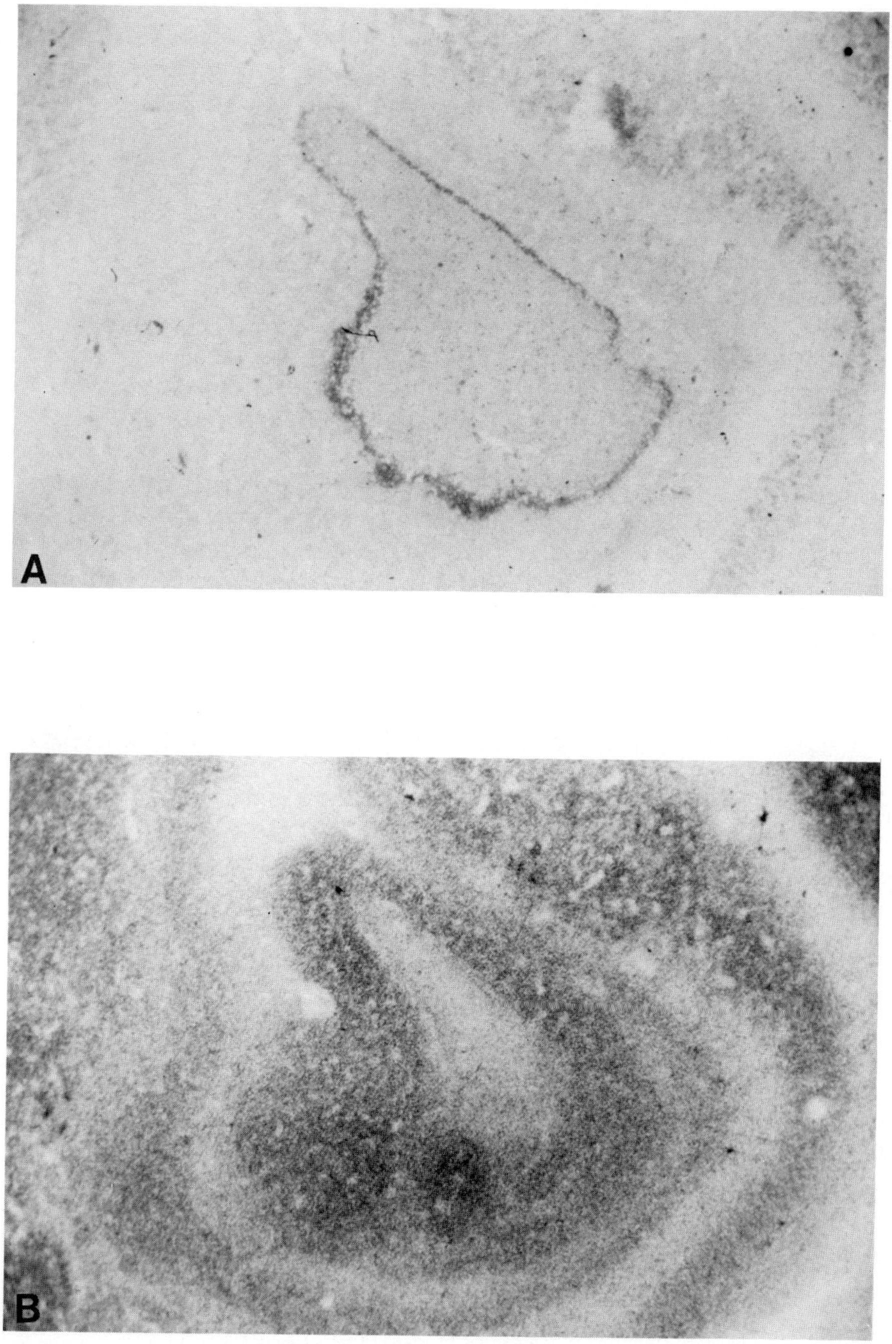

**Fig. 2.** *Mirror sections of postmortem human hippocampal tissue illustrating (a) cells stained with cresyl violet, (b) benzodiazepine receptor autoradiogram generated by incubation of tissue section in 2 nM [$^3$H]flunitrazepam followed by exposure to tritium-sensitive film. The combination of cell counting and quantitative autoradiography allows the determination of whether altered receptor density (if present) in temporal lobe epileptics is related to a change in cell number (see text).*

much effort has been devoted to the search for effective therapeutic intervention or improved management [Terry and Katzman, 1983]. Numerous studies on Alzheimer's disease have focused upon cholinergic mechanisms in the brain [Davies and Maloney, 1976; Perry et al., 1977; White et al., 1977], and different methods of acetylcholine replenishment have been attempted in these patients, with varying success [reviewed by Terry and Katzman, 1983]. Not much attention has been given to the benzodiazepine/GABA system, which probably does not play a major role in Alzheimer's disease. Nevertheless, in one study, [$^3$H]GABA binding was performed on tissue homogenates of brain regions obtained postmortem from six patients dying with Alzheimer's disease, and compared with similarly obtained tissue from age-matched controls [Reisine et al., 1978]. [$^3$H]GABA binding was reduced by 43% in the caudate nucleus, and by 50% in frontal cortex of Alzheimer's disease patients when compared to controls, suggesting a loss of GABA-modulated cells in these regions [Reisine et al., 1978].

### Parkinson's Disease

Although degeneration of the nigrostriatal dopaminergic pathway is now widely recognized to be one of the hallmarks of Parkinson's disease [Bernheimer et al., 1973], the fact that dopamine replacement is only partially effective in controlling the symptoms of these patients has inspired the search for other neurotransmitter alterations. In one such study, [$^3$H]GABA binding was found to be decreased in tissue homogenates of substantia nigra and hippocampus from patients dying with Parkinson's disease when compared to control tissue [Lloyd et al., 1977b]. Whereas the loss of [$^3$H]GABA binding in substantia nigra would imply the presence of GABA receptors in the degenerated nigral neurons or dendrites, diminished [$^3$H]GABA binding observed in the hippocampus could not be explained. However, since dementia is seen in a subset of patients with Parkinson's disease, it would be of interest to determine whether receptor binding properties (not necessarily limited to the benzodiazepine/GABA system) differ between the two subsets. Such a study, to our knowledge, has not been performed.

## POSTMORTEM CHANGES OF BENZODIAZEPINE/GABA RECEPTORS

Ever since the measurement of neurotransmitter receptor binding properties became feasible, there has been concern as to whether these measurements accurately reflected the in vivo properties of the receptor, or whether the storage and handling of tissue affected binding characteristics. This consideration is particularly important in receptor binding studies of human disease where postmortem tissue is utilized, since, for practical and ethical reasons, it is almost impossible to ensure rapid freezing of the tissue after death. Thus, depending widely upon the clinical setting, there is usually a variable exposure period (often several hours) of the cadaver to room temperature, followed by another variable period (hours to days) of morgue temperature storage (4°C) before deep freezing becomes possible. Some brain banks will have available a limited number of brains where the autolysis period was shorter (12–18 hours) between death and deep freezing, but even this is not optimal.

It comes somewhat as a surprise that only a few studies have been performed to investigate the nature and degree of receptor changes under various postmortem conditions. Enna et al. [1976a], measuring [$^3$H]GABA binding in brains from patients with HD, performed additional studies in rats where they were unable to demonstrate large differences in [$^3$H]GABA binding between brain tissue exposed to room temperature for 6 hours and brain frozen immediately after death. Some investigators have attempted to circumvent the problem by taking extreme care to ensure that both control and experimental tissue were collected with no appreciable differences between time to freezing and other handling [Reisine et al., 1977]. Other investigators, in contrast, only utilized brains with acceptable and specified autolysis times [Enna et al., 1977; Speth et al., 1978].

Only a few studies have directly focused upon postmortem effects on receptor binding characteristics. In one such study [Lloyd and Dreksler, 1979], it was shown that, at least for [$^3$H]GABA binding, there was little variation within the range of 4–33 hours between death and freezing of the human brain. In addition, binding was unaffected by deep-frozen storage for up to 64 months. Recently, Whitehouse et al. [1984] systematically studied the effects of postmortem delay and temperature on the binding of several ligands, including [$^3$H]flu, to rat brain tissue. The postmortem handling procedure in this study was designed to simulate the human autopsy process. It was discovered that, whereas the binding of several ligands decreased with increasing autolysis times, [$^3$H]flu binding actually increased, becoming significant at 48 hours postmortem delay. Scatchard analysis revealed that the increased binding represented increased affinity, with no change in receptor density. The authors postulated this increase in affinity to be a result of residual tissue GABA, which has been shown to increase the affinity of the benzodiazepine receptor for its ligands [Reisine et al., 1980]. In support of such an interpretation is the earlier observation that at least in dog brain, brain GABA levels increase during autolysis [Tews et al., 1963]. Again, however, such "artifacts" can be avoided by using tissue samples from which the endogenous GABA has been removed by thorough washing. In our laboratory [unpublished results], no difference was found in [$^3$H]flu binding for rat brain incubated at 22°C for 16 hours after death before homogenization and that prepared immediately after sacrifice, assaying washed membranes. In autoradiography, endogenous GABA can also affect [$^3$H]muscimol and [$^3$H]flu binding [Unnerstall et al., 1981], but this can be minimized by 20–30-minute preincubations of the tissue slices before assay (see Wamsley et al., Chapter 16, this volume).

Although the [$^3$H]flu and [$^3$H]muscimol binding studies being performed in our laboratory on epileptic tissue (vide supra) have the advantage of being deep-frozen in the surgical suite immediately after resection, our control tissue is nevertheless subject to the same problems of autolysis described above. One strategy that was employed in a recent study [Walker et al., 1984] to control for postmortem changes of receptors is to use binding ratios between structures within the same brain, since any postmortem receptor alterations are apt to be generalized and not specific to particular regions. This approach, which is particularly suited for quantitative autoradiography, will undoubtedly be used widely in future studies of receptor alterations in disease.

## INTERPRETATION OF BENZODIAZEPINE/ GABA RECEPTOR CHANGES IN DISEASE

In interpreting ligand binding studies of benzodiazepine and GABA (and indeed all other) receptors, a number of issues arise that must be carefully considered in order not to overinterpret measured differences between disease and control groups. Many of the discrepancies that are existent in the currently available data on receptors can be explained by differences in techniques employed by different laboratories in tissue handling and receptor assays. Consequently, it is important to stress the need for uniformity of these procedures. In addition to these technical issues, several other factors that play an important part in the interpretation of ligand binding studies deserve attention, and will now be discussed.

### Cell Loss

If a population of cells were to degenerate for any reason, the receptors which are an integral part of these cells will also be lost, thus resulting in decreased binding. For example, in Parkinson's disease, decreased

[$^3$H]GABA binding in the substantia nigra is readily explained by cell loss in that structure [Lloyd et al., 1977b]. Cell loss in one structure may also lead to receptor changes at a different site. Thus, increased [$^3$H]GABA binding in the substantia nigra of patients with HD is thought to be a result of striatal cell loss leading to degeneration of striatonigral GABAergic pathway, and a compensatory increase in GABA receptors in the substantia nigra [Enna et al., 1976b]. An apparent increase in binding density may occur in cases of tissue shrinkage, gliosis, or degeneration of neurons that do not contain the receptors being measured. Thus the anatomy of the structure involved is also relevant.

It is, therefore, important to combine receptor binding studies with histological studies in the same specimen when one is studying receptor alterations in disease. Such an approach is being taken in our laboratory where human epileptic tissue removed by temporal lobectomy is subjected to both receptor binding studies using quantitative autoradiography and histological cell counts in adjacent sections. It was also due to a combination of these techniques that led Walker et al. [1984] to hypothesize that dysfunction of striatal neurons prior to cell death may occur early in HD, since in those studies it appeared that increased [$^3$H]muscimol and [$^3$H]flu binding in lateral pallidum and decreased binding in putamen occurred prior to notable cell loss in these structures.

### Effect of Drugs

One of the most frequently encountered difficulties in the interpretation of receptor ligand binding studies on postmortem human tissue is whether the observed alterations are drug-induced, since most of the tissue utilized in such studies would have been obtained from patients taking one or more medications.

In the HD studies described earlier in this chapter, an important consideration in interpreting the increased [$^3$H]GABA binding in substantia nigra was whether those changes were induced by chronic administration of haloperidol or other dopamine blockers commonly prescribed for this disorder. This is particularly crucial in light of the finding of enhanced [$^3$H]GABA binding in the substantia nigra of rats treated chronically with dopamine receptor blockers [Gale, 1980]. Likewise, if drug-induced receptor changes occur, they may be observed in several brain regions. These issues can sometimes be clarified by employing animal models; for example, in the kainic acid model of HD, enhanced [$^3$H]muscimol binding in the substantia nigra of rats occurred in the absence of pharmacological dopamine blockade [Pan et al., 1983].

The issue of drug-induced receptor changes is also of importance in benzodiazepine and GABA receptor studies of human epileptic tissue, since these patients are commonly prescribed a variety of anticonvulsants known to affect the benzodiazepine/GABA receptor complex. Fortunately, there already exists a fairly large body of data dealing with the effects of various anticonvulsants on the different components of the benzodiazepine/GABA receptor complex. For example, in rats, the acute effect of diazepam on [$^3$H]flu binding is to cause an increase in receptor density, but a decrease in receptor affinity for the ligand [Speth et al., 1979]. In contrast, chronic flurazepam, in higher than usual clinical doses, decreases the number of binding sites for [$^3$H]diazepam with no change in receptor affinity [Rosenberg and Chiu, 1979]. No decrease was observed for clinical doses of flurazepam [Möhler et al., 1978; Braestrup et al., 1979]. Phenytoin, another frequently prescribed anticonvulsant, when given in vivo to rats, enhances [$^3$H]diazepam binding in brain homogenates by increasing the affinity [Tallman and Gallager, 1979]. Similarly, a number of direct GABA agonists have been shown to decrease GABA receptor binding when administered chronically [Ticku et al., 1983]. Nevertheless, whenever indicated, it is important to conduct parallel studies in laboratory animals to investigate the effects of drugs on receptor binding properties.

Finally, it is also possible that certain drugs

may indirectly affect a receptor without altering receptor density or affinity. This point is best illustrated by the study of Gallager et al. [1984a], in which chronic benzodiazepine treatment did not affect [$^3$H]flu binding in terms of receptor density or affinity, but resulted in decreased ability of GABA to enhance benzodiazepine binding [Gallager et al., 1984a]. The physiological consequence of such an abnormal benzodiazepine–GABA interaction would be equivalent to decreased receptor density. The affinity of GABA receptor binding in the tolerant animals was increased, on the other hand [Gallager et al., 1984b]. This has been interpreted as evidence that high-affinity GABA receptors represent a nonfunctional, possibly desensitized state, while low-affinity GABA receptors are more likely to represent synaptic function as well as association with benzodiazepine receptors [Unnerstall et al., 1981; Fischer and Olsen, Chapter 12, this volume]. The low-affinity GABA receptors are more difficult to measure but can be assayed with [$^3$H]bicuculline methochloride [Wamsley et al., Chapter 16, this volume]. Thus, with reference to studies on surgically resected human epileptic tissue, normal benzodiazepine and high-affinity GABA receptor binding in the epileptic focus of patients taking benzodiazepines would not necessarily imply normal inhibitory processes at the level of the receptor. It is clear that the interaction between benzodiazepines and GABA must also be investigated in the same tissue before such an inference can be made.

### Effect of Specific Phenomena

In certain conditions alterations of receptors may be the result of a particular phenomenon, such as a seizure or other stressful event, occurring just prior to death. For instance, in the kindling model of epilepsy, some investigators [McNamara, 1984] believe that alterations in GABA and benzodiazepine receptor binding occur transiently as a result of seizure and therefore do not reflect a change in seizure susceptibility induced by kindling. Similarly, acute stress in rats can cause an immediate decline of benzodiazepine receptor binding in cerebral cortex and hippocampus, an effect which can be blocked with chronic diazepam administration [Medina et al., 1983]. Even isolation of mice is associated with a decline of type I benzodiazepine receptors in cerebral cortex, and type II benzodiazepine receptors in diencephalon and cerebellar cortex [Essman and Valzelli, 1981]. Since receptor changes can occur within minutes [Paul and Skolnick, 1978], it is conceivable that certain agonal events may affect benzodiazepine and GABA receptor binding. This is particularly important in receptor studies on human tissue, where resuscitative measures may be undertaken prior to death. Every effort must therefore be made to match for these possible differences in choosing control specimens, and, in the case of animal studies, whenever possible, there should be reasonable standardization of animal housing and handling among laboratories.

### Age Differences

Some receptors change in properties and number with age; therefore, in these cases it becomes important to match specimens from diseased patients with age-matched controls. Such is perhaps the case for benzodiazepine receptors where, in the senescent rat, [$^3$H]diazepam binding capable of being displaced by clonazepam was diminished compared to younger rats [Komiskey and MacFarlan, 1983; but see Pedigo et al., 1981].

### Genetic Alterations

In some instances, differences in receptors occurring between different strains of a species are believed to reflect phenotypic differences, accounting for unique properties of the strain. For example, decreased benzodiazepine receptor density in the midbrain of seizure-susceptible Mongolian gerbils, compared to seizure-resistant strains, may be related to seizure susceptibility [Olsen et al., 1985a]. Other examples of genetic differences in benzodiazepine and GABA receptors have already been cited in this chapter. Depending on the degree of genotypic difference between compared strains, receptor differences may or may not be linked to other phenotypic variations such as seizure susceptibility.

### Statistical Considerations

Olsen et al. [1980] and Van Ness et al. [1982] demonstrated that GABA receptor binding shows a considerable variability in samples from different individuals (animals and humans) and different preparations and storage time. Furthermore, there is a heterogeneity in affinity and resulting curvilinear Scatchard plots, making accurate determinations of the binding parameters (density and affinity) difficult. Therefore, samples to be compared between normal and diseased need to be treated and assayed as similarly as possible. In addition, a suitably large sample number needs to be studied to achieve statistical significance.

## IN VIVO STUDIES OF BENZODIAZEPINE AND GABA RECEPTORS

There have been relatively few studies performed which employ in vivo labeling of benzodiazepine or GABA receptors. Chang and Snyder [1978] administered [$^3$H]flu intravenously to mice. Peak brain radioactivity was achieved 20 minutes after injection. However, at that same time, only 50% of soluble radioactivity represented unmetabolized flunitrazepam. In another study [Williamson et al., 1978a, b], [$^3$H]diazepam was given intravenously to adult rats and homogenates from different brain regions assayed for radioactivity. Maximum binding was observed in the cortex, and lowest activity in the pons and medulla, a rank order in complete agreement with in vitro studies. The ratio of specific to nonspecific binding in these studies was 4:1 [Williamson et al., 1978b], nonspecific binding being determined by pretreatment with pharmacologically active unlabeled benzodiazepines.

Enhancement of benzodiazepine binding by GABA or GABA analogues, a well-known phenomenon in in vitro studies [Martin and Candy, 1978; Tallman et al., 1978], can also be demonstrated in vivo. In one such study, pretreatment of rats with muscimol or aminooxyacetic acid (AOAA), an inhibitor of GABA catabolism, 1 hour prior to intravenous [$^3$H]diazepam led to significant enhancement of total and specific [$^3$H]diazepam binding [Tallman et al., 1979].

Since the benzodiazepines are widely used clinically, and their in vivo binding to brain tissue could now be studied, a number of investigators attempted to correlate in vivo benzodiazepine receptor occupation with pharmacological effect. Paul et al. [1979] showed that in mice the anticonvulsant effect of diazepam against pentylenetetrazol-induced seizures correlated well with occupancy of benzodiazepine receptors by [$^3$H]diazepam. Moreover, total efficacy of diazepam as an anticonvulsant was achieved with only a small fraction (<30%) of benzodiazepine receptors being occupied. In another study of in vivo benzodiazepine receptor binding [Duka et al., 1979], the $ID_{50}$ values of a number of benzodiazepines ($ID_{50}$ being defined as the dose of injected drug required to reduce in vivo [$^3$H]flu binding by 50%) correlated well with their anticonvulsant properties in pentylenetetrazol or picrotoxin-induced seizures. Braestrup and Nielsen [Chapter 7, this volume] summarize their studies correlating the potencies of various benzodiazepines in in vivo binding with anticonvulsant and anxiolytic activity. Recently, in vivo benzodiazepine receptor binding studies have been advocated as a screening procedure for testing the anticonvulsant activity of newly synthesized benzodiazepines [Chweh et al., 1983].

However, in vivo benzodiazepine binding studies have also brought out a number of puzzling issues. For example, there seems to be a discrepancy between the brain concentration of unlabeled flunitrazepam required to displace 50% of [$^3$H]flu binding in vivo and that required in vitro ($IC_{50}$), the former being at least 100-fold greater than the latter. In order to explain this discrepancy, a number of possibilities have been suggested, such as nonspecific binding to plasma proteins, diminished receptor access to benzodiazepines in vivo compared to in vitro, and competition with endogenous ligands or modulators of benzodiazepine receptors in vivo, these being washed out during in vitro studies [Duka et al., 1979]. A major factor is the temperature of assay, since [$^3$H]flu binding is assayed in vitro at 0–4°C and has about a 100-fold lower affinity at 37°C [Speth et al., 1978].

The feasibility of labeling benzodiazepine receptors in vivo has stimulated considerable interest in the development of suitable isotopes for in vivo quantitative imaging of benzodiazepine and GABA receptors in humans, using newly available techniques such as positron emission tomography (PET) [Phelps et al., 1982]. The potential use of this approach is illustrated in the study by Young et al. [1984], where intravenously administered [$^3$H]flu to rats 12 weeks after striatal lesions revealed decreased binding in striatum, and increased binding in globus pallidus and substantia nigra pars reticulata, a pattern similar to that observed by using in vitro [$^3$H]flu quantitative autoradiography. In preliminary studies performed in our laboratory, we have compared autoradiograms generated from in vitro incubation of rat brain slices in solutions containing [$^3$H]flu with those obtained after intravenous administration of [$^3$H]flu. As can be seen in Figure 3, the two images are almost identical, supporting the feasibility of in vivo benzodiazepine receptor imaging.

In screening for appropriate ligands to be used in quantitative receptor imaging with PET, a number of criteria must be satisified. The ligand must:

1. have high lipid solubility and low protein binding so that it crosses the blood–brain barrier easily;
2. bind selectively to the receptor with high affinity;
3. have relatively low nonspecific binding (specific to nonspecific binding ratio should be at least 4:1);
4. remain bound to the receptor for a long enough period to allow imaging with the positron tomograph (i.e., it should have a relatively low dissociation rate);
5. not be significantly metabolized prior to tomographic imaging, or the metabolite(s) must not cross the blood–brain barrier;
6. be relatively simple to synthesize in the

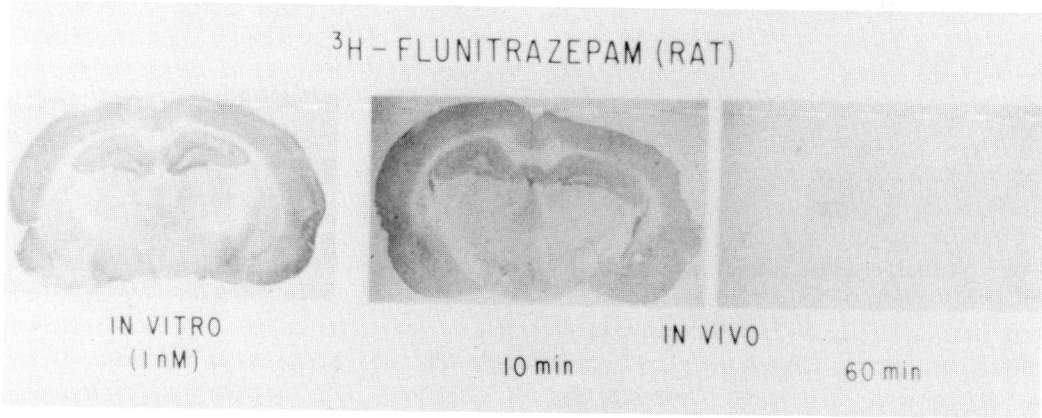


**Fig. 3.** *Autoradiographic comparison of in vitro and in vivo [$^3$H]flunitrazepam binding in the rat. The in vitro autoradiogram was generated by incubation of rat brain sections in 2 nM [$^3$H]flunitrazepam, followed by exposure to tritium-sensitive film. In vivo autoradiograms were generated after intravenous administration of 50 uCi [$^3$H]flunitrazepam. The animals were sacrificed at different time intervals after [$^3$H]flunitrazepam administration and brain sections exposed to tritium-sensitive film. Note the similarity of regional [$^3$H]flunitrazepam binding between in vitro and in vivo (10 min) autoradiograms, and the absence of any labeling at 60 min.*

rapid times required with short half-lived positron isotopes;

7. have relatively high specific activity.

Many of these features can be studied in the laboratory using in vivo autoradiography in animals, so that a thorough understanding of the properties of the selected ligand can be used in developing tracer kinetic models to allow quantitative interpretation of receptor images.

When a ligand satisfying the above criteria becomes available, larger animals such as baboons can initially be studied with the positron tomograph similarly to humans, with the exception that anesthesia or muscle paralysis is often required in the former. One such study [Comar et al., 1979] utilized [$^{11}$C]flunitrazepam (20–25 mCi administered intravenously) of 300–900 Ci/mmol specific activity. Unlabeled lorazepam (1.3 $\mu$mole/kg) was used in parallel experiments to determine nonspecific binding. Due to the poor resolution of the images it was not possible to identify discrete anatomical regions, but a high uptake of the labeled drug in brain compared to extracerebral tissue was found. Approximately 35% of the brain radioactivity was displaced by the lorazepam. The investigators also found that demethylation of the parent compound, which was evident by the presence of labeled $CO_2$ in expired air, ranged from 3–14% of the injected dose. The demethylated product, being pharmacologically active, tended to lead to an underestimation of receptor visualization, since it could no longer be detected by the positron tomograph.

In another study, also in baboons, [$^{11}$C]Ro15-1788, a mixed benzodiazepine agonist–antagonist was utilized [Hantraye et al., 1984]. The specific activity of the compound varied between 500 and 1800 Ci/mmol. The images obtained revealed a high degree of binding in brain regions (temporal, frontal and visual cortices, and the cerebellum) known to be rich in benzodiazepine receptors. Furthermore, the nonspecific binding was low, on the order of 12%, and no demethylation was detected. However, a tracer kinetic mathematical model for regional quantification was not developed. Further PET studies in the baboon have shown that after [$^{11}$C]Ro15-1788 administration, maximum radioactivity in brain is seen at 10 minutes and lasts for approximately another 10 minutes before gradually declining. The cerebral radioactivity is rapidly displaced by unlabeled Ro15-1788 [Maziere et al., 1984]. Finally, other positron emitting isotopes, such as the [$^{75}$Br]1,4-benzodiazepines are also being considered as potential ligands to be employed in PET [Scholl et al., 1983].

At our own institution, ligands labeled with [$^{18}$F] are preferred because of their longer half-life (1.7 hr). We are currently interested in developing [$^{18}$F]fluorodiazepam for benzodiazepine receptor imaging with PET. Apparently, the fluorine substituent in the metabolically active C-3 position minimizes enzymatic degradation and yet retains pharmacological activity. Preliminary studies in our laboratory using unlabeled fluorodiazepam have revealed an $IC_{50}$ of 15 nM compared to 16 nM for diazepam in displacing [$^{3}$H]flu binding at 2 nM concentration. Further work will undoubtedly yield a suitable ligand with high specific activity to allow imaging of the benzodiazepine receptor in humans with PET. In addition, the recent development of a number of GABA mimetics which are capable of crossing the blood–brain barrier has also stimulated some interest in imaging GABA receptors. The successful quantitative imaging of receptors in humans may eventually allow early detection and characterization of receptor alterations in some diseases and ultimately lead to improved diagnosis, subclassification, treatment selection, and management of these disorders.

## ACKNOWLEDGMENTS

Harry T. Chugani is the recipient of Teacher-Investigator Developmental Award 1 KO7 NS00886-01 from NINCDS. The authors wish to acknowledge the expertise of Dr. Thomas L. Babb in correlating hippocampal histology with receptor autoradiograms.

## REFERENCES

Babb TL, Brown WJ (1985): Neuronal, dendritic and vascular profiles of human temporal lobe epilepsy correlated with cellular physiology in vbivo. In Delgado-Escueta AV, Ward AA Jr, Woodbury D (eds): "Basic Mechanisms of the Epilepsies." New York: Raven Press (in press).

Bakay RA, Harris AB (1981): Neurotransmitter, receptor and biochemical changes in monkey cortical epileptic foci. Brain Res 206:387–404.

Bernheimer H, Birkmayer O, Hornykiewicz O, Jellinger K, Seitelberger F (1973): Brain dopamine and the syndromes of Parkinson and Huntington: Clinical, morphological and neurochemical correlations. J Neurol Sci 20:415–455.

Bird ED, Iversen LL (1974): Huntington's chorea: Postmortem measurement of glutamic acid decarboxylase, choline acetyltransferase and dopamine in basal ganglia. Brain 97:457–472.

Bird ED (1976): Biochemical studies on gamma-amino butyric acid metabolism in Huntington's chorea. In Bradford HF, Marsden CD (eds): "Biochemistry and Neurology." London: Academic Press, pp 83–92.

Braestrup C, Nielsen M, Squires RFJ (1979): No changes in rat benzodiazepine receptors after withdrawal from continuous treatment with lorazepam and diazepam. Life Sci 24:347–351.

Braestrup C, Schmiechen R, Neef G, Nielsen M, Petersen EN (1982): Interaction of convulsive ligands with benzodiazepine receptors. Science 216:1241–1243.

Braestrup C, Honore T, Nielsen M, Petersen EN, Jensen LH (1984): Ligands for benzodiazepine receptors with positive and negative efficacy. Biochem Pharmacol 33:859–862.

Burch TP, Ticku MK (1980): Ethanol enhances [$^3$H]diazepam binding at the benzodiazepine-GABA-receptor-ionophore complex. Eur J Pharmacol 67: 325–326.

Burnham WM, Niznik HB, Okazaki MM, Kish SJ (1983): Binding of [$^3$H]flunitrazepam and [$^3$H]Ro5-4864 to crude homogenates of amygdala-kindled rat brain: two months post-seizure. Brain Res 279:359–362.

Chang RSL, Snyder SH (1978): Benzodiazepine receptors: Labeling in intact animals with [$^3$H]flunitrazepam. Eur J Pharmacol 48:213–218.

Chweh AY, Swinyard EA, Wolf HH, Kupferberg HJ (1983): Correlations among minimal neurotoxicity, anticonvulsant activity, and displacing potencies in [$^3$H]flunitrazepam binding of benzodiazepines. Epilepsia 24:668–677.

Comar D, Maziere M, Godot JM, Berger G, Soussaline F, Menini C, Arfel G, Naquet R (1979): Visualisation of [$^{11}$C]flunitrazepam displacement in the brain of the live baboon. Nature 280:329–331.

Cowen PJ, Green AR, Nutt DJ, Martin IL (1981): Ethyl beta-carboline carboxylate lowers seizure threshold and antagonises flurazepam-induced sedation in rats. Nature 290:54–55.

Coyle JT, Schwarcz R, Bennett JP, Campochiaro P (1977): Clinical, neuropathologic and pharmacologic aspects of Huntington's disease: Correlates with a new animal model. Prog Neuro-Psychopharmac 1:13–30.

Coyle JT, McGeer EG, McGeer PL, Schwarcz R (1978a): In McGeer EG, Olney JW, McGeer PL (eds): "Kainic acid as a Tool in Neurobiology." New York: Raven Press, pp 139–159.

Coyle JT, Molliver ME, Kuhar MJ (1978b): In situ injection of kainic acid, a new method for selectively lesioning neuronal cell bodies while sparing axons of passage. J Comp Neurol 180:301–308.

Coyle JT, Zaczek R, Slevin J, Collins J (1981): Neuronal receptor sites for kainic acid: Correlations with neurotoxicity. In Di Chiara G, Gessa GL (eds): "Glutamate as a Neurotransmitter." New York: Raven Press, pp 337–346.

Davies P, Maloney AJR (1976): Selective loss of central cholinergic neurons in Alzheimer's disease. Lancet 2:1403.

Davis WC, Ticku MK (1981): Ethanol enhances [$^3$H]diazepam binding at the benzodiazepine-gamma-aminobutyric acid receptor–ionophore complex. Mol Pharmacol 20:287–294.

Duka T, Hollt V, Herz A (1979): In vivo receptor occupation by benzodiazepines and correlation with the pharmacological effect. Brain Res 179:147–156.

Enna SJ, Bird ED, Bennett JP Jr, Bylund DB, Yamamura HI, Iversen LL, Snyder SH (1976a): Huntington's chorea: Changes in neurotransmitter receptors in the brain. N Engl J Med 294:1305–1309.

Enna SJ, Bennett JP, Bylund DB, Snyder SH, Bird ED, Iversen LL (1976b): Alterations of brain neurotransmitter receptor binding in Huntington's chorea. Brain Res 116:531–537.

Enna SJ, Bennett JP Jr, Bylund DB, Creese I, Burt DR, Charness ME, Yamamura HI, Simantov R, Snyder SH (1977): Neurotransmitter receptor binding: Regional distribution in human brain. J Neurochem 28:233–236.

Essman EJ, Valzelli L (1981): Brain benzodiazepine receptor changes in the isolated aggressive mouse. Pharmacol Res Commun 13:665–671.

Fariello RG, Ticku MK (1983): The perspective of GABA replenishment therapy in the epilepsies: A critical evaluation of hopes and concerns. Life Sci 33:1629–1640.

Gale K (1980): Chronic blockade of dopamine receptors

by antischizophrenic drugs enhances GABA binding in substantia nigra. Nature 283:569–570.
Gale K (1984): Role of the substantia nigra in the anticonvulsant actions of GABAergic drugs. In Fariello RG, Morselli PL, Lloyd KG, Quesney LF, Engel J Jr (eds): "Neurotransmitters, Seizures, and Epilepsy II." New York: Raven Press, pp 57–80.
Gallager DW, Lakoski JM, Gonsalves SF, Rauch SL (1984a): Chronic benzodiazepine treatment decreases postsynaptic GABA sensitivity. Nature 308:74–77.
Gallager DW, Rauch SL, Malcolm AB (1984b): Alterations in a low-affinity GABA recognition site following chronic benzodiazepine treatment. Eur J Pharmacol 98:159–160.
Greenberg DA, Cooper EC, Gordon A, Diamond I (1984): Ethanol and the gamma-aminobutyric acid-benzodiazepine receptor complex. J Neurochem 42:1062–1068.
Hantraye P, Kaijima M, Prenant C, Guibert B, Sastre J, Crouzel M, Naquet R, Comar D, Maziere M (1984): Central type benzodiazepine binding sites: A positron emission tomography study in the baboon's brain. Neurosci Lett 48:115–120.
Horton RW, Prestwich SA, Meldrum BS (1982): Gamma aminobutyric acid and benzodiazepine binding sites in audiogenic seizure-susceptible mice. J Neurochem 39:864–870.
Iversen LL, Bird E, Spokes E, Nicholson SH, Suckling CJ (1979): Agonist specificity of GABA binding sites in human brain and GABA in Huntington's disease and schizophrenia. In Krogsgaard-Larsen P, Scheel-Krugen J, Kofod H (eds): "GABA-Neurotransmitters." Copenhagen: Munksgaard, pp 179–190.
Jenzen LH, Petersen EN (1983): Bidirectional effects of benzodiazepine receptor ligands against picrotoxin- and pentylenetetrazol-induced seizures. J Neural Transmission 58:183–191.
Joy RM, Albertson TE, Stark LG (1984): An analysis of the actions of progabide, a specific GABA receptor agonist, on kindling and kindled seizures. Exp Neurol 83:144–154.
Kanazawa I, Bird ED, O'Connell R, Powell D (1977): Evidence for a decrease in substance P content of substantia nigra in Huntington's chorea. Brain Res 120:387–392.
Komiskey HL, MacFarlan MF (1983): Aging: Effect on neuronal and non-neuronal benzodiazepine binding sites. Neurochem Res 8:1135–1141.
Kraus VMB, Dasheiff RM, Fanelli RJ, McNamara JO (1983): Benzodiazepine receptor declines in hippocampal formation following limbic seizures. Brain Res 277:305–309.
Liebowitz NR, Pedley RA, Cutler RWP (1977): Release of gamma-aminobutyric acid from hippocampal slices of the rat following generalized seizures induced by daily electrical stimulation of entorhinal cortex. Brain Res 138:369–373.
Lloyd KG, Dreksler S, Bird ED (1977a): Alterations in [$^3$H]GABA binding in Huntington's chorea. Life Sci 21:747–754.
Lloyd KG, Shemen L, Hornykiewicz O (1977b): Distribution of high affinity sodium-independent [$^3$H]gamma-aminobutyric acid ([$^3$H]GABA) binding in the human brain: Alterations in Parkinson's disease. Brain Res 127:269–278.
Lloyd KG, Davidson L (1979): [$^3$H]GABA binding in brains from Huntington's chorea patients: Altered regulation by phospholipids? Science 205:1147–1149.
Lloyd KG, Dreksler S (1979): An analysis of [$^3$H]gamma-aminobutyric acid (GABA) binding in the human brain. Brain Res 163:77–87.
Lloyd KG, Munari C, Bossi L, Morselli PL (1984): GABA hypothesis of human epilepsy: Neurochemical evidence from surgically resected identified foci. In Fariello RG, Morselli PL, Lloyd KG, Quesney LF, Engel J Jr. (eds): "Neurotransmitters, Seizures, and Epilepsy II." New York: Raven Press, pp 285–293.
Löscher W, Frey HH, Reiche R, Schultz D (1983): High anticonvulsant potency of gamma-aminobutyric acid (GABA)mimetic drugs in gerbils with genetically determined epilepsy. J Pharmacol Exp Ther 226:839–844.
Löscher W (1984): Evidence for abnormal sensitivity of the GABA system in gerbils with genetically determined epilepsy. In Fariello RG, Morselli PL, Lloyd KG, Quesney LF, Engel J Jr (eds): "Neurotransmitters, Seizures, and Epilepsy II." New York: Raven Press, pp 179–190.
Marangos PJ, Martino AM (1981): Studies on the relationship of gamma-aminobutyric acid-stimulated diazepam binding and the gamma-aminobutyric acid receptor. Mol Pharmacol 20:16–21.
Martin JB (1984): Huntington's disease: New approaches to an old problem. Neurology 34:1059–1072.
Martin IL, Candy JM (1978): Facilitation of benzodiazepine binding by sodium chloride and GABA. Neuropharmacol 17:993–998.
Maziere M, Hantraye P, Prenant C, Sastre J, Comar D (1984): Synthesis of ethyl 8-fluoro-5,6-dihydro-5-[$^{11}$C]methyl-6-oxo-4H-imidazo[1,5-a][1,4]benzodiazepine-3-carboxylate (Ro15.1788-[$^{11}$C]: A specific radioligand for the in vivo study of central benzodiazepine receptors by positron emission tomography. Int J Appl Radiat Isot 35:973–976.
McNamara JO, Peper AM, Patrone V (1980): Repeated seizures induce long-term increase in hippocampal benzodiazepine receptors. Proc Nat Acad Sci USA 77:3029–3032.
McNamara JO (1984): Role of neurotransmitters in seizure mechanisms in the kindling model of epilepsy. Federation Proc 43:2516–2520.

McNamara JO, Galloway MT, Rigsbee LC, Shin C (1984): Evidence implicating substantia nigra in regulation of kindled seizure threshold. J Neurosci 4:2410–2417.

Medina JH, Novas ML, De Robertis E (1983): Changes in benzodiazepine receptors by acute stress: Different effect of chronic diazepam or Ro15-1788 treatment. Eur J Pharmacol 96:181–185.

Meldrum BS (1975): Epilepsy and gamma-aminobutyric acid-mediated inhibition. International Review of Neurobiology 17:1–36.

Möhler H, Okada T, Enna SJ (1978): Benzodiazepine and neurotransmitter receptor binding in rat brain after chronic administration of diazepam or phenobarbital. Brain Res 156:391–395.

Morselli PL, Bossi L (1984): Antiepileptic efficacy of GABA-agonists in humans. In Fariello RG, Morselli PL, Lloyd KG, Quesney LF, Engel J Jr (eds): "Neurotransmitters, Seizures and Epilepsy II." New York: Raven Press, pp 253–261.

Niznik HB, Kish SJ, Burnham WM (1983a): Decreased benzodiazepine receptor binding in amygdala-kindled rat brains. Life Sci 33:425–430.

Niznik HB, Kish SJ, Burnham WM (1983b): [$^3$H]Flunitrazepam binding to amygdala-kindled rat brains: Washed vs. unwashed tissue. Soc Neurosci Abstr 9:486.

Nutt D (1981): Changes in seizure threshold following ECS and bicuculline-induced convulsions. In Morselli PL, Lloyd KG, Löscher W, Meldrum B, Reynolds EH (eds): "Neurotransmitters, Seizures, and Epilepsy." New York: Raven Press, pp 117–128.

Oakley NR, Jones BJ (1980): The proconvulsant and diazepam-reversing effects of ethyl beta-carboline-3-carboxylate. Euro J Pharmacol 68:381–382.

Olsen RW, Van Ness PC, Tourtellotte WW (1979): Gamma-aminobutyric acid receptor binding curves for human brain regions: comparison with Huntington's disease and normal. In Chase TN, Wexler NS, Barbeau A (eds): "Huntington's Disease" (Adv Neurol, Vol 23). New York: Raven Press, pp 697–704.

Olsen RW, Van Ness P, Napias C, Bergman M, Tourtellotte WW (1980); GABA receptor binding and endogenous inhibitors in normal human brain and Huntington's disease. In Pepeu G, Kuhar MJ, Enna SJ (eds): "Receptors for Neurotransmitters and Peptide Hormones." New York: Raven Press, pp 451–460.

Olsen RW (1981): GABA-benzodiazepine-barbiturate receptor interactions. J Neurochem 37:1–13.

Olsen RW, Wamsley JK, Lee R, Lomax P (1985a): The benzodiazepine/barbiturate/GABA receptor-chloride inophore complex in a genetic model for generalized epilepsy. In Delgado-Escueta AV, Ward AA Jr, Woodbury D (eds): "Basic Mechanisms of the Epilepsies." New York: Raven Press (in press).

Olsen RW, Jones I, Seyfried TN, McCabe RT, Wamsley JK (1985b): Benzodiazepine receptor binding deficit in audiogenic seizure-susceptible mice. Fed Proc 44:1106.

Pan HS, Frey KA, Young AB, Penney JB Jr (1983): Changes in [$^3$H]muscimol binding in substantia nigra, entopeduncular nucleus, globus pallidus, and thalamus after striatal lesions as demonstrated by quantitative receptor autoradiography. J Neurosci 3:1189–1198.

Paul SM, Skolnick P (1978): Rapid changes in brain benzodiazepine receptors after experimental seizures. Science 202:892–894.

Paul SM, Syapin PJ, Paugh BA, Moncada V, Skolnick P (1979): Correlation between benzodiazepine receptor occupation and anticonvulsant effects of diazepam. Nature 281:688–689.

Pedigo NW, Schoemaker H, Morelli M, McDougal JN, Malick JB, Burks TF, Yamamura HI (1981): Benzodiazepine receptor binding in young, mature and senescent rat brain and kidney. Neurobiol Aging 2:83–88.

Penney JB Jr, Young AB (1982): Quantitative autoradiography of neurotransmitter receptors in Huntington disease. Neurology 32:1391–1395.

Perry EK, Perry RH, Blessed G, Tomlinson BE (1977): Necropsy evidence of central cholinergic deficits in senile dementia. Lancet 1:189.

Perry TL, Hansen S, Kloster M (1973): Huntington's chorea: Deficiency of gamma-aminobutyric acid in brain. N Engl J Med 288:337–342.

Phelps ME, Mazziotta JC, Huang S-C (1982): Study of cerebral function with positron computed tomography. J Cereb Blood Flow Metabol 2:113–162.

Reisine TD, Fields JZ, Yamamura HI (1977): Neurotransmitter receptor alterations in Parkinson's disease. Life Sci 21:335–344.

Reisine TD, Yamamura HI, Bird ED, Spokes E, Enna SJ (1978): Pre- and postsynaptic neurochemical alterations in Alzheimer's disease. Brain Res 159:477–481.

Reisine TD, Beaumont K, Spokes EGS, Bird ES, Yamamura HI (1979a): Huntington's disease: alterations in neurotransmitter receptor binding in the human brain. In Chase TN, Wexler NS, Barbeau A (eds): "Huntington's Disease" (Adv Neurol, Vol 23). New York: Raven Press, pp 717–726.

Reisine TD, Wastek GJ, Speth RC, Bird ED, Yamamura HI (1979b): Alterations in the benzodiazepine receptor of Huntington's diseased human brain. Brain Res 165:183–187.

Reisine TD, Overstreet D, Gale K, Rossor M, Iversen L, Yamamura HI (1980): Benzodiazepine receptors: The effect of GABA on their characteristics in human brain and their alteration in Huntington's dis-

ease. Brain Res 199:79–88.
Ribak CE, Harris AB, Vaughn JE, Roberts E (1979): Inhibitory, GABAergic nerve terminals decrease at sites of focal epilepsy. Science 205:211–214.
Rosenberg HC, Chiu TH (1979): Decreased [$^3$H]diazepam binding is a specific response to chronic benzodiazepine treatment. Life Sci 24:803–808.
Ross SM, Craig CR (1981): Gamma-aminobutyric acid concentration, L-glutamate 1-decarboxylase activity, and properties of the gamma-aminobutyric acid postsynaptic receptor in cobalt epilepsy in the rat. J Neurosci 1:1388–1396.
Scholl H, Kloster G, Stocklin G (1983): Bromine-75-labeled 1,4-benzodiazepines: Potential agents for the mapping of benzodiazepine receptors in vivo: concise communication. J Nucl Med 24:417–422.
Shin C, Pedersen H, McNamara JO (1983): Parallel alterations in GABA and benzodiazepine receptors in fascia dentata in the kindling model of epilepsy: A radiohistochemical study. Soc Neurosci Abstr 9:487.
Sperk G, Schlogl E (1979): Reduction of number of benzodiazepine binding sites in the caudate nucleus of the rat after kainic acid injections. Brain Res 170:563–567.
Speth RC, Wastek GJ, Johnson PC, Yamamura HI (1978): Benzodiazepine binding in human brain: Characterization using [$^3$H]flunitrazepam. Life Sci 22:859–866.
Speth RC, Bresolin N, Yamamura HI (1979): Acute diazepam administration produces rapid increases in brain benzodiazepine receptor density. Eur J Pharmacol 59:159–160.
Stahl WL, Swanson PD (1974): Biochemical abnormalities in Huntington's chorea brains. Neurology 24:813–819.
Sytinsky IA, Guzikov BM, Gomanko MV, Eremin VP, Konavalova NN (1975): The gamma-aminobutyric acid (GABA) system in brain during acute and chronic ethanol intoxication. J Neurochem 25:43–48.
Tallman JF, Thomas JW, Gallager DW (1978): GABAergic modulation of benzodiazepine binding site sensitivity. Nature 274:383–385.
Tallman JF, Gallager DW (1979): Modulation of benzodiazepine binding site sensitivity. Pharmacol Biochem Behav 10:809–813.
Tallman JF, Thomas JW, Gallager DW (1979): Identification of diazepam binding in intact animals. Life Sci 24:873–880.
Terry RD, Katzman R (1983): Senile dementia of the Alzheimer type. Ann Neurol 14:497–506.
Tews JK, Carter SH, Rao PD, Stone WE (1963): Free amino acids and related compounds in dog brain: Postmortem and anoxic changes, effects of ammonium chloride infusion, and levels during seizures induced by picrotoxin and by pentylenetetrazol. J Neurochem 10:641–653.
Ticku MK (1979): Differences in gamma aminobutyric acid receptor sensitivity in inbred strains of mice. J Neurochem 33:1135–1138.
Ticku MK, Burch T (1980): Alterations in $\gamma$-aminobutyric acid receptor sensitivity following acute and chronic ethanol treatments. J Neurochem 34:417–423.
Ticku MK, Fariello RG, Golden GT (1983): Effects of chronic treatment with some direct GABA agonists on GABA receptor binding. Soc Neurosci Abstr 9:403.
Tower DB (1976): GABA and seizures: Clinical correlates in man. In Roberts E. Chase TN, Tower DB (eds): "GABA in Nervous System Function." New York: Raven Press, pp 461–478.
Tran VT, Snyder SH, Major LF, Hawley RJ (1981): GABA receptors are increased in brains of alcoholics. Ann Neurol 9:289–292.
Tuff LP, Racine RJ, Mishra RK (1983a): The effects of kindling on GABA-mediated inhibition in the dentate gyrus of the rat. II. Receptor binding. Brain Res 277:91–98.
Tuff LP, Racine RJ, Adamec R (1983b): The effects of kindling on GABA-mediated inhibition in the dentate gyrus of the rat. I. Paired-pulse depression. Brain Res 277:79–90.
Unnerstall JR, Kuhar MJ, Niehoff DL, Palacios JM (1981): Benzodiazepine receptors are coupled to a subpopulation of gamma-aminobutyric acid (GABA) receptors: evidence from a quantitative autoradiographic study. J Pharmacol Exp Ther 218:797–804.
Valdes F, Dasheiff RM, Birmingham F, Crutcher KA, McNamara JO (1982): Benzodiazepine receptor increases after repeated seizures: Evidence for localization to dentate granule cells. Proc Natl Acad Sci USA 79:193–197.
Van Ness PC, Watkins AE, Bergman MO, Tourtellotte WW, Olsen RW (1982): Gamma-aminobutyric acid receptors in normal human brain and Huntington disease. Neurology 32:63–68.
Volicer L, Biagioni TM (1982): Effect of ethanol administration and withdrawal on benzodiazepine receptor binding in the rat brain. Neuropharmacology 21:283–286.
Waddington JL, Cross AJ (1980): Characterization of denervation supersensitivity in the striatonigral GABA pathway of the kainic acid-lesioned rat and in Huntington's disease. Brain Res Bull 5 (suppl 2):825–828.
Walker FO, Young AB, Penney JB, Dovorini-Zis K, Shoulson I (1984): Benzodiazepine and GABA receptors in early Huntington's disease. Neurology 34:1237–1240.
Wastek GJ, Speth RC, Reisine TD, Yamamura HI

(1978): The effect of gamma-aminobutyric acid on [$^3$H]flunitrazepam binding in rat brain. Euro J Pharmacol 50:445–447.

White P, Goodhardt MJ, Keet JP, et al. (1977): Neocortical cholinergic neurons in elderly people. Lancet 1:668–671.

Whitehouse PJ, Lynch D, Kuhar MJ (1984): Effects of postmortem delay and temperature on neurotransmitter receptor binding in a rat model of the human autopsy process. J Neurochem 43:553–559.

Williams M, Risley EA (1979): Characterization of the binding of [$^3$H]muscimol, a potent gamma-aminobutyric acid agonist, to rat brain synaptosomal membranes using a filtration assay. J Neurochem 32:713–718.

Williamson MJ, Paul SM, Skolnick P (1978a): Labeling of benzodiazepine receptors in vivo. Nature 275:551–553.

Williamson MJ, Paul SM, Skolnick P (1978b): Demonstration of [$^3$H]diazepam binding to benzodiazepine receptors in vivo. Life Sci 23:1935–1940.

Wood JD (1975): The role of gamma-aminobutyric acid in the mechanism of seizures. Progress in Neurobiology (Pt. 1) 5:77–95.

Young AB, Pan HS, Ciliax BJ, Penney JB (1984): GABA and benzodiazepine receptors in basal ganglia function. Neurosci Lett 47:361–367.

# Appendix

Abbreviations and Chemical Names or Structures of Compounds Mentioned in this Volume (partial list)

| Common Name | Further Identification |
|---|---|
| GABA | 4-amino-*n*-butyric acid |
| TBPS | 4-*t*-butyl-2,6,7-trioxabicyclo [2.2.2]-octane-1-phosphorothionate[1] |
| RO15-1788 | ethyl-8-fluoro-5,6-dihydro-5-methyl-6-oxo (*4H*)-imidazo [1,5-a] [1,4]-benzodiazepine-3-carboxylate |
| RO5-3663 | 1,3-dihydro-5-methyl (*2H*)-1,4-benzodiazepin-2-one |
| RO5-4864 | 5-(4-chlorophenyl) diazepam |
| Beta-carbolines | |
| βCCE | beta-carboline-3-carboxylate ethyl ester[1] |
| βCCM | beta-carboline-3-carboxylate methyl ester[1] |
| PrCC | beta-carboline-3-carboxylate propyl ester[1] |
| DMCM | methyl-6,7-dimethoxy-4-ethyl-β-carboline-3-carboxylate[2] |
| ZK 93423 | Structure provided[2] |
| ZK 93426 | Structure provided[2] |
| ZK 91296 | Structure provided[2] |
| FG 7142 | Structure provided[2] |
| CGS 8216 | 2-phenylpyrazolo [4,3-c] quinolin-3(*5H*)-one[1] |
| CL 218872 | Structure provided[1] |
| EMD 28422 | $N^6$-2-(4-chlorophenyl)-bicyclo [2.2.2] octyl-(3)-adenosine |

[1]See chapter by Lawrence et al., Fig. 1.
[2]See chapter by Braestrup and Nielsen, Fig. 7.

# Index